ONE WEEK LOAN

Urban Transportation Systems

Urban Transportation Systems

CHOICES FOR COMMUNITIES

Sigurd Grava

McGraw-Hill

New York   Chicago   San Francisco   Lisbon   London   Madrid
Mexico City   Milan   New Delhi   San Juan   Seoul
Singapore   Sydney   Toronto

Library of Congress Cataloging-in-Publication Data

Grava, Sigurd.
  Urban transportation systems: choices for communities/Sigurd Grava.
    p. cm.
  Includes index.
  ISBN 0-07-138417-0
  1. Choice of transportation.   2. Urban transportation.   I. Title.
HE336.C5 G73 2002
388.4—dc21   2002070289

McGraw-Hill

A Division of The McGraw-Hill Companies

1 2 3 4 5 6 7 8 9 0   DOC/DOC   0 9 8 7 6 5 4 3 2

ISBN 0-07-138417-0

The sponsoring editor for this book was Larry S. Hager and the production supervisor was Pamela A. Pelton. It was set in Matt Antique by North Market Street Graphics.

Printed and bound by RR Donnelley.

 This book was printed on recycled, acid-free paper containing a minimum of 50% recycled, de-inked fiber.

McGraw-Hill books are available at special quantity discounts to use as premiums and sales promotions, or for use in corporate training programs. For more information, please write to the Director of Special Sales, Professional Publishing, McGraw-Hill, Two Penn Plaza, New York, NY 10121-2298. Or contact your local bookstore.

Contents

Foreword

Dr. Sigurd Grava wraps 40 years of graduate teaching and a like period of private-sector experience in planning, environmental studies, economic analyses, and general feasibility studies into a unique approach to transportation decision making. He focuses individually on each mode of transport in the urban area—from the foot on the pavement, to the rubber on the road, to the steel wheel on steel rail in dedicated rights-of-way. When finished with this interesting and technically precise text, which does include much common sense and some humor, the reader will have an excellent feel for 12 modes of mobility and what factors are pertinent to evaluating choices. The reader will not yet have—and hopefully will realize that he or she does not yet have—sufficient technical knowledge for detailed system planning or economic analysis, but should be able to take several steps toward the rational selection of appropriate modes.

However, if we recognize that in the area of transportation every major decision is a political decision; that in this age of public involvement, every commuter (like the undersigned) believes that he or she is an expert in his or her daily mode; and that a plethora of agencies and organizations believe that they (and they alone) represent the public, then this text provides a superb basis for educating the participants, elevating the level of knowledge and discussion, and hence providing the basis (and hopefully guiding the outcome) for a more informed decision-making process. Also helpful in the education process are field visits to successful and new-start properties with different technologies to get realistic inputs on cost, duration, the implementation and startup process, construction impact, maintenance costs, and the quality of service and its responsiveness to user needs. (Suggestions for such visits are found in App. C, Suggested Places for Field Inspection.)

In addition to being very political, transportation decisions tend to be very emotional issues, ranging from an attitude of "not in my back yard," to a perception of economic gain or loss, to strong preferences for a particular mode. Every town, for example, has a "Monorail Max" who once visited Disneyland and now thinks monorail lines are the only solution. For example, following a presentation on the pros and cons of light rail at a public hearing in Salt Lake City, a gentleman in the audience stood up and made an impassioned plea for a monorail system. After the presentation, I asked him where he had seen this monorail; he replied that he had seen it "up in the air in Atlanta, Georgia." He was totally disinterested when I pointed out that he had seen a heavy rail system on an elevated aerial structure. His mind was made up.

In terms of emotional appeal, the preference scale seems to be monorail, heavy rail, light rail, commuter rail, and buses, in that order. Buses often get penalized because many cities want their own rail lines (just as they want their own professional football teams), and the image of buses is damaged by poor maintenance and poor service on existing systems. A few years ago I listened to a radio talk show featuring a transit general manager who wanted to answer questions on a new rail line. Every one of the callers had a complaint about the bus system or a request to change a specific bus route. Not one call or question was on the billion-dollar rail line! The new rail line location was both driven and limited by political boundaries, which eliminated planned rail lines in counties that voted down local taxation. Politics also required that construction be strictly balanced—a mile to the east required a mile to the west—in the counties that had approved the sales tax financing.

Realistically, the preferred system will vary with the specific local situation, based on various factors, including funding source, population density, political view, and public opinion, among others. Most new-start rail lines were initiated by a strong, united business community, which had a positive impact on local political unity. Every evaluation of a preferred mode should include the alternatives of transit system management (improvements to any existing transit system in conjunction with traffic improvements) as well as an expanded bus system and rapid bus, which are the least costly and most flexible options. If ridership, current or projected, is sufficient, light rail with the flexibility of a mall- or street-running concept deserves special attention. New intelligent transportation system (ITS) technology significantly improves the viability of street-running light rail and rapid bus transit. Of course, commuter rail is a viable option if it fits into existing rail corridors where either a dedicated, single-use line or mixed traffic with Amtrak or freight are options. New-start heavy rail (metro or premetro transit) systems are unlikely in the future; existing systems will continue to expand, but not necessarily with heavy rail technology. The Los Angeles Red Line may well be the last heavy rail subway to be built in the United States for many years. I do not list high-speed rail, since upgrading of existing rail lines in the few high-density corridors which could economically support high-speed rail is the most likely future development in the United States. High-speed rail development in Japan, France, Germany, and Taiwan, with a strong governmental role, does not fit the U.S. political model. The current funding for high-speed rail new-start lines will fracture on the reality of economic feasibility unless the U.S. government changes its "corporate welfare" view of financial support for transit services.

Transit infrastructure is unique in that rail lines such as those in Boston, Chicago, and New York, which were built over 100 years ago, are still in operation today. But, the technology of the rolling stock and the power, propulsion, train control, and communication systems is much more advanced. Safety, ventilation, and regulatory requirements are changing every year, at an accelerating rate since the events of September 11, 2001. The influence of transportation on land use has never been greater, while politically it has become more difficult to consider the future effects of present decisions rather than focusing on solving today's problems today. A few years ago, a commuter rail agency proposed to run a commuter rail line for a one-year test to evaluate the impact on commuter highway traffic and on suburban development patterns without recognizing that no one would move their residence or business facility for a one-year trial. When the agency finally recognized the reality and abandoned the project, one of the board members asked me what their minimum obligation should be. I replied that as a minimum, they should preserve the rail right-of-way in order to retain future transportation options. Fortunately, this was done. Had those board members read this text, they would have had a much better appreciation of options and a more rational approach to the mobility decision process. Too often when new systems or even new lines are opened, the measure of success (in the press particularly) is opening-day ridership, but the real measure of success is how well the system meets its specific goals in the long run—be they changed residential or density patterns, reduced highway congestion, improved social justice, or other goals.

Finally, the readers of this text will understand that one mode does not solve all urban transportation problems and that every mode has a place in the mobility spectrum and is influenced by a variety of factors such as cost, flexibility, density, development patterns (current and desired), and changing technology, to mention a few. If one mode will receive 90 percent federal funding and another mode must be built only with local or state support, this will obviously have a significant impact. Each new system is also influenced by the perceived success or failure of other systems just opened or in construction. As an aside, just as an informed decision-making process is key in mode selection, a timely, responsive, and informed decision-making process is the most important single success factor in the development and construction of *any* selected mode.

After reading this book, public and private decision makers will know better what questions to ask, what criteria are most relevant to their unique environment, and, hence, how to conclude their planning process with a better mode decision. And, I believe that is the real objective of Dr. Grava—and his text succeeds admirably.

—**James L. Lammie**
Director Emeritus
Parsons Brinckerhoff Inc.

Jim Lammie is a former CEO and director of Parsons Brinckerhoff (PB). After 21 years in the U.S. Army Corps of Engineers, he joined PB in 1975 as consultant project director on Atlanta's MARTA heavy rail construction project. Serving in various capacities during the past 20 years, he has been involved in many of the most complex highway, transit, and rail projects in the United States and overseas.

Preface

As we enter the new millennium, there is a temptation to speak of crossing thresholds and starting a new era. As a matter of fact, we are fairly deep in the Postindustrial Era regarding city development already, but there was no sharp break. The changes underway are swift, but not abrupt. Everything happens today at an accelerated pace, but there is no revolution. In urban transportation, there are new possibilities. The hardware is mostly familiar, but we have new capabilities to use it more effectively and selectively. The accumulated experience of the past is a strong foundation for what we do next—assuming that we can do it wisely and constructively.

The private automobile dominates the American transportation scene, and, while it is a splendid device, it cannot be, nor should it be, the only means of urban mobility available to communities. Everybody knows its dark sides, yet we continue to use it with abandon. The car will not be *re*placed in the foreseeable future, but it can be *dis*placed to some extent since there are other choices available. There are, as we have learned the hard way, high-density, truly urban situations where the automobile does not work at all.

There were periods in urban history when only a single transportation mode was reasonably available—horses, steam locomotives, or streetcars—and, thus, evaluation and selection were not much of an issue. After two centuries of intensive technological development, we find today that there is quite a range of options. How then are we to decide what is best for our community?

The aim of this book is to describe the various means of urban transportation, to outline their strengths and weaknesses, but primarily to suggest under what conditions one or another might be suitable. The basic contention here is that, if a community can define precisely what its mobility and accessibility objectives are, then an informed and purposeful selection can be made from the modal options currently available. It is even possible to mix and match, to tailor service systems to the particular needs and desires of any set of consumers. Indeed, there is more available than we can possibly use. While new technological developments are usually welcome, there is no urgent need today for inventors and engineers to come up with additional advanced hardware. Effective choices can be made from what already exists and has been tested in the field.

The information provided here is not intended for technical specialists, but rather for students of the contemporary urban condition and decision makers, whether they are elected officials or service users, and for people interested in the complex mysteries of transportation capabilities. It is hoped that technical specialists, who tend to be experts in if not advocates of only specific modes, may find useful information outside their own fields. The emphasis is definitely on what each mode can do for us.

The working title of this book, while it was being drafted, was *You Can Get There from Here*. Wise people suggested, however, that this might not be a properly dignified title for a semitechnical work. The underlying thought remains, nevertheless, that good possibilities certainly exist to cope with urban transportation issues under various sets of circumstances. An attempt has also been made to keep the material readable by eschewing technical jargon, which is useful but only for card-carrying members of the transportation fraternity, while keeping the technical information reliable and unimpeachable. To show that transportation is not a dry technical subject, but affects deeply the daily lives and concerns of all of us, references and occasional excerpts from literature and the popular arts are included.

The fundamental premise of this review is that *all modes are good,* but only in their proper place. Each of them has a niche, which may be rather broad or quite limited; some may respond to a wide range of needs, others can cope only with very specific demands; some may be effective in general urban environments, others work well only in targeted situations. Aspects of implementation, financing, and acceptance certainly differ, but all modes can do a job in a setting appropriate to its capabilities. They may have done so in the past and may do so again in the future. We are fortunate, in comparison to other periods in urban history, to have so many choices available.

The organization of the book is quite simple: each chapter is devoted to a single mode, following approximately the same format. Some chapters are much shorter than others because the means of transportation under discussion may not be particularly promising as an urban service at this time, it may have a very circumscribed range, or it may respond only to very specific needs. Unlike most works on urban transportation, the material is not integrated page by page, except at the end, to allow a concentrated and uncluttered review of each mode. This is done in the belief that most readers will find it useful to learn about a single mode or a few modes at a time when the need arises. This publication is not a novel with a continuous story line; it has more the format of an almanac. It can be asserted at the beginning, as a given, that modes have to be integrated among themselves to provide a truly responsive service to city residents and workers, but this truism does not have to be repeated on every second page.

There is another difference between this book and several others currently available. Concentrated efforts have been made by other investigators and authors to introduce precision and accuracy in the analysis of urban traffic and transit behavior, going in the direction of operations research. This has been a very commendable trend, giving much insight and understanding. Yet, without any intention of disparaging this research and theoretical work, it also has to be observed that the actual conditions out in the field are frequently capricious, subject to numerous external forces, including the weather and human behavior. Sometimes they border on being chaotic. It is actually quite difficult to count cars accurately on a busy highway, to keep a bus on schedule on a congested street, and to make riders enter and leave a metro car with dispatch. It is by no means suggested that we should revert to an impressionistic attitude in the development and management of transportation systems or that we should not take advantage of the detailed information being generated, but there is also a very great need for flexibility and an ability to respond continuously to changing situations. Traffic and demand conditions are never the same from one day to the next, and commonsense approaches may be often more useful than elaborate and apparently precise solutions based on quantitative analyses. In a longer-range perspective, as city structures and lifestyles change, so does the demand for mobility and types of service that can satisfy contemporary needs.

The modal chapters start with some introductory material as *background* that describes the overall context within which each mode exists. This is followed by a *development history* that is always fascinating in its own right, and much discipline has been required to keep it to a reasonable length and detail. This discussion is important, however, to show how a mode has evolved in response to contemporary technical capabilities and service demands. It should point to the future.

Factual information, usually supported by tables of physical data, is contained in the section devoted to an examination of the *types of operations* that can be found in the field today. Various classification systems are employed whenever possible to provide some systemic order in areas that can be quite fluid and frequently overlap with other modes. A strict and neat segregation of transportation activities by modes does not always work well. As a way of summary, *reasons to support* specific modes or *to exercise caution* are listed, relying on actual recent experience and a recognition of contemporary needs.

The next item of discussion—the *social status* of any given mode—is usually not found in investigations of urban transportation. These concerns touch some very sensitive issues related to policy statements and implicit attitudes in American communities today. Yet, this is where the true and sometimes determining reasons are to be found regarding why some modes will be popular and others will be neglected.

Application scenarios attempt to identify under what conditions and in which settings any specific mode might be suitable and successful. This is amplified by a review of the specific *components* that constitute any given mode, as well as associated *scheduling* and *capacity* considerations. The intention here is to give the technical and functional ranges within which each mode can provide useful service. That leads to concerns about *costs,* which often are the critical determinant of feasibility. Given the great variations in this sector from place to place and differences in the details that are built into any system, as well as continuous changes in prices, this discussion cannot be definitive but only indicative to serve in early evaluations. For anything beyond that, expert estimators will have to be called in.

Transportation systems of high capacity and with substantial capital investment are expected to have a measurable *land use* impact—i.e., they generate development that takes advantage of enhanced access and mobility within service areas. For urban planners, this is a major concern and opportunity. Regrettably, cause-and-effect relationships are often difficult to establish, but a discussion of the implications is nevertheless necessary.

Possible action programs represent introductory material that could achieve implementation if a specific mode is deemed appropriate in a specific instance. This very quickly leads, however, to financing options, government assistance programs, institutional and ownership/management arrangements, labor rules, and other aspects that are generally beyond the scope of this book. Material on such matters can be found elsewhere, keeping in mind the caveat that things change from year to year. Up-to-date information in these fields is always needed to make purposeful progress. Certainly, options and restrictions are different from country to country, with the United States remaining the focus of this discussion.

Each modal chapter has a *conclusion* that attempts to summarize the salient findings regarding choices and opportunities.

There is an annotated *bibliography* for each mode that has been made as lean as possible, but still encompass all the relevant material, particularly references that provide a comprehensive coverage of any mode. Preference has been given not only to recent publications, but also to those that are readily available through publicly accessible sources. Suggestions toward more extensive and more detailed information are contained in the footnotes.

It is not a very wise policy to start explaining what a book is *not,* but for the sake of fairness and to preclude misunderstandings, it should be emphasized that only urban passenger modes are covered, i.e., how people move every day within their own communities and metropolitan areas. What happens with freight or how intercity travel is accomplished has to be found elsewhere. The urban transportation problem and the comprehensive planning process are not discussed either, because much good material has been produced lately, and there is no need to repeat it here. Institutional, labor, financial, and administrative issues are only touched on—not that they are not important, but mostly because they cross modal boundaries and are characteristic of a time and a place, not a type of service. A review of detailed engineering elements of engines and motors, mechanical and structural systems of vehicles, signal and power supply arrangements, the chemistry of fuels, and routine or advanced maintenance procedures would be out of place in this book, useful only to readers with an appropriate background and orientation.

One of my reviewers observed that the information given in this book may be too extensive for transportation novices and would only get them in trouble yet may be only detailed enough to amuse specialists. So be it; that would appear to be a fair and workable balance.

We can note in many instances that situations have been reached in our communities in which *physical* transportation can now be replaced by *virtual* transportation—i.e., communications links substitute for actual travel within a city. Telecommuting is a significant factor already; teleshopping, teleconferencing, and teleentertainment have started to affect urban life profoundly. Watching television at home and having pizza delivered are such common practices that we tend not to fully acknowledge their prophetic nature. These types of activities, however, are not yet to be regarded as full-fledged modes deserving a separate chapter in this book. If there is ever a second edition, that decision quite possibly may have to be modified.

One of the reviewers of early drafts observed that the information given in this book may be too extensive for transportation novices and would only get them in trouble yet may be only detailed enough to amuse specialists. So be it; that would appear to be a fair and workable balance.

Even a diligent reader of this book should not assume that he or she will now be ready to design operating systems—that will take much more knowledge and skill than can be presented here. But there should be considerable confidence in being able to evaluate the options in any given community and to move forward.

Acknowledgments

A book of this scope is not produced by a single person within a short time period. As the author responsible, I wish to utilize briefly the first-person singular form to acknowledge my indebtedness to organizations and individuals. Teaching transportation planning courses at Columbia University for many decades has given me an opportunity (nay—a necessity) to keep abreast of the field, and I thank my students for their questions and concerns, which have shaped and steered my own attitudes more than they know. Being active in a field for a long time does not necessarily bring wisdom, but it does generate experience and a sense of reality and opportunity as situations keep changing. My academic involvement has been complemented, rather extensively, I think, by work on numerous transportation projects in some 25 countries. I have done this primarily as the technical director for planning and vice president of Parsons Brinckerhoff, a consulting firm with an international standing in the field of transportation, but also in association with other organizations, such as various municipal governments and the United Nations. I remain an unabashed New Yorker, my permanent home by choice, even though I have the opportunity now to be active also in my native country, Latvia.

My consulting and academic work has had the added informative but tiring dimension of extensive travel. I have seen and at least "touched"[1] just about all the systems and places mentioned in this book and have taken many pictures. Even a short visit, while it does not generate full insights on how well a service performs, it is useful as a base for comparisons.

I am fortunate to count among my friends and colleagues many individuals who are experts in various sectors of transportation and are so recognized in the industry. I have taken advantage of this situation, and, besides benefiting from working together with them in many situations, I have asked several of them to review various chapters of this book. (What are friends for?) Most of them come from the Parsons Brinckerhoff family, and, since each of them knows much more about his mode than I do, I have taken their comments seriously. This has been more than just another peer review. My colleagues, every one of them, have taken the pains to review the draft thoroughly and to set matters straight in terms of how systems operate and how elements behave. There is much comfort, certainly on my part, in knowing that real experts who work with a given mode every day have scrutinized the material. I can only hope that I have been able to respond to the factual suggestions adequately. As is always the case, the reviewers are not responsible, however, for the opinions and interpretations that permeate this discussion. I am sure that much of a personal attitude shows through, and I solely take the credit or blame for that.

Specifically, the principal reviewers have been:

Greg Benz	Chap. 2, Walking
Stephen Faust	Chap. 3, Bicycles
Elliott Sclar	Chap. 5, Automobiles
Bruce Schaller	Chap. 7, Taxis
David Miller	Chap. 8, Buses
Charles Fuhs	Chap. 9, Bus Rapid Transit
David McBrayer	Chap. 9, Bus Rapid Transit
Mark Walker	Chap. 11, Streetcars and Light Rail Transit
Winfred Salter	Chap. 13, Heavy Rail Transit (Metro)
Robert Olmsted	Chap. 13, Heavy Rail Transit (Metro)

[1] For rapid transit, I define this as riding every line from beginning to end, past every station.

Foster Nichols	Chap. 14, Commuter Rail
Peter Marcuse	Chap. 14, Commuter Rail
Joseph Silien	Chap. 15, Automated Guideway Transit
Vern Bergelin	Chap. 16, Waterborne Modes
Alan Olmsted	Chap. 16, Waterborne Modes

My firm—Parsons Brinckerhoff—has very generously extended technical assistance in the preparation of this book. This encompasses not only the use of office facilities and equipment, but also free access to its rich archival material, and—most important—the preparation of the graphics, which is a major component of the entire effort. Judith Cooper managed this activity with enthusiasm and extended competent personal advice in the editorial field. Kevin Peterson is responsible for a number of the illustrations—those that show the touch of a fine hand. Pedro P. Silva did all the technical work on graphics, in most thorough fashion. James Anderson was consistently helpful in navigating through the more advanced procedures of electronic processing of words, images, and numbers. I took all the photographs myself, except for a few as noted.

Special thanks go to my research assistant, Shane Taylor, who coped successfully with libraries, electronic data sources, and agency and manufacturer contacts, and also reacted with diligence and intelligence to the material as it was being assembled or generated.

Contemporary urban transportation problems are also experienced by communities that are regarded as being ahead of the game. This article was originally published in *The Economist,* July 14, 2001.

Going Nowhere

And Certainly Not Fast

SEATTLE—Each day, Sally del Fierro and a neighbour share a car into Seattle, 26 miles from where they live. Even in the lanes reserved for cars with two or more occupants, the drive generally takes more than an hour. In bad weather, or if an accident blocks part of Interstate 5, it takes two hours. In the three years that Ms del Fierro has been making the journey, the traffic has become noticeably worse. Moreover, 600 new houses are planned for her town, each of which will dump another car or two on narrow local roads.

A report in May from the Texas Transportation Institute, which keeps an eye on such things, confirmed that Seattle has the second-worst traffic in the country, equal with Atlanta and just behind Los Angeles. Commuters in the area lose 53 hours a year stuck in traffic over and above their normal commuting times, burning 81 gallons (307 litres) of petrol while their cars idle away. Businesses find it increasingly difficult to move supplies and people from one place to another.

What to do? A light-rail system for Seattle, which at present relies largely on overworked buses for public transport, has been in the works since 1993. But, without a single spadeful of earth turned, the project is $1 billion over budget and years behind schedule. A citizens' initiative passed in 1999 lopped $2 billion from the state transport budget by limiting fees levied on vehicle licence-plates. A sharp earthquake in February complicated matters by weakening a freeway viaduct along the Seattle waterfront; it almost certainly needs replacing, at a probable cost of $600m. And there are scores of rickety roads and bridges all across the state.

When the state legislature convenes on July 16th for its fourth session this year, its chief task will be to vote on a $17 billion package proposed by Gary Locke, the Democratic governor, that is designed to improve the state's roads as well as its rail service and its extensive ferry system. But winning approval will not be easy. The lower house is evenly divided, 49-49, with neither side willing to approve anything for which opponents might take credit. The state Senate is on the whole readier to spend big money on transport. But politicians from eastern Washington—where a storm on June 26th caused huge damage to cherry and apple crops—are reluctant to raise constituents' taxes to help fuming motorists in Seattle.

Mr Locke, an able but low-key technocrat, has been pushing his package hard, working on lobbyists and touring the state to muster support. But even he is reluctant to admit what everyone knows—that improving transport will require an increase in the state petrol tax. (European readers may insert bitter laughter here—a gallon of gasoline in Washington costs a trifling $1.60.) Anyway, both Mr Locke and the legislators seem inclined to let the voters choose for themselves what they want to build or not build, rather than take a risk by deciding on their behalf.

The irony, notes David Olson of the University of Washington, is that the state was once a leader in almost every form of transport. Roads, railways and ports figure prominently in the region's early history, as the first settlers fought against their geographical isolation. Shipping to Alaska made Seattle rich in the 1890s. William Boeing left a career in timber to go into aviation, and never looked back. Seattle's port was among the first in the country to be publicly owned.

In recent years, however, transport has been seen as a necessary evil. Voters abhor gas taxes and tollways, and local towns bitterly resist proposals to enlarge nearby freeways. A proposal to add a third runway at Seattle-Tacoma International Airport has resulted in 12 years of lawsuits and disputes. The reason for this change of heart is the region's economic success. The region's new techno-elite is less interested in building roads than in preserving nice, green views from its windows.

Only a big recession, Mr Olson thinks, is likely to make these people change their minds. So far the area's technology firms have been rather less clobbered by their industry's slump than their peers in Silicon Valley—a fact to which the roads around Redmond, full of commuting Microsoft programmers, bear ample witness.

CHAPTER 1

Overview—Criteria for Selecting Modes

Life in cities—i.e., in organized human settlements, which are mostly referred to as *communities* in this book—is possible only if people have *mobility*[1] on a daily basis—the ability to move around so that they can do what they have to do or like to do. One characterization of a city is that it consists of specialized, frequently clustered, activities that perform discrete functions. Residences are separate from workplaces, major shopping is concentrated in identifiable centers, and larger entertainment and relaxation facilities are found at specific locations. They have to have *accessibility*.[2] Unlike in a village, very few of these destinations are reachable on foot; at least, they tend not to be within a *convenient* walking distance.

The large ancient and medieval cities were actually conglomerations of neighborhoods in which daily life could take place

---

[1] *Mobility* is here defined as the ability of any person to move between points in a community by private or public means of transportation. The usual obstacles to mobility are long distances, bad weather, steep hills (all constituting friction of space), but, above all, the unavailability of services, high fares, and possibly other forms of exclusion.

[2] *Accessibility* is here defined as the possibility of reaching any activity, establishment, or land use in a community by people (or by conveyances of goods or information) who have a reason to get there. It is a measure of the quality and operational effectiveness of a community.

within a short radius; only occasionally was a longer trip to a major event necessary. Industrialization during the nineteenth century caused a true urban revolution by disaggregating the small-scale pattern into metropolitan structures with strong and intensive production and service zones. Assisted transportation became mandatory, and was, indeed, quickly invented—horse cars, steam railroads, electric streetcars, and eventually underground metro (electric heavy rail) systems.

The twentieth century brought further development of the rail modes, and introduced individual motor (gasoline- and diesel-powered) vehicles—buses and automobiles. The latter came to dominate the transportation field, at least in North America, and dispersed the urban pattern further into sprawl. We are all familiar with this situation, since this is our environment, and it has been examined endlessly by scholars, journalists, and concerned citizens. What is not quite so apparent is that urban life and spatial patterns are entering a new, postindustrial, period, which is characterized by the emergence of many dispersed special-purpose centers (not just the historic single all-purpose center), overall low densities, and movement in many different directions at any given time with diverse trip purposes. Electronic communications systems play an increasingly large role. All this makes it more difficult to operate effectively the traditional transportation modes that served us well under more structured conditions. Everything has not changed, but the task of providing responsive transportation services is now more challenging. Also, the expectations are higher.

There is a large inventory of available means of mobility today, most of them tested under various conditions in various places. In the United States, it is not just a question of how to cope with the automobile—admittedly a very seductive mode—but rather of how to equip our communities with a reasonable array of transportation choices, so that the best aggregate level of mobility is offered to all people. Never before has any other culture enjoyed the same freedom of movement, but there are deficiencies: not everybody can take full advantage of the current car-based transportation capabilities, and the systems that we do have are not necessarily (quite unlikely, in fact) the best, the most economical, the cleanest, and the most responsive options that could be provided. The vehicular pollution problem is perhaps on the verge of being solved, if some serious additional effort is applied, but plenty of other issues remain.

The trends and problems are global, and while the scope of inquiry of this book is definitely directed to North America, these concerns do not exist in isolation, certainly not as far as transportation technology is concerned. It is common practice to refer to "industrialized countries" as having special needs and capabilities—which is an obsolete concept, because *industry* (i.e., manufacturing) is no longer the determining factor. The search for a proper label has some significance. "Advanced countries" is a pompous and patronizing characterization that does not contribute much to an operational discussion. "Peer countries" has some validity, but only if everything is compared to a U.S. situation. "Developing countries," on the other hand, is a very common term that helps to summarize broad descriptions, but obscures the fact that there is tremendous variety among these countries. Saudi Arabia, Brazil, Kenya, and Indonesia do not fit in the same box easily.

The fact of the matter is that cities and their populations are not homogeneous in different parts of the world, not even within the same country (not even in Sweden). Each city has components that range in their transportation expectations from the most comfortable to the most affordable. There are districts in African cities that expect and can pay for the most advanced services, and there are neighborhoods in American metropolitan areas that are not much different from those found in Third World countries. The relative size of the various user cohorts is, of course, different, but the demands within them are quite similar.

Therefore, for the purposes of examining transportation needs, it can be suggested that we recognize the presence of various economic and social classes (user groups) that react differently to transportation systems and have to be serviced differently. In a perfect world, such distinctions would not have to be made, since everybody is entitled to mobility. Equity is an important concept, and social reforms are undoubtedly needed in many instances, but the duty of urban transportation is to provide service for communities the way we find them today. Purposeful and relevant change comes next, but upgraded mobility systems can only do so much in implementing community reforms.

Thus, to define a base for the discussion of transportation modes, the following distinctions that are present in any society can be made:

- *The affluent elite.* This group is basically separate and only barely visible from the outside. The members live and play in their own enclaves and have their own means of mobility (limousines and private jets). They do not affect the rest of us, except to cause some envy; they do not participate in daily urban operations, and they do not use the subway. They do have much influence in decision making.

• *The prosperous cohort.*   This group has the same expectations from transportation services as everybody else—rapid, comfortable, and secure accommodations—but members of this group can exercise a choice and be selective. They insist on control over their private space, and they might use public transportation, but only if it meets very high standards. The expense of transportation is not a significant barrier; the demand is for individually responsive means of unconstrained mobility. The private automobile does this (most of the time), and there is an open question as to what proportion of Americans falls in this group of dedicated motorists who have no other choices in mind.

• *The middle class.*   This group has largely the same attitudes as the previous group, except that they operate with more frugal means. They include among their members proportionally more individuals who will favor public transportation as a matter of principle and the proper thing to do. It has always been the case that the professional and educated classes lead the public debate, start revolutions, and demand reforms. They have to be counted on as the formulators of public opinion, and they will determine policy directions in places where they constitute a vocal presence. It is a fact that members of this group, whether they are Argentines, Egyptians, Belgians, or Americans, will act and behave in the same way and demand the same type of services and facilities. They all read the same books and drive the same cars. The only differences among them are their relative proportion of the populace in any given society and some cultural variations. Europeans, for example, cherish their old city districts; Americans regard them as quaint "theme" areas; and members of emerging societies are still frequently embarrassed by them.

• *The surviving cohort.*   This group consists of working people of modest means whose principal preoccupation is basic existence. They have little influence on decisions and political processes—except in instances where they constitute the overwhelming majority and are politically organized. They need and deserve transportation services, but they cannot afford high charges, and their choices tend to be limited. Some degree of subsidy will almost always be necessary to attain acceptable service levels.

- *The disadvantaged class.* This group includes the poor and those who have some personal handicap and insufficient resources to purchase proper services. Poverty always comes at different levels, but the problems are universal and unforgiving. This group represents the largest challenge to public agencies and institutions in achieving basic mobility for all. No social assistance program really works unless physical accessibility is ensured. Communities in the United States are certainly not immune from these requirements, and the current "welfare-to-work" effort is only one example of the initiatives needed.

The preceding is not by any means intended to be a sociological analysis of contemporary societies, but only a hypothesis of how different populations react to mobility needs and services provided. More specifically, the adequacy of operations can be looked at from three perspectives, which eventually leads to the selection of a proper response or transportation mode:

- The point of view of the *individual,* which will stress personal attitudes and emphasize usually humanly selfish considerations
- The policy of the *community,* which has to stress the common good and long-range capabilities
- The concerns related to *national* efficiency and well-being

The personal concerns will encompass the following:

- *Time spent in travel.* This includes time spent to reach the vehicle or access point, to possibly wait, to actually travel, to possibly transfer, and to reach the final destination (probably on foot).
- *Costs incurred.* These include primarily the out-of-pocket expenses on any given trip (including possible tolls and purchase of fuel), but there are also considerations of previous investment (buying a car) and the sunk costs (investment in equipment and insurance).

- *Operational quality.*   This concerns reliability, safety (from accidents), and smoothness of motion.
- *Human amenities.*   These include security (from criminal activity), privacy, sanitation, climate control, seats, visual quality, and social standing.

The communal concerns should include the following:

- *Efficient networks and services.*   They should have the ability to support economic and social life, and cause minimal disruptions and delays in normal urban operations.
- *Efficient urban patterns.*   To the extent that transportation systems can help to achieve more compact settlement forms, the configurations and activity locations should be deliberately shaped.
- *High degree of livability.*   Transportation modes should provide access to all places and establishments and have minimal local environmental and visual impacts.
- *Economic strength.*   Economic development, tax revenues, and local jobs should be boosted due to good transportation.
- *Fiscal affordability.*   Services should result in limited drain on local resources, maximum use of external assistance, minimal indebtedness, and low annual contributions.
- *Institutional peace.*   There should be minimal need to change ordinances or regulations, modify labor rules, displace families and establishments, disturb existing institutions, etc.
- *Civic image and political approval.*   Services should include features that are admired by outsiders and endorsed by local residents (voters) and businesses.

The national concerns exist at a higher and overarching level, and they might not always be achieved if left to local initiatives:

- *Use of national wealth.*   This involves the implementation and operation of the most cost-effective systems, particularly as seen from the perspective of the national budget.
- *Conservation of fuel resources.*   This particularly concerns those derived from petroleum.
- *Environmental quality.*   Air quality over large areas and regions demands specific attention.

- *Equity.* This is a concern to ensure that the needs of the less-privileged members of society are specifically addressed.

- *National technological capability.* Those systems that enhance technological advancement and production capacity within the country should be emphasized.

- *Well-functioning, well-equipped, and balanced communities.* Such built environments should be created in all parts of the country and within all metropolitan areas.

Recognizing the fact that no proposed or existing transportation system can satisfy equally well three separate sets of criteria, there is a need to amalgamate the preceding lists, perhaps even to make some compromises. There is also the practical consideration that the discussion here has to move toward workable guidelines for the *selection of appropriate modes* in any given urban setting. This means that some of the considerations are so overarching and basic that they simply have to be accepted as given; others make no distinction among modes and, therefore, are not operative in the evaluation process. Attention has to turn to functional aspects. All services and systems eventually exist and perform at the *local* level in communities.

Trip Purpose and Clientele

Most transportation modes can make a reasonable claim to be able to satisfy all trip purposes within a community. They have to, because no city can provide too many overlapping services. There are, however, modes that respond best to selected situations with identifiable needs. These usually encompass paratransit and various high-technology modes (shuttles and district services). With respect to user groups, the options are more complicated, because people tend to have differing expectations. These range from placing comfort features first to a single-minded emphasis on affordability. Concerns with equity very much enter into these evaluations.

Geographic Coverage and Grain of Access

The more capital-intensive modes best serve concentrated corridors, and door-to-door accessibility has to be added by feeder services. The grain of the former has to be rather coarse, i.e., not able to reach many dispersed points directly. Any mode that attempts to do the latter as communal transit for the sake of user convenience will not be in a position to provide rapid service, because of the many stops that will have to be made. To a large extent, this consideration explains the popularity of the private automobile.

Carrying Capacity

Transportation modes available today cover a wide spectrum in their ability to do work, i.e., carry people. A fundamental and not-too-difficult selection task is choosing a proper mode to respond to estimated demand volumes. If the users from a district number a dozen or so during a day, only individual street-based vehicles (perhaps in joint use) can be considered; if they number several tens of thousands, a subway will have to be built. The suitable responses at the extreme ends of the scale will be expensive in one way or another.

Speed

Time distances, not *physical* distances, are of concern here. For any given traveler in an urban situation, the maximum speed that a vehicle or train can attain on an open channel is of little interest; what matters is the total time consumed from the origin point to the destination and the inconveniences of transfers along the way. The private automobile is a formidable competitor again, except on truly congested street networks. The aggregate rapidity of movement is also a communal concern to the extent that time spent in travel is unproductive and tiresome to the participants.

Passenger Environment

In a prosperous society, personal comfort and convenience features are increasingly significant. If certain levels in quality of life have been attained in residences and workplaces, greatly inferior conditions will not be tolerated during travel. These features encompass the smoothness of the ride, privacy (or at least some distance from strangers), sanitation, climate control, availability of seats, visual quality, and anything else that registers through human senses. The challenging task in communal transit is to measure up to what private cars provide.

Reliability

Life in contemporary cities is stressful enough, and our society (as well as our employer) expects punctuality. Delays in traffic and travel are acceptable only as rare occurrences. There are modes that are more immune to traffic overloads and bad weather (rail-based, mostly), and there are others that are quite vulnerable to urban disruptions (street-based, mostly).

Safety and Security

Residents in cities are well sensitized, through continuous media attention, toward issues of personal safety and security—for good reasons. This is mostly a matter of the overall level of civilized behavior in a community and police protection, but there are modes that are perceived to be more susceptible to antisocial action and physical breakdown than others.

Conservation of the Natural Environment and Fuel

The attention paid lately to the quality of air and water around us and the concerns with resource depletion enter in the planning and design of many urban systems, particularly so with trans-portation. While these are national issues with national man-dates, solutions can be achieved only through work at the local level, even if the consequences of any individual small action may be seen as marginal. Generally speaking, transit is benign, and low-occupancy automobile use is damaging.

Achievement of a Superior Built Environment

We can continue to expect that major transportation systems that significantly enhance the accessibility of specific nodes or corri-dors will generate a positive effect on land use and distribution of activities. This feature has potential for organizing the urban pat-tern, but evidence shows that this does not happen in all instances and it does not happen automatically—unless other constructive organizing programs are also implemented.

Costs

The expenses associated with transportation improvements and management can be broken down in considerable detail, but the commonly listed elements are right-of-way acquisition, construc-tion of the channel (roadway or guideway) and facilities, purchase of rolling stock, and annual operation and maintenance expenses, which include compensation for the work force, purchase of fuel or power and supplies, maintenance of equipment and facilities, and managerial expenditures. Nothing is cheap, but some modes involve massive capital investments, while others consume large amounts of resources to run services and maintain hardware. It should not matter in the long run whether the funds come from municipal, state, or federal budgets since they are all drawn from the wealth of the entire society and country, but it does matter when decisions have to be made with respect to any specific system. The costs, either in their entirety or by separate components, are frequently, as might be expected, the life-or-death factors for any transportation project.

Implementability

This concern refers to elements that are complex, not always well defined, and frequently obscure to the general public in the political and institutional realms, sometimes reflecting established practices and habits. They can be critical items if progress with any project is expected, and they may sometimes represent insurmountable barriers. The engineers have an equivalent term— *buildability*—in public works construction. But that is a comparatively easy task since it refers to the physical ability to get something done. Implementability encompasses social, administrative, and political arrangements and habits, often unique to a specific community. Transportation systems affect much more than tangible artifacts and their operation. These factors operate at the local and state levels primarily, and no generalizations will be made here, except to call for serious attention and understanding well before any irreparable damage is done due to neglect or ignorance.

Image

Transportation systems and services are the public face of a community. Everybody comes in contact with them, and they are usually the first thing that a visitor from the outside experiences. They are elements of civic pride in many instances, and they show the seriousness that is applied to the creation of a livable and efficient community. But pride can also be a sin, and there are instances on record in which transportation solutions have been implemented for reasons other than functional necessity. This should not happen with full knowledge of the capabilities and potential of transportation modes in the contemporary city. There are legitimate reasons to applaud service systems that respond to the needs and capabilities of a community, to take pride in something that works well.

We should be ready now to apply the preceding criteria as a screen in reviewing the many transportation modes available for service. We shape our service systems, they do not shape us, but they do have a fundamental role in defining the structure of communities and how we live and operate in cities and metropolitan areas. Transportation systems and land use are two sides of the same coin. To achieve the exact built environment that we wish to have, work with both of them in a mutually supporting fashion is indicated. The record from the past has not always been inspired; we have the means, the methods, the choices, and, let us hope, the knowledge today to do better.

Thirty years ago, Lewis Mumford articulated his vision as to what urban transportation should be. It is valid today, although it still remains a vision.

Transportation: "A Failure of Mind"

Lewis Mumford

CAMBRIDGE, Mass.—Many people who are willing to concede that the railroad must be brought back to life are chiefly thinking of bringing this about on the very terms that have robbed us of a balanced transportation network—that is, by treating speed as the only important factor, forgetting reliability, comfort and safety, and seeking some mechanical dodge for increasing the speed and automation of surface vehicles.

My desk is littered with such technocratic fantasies, hopefully offered as "solutions." They range from old-fashioned monorails and jet-propelled hovercraft (now extinct) to a more scientific mode of propulsion at 2,000 miles an hour, from completely automated highway travel in private cars to automated vehicles a Government department is now toying with for "facilitating" urban traffic.

What is the function of transportation? What place does lomocotion [sic] occupy in the whole spectrum of human needs? Perhaps the first step in developing an adequate transportation policy would be to clear our minds of technocratic cant. Those who believe that transportation is the chief end of life should be put in orbit at a safe lunar distance from the earth.

The prime purpose of passenger transportation is not to increase the amount of physical movement but to increase the possibilities for human association, cooperation, personal intercourse, and choice.

A balanced transportation system, accordingly, calls for a balance of resources and facilities and opportunities in every other part of the economy. Neither speed nor mass demand offers a criterion of social efficiency. Hence such limited technocratic proposals as that for high-speed trains between already overcrowded and overextended urban centers would only add to the present lack of functional balance and purposeful organization viewed in terms of human need. Variety of choices, facilities and destinations, not speed alone, is the mark of an organic transportation system. And, incidentally, this is an important factor of safety when any part of the system breaks down. Even confirmed air travelers appreciate the railroad in foul weather.

If we took human needs seriously in recasting the whole transportation system, we should begin with the human body and make the fullest use of pedestrian movement, not only for health but for efficiency in moving large crowds over short distances. The current introduction of shopping malls, free from wheeled traffic, is both a far simpler and far better *technical* solution than the many costly proposals for introducing moving sidewalks or other rigidly automated modes of locomotion. At every stage we should provide for the right type of locomotion, at the right speed, within the right radius, to meet human needs. Neither maximum speed nor maximum traffic nor maximum distance has by itself any human significance.

With the overexploitation of the motor car comes an increased demand for engineering equipment, to roll ever wider carpets of concrete over the bulldozed landscape and to endow the petroleum magnates of Texas, Venezuela and Arabia with fabulous capacities for personal luxury and political corruption. Finally, the purpose of this system, abetted by similar concentration on planes and rockets, is to keep an increasing volume of motorists and tourists in motion, at the highest possible speed, in a sufficiently comatose state not to mind the fact that their distant destination has become the exact counterpart of the very place they have left. The end product everywhere is environmental desolation.

If this is the best our technological civilization can do to satisfy genuine human needs and nurture man's further development, it's plainly time to close up shop. If indeed we go farther and faster along this route, there is plenty of evidence to show that the shop will close up without our help. Behind our power blackouts, our polluted environments, our transportation breakdowns, our nuclear threats, is a failure of mind. Technocratic anesthesia has put us to sleep. Results that were predictable—and predicted!—half a century ago without awakening any response still find us unready to cope with them—or even to admit their existence.

Walking

Background

We are all pedestrians; any trip by any means includes at least a small distance covered on foot at the beginning and end of each journey. Walking is the basic urban transportation mode that has allowed settlements and cities to operate for thousands of years. It is still very much with us, but its role has been eroded with the introduction of mechanical means of transportation, drastically so in American communities, with the dominant presence of the private automobile in the last half century.

The principal transportation mode in the developing world, even in large cities, is still walking because of constraints on the resources needed to operate extensive transit systems. People cover long distances on foot every day and expend human energy that they can scarcely spare. Walking under those conditions is an unavoidable chore that consumes productive capability. In North America and Western Europe, the attitude and policies are just the opposite: walking is efficient, healthful, and natural. We should do more of it—almost everybody agrees—and some of the current trends should be reversed. Ironically, among the most popular exercise machines in health clubs and in homes are treadmills that simulate walking, which could be otherwise accomplished with a transport purpose on the street.

Admittedly, because of the size of contemporary metropolitan areas, with origin and destination points far apart, the need to save time consumed in routine travel, and the desire for basic comfort and avoidance of severe weather conditions, walking as a transportation mode has limitations. But the niche that it can fill is still rather large, and the opportunities are by no means fully exploited. Just the reverse is happening today, and some proactive programs will be necessary to restore reasonable balance.

The trend in the percentage of commuters who walk to work in the United States[1] has been negative:

1960	9.9 percent
1970	7.4 percent
1980	5.6 percent
1990	3.9 percent
1999[2]	3.1 percent

Much of this can be explained by the fact that land use patterns have become more coarse-grained (i.e., greater segregation of job places and commercial activities from residences), and trips have become longer overall, but there is also the greater propensity to use the car for any purpose, even just to go around the corner. Working at home has increased slightly, but not enough by far to explain the drop in walking to and from workplaces. Appeals to reason and civic responsibility will not alter the prevailing attitudes much; programs to make walking *attractive* to individuals will have to be expanded and implemented. The contemporary built environment in North America is not always fully enabling toward pedestrians. Not all new streets have sidewalks, they are not always structured into coherent networks, and they frequently lack proper amenities (good pavement, lighting, rest areas, etc.).

[1] U.S. Census data.
[2] Since the 2000 U.S. Census data were not yet available, information from the American Household Survey was used for 1999.

Development History

> *What is a featherless biped?*
>   —*A plucked chicken or a pedestrian.*
> *What is a pedestrian?*
>   —*A driver who has found a parking space.*

The historical review of walking could begin some 20 million years ago, when certain animals—our ancestors—started to move around on their hind legs.[3] That would not be a very profitable discussion; even the last 6000 years (save the last 150 years or so) can be quickly summarized to arrive at the conditions that prevail today.

For thousands of years, settlements and urban groupings, eventually evolving into cities, were almost entirely walking environments. Some deliveries were made by pack animals and carts, some people were carried by one device or another, and soldiers and chiefs liked to ride, but most movement and linkages inside cities were accomplished on foot, even the carrying of heavy bundles and parcels. Cities had to be of a walking scale, and they were—almost all of them could be easily traversed on foot within a quarter of an hour. Even the few very large ones (imperial Rome, Beijing, Paris, London) were assemblages of neighborhoods that each contained the daily life of the residents, including their workplaces. Extensive wheeled and animal traffic, however, was present in the larger cities as a part of production and distribution activities. Street congestion on the narrow streets was known even in ancient cities.

The streets, often just the linear spaces left between building lines, usually made no provisions for separate types of movement. People, carts, and animals used the same channels, mixed freely, and were all impeded by the many activities that spilled out on the street and have traditionally been a part of the urban scene: ped-

---

[3] Needless to say, any dates with prehominid hominoids are uncertain. Progress was slow and gradual, and new archaeological discoveries are always making adjustments to the dates. The evolution toward erect locomotion apparently is not yet quite complete either, judging from the fact that many of us tend to get chronic back pain.

dlers, vendors of food, purveyors of various services, entertainers, musicians, preachers, children, thieves, and beggars. Depending on the organizational level of any society and the attention paid to public works by any city administration, the streets may have been paved,[4] but usually were not (except for major avenues), and there may have been provisions for drainage, which can be seen even in some of the very ancient cities. Principal streets in the cities of Mesopotamia and the Indus Valley show evidence of drains and pedestrian lanes. Significant Roman cities provided raised sidewalks along both sides of the street, leading to adjoining store and housing entrances. There were stepping stones across the vehicle channels, spaced to allow carts with wheels a standard distance apart to move along.

Yet, the practice of providing sidewalks became lost for many hundreds of years. Water and liquids found their own way, not infrequently turning street surfaces into malodorous and pestilent bogs. European cities, as a rule, had no sidewalks during the medieval and Renaissance periods; they appeared in the second half of the eighteenth century, at least in the more prominent cities.[5] Available evidence indicates that walking on urban streets, up to the Age of Enlightenment, was a dangerous and dirty practice. Despite certain images based on romantic nostalgia, people walked when they had to, but not for pleasure and recreation. That became possible only considerably later, when some protected and designated spaces were developed and opened to the public—promenades, public gardens, and parks.

In the nineteenth century, sidewalks were always present along the sides of improved streets, with a curb and a gutter, in all the cities of the Old World, as well as in colonial towns. The prime pedestrian environments were the grand boulevards, not in Paris alone.[6] In American cities, the nineteenth-century "parkways" in their early form extended the landscaped park environment into the city itself and were intended for leisurely strolls and carriage rides.

[4] *Paving* means that a reasonably watertight surface is created that will keep its shape with no ruts under the pressure of wheels, hooves, and feet, and that there will be little dust during dry periods and no mud on wet days.

[5] This is a conclusion drawn from scanning many contemporary images of cities, assuming that they are reliable in the details.

[6] See J. Cigliano and S. B. Landau (eds.), *The Grand American Avenue: 1850–1920* (Pomegranate Artbooks, 1994, 389 pp.).

The functional purpose of a sidewalk and a curb was to protect pedestrians from the consequences of horse-drawn wheeled traffic, which had grown immensely in volume and impact. Crossing major streets became a dangerous adventure, horses and heavy wagons were frightening to most people, and—most important—sanitary conditions on the streets were abysmal. Litter and garbage were not collected with particular diligence, nor was the excrement of hundreds of horses. On a wet day, ankle-deep slurry covered street surfaces, as it has been described by some earthy contemporary authors. With a raised sidewalk in place, the tides could be held back, a reasonably solid surface could be provided, the adjoining property owners could at least sweep their own frontages, and some positive drainage could be achieved. Nevertheless, a gentleman escorting a lady was expected to walk on the curb side to shield her from likely splashes. Gallantry aside, it was presumably easier to clean a pair of trousers than multiple voluminous and frilly skirts.

Covered sidewalk in a Renaissance city (Vicenza, Italy).

A specific building form favorable to pedestrians was the *arcade, colonade,* or *loggia* along the street fronts of buildings. They are encountered in various historical periods at many locations. They were built primarily to gain more space on the upper floors by protruding into the street, but they also offered thereby a sheltered path for walkers (provided that the arcades were not cluttered up with other activities), since wagons and carts could not conveniently enter. In very hot climates and in places with extreme rainfall, they are a practical means of minimizing the disruptions of urban life. Unfortunately, today, in crime-prone American cities, street-front arcades are sometimes banned because of the fear that muggers may hide in their shadows.

Another similar device was the *passage* or *galleria* (also called an *arcade* in American English)—a building perpendicular to the street that provides accommodation for many stores on a single or multiple levels, with an open central circulation space. The first of these probably was the *Galleries de Bois* in Paris, opened in 1786. The most famous one is the *Galleria Vittorio Emanuele I* in Milan (1865), but there are many others throughout Europe—

Multistory nineteenth-century arcade in Cleveland, Ohio.

practically in every large city, since they were seen as prestigious retail venues in the nineteenth century. America has its surviving examples as well—in Cleveland, Providence, and Los Angeles. This concept has emerged again in today's planning for pedestrian spaces, as is discussed subsequently.

In the second half of the nineteenth century, the movement toward park creation in American cities resulted in attractive walking environments as well, but for recreational purposes only. The *City Beautiful* efforts in the very early twentieth century extended the concept of formal boulevards, with landscaped walkways, enhancing the prestige of any city at that time.

The appearance of modern architecture some decades later brought with it new concepts in pedestrian space. For the purposes of this discussion, the principal characteristic was the separation of pedestrian walkways from vehicular traffic, which by that time had become a threat to safety and, due to its speed, incompatible with the pace of human locomotion. The results were *superblocks* (towers in a park), with motor vehicles kept on the periphery and pedestrians able to follow separated paths. Le Corbusier—perhaps the most influential architect of the twentieth century—was a principal proponent of this design form through his many conceptual plans and several projects that were implemented, starting in the 1920s. Independent walkways were not exactly a completely original idea, but they were brought into the city fabric as a major design-governing feature. Superblocks were advocated for lower-density neighborhoods as well (for example, Radburn, New Jersey, in 1929). Not all the efforts were attractive or successful (for example, the massive and monotonous public housing projects in large cities), and frequently problems with maintenance, policing, and privacy came to the fore.

Nevertheless, the separated pedestrian walkway at the local level leading to service facilities, institutions, and transit stops remains a strong concept in the inventory of planning and design options. Within the last few decades, however, a different approach has also emerged under the label of *new urbanism* (or the *neotraditional planning, pedestrian precinct,* or *transit village* concept), which attempts to restore primacy to pedestrians within residential and commercial districts.[7] It recognizes that motor vehicles create conflicts, but, instead of seeking to segregate people in "safe" environments, asserts that automobiles should be made to behave and that streets should accommodate all movement, particularly walking, with no interference or dangers. The street traditionally serves all; there should be no reasons to drive fast or irresponsibly, and, by creating reasonably high densities and clustering activities, most destinations should be accessible on foot. Whether residents living in these districts do actually always walk and whether developers will embrace the concept and will be able to create a mass market remains to be seen. The new urbanists also believe that contemporary city folk want to recapture a sense of belonging to a community and basic neighborliness through walking to accomplish daily chores and sitting on a traditional front porch facing the public street. The intent is to be applauded; social scientists, however, continue to record a growing deliberate alienation by most residents from the public realm and a retreat into their own private spaces (in the enclosed car and in front of the TV set).

Similar community-building aims are expressed under *traffic calming* programs as approaches toward a safe and attractive pedestrian environment, which are outlined further in Chap. 5, Automobiles.

No discussion of pedestrianization in the United States can avoid suburban shopping centers. This purely American building form has at its core the presence of a pedestrian space—deliberately designed to attract people, make them linger, and encourage spending. While these service clusters may be almost impossible to get to on foot in low-density suburbia, once the customers park the car, everything that has been learned about keeping people happy and comfortable is applied here, deliberately and through countless repetitions. This encompasses weather-protected space (usually enclosed, always at the same temperature), widths of corridors that allow show windows on both sides to be seen, plenty of sitting and rest areas out of the main flow, colorful (if some-

---

[7] Much literature has been generated recently regarding this concept. See A. Duany and E. Plater-Zyberk, *Towns and Town-Making Principles* (Rizzoli, 1991, 119 pp.) and P. Calthorpe and W. Fulton, *The Regional City* (Island Press, 2001, 304 pp.).

times garish) décor with historical or popular design references that are easily understood, food courts that provide for all tastes and keep customers on the premises, no interference by (not even visibility of) goods and maintenance operations, sufficient escalators and elevators if operations take place on several levels, exemplary cleanliness, and complete security.

The ancestry of American shopping malls can be traced back to the first half of the twentieth century, to a few examples of clustered stores in model developments (Roland Park, for example, in Baltimore), and particularly to the Country Club Plaza in Kansas City (1922) as the first automobile-oriented center. The real model, however, was more likely the series of farmers' markets built in Los Angeles in the 1930s, with off-street parking lots and central pedestrian circulation areas. A few shopping malls appeared before World War II, but the real boom started in the 1950s, accompanying the explosive trends of suburbanization and movement of families to open peripheral territories. Thousands were built each year during the peak period in the 1960s and 1970s, reaching a total inventory of more than 26,000 malls across the country by 1990.[8] The first regional mall was Northgate in Seattle (1950), followed by Northland in Detroit (1954). Since the 1960s, almost all the large malls have been enclosed and equipped with full climate control. The size records were set by the West Edmonton Mall in Canada (3.8 million ft^2 [350,000 m^2] of retail space, 1981) and the Mall of America outside Minneapolis (4.2 million ft^2 [390,000 m^2], 1992).

While shopping malls are not specific transportation facilities, they are the prime popular examples of pedestrian environments in North America today. They can be seen as laboratories where human walking

Interior of a contemporary shopping mall in Newark, New Jersey.

[8] See M. D. Beyard and W. P. O'Mara, *Shopping Center Development Handbook* (Urban Land Institute, 1999, 3d ed.) and W. Rybczynski, *City Life: Urban Expectations in a New World* (Scribner, 1995, 256 pp.), Chap. 9.

behavior can be observed and conclusions reached as to which features are favored by real people and which are not, even if the findings are not always encouraging. (The counteraction of old city and village centers against shopping malls to recapture lost business by rebuilding traditional commercial cores into pedestrian-friendly environments is outlined under "Automobile-Free and Automobile-Restricted Zones" in Chap. 5, Automobiles.)

Another approach toward expediting pedestrian operations—not particularly new, either—has been the creation of several levels, thus giving crowds adequate space. The issues associated with multiple pedestrian levels are discussed on subsequent pages, but successful examples are found either as entire underground networks or as connecting mezzanines under major street intersections. Bringing pedestrians one level up is also a possibility, but the record with this idea is rather spotty. Efforts to do that in the core areas of London (Barbican), Stockholm, Bogota, San Juan, and other places have largely not fulfilled expectations. Second-level *skyways* are an American invention, and they work well and are popular as weather-protected connections in cold climates between garages and commercial establishments.[9] They are found in Minneapolis–St. Paul, Rochester, and Cincinnati, and as short linkages at many other places. While they can penetrate and enter buildings, they operate mostly as corridors, not so much as shopping streets, and thus can be clearly classified as components of the local transportation networks.

The situation is different with underground pedestrian systems. While some of them take the form of interconnected corridors (as in Houston), they can accommodate activities along the sides. The most extensive network has been developed in Montreal, where spacious pedestrian passageways under the street level connect many key buildings and transit access points. It is truly an underground system. Very large pedestrian networks have been developed in the larger cities of Japan, as well as in Seoul, Korea. Extensive, albeit smaller, arrangements are found in Toronto and New York. The World Trade Center, for example, had an extensive below-street environment with a large retail compo-

[9] For a full analysis see K. A. Robertson, "Pedestrian Walking Systems: Downtown's Great Hope or Pathways to Ruin?" *Transportation Quarterly,* July 1988, pp. 457–484.

Nicollet Transit Mall in Minneapolis, Minnesota, with walkways and skyways.

nent. It was about to be rebuilt and improved before its destruction. In many instances, connections from underground transit stations extend into the surrounding blocks and offer direct entries to stores.

Localized additions to the overall pedestrian system in high-density districts are underground crossing mezzanines under major street intersections. The best examples are equipped with escalators, offer a wide inventory of convenience shopping and grazing (fast food), have public toilets, and can even be effectively used for safe loitering. Vienna, London, and Prague, among other cities, have a number of successful cases.

To complete this review of options, mention has to be made of contemporary enclosed commercial and activity spaces as a single building or combination of buildings, as represented by enclosed urban malls. They are not transportation paths, although pedestrian movement crosses them, and they frequently offer attractive alternatives to the outdoor sidewalks. Indeed, the latter point can be regarded as a criticism, because they tend to siphon life away from the traditional walkways. At this time it is hard to think of any city, particularly in Europe, that would not have one or more examples. The first such project in the United States was the Midtown Plaza in Rochester (1962), and the more visible examples today are the ZCMI Center in Salt Lake City, the Gallery in Philadelphia, Eaton Center in Toronto, the Embarcadero Center in San Francisco, Peachtree Center in Atlanta, Water Tower Center in Chicago, and many more. There are also interesting projects in converting historical buildings into such environments (for example, the Old Post Office in Washington and the recently refurbished Grand Central Terminal in New York).

Besides examining physical improvements encompassing pedestrian spaces and paths, there is the question of the extent to which free human behavior, as represented by walking as a repeated daily activity by everybody, can be analyzed systematically, provided for, and planned accurately through quantified methods. For thousands of years, pedestrian spaces happened (a few were deliberately designed), and they either worked or they didn't. If these environments had problems, they were progressively changed; i.e., they became adjusted to needs and expectations over the years. Or they were scrapped. Planning and design depended on the instincts, good sense, talent, and experience of designers and builders. Layouts and dimensions were arrived at impressionistically. Some great successes were thereby achieved, but in most cases a trial-and-error process ensued.

That situation has changed, at least to the extent that design and planning tools have been created that bring considerable specificity and reliability to estimates of space needs and the structuring of safe and efficient flow paths.[10] No claims are to be made that this will automatically produce superior designs, but any design can now be tested as to its functional adequacy and be at least sized accordingly. The principal point is that the planning of pedestrian facilities (spaces, paths, stairs, and sidewalks) can be supported by documentable analyses, thus bringing much more reliability to a task that has otherwise uncertain dimensions.

Types of Walking Practice

The behavior of human beings when they are in motion in public spaces is affected by the purpose of such action for each individual at any given time. As familiar as walking is to all of us, several distinct situations can be identified.

Walking Briskly

This represents the need to move expeditiously from Point A to Point B. The principal purpose is to overcome distance quickly, and to do this by mostly ignoring all distractions and not being diverted by other destination or action possibilities. The best example is going to work in the morning under time pressure to reach the desk, or attending any significant event with a precise

---

[10] The seminal work was done by J. J. Fruin of the Port Authority of New York and New Jersey (published as *Pedestrian Planning and Design* (Metropolitan Association of Urban Designers and Environmental Planners [MAUDEP], 1971, 113 pp.), later joined by G. Benz (several articles jointly with Fruin and *Pedestrian Time-Space Concept* (Parsons Brinckerhoff Quade & Douglas, 1986). Subsequently, other analysts have expanded this subfield, and it is now a regular component of traffic engineering.

starting time. The shortest time distance is chosen; safety on the path and reliability do count, but aesthetics and human amenities can be largely ignored. The walker in a real hurry will slosh through puddles, jump traffic signals, elbow others out of the way, avoid stairs, and take any possible shortcut to keep to a straight line. This is definitely a form of transportation, with the trip characteristics easily identifiable (origin and destination points, time, mode, purpose, etc.). The action is rational within the operational context, and it is predictable as to its execution. Volumes, levels of occupancy, and speeds on connecting paths and within spaces along the way can be estimated, if the number of trip ends associated with origin and destination points are known. Sidewalk loading conditions and the utilization of spaces can be calculated. This is the pedestrian equivalent of channelizing motor traffic flow and building highway lanes (in concept only, needless to say). The precise pedestrian flow analysis methods, outlined in following sections, are very much applicable to this situation and give good results.

Meandering[11]

But we are not always in a rush. People employ their senses to enjoy the surroundings, to look at interesting things that catch their attention, to gawk at other people, and to be a part of the street scene as they walk along. Yes, it is still movement from Point A to Point B, but not necessarily in a straight line and not at a constant speed. Indeed, this type of action embodies the attractiveness of being in a city, and every opportunity should be taken to enjoy the walk and be distracted. Walking should be an unconstrained and positive experience, as long as we are not late for the next appointment. How much time any such journey will take is a function of the time available to the walker, the interest level of the surroundings, and weather conditions. In other words, this is not a steady-state situation, not even for the same traveler on a regular schedule, and the level of predictability regarding the exact path and time consumed decreases. If a designer lays out an exciting path for walking and meandering, will all of us move in the same way and consume the same amount of time? Most likely not. It is a dynamic and changing situation—buildings, spaces,

---

[11] The English language offers a few other possible descriptors for this type of locomotion: *rambling, peregrinating, sauntering,* or *strolling.*

other people, and urban elements are seen in motion, causing occasional slowdowns.

Technical planners have great problems with these conditions, being able at best to use statistical averages on human behavior.

Tarrying[12]

People are also found in pedestrian spaces and on paths with really no urgent intention to get anywhere, but simply for the purpose of enjoying the scene, meeting others, or having an outdoor lunch. A pedestrian leaves Point A, goes to a few other locations, and quite frequently comes back to Point A. He or she is a part of the urban street theater, sometimes blocking those who are in a hurry, consuming space and time for the best possible reasons—personal enjoyment.

We all do it when we have the time and when the weather is good, but this is certainly not transportation. Thus—reluctantly—we have to stop further discussion of this pedestrian situation, and leave it for urban designers and urban recreation specialists. Technical pedestrian space planners, however, have to recognize the presence of this element as a possible and likely use of space and facilities. Systematic observations of what people do in public spaces have brought insights on their behavior, and people's behavior is no longer a complete mystery.[13] Designs can be guided fairly well, and it is possible to predict what will work and what will be avoided, where people will congregate and what they will pass by. It is also possible to estimate how long persons will be present, on the average, in such spaces.

In summary, our ability to be analytically precise and conceptually purposeful with the walking mode is still limited and only selectively reliable, particularly when pure functional linkages do

---

[12] Again, other choices for a term are available: *lingering, poking, dallying, dawdling, delaying,* and—most important—*loitering.* Unfortunately, all of them carry a negative connotation in English, as if such activities were deplorable. That is not so, and the rather common signs reading "Do Not Loiter" should be used sparingly, if at all.

[13] The most perceptive and productive student of these situations was William H. (Holly) Whyte, whose findings should be required reading for anybody dealing with pedestrian issues. Particularly recommended are his *Social Life of Small Urban Spaces* (The Conservation Foundation, 1980, 125 pp.) and *City: Rediscovering the Center* (Doubleday, 1988, 386 pp.).

Central pedestrian street in Munich, Germany.

not dominate. Space-time occupancy, however, can be defined, movement paths can be identified, and restful places and enclaves can be structured. Rationality and understanding gained from controlled previous and on-going research can be applied. The pedestrian situation is most complex because people do not stay in marked lanes as cars do, and in many instances all three types of walking, as previously outlined, will exist concurrently.

To create any successful pedestrian environment, ideas and original concepts generated by first-rate designers are still crucial. Nothing very good is likely to happen without a creative spark, but beyond that plans do not have to be guided by intuition alone. The range of promising possibilities can be narrowed effectively and pilot projects can be tested systematically by applying the study techniques developed recently. The design of pedestrian spaces, movement networks, and nodes of convergence can benefit greatly from a rigorous analysis.

Reasons to Support Walking

Economy

The walking mode involves very little expense, either public or private. The paths themselves, usually sidewalks, are usually built together with normal street construction, and the specific expense is rather minimal. On a local street, adding sidewalks to the other cost items would account for only a small fraction of the total cost. The pavements can be of a rather light construction since they do not have to carry heavy loads; however, they have to be strong enough to support an occasional maintenance or service vehicle. Separate walkways that traverse parks and open spaces have to be able to accommodate police cruisers and service trucks.

Cost of operation is not a concept associated with the walking mode. Each person is responsible for his or her own equipment (shoes, mostly), which do wear out, but it is a normal personal expense. An interesting issue is the consumption of energy. A 155-lb (70-kg) person will burn 280 calories walking briskly for 1 hour[14]—which is a benefit in an overfed society in need of exercise, but may be a significant consideration for undernourished populations. Walking slowly, only half of that amount is consumed; climbing stairs doubles the energy expenditure as compared to normal walking.

Health

As mentioned previously, an argument can be made that the obvious health benefits of this, the most basic, form of exercise should constitute a major reason to equip American communities with the best possible walkway systems. This is a question of not only offering a rich inventory of physical facilities, but also a matter of structuring districts so that walking is a logical modal choice, and the experience is safe and attractive.

Availability

There is no need to wait for a transit vehicle or even to turn on the ignition; the mode is always present and ready for use (within reason). Most cities, particularly in the industrialized countries, have done quite a lot to make the sidewalk network navigable to people with mobility impairments (curb cuts with ramps throughout), thus making the walkway system more free of obstacles to use than any other transportation mode.

Cognition

A pedestrian is in direct contact with the surrounding environment. The act of walking is automatic and does not require deliberate attention or even too much care to avoid obstacles and dangers. The senses and the mind can be employed to appreciate the streetscape or the landscape and to pursue independent thought.

Environmental Protection

Walking is the ultimate environmentally friendly transportation mode. (No need to worry about heat generated by bodies and evaporated perspiration.)

[14] *Source:* NutriStrategy Web page.

Reasons to Exercise Caution

Walking, as pleasant and advisable as it might be, is not an all-purpose transportation mode because there are functional limitations.

Distance

The human animal does become tired, rather quickly. We also have an advanced brain that is constantly searching for the paths of least resistance and seeks to conserve personal energy (otherwise known as *self-preserving laziness*). The question that has preoccupied transportation planners for some time is the reasonable walking range that people will accept, particularly Americans—well-known as car-obsessed individuals. The examples range from some motorists who will circle a shopping center parking lot endlessly to find a space closest to the mall, to dedicated hikers who take great pleasure in long and challenging walks. On-street parking spaces are deemed to be good only if they are within 200 to 300 ft (60 to 90 m) of the door.

A specific planning concern is acceptable access distance to transit stations or stops on foot. The general consensus today is that a quarter mile (1320 ft; 400 m) is a range within which just about everybody will walk; within a half mile (2640 ft; 800 m), the number of walkers may be cut by 25 or 50 percent; a few, but only a few, will walk a mile to any destination or transfer point. Eighty percent of walking trips are less than 3000 ft (0.9 km). It is a curious fact that residents of large cities are more likely to embrace walking than those living in smaller places. New Yorkers in particular tend to be at the top of this list. Commuters from the Port Authority Bus Terminal or any of the major rail stations will readily walk 3000 ft or more to and from their Midtown offices.

Major events and festivities represent an exception to walking limitations—a mile is quite acceptable under those circumstances. (Any National Football League game, by definition, is a major event.) Even school children are pampered in this country: most school districts have a rule that any pupil or student is entitled to a school bus pickup if he or she lives beyond a 1-mi radius.

Speed

A human being is not particularly fast. A regular walking pace is 15 minutes to the mile, which may be extended to 20 minutes. This translates into 4 or 3 mph (6.4 or 4.8 kph). For short distances,

Distances That Can Be Traveled in 30 Minutes

Mode	Miles	Kilometers
Pedestrian walking leisurely	1.5	2.4
Pedestrian walking briskly	2	3.2
Race walker	4.5	7.2
Jogger	3	4.8
World-class runner	6.5	10.5
Bicycle at normal pace	5	8.0
Bicycle in 1-h race	15	24.0
Local bus in dense city traffic	3	4.8
Bus on suburban streets	8	13.0
Express bus (suburb to central business district)	15	24.0
Streetcar in mixed traffic	4	6.4
Light rail service	8	13.0
Subway in regular service	12	19.5
Commuter rail in regular service	18	29.0
Regional express train	22	35.4
Metroliner	45	72.5
French TGV (*train à grande vitesse*)	80	130
Private car in a badly congested city district	1	1.6
Private car moving at normal urban speed limit	12	19.3
Private car on an expressway at 55 mph	27	43.5
Indianapolis 500 race car	90	145.0

since walking does not involve a startup or terminal time loss, the slow pedestrian speed does not matter, but it becomes a factor with longer trips. There are some athletes who can cover a mile in less than 4 minutes, but even that is only 15 mph (24 kph), hardly comparable to any motorized mode. A very good marathon runner can maintain a 13-mph speed. The best that any human being has done is 9.79 seconds in the 100-m dash (22.8 mph; 36.8 kph), which cannot be sustained for any distance, either.

Traffic engineers, who are responsible for adjusting traffic signals so that there is enough time for every pedestrian to cross a street, are particularly concerned with velocities over short distances.[15] The usual design assumption is 4 ft/s (1.2 m/s), or 2.7 mph (4.4 kph), which accommodates almost everybody. Young

[15] Chapter 11.6 in J. D. Edwards, *Transportation Planning Handbook* (Prentice Hall/ITE, 1992), p. 396 ff.

people walk faster, and 5 ft/s (1.5 m/s), or 3.6 mph (5.8 kph), would be workable for them. However, with the presence of a significant number of elderly persons in the flow, 3 ft/s (0.9 m/s), or 2.0 mph (3.3 kph) may be used.

Some tentative studies suggest that Americans generally walk faster than all other nationals, except the Japanese. Also, that residents of the larger cities are more in a hurry, especially in Boston and New York, than in smaller places. This is an intriguing hypothesis, which should be tested further.

Change in Elevation

People are reluctant to change elevations, because we know instinctively that this involves significant energy expenditure, as compared to level walking. Thus, in any pedestrian designs for sizeable flow conditions, this aspect requires particular attention. Changes in level should be avoided if at all possible; mildly sloping ramps that are not particularly prominent visually, escalators, or elevators have to be provided. There is a natural reluctance to use overpasses and underpasses (the latter also because of safety reasons).

Weather Conditions

Adverse weather, whether it is rain, snow, high wind, or broiling sun, will reduce considerably any propensity for walking. This is one of the principal reasons why indoor pedestrian environments have been so successful. In almost all climatic regions, however, any outdoor pedestrian space or path can benefit from full or partial shelters, particularly because walking trips will otherwise shift to motorized modes at certain times, thus either overloading transit systems or requiring them to have standby capacity. In most instances deciduous trees are suitable additions since their leaves provide shade, but their bare branches let the sun penetrate in the cold months when some warmth is welcome.

Carrying Goods

Again, pedestrians have limitations, including how much weight they are able or willing to carry with them. Briefcases and pocket books are one thing, but even shopping bags may present problems. Being accompanied by small children or pushing strollers are not exactly encumbrances, but such common situations do require attention in the design of walkways.

Impaired Personal Mobility

Any community has a certain percentage of people who have larger or smaller disabilities that will reduce the extent of their participation in the walking mode. These may be temporary, such as a sprained ankle, or they may be permanent, such as missing limbs. There are always the very young and the very old. The task in the design of pedestrian facilities is to achieve accessibility that allows the greatest percentage of the population to use them. Methods range from curb cuts at intersections that allow wheelchairs to move, to audible traffic signals that assist those with visual impairments.

Safety and Security

The technical term *pedestrian–vehicle conflict* refers to the one-sided violent encounter between the soft tissues of a human body and a large, hard, frequently fast-moving, invulnerable metal object. There are no air bags or other protective devices for the walker; the only safeguards are to stay out of the way and to ensure that motorists recognize the presence of pedestrians and know that people always have priority in traffic channels when safety issues are concerned.[16]

In the United States, pedestrian accident rates are considerably lower than those in most other countries, but this may simply be due to the fact that fewer people walk here. The rates are too high nevertheless, no matter what they may actually be. This is a national concern, particularly if walking and jogging are to be encouraged as an overall policy, and programs have been developed both to educate and instruct drivers and walkers and to provide physical safety elements that would minimize, if not preclude, such occurrences. There were 5307 pedestrians killed in 1997, which represents a 43 percent decrease in the rate per 100,000 population since 1975. During this period, 13 to 17 percent of all victims of fatal accidents associated with motor vehicles were pedestrians.[17] By 2000, the fatalities had decreased further to 4739. The largest causes were "walking,

---

[16] A reasonably complete summary of pedestrian safety issues and programs is found in J. L. Pline, *Traffic Engineering Handbook* (Prentice Hall/ITE, 1992), p. 19 ff.

[17] U.S. Department of Transportation, Fatality Analysis Reporting System.

playing, or working in roadway" and "improper crossing of roadway at intersection."

Unfortunate associated issues are that many people do not observe traffic regulations scrupulously when walking, and enforcement frequently tends to be rather lax with respect to pedestrian behavior. There are places where local residents have a rather cavalier attitude toward "minor" regulations and will jaywalk and ignore "Walk/Don't Walk" signals (New York, Paris, and Bangkok come to mind),[18] leaving aside the general urban situation in Third World cities where traffic and crossing conditions can be quite chaotic. There are other locations—primarily in Western Europe and Japan—where rules are respected diligently. All this argues for the implementation of hard control devices (barriers, for example) that leave little discretion, even though they can be seen as constraints on free choice.[19]

Most pedestrian accidents occur when people have to cross vehicular traffic streams, as was pointed out before. The number and intensity of incidents vary widely under different street conditions and among various population groups. For example, there are many more accidents with cars turning left than with those turning right. This is due largely to the impaired field of vision of the driver and some confusion as to who should yield the right-of-way. Zebra stripes at crosswalks are highly recommended traffic markings, but they do not necessarily reduce accidents because many pedestrians ignore them, and motorists have the right to expect that walkers will stay within the designated space. A common cause of serious accidents is children who run into the street from between parked vehicles. Elderly people also experience higher accident rates, presumably due to being less agile in getting out of the way and moving more slowly than anticipated by some impatient motorists.

The most dangerous, and most deplorable, situation is the absence of sidewalks at all in many low-density suburban areas,

---

[18] The author tends to brag about the two jaywalking tickets that he has received on the streets of New York as proof of his personal independence, recognizing full well that this represents a juvenile attitude.

[19] A loud controversy was generated in New York in late 1997 when Mayor Giuliani caused short fences to be erected at high-volume pedestrian intersections near Rockefeller Center, limiting the crossing of avenues to the upstream side only. This was seen as an effort to primarily expedite vehicular movement, although undoubtedly there is a strong pedestrian safety feature as well. They are now grudgingly accepted.

forcing the few walkers to use the street pavement or the narrow shoulder. Joggers on busy streets are particularly vulnerable.

The other dimension of concern is *personal security*—the potential for criminal or threatening behavior by other people within pedestrian environments. Again, walkers, especially when they are alone, are vulnerable in certain situations. The possible countermeasures, besides the visible presence of police, are good lighting, clear visibility in all directions, absence of hiding places and secluded enclaves, and—above all—many people on the walkways at all times who maintain "eyes on the street."

Application Scenarios

A strong argument can be advanced that a basic walkway network should extend over the *entire community* where people live, work, and use various facilities. All points should be accessible on foot, with some convenience and safety. The only exceptions might be very local streets (cul-de-sacs and loops) where motor traffic is minimal and people have obvious priority—streets that are formally or in effect traffic-calmed. The desirable system, then, would be a completely interlinked network of sidewalks and walkways, with adequate dimensions and surface quality and equipped with proper safety arrangements at all crossings of significant vehicular movement.

There will be always places within any city where the existing or desirable pedestrian flows are sufficiently intense to apply improvement programs beyond the standard provision of sidewalks. These are opportunities to structure an urban environment at an enhanced level of livability, convenience, and attractiveness. Such obvious nodes are commercial and service districts, entertainment and sports centers, stadiums, major education and cultural institutions, transit stations, and similar venues where people congregate. These patrons deserve the means to move to and from them on foot, which suggests the development of special or improved walkways leading to and from the surrounding districts and principal access points (stations and parking lots). In many instances, interior circulation networks within special districts and campuses are likewise candidates for careful planning and construction of walkways.

Any pedestrian plan has to recognize certain external, rather obvious, factors that significantly influence the demand for facil-

The book *The Death and Life of Great American Cities* by Jane Jacobs revolutionized thinking about cities. The key theme was a return to the traditional street as the focus of urban life where the active presence of local residents ensures safety and a sense of community.

Streets in cities serve many purposes besides carrying vehicles, and city sidewalks—the pedestrian parts of the streets—serve many purposes besides carrying pedestrians. These uses are bound up with circulation but are not identical with it and in their own right they are at least as basic as circulation to the proper workings of cities.

A city sidewalk by itself is nothing. It is an abstraction. It means something only in conjunction with the buildings and other uses that border it, or border other sidewalks very near it. The same might be said of streets, in the sense that they serve other purposes besides carrying wheeled traffic in their middles. Streets and their sidewalks, the main public places of a city, are its most vital organs. Think of a city and what comes to mind? Its streets. If a city's streets look interesting, the city looks interesting; if they look dull, the city looks dull.

More than that, and here we get down to the first problem, if a city's streets are safe from barbarism and fear, the city is thereby tolerably safe from barbarism and fear. When people say that a city, or a part of it, is dangerous or is a jungle what they mean primarily is that they do not feel safe on the sidewalks. But sidewalks and those who use them are not passive beneficiaries of safety or helpless victims of danger. Sidewalks, their bordering uses, and their users, are active participants in the drama of civilization versus barbarism in cities. To keep the city safe is a fundamental task of a city's streets and its sidewalks.

This task is totally unlike any service that sidewalks and streets in little towns or true suburbs are called upon to do. Great cities are not like towns, only larger. They are not like suburbs, only denser. They differ from towns and suburbs in basic ways, and one of these is that cities are, by definition, full of strangers. To any one person, strangers are far more common in big cities than acquaintances. More common not just in places of public assembly, but more common at a man's own doorstep. Even residents who live near each other are strangers, and must be, because of the sheer number of people in small geographical compass.

The bedrock attribute of a successful city district is that a person must feel personally safe and secure on the street among all these strangers. He must not feel automatically menaced by them. A city district that fails in this respect also does badly in other ways and lays up for itself, and for its city at large, mountain on mountain of trouble.

This is something everyone already knows: A well-used city street is apt to be a safe street. A deserted city street is apt to be unsafe. But how does this work, really? And what makes a city street well used or shunned? . . . What about streets that are busy part of the time and then empty abruptly?

A city street equipped to handle strangers, and to make a safety asset, in itself, out of the presence of strangers, as the streets of successful city neighborhoods always do, must have three main qualities:

First, there must be a clear demarcation between what is public space and what is private space. Public and private spaces cannot ooze into each other as they do typically in suburban settings or in projects.

Second, there must be eyes upon the street, eyes belonging to those we might call the natural proprietors of the street. The buildings on a street equipped to handle strangers and to insure the safety of both residents and strangers must be oriented to the street. They cannot turn their backs or blank sides on it and leave it blind.

And third, the sidewalk must have users on it fairly continuously, both to add to the number of effective eyes on the street and to induce the people in buildings along the street to watch the sidewalks in sufficient numbers. Nobody enjoys sitting on a stoop or looking out a window at an empty street. Almost nobody does such a thing. Large numbers of people entertain themselves, off and on, by watching street activity.

In settlements that are smaller and simpler than big cities, controls on acceptable public behavior, if not on crime, seem to operate with greater or lesser success through a web of reputation, gossip, approval, disapproval and sanctions, all of which are powerful if people know each other and word travels. But a city's streets, which must control the behavior not only of the people of the city but also of visitors from suburbs and towns who want to have a big time away from the gossip and sanctions at home, have to operate by more direct, straightforward methods. It is a wonder cities have solved such an inherently difficult problem at all. And yet in many streets they do it magnificently.

ities. People walk much more in high-density areas than in lightly developed places, because there are more destination points within walking range. Likewise, communities with a large population of older residents and children will show more people on foot than those with an average demographic composition. Males 25 to 54 years old are the least active participants. Almost all traffic to a corner grocery store (if it still exists) in a dense city will be on foot; almost nobody can or will walk to a regional shopping center. Except for movement associated with commuting, most walking will occur in the middle hours of the day.

(The development of walkway systems with a purely recreational purpose within open spaces and along water bodies is a most desirable action as well, but it is not exactly transportation; therefore, it is not included in this discussion. Similarly, the creation of civic spaces is a subject for other analyses, recognizing that they too generate major pedestrian presence and flows.)

Components of Walking Systems

The presumably ubiquitous pedestrian system in any community consists of only a few rather simple physical elements, but—given the importance of this network—a closer examination of each is warranted.

Sidewalks and Walkways

Sidewalks are normally placed within the public right-of-way on either or both sides of the central vehicular channel, within the marginal reserved strips, which will usually be at least 7 ft (2 m) wide (with a 50-ft [15-m] right-of-way and a 36-ft [11-m] pavement). The sidewalk itself should be *at least* 5 ft (1.5 m) wide, where the governing consideration is not the size of persons, but rather the ability of two baby carriages (or mail carts or wheelchairs) to pass each other. In higher-intensity areas, the sidewalks are frequently 8 ft (2.4 m) wide; in commercial districts of large cities, they may be 15 ft (4.6 m) wide or more.

Sidewalks may be placed directly adjacent to the curb, thereby allowing some savings in construction costs, but with the drawback that pedestrians will be in very close proximity to moving vehicles, and the opening of doors from parked cars may create obstructions to pedestrians. A preferred approach is to place sidewalks directly or almost adjacent to the outside right-of-way line, which creates a buffer strip between the sidewalk and the curb. This strip provides a safety zone that can be landscaped. It is advisable to place utility lines under the unpaved strips, thereby making repair and excavation less costly.

The utility of sidewalks is frequently impaired by various obstructions—sign- and lampposts, hydrants, mailboxes, bus shelters, newspaper vending machines, parking meters, trees, and benches, not to mention protruding outdoor cafes, produce stands, and staircases. All this may be useful and necessary, but the effective sidewalk width will be thereby reduced (as is discussed later under "Capacity Considerations"; also see Figs. 2.1 to 2.3).

In most communities the responsibility for maintaining and cleaning the sidewalk rests with the adjoining property owner, even though it is located in a public right-of-way. The local landlords have to ensure that broken surfaces are repaired to preclude accidents, that snow and debris are cleared to allow passage, that water does not flood the paths, and that the facility is generally of adequate quality. They may receive citations for negligence, and

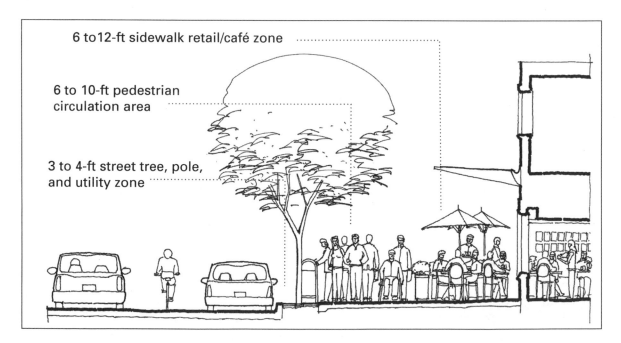

Figure 2.1   Activity street with intensive sidewalk use (green street).

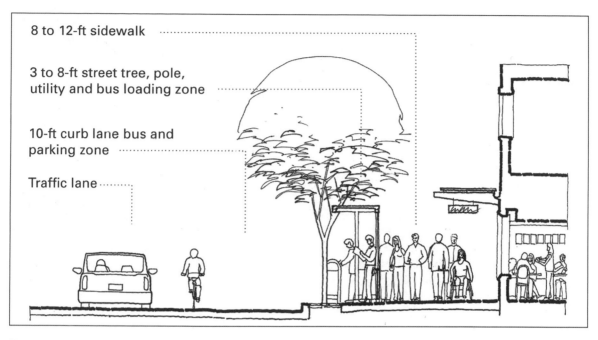

8 to 12-ft sidewalk ....

3 to 8-ft street tree, pole,
utility and bus loading zone ....

10-ft curb lane bus and
parking zone ....

Traffic lane ....

Figure 2.2   Urban street with regular sidewalk.

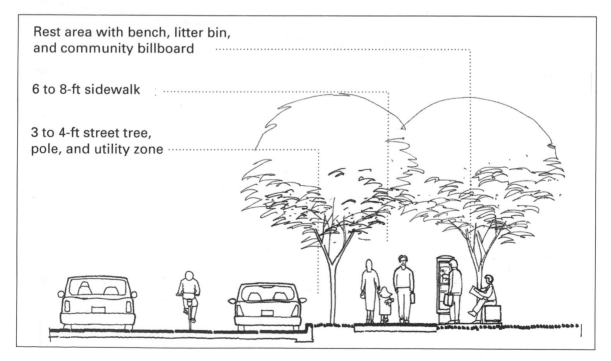

Rest area with bench, litter bin,
and community billboard ....

6 to 8-ft sidewalk ....

3 to 4-ft street tree,
pole, and utility zone ....

Figure 2.3   Suburban street with pedestrian amenities.

injured parties may sue the private owners as well as the municipality. The owners have the duty at least to notify the appropriate city agency about problems and deficiencies.

Surfaces

The type of material used for sidewalks and the resulting surface finish have at least three areas of concern: ease of walking, permanence, and visual attractiveness.

The most common type of construction is poured concrete with wire mesh reinforcement. It is easy to build, since not much subsurface preparation is necessary (no heavy loads will have to be carried); the surface is nonslip and watertight, and the pavement is durable. Some shortcomings are that the surface is not particularly interesting, the repair effort is rather extensive when the slabs crack or break, and the surface is hard on the feet (in every possible sense).

Bituminous blacktop pavement is usually cheaper yet, and construction is very easy. It has a great advantage in that the material is somewhat resilient, thus offering very good walking quality. However, this softness (the surface may even melt under very hot sun) makes blacktop unsuitable for real urban application—sharp heels and small hard wheels will destroy the surface rather quickly. While repairs are easy, the patched patterns and the overall "common" appearance of blacktop walkways make them suitable only for long recreational paths, where most users will wear shoes with soft soles.

The most attractive surfaces in general use are provided by special paving blocks and brick. Various sizes and shapes are available, and interesting geometric designs can be achieved—pleasing to the eye as well as to the feet. These materials are more expensive, and great care has to be taken in construction to maintain the integrity and evenness of the surface. Individual elements may become dislodged, and such surfaces become uncomfortable to walk on because the ankles are continuously twisted. A protrusion of even half an inch may trip some people.

At the top of the line are stone slabs and polished terrazzo. Undoubtedly, they give the best impression due to the rich quality of the material, and they are quite durable. The problems are high cost and the slippery conditions that are quite likely to occur with any moisture.

Drainage and Lighting

Full and uninterrupted utilization of pedestrian facilities depends to a significant extent on overcoming natural constraints, which in this case are too much water and darkness. The need for drainage systems is rather obvious and unavoidable, not only to maintain clear paths at all times, but also to ensure that no structural damage is done to the infrastructure through gradual erosion or sudden wash-outs.

Lighting is not a mandatory requirement, but is to be expected anywhere that urban life continues beyond dusk. It is not only a matter of maintaining adequate and uniform illumination levels along the entire walking system, it is also an opportunity to create and heighten visual interest within the pedestrian environment. Crime prevention is a significant consideration, with well-lit paths and spaces to deter criminal activity (or at least to push it to the dark places).

Traffic Control Devices and Accessibility Concerns

There should be no constraints on the movement of pedestrians, and people should be able to proceed with no unwanted stops, even if they have some handicaps to walking. This may hold for ideal situations only; in real life, adjustments need to be made. First, there are the many crossings with vehicular traffic that will be present in any pedestrian network. The needs and choices are quite well understood and worked out by this time, and various levels of controls are available—including simple stop signs for vehicles, regular traffic signals, and controls with special phases for pedestrians. However, in American communities reminders are needed that walking is a legitimate transportation mode that should be encouraged and that its participants need some protection. Their presence should be recognized in the timing and deployment of signals and traffic controls, which routinely tend to take into account the speed of cars, not of pedestrians.[20]

It is now the law of the land that people with physical or mental impairments should receive every consideration when systems

[20] Manhattan avenues, for example, have signals timed to move cars in a green wave over long distances. The experience of the author over many years shows that a pedestrian moving at a normal pace, either with or against the traffic flow, will face a red signal at every intersection.

are designed and built so that they are not excluded from participation (as much as is reasonably possible). This certainly applies to walking, and includes the requirements that public walkways must not encompass steep grades (no slopes along the path in excess of 1:12),[21] and that there must be no steps or curb faces along the way. The latter requirement results in curb cuts at intersections in all the directions that people may travel. Curb cuts not only allow wheelchairs to move without great difficulty, they also assist all others who find climbing up and down somewhat of a chore. Another set of desirable improvements address the needs of people who have impaired vision. These improvements include tactile surface treatments that can be sensed along edges and at points where care has to be exercised (at intersections, for example), traffic signals accompanied by audible sounds indicating Walk or Don't Walk conditions, and the clearance of obstructions that may be difficult to see or sense otherwise (Figs. 2.4 and 2.5).

Grade Separation and Multiple Levels

At places where vehicular traffic is heavy and serious conflicts may occur, grade-separated paths are most desirable. There have been thousands of such idealistic or practical designs, but not too many have actually been implemented. The concern is not one of cost only; there is always the potential problem of whether people will actually use them. If the vehicular flow is not an absolutely solid wall and the protected passage consumes longer time or effort, many walkers will take a risk and follow any shortcut.

Pedestrian *overpasses* bridging heavy motorways are the easiest type to implement because they can be relatively light structures (carrying only people), even though they may have to be built rather high to provide enough headroom for large trucks below. This fact also embodies the principal problem: users face a high climb upward, which they are most reluctant to undertake, even knowing that coming down on the other side will be easier. Another problem is that in some places the younger members of the community find great sport in bombarding the cars below

---

[21] This means that the centerline rises or falls vertically no more than 1 unit within a 12-unit distance horizontally. Grades (or gradients) are also expressed as percentages, i.e., measuring the vertical change over 100 units horizontally. The 1:12 ramp corresponds to an 8.3 percent grade.

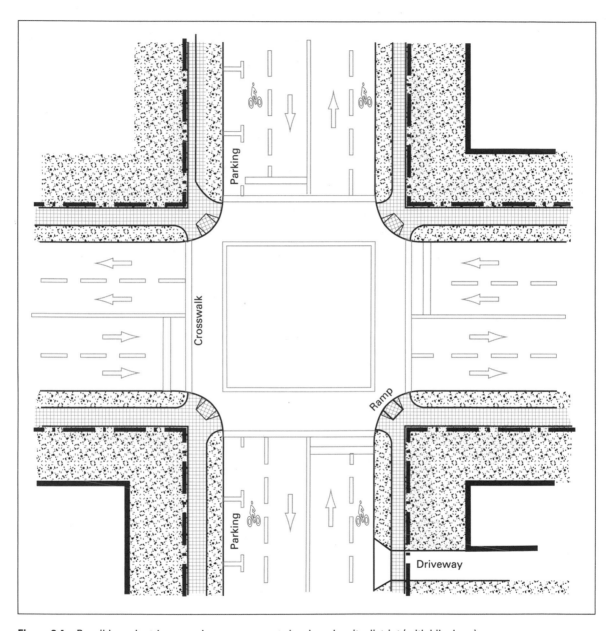

Figure 2.4   Possible pedestrian crossing arrangements in a low-density district (with bike lane).

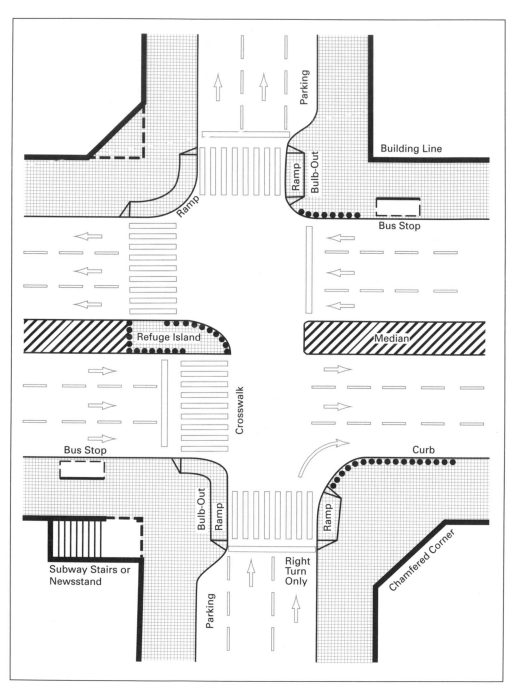

Figure 2.5   Possible pedestrian crossing arrangements in a high-density district.

from the overpass. An early famous example was the footbridge across Broadway at Fulton Street in Manhattan (in the 1860s), which was lightly used, became a nuisance, and was soon removed.

The actual results, as frequently seen in cities, are rather unsightly caged tubes with formidable staircases on both sides—effective only if all possible at-grade shortcuts are fenced off. They will be built when they have to be built. Good architectural design may be of some help, particularly because they usually have a very prominent visual location. Overpass structures may also serve as gateways to districts and as the scaffolding for signs, information, and artwork. The real solution would be to provide escalators, at least up, but they are expensive, particularly because they do not operate reliably if left exposed to the weather. There are some fortuitous instances in which an overpass with elevators and escalators can be built in conjunction with a transit facility, even if it is only a major bus stop. A fully acceptable overpass should be able to accommodate wheelchairs, which means a ramp extending for a considerable distance or taking the form of a helix is needed.

Underpasses (pedestrian tunnels under roadways), on the other hand, offer more attractive possibilities, even if they are more expensive than light bridges. The vertical clearances are less (8 ft for pedestrians), and the walking entry is always downward (people do not always consciously think of the need to climb up a short time later). The more successful examples have gently sloping ramps with well-landscaped or otherwise attractive sides. They tend to be expensive to build, particularly if utility connections have to be maintained under the street above, and care with drainage systems is always needed to keep them dry.

The principal problem with underpasses is real or perceived threats to personal security—all of us have seen enough Hollywood movies that show what can happen when good people get trapped in tight tubes. Therefore, they can only work if high levels of illumination are main-

System of second-level overpasses in Shanghai, China.

Pedestrian underground mezzanine in Vienna, Austria.

Second-level walkways in central part of Hong Kong.

tained, the tunnel is short, and every part is completely visible. This means that any user must be able to see fully to the exit end before he or she enters the underpass.

All the preceding concerns also apply when *multilevel* pedestrian environments are considered. The issues of populating such paths and spaces with sufficient volumes of users have been mentioned previously.

Roofs and Shade

The issue of providing weather protection on an open pedestrian network has been touched on already. In extreme climates, it is almost a must if reasonable volumes of walkers are to be attracted, and in temperate zones there are obvious advantages as well. There is no need for a continuous canopy, but occasional shelters are most welcome. Frequently, advantage can be taken of nearby buildings to provide awnings and other devices against rainfall, sleet, snow, wind, and extreme sun conditions. Whenever they are uncomfortable, people will seek protection indoors.

Landscaping and Amenities

Pedestrians appreciate visually attractive environments; indeed, it cannot be expected that paths and spaces will be popular if significant attention is not devoted to such features. Also, walkers need resting places from time to time. Landscaping, coordinated graphics, and attractive street furniture are all a part of the design challenge. Since these tasks enter into the realm of architecture and urban design,[22] the discussion here has to be limited to only a listing of desirable elements. These include benches, water fountains, telephone booths, signs, information kiosks, public toilets, litter baskets, and possibly quite a bit more. Sidewalk cafes, vendors, pushcarts, musicians, and entertainers may also be present— adding lively activity, color, and possible obstructions to walkers in a hurry.

Mechanical Movement Assistance

Those parts of the pedestrian system that have to accommodate high volumes of walkers can benefit significantly from the provision of mechanical auxiliary means of mobility. Obviously, this will increase the costs, which are not recoverable by user charges if the full purpose of pedestrianization is to be achieved. Some assistance may be obtained from private sponsorship; however, this is feasible only in commercial or institutional areas.

The choices include moving sidewalks (horizontal or inclined belts or escalators) and small tractor-pulled trains (for which there is no good name that would be understood by most people).

Moving sidewalks, widely used in airport terminals and other protected spaces, can also be employed on pedestrian malls to allow the spanning of distances for those who are reluctant or unable to walk extensively. Several manufacturers serve this market, and construction is a routine task. However, there are a few caveats. First, in the United States, for safety and insurance reasons, the speed is somewhat less than normal walking pace (90 ft/min [27 m/min] as compared to a normal walking pace of 200 to 250 ft/min [60 to 76 m/min], and some people exercise great care boarding a moving belt. There is also the potential problem that a dangerous pileup may occur if users do not disperse quickly at the exit end. Belts cannot be boarded from the side or crossed, again because of safety reasons; thus they have to operate as a series of separate segments over longer distances. A two-lane moving sidewalk will consume a width of about 5 ft (1.5 m); moving at 2 ft/s (0.6 m/s), it can carry about 8000 people per hour.

---

[22] Numerous good references are available; these include: C. C. Marcus, *People Places* (Van Nostrand Reinhold, 1998, 376 pp.); R. Hedman, *Fundamentals of Urban Design* (APA Planners Press, 1985, 146 pp.); and N. K. Booth, *Basic Elements of Landscape Architectural Design* (Elsevier, 1983, 315 pp.).

Network of moving belts at the World's Fair in Osaka, Japan.

To overcome the problem of slowness, there have been several efforts to design a *variable-speed* belt—i.e., a device that moves slowly at both ends but picks up speed in the middle portion. While there are several imaginative design ideas on how to do that, there are no operational examples yet. (Such a device may be introduced on the Paris metro system, and explorations are under way again in New York.) Side entry/exit on a moving sidewalk are intriguing possibilities, but are not likely to be attempted by any agency concerned with personal injury lawsuits in the United States.

The problem with all moving belts is that in these devices most of the infrastructure has to be in motion, not just individual vehicles, and they have to be activated along their full length whether there are many users or only one at any given time, thus generating considerable operational costs. Furthermore, a parallel walkway is still necessary, because some people will not or cannot use these mechanical devices, which occasionally will be out of service for maintenance anyway.

Beyond closed environments, moving belts have been employed successfully in comprehensive, separated (second-level) walkway systems in several world's fairs and major exhibition sites. They operate in the Tacoma business center, at the Hollywood Bowl, and in some European city cores in a largely open public environment.

The other possible device for mechanical assistance in pedestrian areas is a low-speed train of small open cars pulled by a tractor or "locomotive." These are quite common in pedestrian malls, amusement parks, and large parking lots. They are usually open so that riders can enter and exit quickly, and they operate as a convenience service; i.e., they usually charge no fares. Frequently, to maintain a local design theme, they are decorated in appropriate style, or at least made colorful and inviting. The principal functional characteristics of these trains are that they are small enough and slow enough that they can mix freely with pedestrian traffic, not being seen as a threat to walkers. Examples are found in pedestrian malls in Miami Beach and Pomona, in parking lots at Disneyland and Disney World, in the San Diego and Bronx zoos, along the boardwalk of Atlantic City, and at a number of other places with intensive pedestrian flows in North America.

Tractor-train on the boardwalk to assist pedestrians in Wildwood, New Jersey.

Capacity Considerations

The estimation of walkway capacities is not an easy task, because people on public paths and in pedestrian spaces do not follow fully predictable patterns of movement, and sometimes do not move at all, as discussed earlier. To bring some rationality to this situation, capacity analyses are separated into two distinct types:[23]

- *Linear paths,* where there is mostly purposeful motion, almost entirely in two directions (such as sidewalks and corridors)

- *Area situations,* where movement may be in any direction, and participants may simply occupy space for varying periods of time within the confines of a designated floor area (such as station mezzanines and platforms, outdoor rooms used for any purpose, and public squares)

[23] These study methods, following the aforementioned work of Fruin and Benz, are outlined in full detail in several articles and reference books. The basic reference for all capacity investigations is the *Highway Capacity Manual 2000* (Transportation Research Board, Washington DC, 2000), with pedestrian issues covered in Chaps. 11 and 18. Shorter discussions are found in the Institute of Transportation Engineers *Transportation Planning Handbook,* op. cit., pp. 227–230 and 440 ff. and *Traffic Engineering Handbook,* op. cit., various sections.

While human beings come in different sizes and shapes, for the purposes of transportation analyses, it is assumed that all of them fit into an ellipse with 18- × 24-in (45- × 60-cm) dimensions as the basic design unit. This represents a base density of 3 ft² per person (0.28 m² per person or 3.6 persons per m²), but not "crush" conditions, which are a matter of cultural tolerance or dire necessity. On some overloaded trains in developing countries, 0.9 ft² per person (12 persons per m²) have been observed, which would cause broken ribs, except that some of them are hanging on to the outside and some are sitting on the roof. In the industrialized countries, crowding at the 2-ft² (0.2-m²)-per-person level may be acceptable only on trips with very short duration (in elevators, for example; floor space needs in vehicles are discussed further in various chapters under "Capacity").

On the other hand, people feel comfortable only if they are able to preserve their own "private space bubble," which tends to be 8 or 10 ft² (0.8 m²) in area—the approximate size of an open umbrella.[24]

All the preceding refers to people when they are standing. Since this is a discussion of *walking,* the moving state is more important, and it is more complex. Basically, the space needs are larger because a walker has to step forward into an area that should be open. The fundamental relationships in determining pedestrian flow capacity are *speed–volume–density.* An overriding consideration in these analyses is the concept of *level of service* (LOS). This is a qualitative measure that characterizes various situations with reference to the ease with which the operations can be performed. For example, LOS A means that motion by any individual is not constrained by anybody else, while LOS F describes an unacceptable situation in which extensive contact with others will be necessary to make any progress. Each level is associated with a specified space requirement (see Figs. 2.6 and 2.7). For example, if only 12 ft² per walker are available (LOS E), continuous conflicts will be experienced, and speed will be retarded, but considerable capacity will be achieved. At about 8 ft² per person (0.7 m²) friction becomes excessive, and all movement will stop at 2.5 ft² (0.2 m²).

---

[24] Tolerated or welcome intrusions in each personal space will not be discussed here.

LOS A

Pedestrian Space > 60 ft²/p   Flow Rate ≤ 5 p/min/ft

At a walkway LOS A, pedestrians move in desired paths without altering their movements in response to other pedestrians. Walking speeds are freely selected, and conflicts between pedestrians are unlikely.

LOS B

Pedestrian Space > 40–60 ft²/p   Flow Rate > 5–7 p/min/ft

At LOS B, there is sufficient area for pedestrians to select walking speeds freely, to bypass other pedestrians, and to avoid crossing conflicts. At this level, pedestrians begin to be aware of other pedestrians, and to respond to their presence when selecting a walking path.

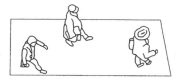

LOS C

Pedestrian Space > 24–40 ft²/p   Flow Rate > 7–10 p/min/ft

At LOS C, space is sufficient for normal walking speeds, and for bypassing other pedestrians in primarily unidirectional streams. Reverse-direction or crossing movements can cause minor conflicts, and speeds and flow rate are somewhat lower.

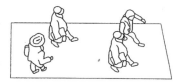

LOS D

Pedestrian Space > 15–24 ft²/p   Flow Rate > 10–15 p/min/ft

At LOS D, freedom to select individual walking speed and to bypass other pedestrians is restricted. Crossing or reverse-flow movements face a high probability of conflict, requiring frequent changes in speed and position. The LOS provides reasonably fluid flow, but friction and interaction between pedestrians is likely.

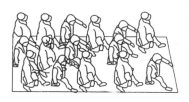

LOS E

Pedestrian Space > 8–15 ft²/p   Flow Rate > 15–23 p/min/ft

At LOS E, virtually all pedestrians restrict their normal walking speed, frequently adjusting their gait. At the lower range, forward movement is possible only by shuffling. Space is not sufficient for passing slower pedestrians. Cross- or reverse-flow movements are possible only with extreme difficulties. Design volumes approach the limit of walkway capacity, with stoppages and interruptions to flow.

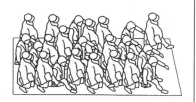

LOS F

Pedestrian Space ≤ 8 ft²/p   Flow Rate varies p/min/ft

At LOS F, all walking speeds are severely restricted, and forward progress is made only by shuffling. There is frequent, unavoidable contact with other pedestrians. Cross- and reverse-flow movements are virtually impossible. Flow is sporadic and unstable. Space is more characteristic of queued pedestrians than of moving pedestrian streams.

Figure 2.6   Illustrations of pedestrian walkway levels of service. [From Transportation Research Board, *Highway Capacity Manual 2000* (National Research Council, 2000), p. 11-9. Copyright © 2000 by the Transportation Research Board, National Research Council, Washington, D.C.]

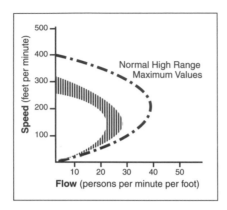

Figure 2.7   Speed–flow relationship. [From data in Transportation Research Board, *Highway Capacity Manual 2000* (National Research Council, 2000), p. 11-3.]

The highest observed rates of pedestrian flow on busy sidewalks are about *25* persons per minute per foot of sidewalk width. This occurs with a speed of 150 ft/min (2.5 ft/s; 0.76 m/s, with a density of 5 to 9 ft² [0.5 to 0.8 m²] per person) and it represents conditions close to LOS E (see Table 2.1). In special instances, such as in controlled pedestrian corridors and tunnels, *35* persons per minute per foot can be reached and even exceeded for short periods. Actual pedestrian counts in the centers of the larger cities in North America (Chicago, New York) have recorded averages of 250 to 350 people moving past a point on a regular city sidewalk within a minute. Volumes during the peak 15 minutes have approached 500 per minute.

The preceding paragraph introduces another element that requires some explanation—the *effective width* available for walking, or the *effective space* available for circulation. This recognizes the obvious fact that people will not walk with their shoulders rubbing against lateral obstructions, such as buildings, trees, parking

Table 2.1   Pedestrian Flow Characteristics on Walkways

Characteristic	Level of service*					
	A	B	C	D	E	F
Flow rate, pedestrians per min per ft	Less than 5	5–7	7–10	10–15	15–23	Variable
Spacing, ft² per pedestrian	More than 60	40–60	24–40	15–24	8–15	Less than 8
Walking speed, ft/min	More than 255	250–255	240–250	225–240	150–225	Less than 150

* Pedestrian specialists have not yet agreed on the exact space standards. Several sets exist; the norms shown here are the ones most frequently used. Research and discussions continue, and a commonly accepted set is promised soon.
Source: Transportation Research Board, *Highway Capacity Manual 2000* (National Research Council, 2000), p. 18-4.

meters, and other vertical elements. Consequently, the physical width may have to be reduced by 6 to 18 in (0.2 to 0.5 m) on each side, depending on the severity of the marginal obstructions, if any.

Linear Capacity

The simplest cases are situations in which pedestrians are in continuous motion along a defined channel. If the movement is in two directions, there will be some internal turbulence and friction as people deviate from a straight line, but by and large they will walk on the right-hand side and avoid contacts rather adroitly. Given sufficient time, they will establish reasonably uniform densities within the total mass. Nevertheless, individuals may meet

Linear Pedestrian Capacity Example—City Sidewalk

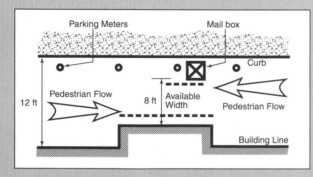

Effective width = 8 ft − 2.5 ft (side obstructions) = **5.5 ft**

Selected level of service = **C/D** (the break-point between LOS C and LOS D)

Flow rate = **10** pedestrians per minute per foot of width

Capacity = 10 × 5.5 = **55 walkers per minute**   or 3300 per hour

Or:

If there are 1400 walkers per hour westward and 1600 eastward, they will operate at level of service C.

$$\text{Walkers per minute} = \frac{1400 + 1600}{60} = 50$$

$$\text{Walkers per foot of width} = \frac{50}{5.5} = 9.1$$

which corresponds to LOS C (see Table 2.1)

Center of Rome under an automobile ban.

and stop for a chat in the middle of the stream,[25] and other events may occur that reduce the base capacity performance.

In a city, pedestrian flows, regrettably, do not follow straight lines for very long distances. There are crossings and intersections that interrupt smooth progress. As a matter of fact, the governing capacity conditions are found at street corners where people accumulate, awaiting the next opportunity to proceed. In busy places, the complexities intensify, because other pedestrians walking in a perpendicular direction have to penetrate the crowd already assembled at the edge of the sidewalk; other business or personal actions may take place at the corners; and turning cars may impede the flow of people across the street even if they have a Walk signal in their favor. If the volumes are large, two clusters of walkers will clash in the middle of the street, and numerous evasive maneuvers will take place.[26]

The consequences of all this are that pedestrian flows frequently become organized in platoons that act as a single body, that crosswalks will require special analyses because many users will stray from designated paths, and that street corners merit special attention in terms of analyses (see next subsection, "Area Capacity"). Frequently, in high-volume cases, street corners may have to be redesigned to gain sufficient pedestrian space by cleaning out all clutter, possibly chamfering building corners and "bulbing out" the curb line. (The latter can only be done at the expense of vehicular movement or parking lanes, but it also has the benefit of shortening the crossing distance from curb to curb.

---

[25] This is called a *Holly Whyte encounter*—showing a complete disregard for the movement space of others, a phenomenon documented by W. H. Whyte.
[26] These complex situations require the attention of specialists and are not discussed in detail here.

Area Capacity

The method of analysis for circulation spaces is the next step, and it recognizes that people do not only walk in pedestrian spaces, but that they also linger and do other things. Thus, another dimension has to be introduced—time—which takes into account the fact that each user occupies the available space for differing durations. The demand for time-space is compared to the total time-space supply available. Each user generates a product of square feet occupied multiplied by seconds of presence in the space. The information needed to judge the functional adequacy of any layout is not only the number of users, but also the time that they take to cross the space (or linger), considering the level of service to be achieved. (Any space can accommodate differing numbers of people in a more crowded or in a more comfortable state.)

The type of pedestrian activity (walking, meandering, or tarrying) and the amount of each will depend on the general purpose of the space, the time of day, weather conditions, and other external factors. For example, spaces that have a significant recreational dimension will be at the highest use level in the middle of the day or in evenings and on weekends, and the users will not be in great hurry. Paths and spaces associated with transit facilities will have intensive loads during the two daily peak periods when patrons will be walking briskly through them, waiting for various reasons, or being processed (buying tickets, reading maps and schedules, or picking up a newspaper).

Possible Action Programs

The presence of pedestrians and the need to provide for sufficient spacious, safe, and attractive walkways and spaces is reasonably accepted in American communities. Nobody will argue against the concept, and this is the politically correct thing to do. However, there may be general neglect of this sector, and priorities are frequently placed on other needs and programs. Recent history shows that effective plans and campaigns to achieve better pedestrian conditions are often mounted by public interest groups concerned with environmental quality, health, traffic safety, or general livability. The triggering mechanisms may be the occurrence of serious traffic accidents at a given location, which calls for remedial mea-

Area Pedestrian Capacity Example—Subway Mezzanine

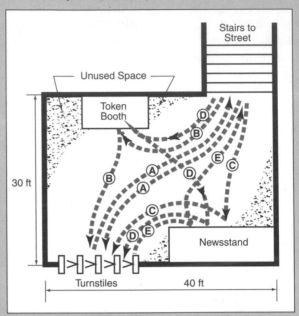

Types of Movement (per hour):

A. Stairs to turnstiles (or vice versa)	3200 patrons,	10 seconds
B. Stairs to token booth to turnstiles (assuming no excessive queues at the booth)	600	30
C. Stairs to newsstand to turnstiles	50	45
D. Stairs to token booth to newsstand to turnstiles	200	60
E. Turnstiles to newsstand to stairs	100	45

Note: Preceding list of space users does not include:

- Anybody who comes from the street to buy a newspaper and then returns to the street

- Anybody who waits within the mezzanine for any reason

- Any exiting passenger who stops at the token booth

While the total mezzanine area is 1200 ft^2, only approximately two-thirds, or 800 ft^2, is usable for circulation.

If a level of service C is to be maintained, each person will occupy, let us say, 30 square feet. (Note: when standing at the token booth or newsstand, less space will be consumed, but, instead of applying this correction, it can be regarded as a conservative safety factor.)

Time-space **demand** per hour:

A. $3200 \times 10 \times 30 =$ 960,000 ft²·s

B. $600 \times 30 \times 30 =$ 540,000

C. $50 \times 45 \times 30 =$ 67,500

D. $200 \times 60 \times 30 =$ 360,000

E. $100 \times 45 \times 30 =$ 135,000

 Total   2,062,500 ft²·s

Time-space **supply** per hour:

$1200 \text{ ft}^2 \times 0.67 \times 3600 \text{ s/h} = \mathbf{2{,}894{,}400 \text{ ft}^2 \cdot \text{s}}$

Thus, the demand can be accommodated comfortably. Indeed, the overall level of service would be B, with about 42 ft² available to each patron (see Table 2.1).

sures, extending all the way to the advocacy of regionwide walkway systems for the benefit of the entire population.

A pedestrian component should be a part of any overall transportation plan, whether it is prepared at the corridor, citywide, or even regional level. It is particularly important to structure pedestrian systems when district and neighborhood plans are prepared. The development of transit systems, whether they are rail-based or consist of rubber-tired vehicles only, should encompass walkway access extending from stops or stations into the direct tributary areas.

Since pedestrian systems by definition are local facilities and accommodate operations at the district scale, the resources to plan for them and funds to construct paths and spaces are usually also a local responsibility. However, if a more comprehensive approach is taken; systemwide implications are examined; and congestion management, air quality, and community improvement issues are involved, funds from higher levels of government may

The Galeria in Milan as a civic gathering place.

An array of pedestrian amenities on a street corner in Paris.

be available, including specifically targeted federal assistance programs and as components of coordinated multimodal approaches. Work to be done in local business districts is frequently supported by private civic or trade groups dedicated to the well-being of such areas (e.g., merchants' associations and chambers of commerce).

Conclusion

American communities, generally speaking, have neglected their pedestrian systems in the recent past. Attention and resources have been devoted to mechanical transportation options under the overall assumption that walkers will take care of their own needs, and that most people will drive anyway. This attitude has led to the building of many subdivisions with many miles of streets that provide nothing for pedestrians. There is a need to change this attitude—not so much in policy statements, which tend to say the right things anyway, but with real programs and funded projects. It should be possible to walk efficiently and safely, and perhaps with pleasure, in all parts of any community. There are encouraging signs that public opinion is moving in that direction, and many places can proudly point to successful walkway projects. It is just that such achievements should not be perceived as special efforts; they should be routine.

Simple and effective means of creating a pedestrian street in Dublin.

Bibliography

Brambilla, Roberto, and Gianni Longo: *For Pedestrians Only: Planning, Design, and Management of Traffic-Free Zones,* Whitney Library of Design, 1977, 208 pp. A comprehensive review of pedestrian planning issues, with an inventory of facilities in operation by the mid-1970s.

Breins, S., and W. J. Dean: *The Pedestrian Revolution: Streets without Cars,* Random House, 1974, 151 pp. An early survey of pedestrian projects and improvement possibilities.

Hass-Klau, Carmen: *The Pedestrian and City Traffic,* Belhaven Press, 1990, 277 pp. Emphasis on pedestrianization efforts during the twentieth century, particularly in Europe.

Pushkarev, Boris S., and Jeffrey Zupan: *Urban Space for Pedestrians,* MIT Press, 1975, 212 pp. A documentation of the pedestrian situation in New York City, defining many of the parameters used subsequently.

Rudofsky, Bernard: *Streets for People: A Primer for Americans,* Doubleday & Company, Inc., 1969, 351 pp. The seminal review of pedestrian spaces and their performance through the ages.

Transportation Research Board: *Highway Capacity Manual 2000,* National Research Council, 2000. Technical handbook on all functional elements related to the planning for pedestrian facilities.

Uhlig, Klaus: *Pedestrian Areas: From Malls to Complete Networks,* Architectural Book Publishing Co., 1979, 152 pp. Concepts of remodeling towns toward pedestrian use; many actual examples.

Warren, Roxanne: *The Urban Oasis: Guideways and Greenways in the Human Environment,* McGraw-Hill, 1998, 196 pp. A brief but recent examination of pedestrian zones.

Whyte, William H., *The Social Life of Small Urban Spaces,* The Conservation Foundation, 1980, 125 pp. An investigation into the behavior of users and the factors that make pedestrian spaces effective.

Bicycles

Background

The most amazing thing about bicycles is that among all machines and animals that move they are the most efficient devices for transporting weight over a distance for a fixed amount of energy consumption (about 0.15 cal/g·km).[1] Apparently, the mechanical assemblage of pedals, a chain, and wheels, in combination with the powerful thigh muscles of the human body, is the best arrangement to achieve forward motion. It beats a salmon swimming, a jet plane or seagull flying, or a train or a horse running. It is curious that none of the life forms on this planet have locomotion systems in any form resembling wheels.

Be that as it may, the bicycle is a respectable contemporary form of transportation, but it is also much more than that. Because it has many dimensions, it is frequently hard to deal with the bike option as a simple utilitarian mode and to develop from it serious service systems. Yet, the bicycle today has no known or acknowledged enemies; it is absolutely politically correct because of its nonpolluting, space-saving, resource-conserving, and health-enhancing characteristics. Politicians endorse it, and there are dedicated support groups that vocally promote bicycle systems as the solution for

[1] S. S. Wilson, "Bicycle Technology," *Scientific American,* March 1973, p. 90.

almost all city mobility problems. The advocates are mostly right, but the world would have to be largely populated by socially responsible and physically fit persons, thinking the right thoughts, to achieve effective pure bike systems in cities.

Just about all of us learned how to ride a bike when we were children,[2] and we derived much pleasure from it. This was our first means to gain an enhanced sense of free movement—a seminal human desire. However, as adults we usually seek more comfortable and less visually flamboyant means of circulation. It could be that one of the strongest subliminal reasons why bicycling is not a universally popular mode of transportation in the United States is our unease with being part of an urban traffic flow dressed in the colorful contemporary attire of the real cyclist—or, conversely, with perspiring in a suit and tie on a public street.

The real reasons (a hypothesis at this point) why bikes are not a stronger transportation mode are that most people are unsure of their cycling skills and do not feel safe in vehicular traffic, and that bikes are a mode quite unlike cars and pedestrians, but there is no reasonable possibility in most instances to give them their own space.

It is difficult but not impossible, even in North America today, to identify types of communities where bicycles could be the dominant urban transportation mode. There are many more places where human-powered subsystems can perform specific functions very well, in support of other services or in filling defined service niches. Let us search for such opportunities, because the potential for bikes to be useful in today's communities is not yet really tapped.

Development History

For a relatively simple device with much basic utility, it is remarkable that the bicycle is only little more than 100 years old.[3] True,

---

[2] This does not necessarily mean that all of us are sufficiently skilled and well enough trained in safe cycling techniques to ride competently in city traffic.

[3] There are a number of sources that cover the history of cycling and the machines. They include: D. B. Perry, *Bike Cult* (Four Walls Eight Windows, 1995), pp. 2–43; S. S. Wilson, op. cit., pp. 81–90; and *Encyclopaedia Britannica,* "Bicycle," vol. 2, pp. 981–984.

A few communities in North America have defined a full set of bike-oriented policies and programs. One such place is Palo Alto, California, as originally published in the *New York Times,* June 29, 1989.

Where Bicycle is King (And the Queen, Too)

JANE GROSS

PALO ALTO, Calif., June 22—In this car-crazy state, Ellen Fletcher's boast is at least eccentric and perhaps bizarre. She fills the gas tank of her battered 1963 Plymouth Valiant only twice a year and has to remind herself to take the car out of the garage once in a while for some exercise.

At 60 years of age, Ms. Fletcher bicycles everywhere.

For excursions to San Francisco, 35 miles to the north, she takes the train and brings her folding bike, thumbing her nose at the hills.

For meetings in San Jose, 15 miles south, she rides her 18-speed touring bicycle, making the trip in 80 minutes, unless a rare south wind slows her down.

The grocery store is easy. She tethers her three-speed to one of the racks that are ubiquitous here, then loads packages in her wire basket.

Ms. Fletcher is a City Council member and author of most of the legislation that has made Palo Alto the "most bicycle-friendly town in the United States," according to a CBS News broadcast last year that drew howls of protest from the community of Davis, Northern California's other cycling mecca.

• • •

Superlatives aside, Palo Alto is the sort of place where everybody who is anybody wears skin-tight black shorts and a mushroom-shaped helmet, where engineering freaks who have made it big shell out $2,000 for the Kestrel carbon fiber bike or $3,000 for a Tom Richey tandem.

With at least 14 percent of the city's 60,000 residents commuting by bicycle, according to the 1980 census, lots of people have snazzy models to ride on weekends and "station bikes" to leave outside the railroad terminal. There are six-month waiting lists for the few dozen bicycle lockers at the two stations.

For at least a decade, the city's comprehensive transportation plan has been built on the twin premises that "needs cannot continue to be met primarily by the private automobile" and bicycles should be "encouraged for nonrecreational activities."

That encouragement comes in many forms, including a 40-mile network of bicycle lanes and paths. The centerpiece of this network is the "bicycle boulevard," a two-mile stretch of Bryant Street, a main east-west artery, where cars have limited access because of a series of barriers and bikes-only bridges.

Then there are laws and zoning ordinances passed in recent years. The city requires large businesses to provide showers for cyclists and parking for bikes. City employees are reimbursed 7 cents a mile for business travel by bicycle, as against 23 cents for cars.

Traffic signals at busy intersections stay green longer so cyclists can get across the street, and there are "smoothness" standards for road repaving. Drive-in businesses, except for car washes, must provide access for bicycles. Junior high schools offer a cycling course. And the Police Department runs a traffic school for juveniles who receive summonses.

Earlier this month, the City Council gave a legal green light to another alternative mode of transportation, the skateboard, by authorizing its use on all but the 25 busiest streets. Ms. Fletcher was one of two City Council members who voted against the ordinance and was accused by others on the council of jealously guarding the streets for her beloved bicycles.

• • •

There are many explanations for Palo Alto's love affair with the bicycle. It adjoins Stanford University, and college towns tend to be cycling hotbeds in California. The weather is balmy here, sunny and dry when San Francisco and Marin County to the north are dank with summer fog. And the terrain is flat, manageable for a novice cyclist. Finally, there is a widespread interest in ecological issues and physical fitness, two of the Bay Area's abiding preoccupations.

Ms. Fletcher says there is also an economic component, with well-to-do people, who predominate here, more likely to ride. At Stanford Research Park, home to a score of high-technology companies, a city survey found that 5 percent of the 22,000 employees bicycled to work. "But the assembly-line people all came in cars," said Ms. Fletcher, with a sniff of disapproval. "I'm not sure they realize the economic impact of depending on a car."

The economic issue is not lost on Ric Hjertberg, both as the owner of Wheelsmith, a local bicycle shop, and as a bicycle commuter. Mr. Hjertberg and his wife both bike to work and thus save "thousands of dollars a year" on gas and repairs, he said.

For long trips and ferrying a 13-month-old baby, the family gets by with a pair of 20-year-old Volkswagens. "Your cars don't have to have the same kind of integrity if you don't drive them," Mr. Hjertberg said. "If I couldn't bike to work, I'd have to get a modern car."

it is possible to find some quaint two-wheeled devices pushed by feet much earlier, but they were of no specific consequence. Throughout the nineteenth century there was a string of incremental technical developments that gradually contributed to the eventual composition of the bicycle: foot treadles and cranks, wheels and spokes, hard rubber tires, springs and slender steel structural members, and so on. This resulted in the French *velocipede* of 1863 (which was largely a curiosity) and—most important—the high-wheeler or "ordinary" bicycle of 1870, which was developed by members of the legendary Starley family of Coventry.

The high-wheeler was a workable machine that caught the popular imagination, and even today all of us know what it looked like. It was a clumsy and dangerous device; considerable skill and fortitude were required to ride one (how do you stop and dismount?), but it could be done. It was good for show-offs, but it also whetted appetite for an apparently attainable higher mobility. The ordinary left not only a useful expression in the English language—to "take a header" (fall forward over the large wheel)— but also a legacy of technical experience and knowledge, and inspiration and confidence to do better. With the invention of the tricycle (which Queen Victoria liked), it became socially acceptable for women to engage in joint recreational activity, a significant breakthrough in that era.

It also has to be observed that the bicycle industry was instrumental in laying the foundations for a number of refined technological developments in the advanced countries. After all, the Wright brothers were bicycle mechanics, and a number of the early automobiles were light vehicles with bicycle-type elements and an engine hooked on.

All the accumulated experience came together in 1885 when the Rover[4] "safety" bicycle came out of the Starley family workshop in Coventry. Two equal-size wheels with wire spokes and a chain drive with pedals set the pattern that has not basically changed since that time. All the subsequent improvements have only consisted of refinements, incremental additions, and the use of more advanced materials. All of us could ride the original safety and not fall off.

This machine sparked an unprecedented change in the lives and recreational orientation of urban residents, of all ages and

---

[4] The name *Rover* lives on in the robust British land vehicle.

A Brief History of the Bicycle

1816	Two-wheeled "pedestrian hobby horse" made in Karlsruhe.
1839	Treadle mechanism developed in Scotland.
1861	Velocipede with foot cranks/pedals built in Paris.
1868	First bicycle races in France.
1871	High-wheeler or "ordinary" machine developed in Coventry by James Starley.
1876	Starley tricycle allows women to become cyclists. Philadelphia Centennial Exhibition displays British wheels.
1877	First American wheel made in Boston. Introduction of tubular frame and ball bearings.
1879	Chain-and-sprocket drive to rear wheel developed.
1880	League of American Wheelmen organized.
1882	Speed record set: 20 mi, 300 yd in 1 hour.
1885	Rover safety bicycle (direct ancestor of today's bike) produced in Coventry; becomes standard machine.
1886	Roads Improvement Association founded in Great Britain.
1888	Pneumatic tire made practical.
1890s	Worldwide bicycle craze.
1899	Derailleur gear-changer invented (external).
1901	Gears for rear hub patented (internal).
1903	Tour de France bicycle race started.
First half of 20th century	Popular attention turns to automobiles and other mechanical devices. Bicycles remain in great demand among children. Cycling as a sport continues.
1950s	English three-speed bikes imported in great numbers.
1958	Ten-speed lightweight bike introduced.
1960s	"Banana" bikes become popular among the young.
1962	Cross-frame bicycle invented.
Early 1970s	Gasoline crises, environmental awareness, and great resurgence in the purchase and use of bicycles in the United States.
1974	International Human Powered Vehicle Association formed.
1979	55-mph barrier broken. *Gossamer Albatross* human-powered airplane crosses the English Channel.
1981	The first mountain bike—the "stump jumper"—introduced.
1986	65-mph barrier broken. DuPont wins prize for human-powered aircraft.
1992	Individual flying-start speed record over 200 m set at 68.74 mph.

genders, except the poorest classes. Everybody tried to acquire a bicycle and use it as much as possible during leisure hours and as a transportation device. The decade of the 1890s was colored by a true cycling craze and single-minded dedication to this specific activity. It was soon discovered that the quality of available roadways in cities and outside was inadequate, and the immediate demand was for better (smooth and watertight) road surfaces. Mud and ruts on streets were no longer acceptable, and a number of strong local and national bicyclist organizations made themselves heard effectively. The leading association was the League of American Wheelmen (now the League of American Bicyclists), which sparked the national Good Roads Movement. Even today, cyclists appear not to have lost purposeful dedication and a ready willingness to do battle for their cause.

It is sad to note that over the following half century the popularity of the bicycle as a mass means of transportation and recreation faded in the United States. It remained a most popular toy for children, and, therefore, there were always American manufacturers of the machines, but the beacon for young people and adults became the automobile. It is fair to say that in the 1950s and 1960s few self-respecting citizens would consider being seen on a bicycle, and the few dedicated riders were looked at somewhat askance. Even American athletes had no visibility at international cycling competitions. The introduction of sleek 10-speed, lightweight racing machines with skinny tires in the late 1950s—mostly of Italian origin—brought some excitement to the field, but not a reversal of popular attitudes. These bikes were chic and fast, and some people looked good leaning far over dropped handlebars, but they were not very suitable for city cycling. The three-speed "English racers" became the machines of choice, replacing the simple and rather tiring one-speed bikes, effective for urban use as well as for long-distance touring.

Bicycles did not lose their popularity everywhere. Europeans, for example, found the bicycle a very appropriate device within a constrained economy after World War II, and many cities became equipped with suitable facilities over the following decades. The legacy some years later varies. In some countries (Denmark, the Netherlands) bike-riding habits remain strongly in force; in others mopeds, scooters, and later cars have surged ahead. In developing countries, provided that climate and topography are not major obstacles, the bicycle is a significant factor, albeit frequently beyond the means of the poorer segments of the population. Cities in China, India, Indonesia, Bangladesh, and a number of other places in Africa and Latin America still depend on human-powered transportation, but those situations are driven by economic necessity and thus are not quite comparable to our transportation concerns. They can teach us much about the physical characteristics and facility operations under extreme loads nevertheless.[5] It is, however, a fact that many cities in the Third World have instituted policies and traffic controls that are inimical to bicycle use because of their lack of a "modern" image. This is particularly true with respect to bicycle rickshaws as human-powered transportation for hire.

Attitudes toward cycling changed again in the United States as an aftermath of the social turmoil of the late 1960s, propelled by a general environmental awareness and most directly by the oil crisis. The bicycle became a status symbol, and, even if the riding cohort did not become very large, everybody had to agree that this was the right thing to do. It is difficult today to find anybody making disparaging remarks publicly about bicycles, not even taxi drivers and car manufacturers. If that is so, then why does not every American city have a full network of bikeways? This question is the principal topic of inquiry for the rest of this chapter.

There has been a significant development regarding rolling stock as well. The introduction of the "mountain" bike in 1981 was not a revolution, but a breakthrough nevertheless. It is basically a redesigned English three-speed machine in a lighter format, with a wider gear range and better brakes. This rather simple but sturdy machine, with thick tires that can absorb surface imperfections with comfort and a configuration with flat handlebars that allows the rider to maintain a natural posture on a wide saddle, was designed for off-road use, but it was soon discovered that it can cope even better with city conditions.

We now have a suitable urban machine, and it is difficult to think of any urgently needed improvements to the hardware. Future advances are likely to be in the use of lighter and stronger composite materials, different types of spokes, stronger tires, upgraded gear shifts and brakes, and most likely an endless stream of electronic gadgetry. Removable electric motors to assist

---

[5] See M. Replogle, *Non-Motorized Vehicles in Asian Cities,* World Bank Technical Paper Number 162, 1992.

human power (which upset the purists) and folding models that help in storage and transport are already available. There are also improvements in lighting, baggage space, and protection devices for the rider and clothing.

Another branch of technical development has been the sector devoted to racing and pure speed—subjects somewhat outside our current discussion. The racing machines, of which there is an endless variety,[6] are stripped-down versions of the standard bike to attain the least possible weight and are exclusively suitable for a specific form of racing. For example, the track bikes used on the closed oval of a velodrome have fixed gears and no brakes.

Much ingenuity and engineering effort have gone into the quest for speed. Given the limited amount of power that a human body can produce (⅓ hp over an extended period and 2 hp in short bursts by the best-trained athletes), attention has to be devoted to lightweight materials and aerodynamics. To reach beyond the speed levels that can be attained by the best racers on reasonably regular bikes (in the 35-mph [56-kph] range), an enclosed teardrop-shaped body becomes necessary, and much experimentation has taken place during the last three decades with rather weird-looking contraptions. Not counting downhill and motor-paced efforts, the speed record (over 200 m, flying start) at the present time is 68.72 mph (110.60 kph), set in 1992.[7] In the meantime, cyclists have also been the engines for human-powered aircraft.

Bike paths of various kinds have been built since the very early years of cycling development, but almost entirely as recreational trails for leisure riding and sport. The first such path is said to be the dedicated lanes along the landscaped Ocean Parkway in Brooklyn, opened in 1895, connecting Prospect Park to Coney Island. Not much can be reported from the middle decades of the twentieth century in the United States, except that trails were created here and there, as the automobile took over most individual transportation tasks. Many of these bike facilities were in parks and protected open spaces (under WPA programs), as well as in association with several parkways built during that period.

---

[6] See D. B. Perry, op. cit., pp. 96–156.
[7] Ibid., p. 518.

In Europe, even with the prevailing high volumes of cyclists after World War II, purposeful improvement programs for bicycle systems really emerged only in the 1960s; they soon gained much popular support, and they blossomed in the 1980s.[8] Even though overall bike use declined with growing prosperity, sufficient reasons were identified in a number of locations to embark on extensive programs creating circulation networks. Denmark, the Netherlands, and West Germany were the pioneers, and they remain the leaders today, generating experience and information that have been useful to others. Cities that can be singled out are: Odense, Copenhagen, Nottingham, Delft, Groningen, Erlangen, Münster, Munich, Winterthur, and Graz. Several new towns have been built with bicycles as a principal transportation mode: Stevenage and Milton Keynes (Great Britain), Västeras (Sweden), and Houten (the Netherlands).

Other European countries have lagged behind or have shown little interest in the development of cycling systems for various social, physical, or economic reasons. Considerable progress, however, has been made in the rest of Scandinavia since the 1970s; several places in Great Britain and France have taken major steps in facility expansion as well.

The experience with city systems in the United States so far has not been very inspiring. As bikes became socially and politically acceptable—and, to a certain extent, icons of a new lifestyle—in the late 1960s, a few places initiated programs, to the amazement of the rest of the country. These were Davis, Palo Alto, and Berkeley in California, soon followed—as could be expected—by Portland, Oregon; Seattle; Pasadena; Madison, Wisconsin; Ann Arbor, Michigan; Boulder, Colorado; and a few other university, resort, and retirement communities. Among large cities, Toronto and Chicago have implemented particularly constructive programs. Since 1966, the bike lane programs of Central Park in New York City have had high visibility because of their thoroughness and the sheer volumes of users they attract. Gradually, several states established financial assistance programs and started to publish guidelines, while demonstration and pilot projects became eligible for federal help—but almost entirely in association with highway development programs.

---

[8] See Part II of H. McClintock's *The Bicycle and City Traffic: Principles and Practice* (Belhaven Press, 1992).

Bicycle trail along an old rail alignment (Connecticut).

Types of Cycling

The bicycle is a versatile machine that can be used in a number of ways, but the most common applications are the following:[9]

Children's Toy

The popularity of bikes among the younger set continues, and it represents an industry by itself, with many models and promotional campaigns. It is, however, not a matter of transportation, and therefore outside the scope of this discussion. Nevertheless, there is a crucial concern with safety and livability at the very local level. There has been a measurable decrease in bike use as American parents have become increasingly cautious. These issues have to be addressed in the context of neighborhood street design and management, particularly associated with traffic calming efforts. Education programs in safe riding are advisable, a major concern being the coexistence of bikes with pedestrians and cars.

Recreational Device

Cycling has evolved during the last few decades into one of the most popular recreational activities for Americans, right after walking and swimming. Its health benefits are unquestionable, and the activity itself is most pleasurable and sociable. Distances are covered, nature is observed, and like-minded people are met. It supports also a very large manufacturing and products distribution industry, ranging from specialized apparel to high-tech devices and components.

The demand in this sector is for appropriate and visually attractive trails and paths. To ride in mixed traffic with automo-

[9] W. C. Wilkinson et al., in their monograph *Selecting Roadway Design Treatments to Accommodate Bicycles,* classify cyclists by skill level rather than activity into Type A (proficient and competent), Type B (casual and not fully skilled/trained), and Type C (children).

biles impairs the purpose and concept of this activity, and therefore the preferred model is a system of separate and exclusive channels that should extend for considerable distances. Inside urbanized areas, this can only be achieved through (preferably linear) parks, along wide boulevards, on the shores of rivers and lakes, and perhaps along underutilized rail and canal rights-of-way, paying much attention to scenic quality. Rest areas and service clusters along the way are welcome amenities. Splendid examples of this type of facility can be found in Boulder; Dallas; Davis, California; Eugene, Oregon; Gainesville, Florida; Madison; Missoula, Montana; Montreal; New York City; Palo Alto; Portland; San Diego; Seattle; Tucson; Washington, DC; and many other cities across the country.

A more complicated effort is being exerted to establish bike trails at the metropolitan or state level. There are jurisdictional and financing issues that will not be covered here, but it is encouraging to note that a number of successful networks can be identified in the United States. Notable programs exist in California, Florida, Illinois, Maryland, Michigan, Minnesota, North Carolina, Ohio, Oregon, and Wisconsin, but much more could be done, and most probably will be done because the demand is strong.

The emphasis on separate bike trails should not obscure the fact that bike routes and lanes on regular streets are also vital components of the recreational system. An efficient and safe network on surface facilities is a part of the effort, as described in more detail in the following text.

There is one issue from the urban operations point of view that is not yet adequately addressed: the transport of the bike itself from a city residence to the system of trails, which in many instances cannot be done in the saddle. To minimize the need to use private cars with outside racks, accommodations in public transit have to be developed. Trains in Europe tend to have designated compartments for bikes; this is not often the case in the United States, but examples exist on several of the North American commuter rail systems. Buses cannot carry bicycles inside, but special racks can be added on the outside ends, which is entirely justifiable when warranted by demand.

Cycling of this type is not, strictly speaking, urban transportation, but rather falls under the planning heading of recreation and open space, and it has its own set of concerns and concepts. Therefore, it will not be discussed further in this review, except by remaining alert to the frequent instances where bike trails can serve a joint purpose of recreation and urban transportation activity. It has to be recognized, of course, that there are many situations where the bike is a convenient and appropriate means of transportation to established recreation facilities, thereby blurring the distinction between use purposes.

Competitive Sport

Cycling has always been a recognized form of sport, ranging from long endurance races to blistering sprints and engaging a large number of athletes. There are well-known races (see the sidebar on the Tour de France), it is part of the Olympics, and there is a long list of local "minor-league" events. Again, racing is not transportation, but cycling as a sport serves to maintain high interest in the activity, develops equipment, and builds pride and support by making the effort newsworthy.

To operate races, streets and roads have to be occasionally closed to create temporary racing circuits, which should create no significant problem for communities even if regular street traffic is interrupted for a few hours on a weekend. It can also be argued that any sizable city should have a velodrome, as most European cities do, to encourage the sport. This is a highly specialized structure housing an indoor banked track, constructed of timber in an oval shape, with seats for spectators and accommodations for racers. The track is usually 200 m long. Many American cities in the early part of the twentieth century had such facilities.

Urban Transport

Commuting to work, going to school, visiting a doctor or travel agent, shopping, or seeing friends are all regular—almost daily— activities that all of us have to undertake, and for which we need some sort of transportation. Finding the proper mode in each instance is the principal focus of this work, and the rest of this chapter is devoted almost entirely to the option of using bicycles for these tasks. A community may be planned with bikes as the dominant system (very rare), it may have an elaborate network of dedicated paths (a few do), or it could have a set of rules and regulations, with some physical elements in place, that assist the use of bicycles for various purposes or as feeder services. Or it may have nothing that takes advantage of this useful mode.

Tour de France

The premier event of the cycling world is undoubtedly the annual Tour de France race, which has been run every year since 1903 except during the World Wars. It is a supreme test of human endurance and tenacity, and, of course, athletic skill. It is also one of the top global sporting festivals, watched by millions, outranked only by the Olympics and perhaps the World Cup. It is run in 21 stages as separate consecutive daily races, where the aggregate time over more than 3600 km determines the winner. As the name suggests, it follows a circuit around France, but it is not a continuous chain, nor has it lately been contained within France alone.

The great interest in the race is enhanced by the fact that there are flat stages where the sprinters dominate, and punishing mountain stages where the climbers reign. There are also experts on the transition stages; however, in order to win, the capabilities of a superman on wheels are called for. The names of Anquetil, Merckx, Hinault, LeMond, and Indurain will live forever (LeMond is actually an American); Armstrong is now added to this list. The highest overall average speed of 39.5 kph (25.5 mph) was achieved by Miguel Indurain in 1992. In recent years the Tour has become an advertising and television extravaganza that involves thousands of people—besides the 200 or so cyclists—and hundreds of vehicles that accompany the race as it unfolds. No opportunity for promoting a product is lost, but this is not all bad because the idea of cycling becomes visible to everybody.

This is a race supposedly of individuals, but teams are in operation where the duty of the secondary members is to assist their leader through various (legal) tactics. The teams are commercial enterprises, but very strong national identities are always present, particularly among the Europeans.

It is not particularly interesting to watch the race in the field, as the riders flash by in a few seconds; the real excitement is brought by television coverage and the continuous review of statistics as the race proceeds. The yellow jersey—the current leader—is the icon to be watched. Stay tuned every July, particularly as the race ends gloriously with a sprint down the Champs Elysées in Paris, even though usually the winner is known well before that.

It is stressed repeatedly that the practical value of the Tour is to test new hardware, equipment, and accessories.

Service Vehicles

In developing countries, bike-based human-powered devices are engaged in a multitude of work tasks. They deliver goods (including large pieces of furniture and live animals), they carry people in bicycle rickshaws, they propel pumps and machinery processing agricultural products, and they deliver urgent messages. All take advantage of the small investment required for the rolling stock, the high efficiency of movement generation, and the extreme maneuverability of the vehicles in constrained urban situations. Unfortunately, as mentioned before, many local governments think that these operations impair their civic image, and officials tend to create obstacles for bicycle-type vehicles, including outright bans.

In communities of developed countries, human-powered vehicles exist as well, and they have useful roles to play. They are almost indispensable in large cities with crowded and congested centers where motor vehicles simply cannot get through and even short-term parking is a major problem.

Perhaps the most common form is delivery vehicles—either bikes with large containers servicing supermarkets or ordinary bicycles used by fast food establishments (for example, pizza and Chinese food). A special application may be mail delivery in moderate-density areas with the proper spacing between houses. Newspaper delivery by bike is an old American tradition. Couriers of messages and small parcels use light ordinary bikes and shoulder bags.

A number of cities (Newark, New Jersey; Seattle; Las Vegas) have created bicycle police squads that are extremely fast and effective in situations where a police car becomes helpless. It could even be suggested that first-response firefighters and emergency medical personnel consider this option.[10] Bike-mounted police were very common in the 1890s; they have reappeared again in many places because of the suitability of the machines for the tasks at hand.

A special case are bicycle messengers, with the Manhattan business district a prime example, where these superbly skilled riders have attained a legendary and threatening aura.[11] This activity expanded explosively in the 1970s when it proved itself as the quickest and most reliable means to deliver documents and written messages between offices. The fleet grew, and the riding habits became increasingly reckless because this is essentially piecework. While one could admire the acrobatic skill of the mes-

---

[10] The author has worked in a large consulting office in Manhattan where twice people have died from heart attacks while waiting for an ambulance to arrive.
[11] This story has been recorded by the *New York Times* since the late 1970s, maintaining a critical attitude. See the *New York Times Index*.

sengers, countless dangerous situations were created since all legal and physical constraints to fast movement tended to be ignored. The messengers themselves suffered the greatest damage (one or two were killed each year), but pedestrians were seriously endangered as well (in some years two or three were killed after collisions with fast-moving bicycles), and a public outcry for control arose in the mid-1980s. Police campaigns were instituted, but, with the widespread introduction of fax and e-mail, the demand for bike messenger services waned. Also, greater attention was being paid by state and federal agencies regarding the payment of taxes and observance of labor rules. With the onset of a general business recession, the operations abated, and the menacing aspects eased. However, that was not the end of the activity. There is a resurgence of bike presence today with the growing practice of ordering things through the Internet and expecting immediate delivery, which the U.S. Postal Service is unable to provide.

The overall transportation planning point to be made about all service bikes is that they need to reach every location in a community, and thus they have to be absorbed in the general traffic flow on existing streets. Use of special bicycle facilities will only be incidental to their general operations. While this would be welcomed by the riders to save time and energy, a fine-grained network cannot be expected, and, therefore, regular surface streets will have to carry this mixed traffic. A major problem for service delivery is the safe—if brief—parking of the bike at any number of destination points.

Reasons to Support Bicycle Systems

Bicycles as a form of urban transportation, leaving aside at this point their capabilities as a means of recreation, have a series of positive characteristics, and in some respects superior aspects as compared to other transportation options. These are well-known features

Bike messenger in the traffic stream of New York City.

because cycling advocates have promoted them loudly for years;[12] however, an overall survey is necessary here to place these characteristics in context.

Direct Access

Except for walking, bicycles, because of their small size and weight, can provide the most direct door-to-door service compared to all other mechanical modes. Indeed, they go beyond the door, if the proper accommodations are made. It is an "intimate," truly personal service, almost an extension of the human body. The bike is actually as close to personal rapid transit (see Chap. 15, Automated Guideway Transit) as we are likely to get.

Low Energy Consumption

Bicycles, of course, do not tap into the fossil fuel pool, and, to the extent that gasoline availability will certainly become a crisis again, this is a significant feature of public concern. At least one government official in defense of the national fuel supply plan has assured the American public that "we will not have to go back to bicycles." That may not be the worst thing that could happen to us.

Cycling does consume energy through the human engine, but—as mentioned at the start—it is the most efficient way to achieve motion. A bike weighs 20 to 30 lb, a car 2000 to 4000. A single-occupancy automobile will consume 1860 calories to generate one passenger mile, and walking will burn 100, while a person on a bike will carry himself or herself 1 mi for 35 calories.[13] The energy aspect has a different interpretation in the prosperous societies, as compared to populations where undernourishment is an issue. In developing countries the added caloric intake necessary to push a bike forward does become a cost consideration in the total equation, but it is less than for walking and cheaper than taking the bus.

[12] The Web carries much of this material under "bicycling," "bike lanes," "bike paths," and similar key words. Particularly extensive as sources of information and advocacy material are the home pages of "transportation alternatives" (of New York City) and the City of Portland, Oregon.

[13] M. C. Holcomb, *Transportation Energy Data Book: Edition 9* (Oak Ridge National Laboratory, 1987).

Absence of Pollution

Unlike motorized modes, bicycles, of course, produce no air pollution, except what trace amounts may result from human exertion. Cycling creates no noise pollution either. This can be a bit of a problem because nobody can hear a fast rider approaching and get out of the way. That is why some bike messengers carry a whistle in their teeth, while the rules require that each machine have a bell.

Healthful Exercise

If the total population is largely overweight, and exercise is a national preoccupation that is highly endorsed, then the widespread use of bikes has an obvious benefit.

Space Conservation

The bicycle is a compact machine. It occupies about 22 ft^2 (2 m^2) when standing and 55 ft^2 (5 m^2) when in motion. This corresponds to 270 ft^2 (25 m^2) and at least 600 ft^2 (55 m^2), respectively, for the automobile. If, let us say, 5 percent of single-occupancy motorists were to switch to bicycles, all street congestion problems would disappear. However, this is surely an impossible dream in today's American communities.

Low Public Investment

Even under the most elaborate plans, a bike system is a low capital investment. There could be right-of-way acquisition, but in most cases recreational bike trails would be fitted into already designated open spaces, and in an urban setting the lanes would be created within existing streets and roadways. The only expenses then are physical elements and paint to demarcate the lanes, signage, and perhaps the construction of some rest areas. There may be a need for new or upgraded pavement since smooth surfaces are required, but this is not a large construction effort. Some drainage inlets may have to be rebuilt to prevent accidents and other localized modifications made.

An added expense item and legal/administrative complication will be the creation of places for bicycle storage at nonhome destination points: in commercial areas, at office buildings and institutions, at transit stations, schools, etc. Some space, either publicly or privately owned, has to be found for this purpose; at least minimal but secure improvements have to be made; some supervision must be provided; and responsibilities (insurance and management) must be defined. Advantage can be taken of underutilized scraps of urban space, and in many instances building owners may be in a position to contribute space or other resources. Existing automobile garages and lots may be willing to participate.

To make cycling really effective in North America, a most desirable feature—second to parking—would be changing rooms and shower facilities at the nonresidential end. As has been done in Portland and a few other places, athletic and health clubs may be brought into the overall system, short of establishing such amenities by individual employers.

Experience shows that the availability of reasonably convenient and secure parking accommodation of bikes during the time their owners accomplish their trip purposes is a critical requirement. To achieve a successful program, this aspect cannot be overstressed.

Low Private Expense

The interesting thing about cycling is that the users provide their own rolling stock. But, unlike automobiles, the investment is low. An acceptable adult bike can be purchased today for about $200, mountain bikes start at $300 and go upward, and an advanced Italian racing machine will be in the $3000 range. Unlike automobiles, there is no fuel to buy or tolls to pay. There is, however, peer pressure to acquire the proper accoutrements of the cycling fraternity. Helmets are a desirable piece of equipment, and they are now mandatory in most places for riders up to 14 years of age. There are a few voices suggesting that prevention of injuries would be higher if the amount spent on headgear could be devoted instead to education programs about safe riding and proper behavior by motorists.

Reasons to Exercise Caution

Bicycles are not suitable for everybody, and they do not fit easily into urban traffic operations as we have them at this time in American communities.

Traffic Compatibility

Bicycles are quite different from motor vehicles in their size, speed, and operational characteristics. They do not coexist easily on the same channels, thereby creating constant friction and occasional danger, mostly to the unprotected cyclist. It can be argued that bicycles are vehicles like all others (state laws say so) and therefore have the same rights regarding road space, but this argument is belied by the physical fact of their vulnerability. (See the following "Bike Lane Debate" section.)

Bicycles are even less compatible with pedestrians, creating serious friction, if not serious accidents, between the two modes. Thus, fast-moving bikes cannot be allowed on sidewalks, and reasonable provisions have to be made at intersections where volumes are significant. Even on low-intensity recreational paths it is advisable to make it clear on which side pedestrians and bicycles are supposed to move.

One solution to this dilemma is the creation of separated facilities and exclusive lanes, but that suggestion raises issues of cost/benefit: is the large expense of allocating space in a tight urban environment justifiable if the usage is not of a high volume? How many real benefits are gained? In many cases, dedicated space can only be found at the expense of something else—established vehicular lanes, for example. In other instances, it is not just a matter of rational cost/benefit analysis, but almost certainly involves attitudes, emotions, and political clout among various user groups.

Human Capabilities and Attitudes

To pedal a bicycle requires a certain amount of strength, stamina, and agility, which are not possessed in equal amounts by all people. Many cannot use a bike at all. Elderly people, for example, can use a tricycle, but this would create further friction in any traffic stream, including bike lanes.

Bus and bike lane within street pavement space (Odense, Denmark).

Separated and buffered major bike lane (Münster, Germany).

Urban networks are stressful environments in any case; bike operations introduce another level of concern that requires constant alertness and individual responsibility in behavior. Children presumably should not be exposed to such conditions; many people will rather forgo this challenge. There are also a certain number of people who will regard cycling as a not very dignified activity, associated with children's games and teenage exuberance.

Since transport by bicycle will always be a voluntary choice, such a system can only be successful if it has sufficient attractive features, compared to other movement options, to generate and maintain a willing and dedicated user group. A nontrivial issue is the availability of good-quality support facilities, such as secure parking, changing rooms, and showers. These are reasonable expectations if a steady and committed ridership volume is to be maintained in any American community.

Safety and Security

The city environment is rife with physical imperfections that become problems for a fragile bicycle: potholes, grooves, and ruts in the pavement; longitudinal or diagonal rail tracks; slippery surfaces; sewer gratings that catch a wheel; deep puddles; debris and broken glass; and similar elements that can cause a crash if cyclists are not continuously alert. There is also the behavior of other participants in normal street activity: drivers of large trucks may not be able to see a bicycle well, doors of parked cars may open suddenly, pedestrians may emerge suddenly on paths, some riders may have little sense of movement etiquette, and it is even rumored that some motorists take pleasure in intimidating cyclists to cede space.

Safety concerns relate also to the behavior of cyclists themselves. Not everybody on the streets is adequately skilled in—or, worse, fully aware of—appropriate riding techniques. Common unsafe practices include riding against vehicular traffic, ignoring traffic signals, using sidewalks, not making proper turn signals, not wearing helmets, riding at night with poor reflectors or no headlights, exceeding appropriate speed, overloading a bike, cutting in front of vehicles, not knowing how to make turns at intersections, and others.

Consequence of a motor vehicle–bicycle conflict (Nigeria).

It is easy to steal a bicycle, and often tempting opportunities are present. If secure and convenient storage places are not available, the carrying of a very heavy chain is a most inconvenient practice and does not necessarily guarantee complete peace of mind.

All these cautions and shortages represent a filter that screens out most potential bike commuters under currently prevailing conditions, particularly in the centers of large cities.

The Natural Environment

Steep topography clearly will not encourage bike riding, except for those who embrace this challenge. Modern lightweight machines and efficient gears reduce the problem, but the difficulty is still there. Any street with a gradient steeper than 3 to 5 percent will bother most regular riders; steep downgrades present a potential safety problem.

Climate is a concern as well. It is most uncomfortable to engage in vigorous physical activity in high temperatures with no air conditioning possible. There are places around the globe and even in this country where cycling is simply not viable as a steady practice because of weather. Cold temperatures, high winds, sleet, snow, and ice do not necessarily preclude cycling, depending on their severity, but little enjoyment will be derived, and high volumes of riders cannot be expected. In China, a tentlike plastic poncho is widely used, but it is doubtful that Americans would find this response acceptable. Space-age fabrics are available to provide protection against rain and cold, but expense and comfort considerations are present nevertheless.

The complication is not the fact that some regions are simply not bike friendly, but rather the problem everywhere of repeated interruptions in operations, such as by a heavy rainstorm. When that occurs, most regular bike commuters will seek a dry substitute mode, which then would overload regular transit and vehicular services. The more successful the bike mode is in a given community, the larger the temporary dislocation will be. Neither of the potential theoretical solutions are particularly attractive: keep a reserve transit capacity available that would be idle most of the time or declare a rain day that would be a forced vacation, except for those who can work at home, causing some business disruption.

Reach and Speed

Since cycling depends on a human engine, which gets tired, certain physical limitations have to be recognized for this mode. There is general consensus that very few people would ride more than 10 mi (16 km) on a regular basis each day, which can be covered in less than 1 hour. Anything below 5 mi (8 km) can be considered a comfortable distance that should represent no difficulty for most cyclists. Such a radius encompasses the territories of small and medium size cities, except for some suburban sprawl around the edges, and thus conceptually a bicycle transportation potential exists almost anywhere.[14] Even in large metropolitan areas most trips fall in the shorter range, keeping in mind that the average commuting trip in the United States was 11 mi (18 km) in 1990.[15]

The regular cycling speed is about three times walking pace—in the range of 10 mph (16 kph) or 6 minutes for each mile. While almost all cyclists can maintain a 12-mph speed over extended distances, many will usually ride at 6 mph (10 kph). A respectable racing speed is 20 to 25 mph (32 to 40 kph), and anything above 30 mph (48 kph) is a sprint in major competition. There is one inescapable physical fact: at 20 mph, 90 percent of the energy produced by a cyclist is consumed in overcoming air resistance, not rolling friction.

This means that for a trip of only a few miles, if the storage of the bike is not cumbersome at either end, cycling will be fully

[14] The *National Personal Transportation Survey* documents that one-quarter of all trips are shorter than 1 mi (1.6 km) and two-thirds are within 5 mi (8 km).
[15] "Journey-to-work" statistics of the 1990 U.S. Census.

competitive with transit time in most instances (including access and waiting times); there will be a time penalty for longer trips. Comparisons with automobiles will depend, of course, on local street congestion and parking conditions. During the subway strikes in New York City, when all streets were completely jammed, bicycles were the only rational mode to use.

Storage

Problems with parking a bicycle at the nonhome end of a trip have already been mentioned. With apartment buildings, particularly old ones, there is also frequently a problem at the residence end: space to store equipment may not exist or be very inconvenient. (Seinfeld keeps his $2000 bike on hooks on the wall of his apartment.)

Established Urban Patterns and Facilities

Besides the question of the overall size of an urban settlement, there is also the issue of the extent that land use distribution and clustering of activities foster or constrain bicycle transportation. The bike is a small device in individual use, and therefore is not quite compatible with large concentrations of intense activities. It serves well on shopping trips to the neighborhood center, but it may have problems dealing with a regional mall or a cluster of hyperstores. It can bring commuters to a group of office buildings, but there may be frictions moving to and within a large industrial district, particularly in a mix with heavy truck traffic. Large urban places tend to have districts with more segregation of uses and more distinct single purposes than smaller cities, and thus a coarser traffic mix, not very compatible with bicycle operations.

Existing massive transportation facilities are quite likely unable to accommodate bicycles and to have no intention of doing so. Fast limited-access highways are clearly not compatible with bike traffic, and few have excess right-of-way width for separated bike paths. Bridges and tunnels frequently have no provisions for bike movements, and retrofitting them may be nearly impossible. Some of the more vocal recent controversies have involved demands by cyclists to gain access across some of the major bridges in this country. Many more recent highways do have generous right-of-way widths within which bike lanes could be accommodated; some states allow cyclists to use interstate highway shoulders.

The critical issue in most urban environments is the fact that we are facing a zero-sum game for space. Additional room for circulation usually cannot be created, and, therefore, if anything is given to bicyclists, it has to be taken away from motorists or pedestrians. The battles over turf can become fierce, and the stakes are high.

Official Attitude

Cycling, particularly for commuting to work, is not among the normal patterns of operation in most communities. Even if there is no outright hostility toward it, the level of activity will be substantially influenced by the attitudes and opinions of political and civic leaders, the position taken by the local media, and the general atmosphere in the community toward bicycles. A positive environment is a prerequisite toward program implementation, and a proactive approach—not just neutrality and pro forma endorsement—will be necessary to achieve results. It usually requires some locally recognizable and respected individuals or groups, not just cycling enthusiasts, to initiate action.

Application Scenarios

If bicycles are not an answer to all urban transportation problems, they have a significant potential in the proper situations nevertheless. Starting with the ideal conditions in order to define one end in the range of applications, several such instances can be identified.

An Entire Community

- Compact, relatively small size (all internal trips below 5 mi [8 km])
- A large population cohort in the 15 to 35 age group (such as a college town or institution-dominated community)
- Flat terrain (or gently rolling at most)
- Mild weather (with not too many instances of extreme temperatures)
- Several destination centers (shopping, recreation, education)

- Wide streets or other rights-of-way (with available space for bike lanes)
- A system of parks and waterway and railroad rights-of-way
- A strong tradition of outdoor activity and environmental awareness

This scenario describes a college town that would have the ability to provide separated bike paths along some major (and scenic) corridors, a network of bike lanes along many streets, and sufficient bike parking at activity centers.

A Residential Neighborhood

- Moderate density (houses close together with narrow side yards)
- Local service centers and establishments within 2 mi (3 km) or less
- No through vehicular traffic
- Transit stations or stops within or adjacent to the area
- Flat terrain
- Mild weather

This scenario describes what neotraditional planners and architects would call a *transit village*. Bike operations may be accommodated on regular interior streets, without any dedicated lanes.

Feeder Service

This includes service to a node that requires daily access, such as a transit station. It would require the same general characteristics as outlined in the two preceding sections, as well as:

- Secure storage facilities at the node
- Preferential lanes in the vicinity of the node

The bicycles may be owned by the individual commuters, or there may be a fleet of public bikes that are picked up (or rented via a meter), used as necessary within the defined neighborhood, and brought home for the return trip next morning. Obviously, there are potential issues with theft and vandalism, but some experience is already available, and the aspirations toward a civil society are not unreasonable. It is also possible to consider the

Facility Definitions

bicycle lane　A portion of a roadway that has been designated by striping, signing, and pavement markings for the preferential or exclusive use of cyclists.

bicycle path　A bikeway physically separated from motorized traffic by an open space or barrier and either within the highway right-of-way or within an independent right-of-way.

bicycle route　A segment of a system of bikeways, not in exclusive use, designated by the jurisdiction having authority with appropriate directional and informational markers.

bikeway　Any road, path, or way that in some manner is specifically designated as being open to bicycle travel, regardless of whether such facilities are designated for the exclusive use of bicycles or are to be shared with other transportation modes.

multiuse (recreational) trail　A pathway or facility on a separate right-of-way or within a public open space for use by a variety of different nonmotorized users, including cyclists, walkers, hikers, inline skaters, skiers, and equestrians. With more intensive use, the separate types of movements would have their own dedicated paths.

shared lane　A portion of the roadway that has been designated by striping, signing, or pavement markings, and may have enlarged width, for the joint use of bicycles and motor vehicles.

reverse arrangement: communal or rental bikes being available at the destination (nonhome) end of the trip.

The Bike Lane Debate

For some time now a fundamental debate has raged within the cycling fraternity as to whether bicyclists are drivers of vehicles (comparable to motorists) or whether they are people with wheels (comparable to pedestrians). This is by no means an academic exercise, because the decision is basic to the placement of bicycles within the operating built environment and the provision of suitable facilities. If cyclists are drivers, then they can and should share regular roadway space in mixed traffic; if not, then they have to be protected and segregated.

Fuel for the fire—so to speak—is added by the rather insistent attitude of some dedicated cyclists who are not only highly skilled (certainly Type A) but also adamant that they have all the rights that a motorist does, and that this standing is not to be impaired.[16] This is entirely true legally, but there are also other cyclists, mostly of the casual recreational type, who would much rather be as safe and separate as possible. Rational discourse is not always employed since historically both ends of the range are characterized by much emotion and stridency ("we will not be discriminated against" versus "we do not wish to have our children killed"). Where does this leave cycling as an urban transportation mode?

In the practical realm, the dispute focuses on bike lanes. The first group sees them as an affront, regards them as dangerous, and insists that all bike riders become skilled enough to participate in regular traffic flow. Some committed cyclists contend that bike lanes are advocated primarily by motorists, thereby hoping to push bicycles out of their way and segregate them on mandatory lanes. There is, indeed, a specific problem at street intersections. Since bike lanes are usually on the right-hand side of the roadway, cyclists have considerable difficulties in making a left turn across the vehicular traffic moving forward, and motorists turning right across the bike lane may not see cyclists going straight. There are ways of coping with this conflict, but none are elegant and foolproof.[17]

The position taken in this review is that the objective of a bicycle urban transportation system is to attract as much ridership as possible, even if all participants are not completely proficient and never will be. They should not only be safe but also feel safe. A tentative rider will become most discouraged by the air turbulence left in the wake of a fast-moving 18-wheel truck passing close by. This should not be experienced ever, and social engineering to achieve high cycling expertise by everybody is not called for either. Bicycles are inherently different from both motor vehicles and from people on foot. A cyclist cannot do much damage to a motorist, but he or she can hurt a pedestrian, and a motorist is certainly easily able to injure both.

---

[16] See any publication or statement by John Forester and the home pages of John S. Allen and others.

[17] The manuals outline the various possible responses, which include merging of right-turning vehicles into the curbside lane, providing separate advanced signal phases for bicycles, designating a crossover reservoir space for bikes at the head of the queue, etc.

Thus, with the full recognition that most of the bicycle miles accumulated within any community will be on regular streets, there is a strong case for providing separate facilities where possible and reasonable. It is fundamentally a matter of recognizing cyclists as legitimate users of public rights-of-way (not being relegated to a secondary status) and providing as much encouragement as possible for nonexpert riders to participate in this activity. Undoubtedly, there is also a larger policy implication here—since in most instances bike space can only be created by taking away some space from motor traffic, priorities are being defined. It may be anathema to some Type A riders to hear this, but motorists need protection from cyclists as well, at least to minimize their nervousness and level of possible road rage. Bike lanes, if they are properly controlled, provide also a more open street environment with better visibility, more maneuverability options, and reservoir space to minimize accidents. But the debate will surely continue.

The City of Cambridge, Massachusetts,[18] as an example, while it does not bar cyclists from any streets or lanes, supports the use of designated bike lanes because they:

- Help define road space
- Provide cyclists with a path free of obstructions
- Decrease stress level
- Signal motorists that cyclists have rights

Components of the Physical Network

A decision to plan a bike-oriented transportation system for a community, as is the case with all other modes and systems, has to be supported with a reasonable assurance that sufficient demand will be generated and that broad-based support exists. In the case of bikes, the preparatory steps should not require elaborate analyses and simulation modeling because the future usage patterns are not so much a matter of cause and effect logic as of local trends and human attitudes. There is no reliable way to predict how far the population will embrace the new transportation option and how large the user pool will be. Rather, there is good reason to suggest

[18] See the city's home page on the Web at www.ci.cambridge.ma.us.

Bicycle Lane Fact Sheet

Minimum lane width with no lateral obstructions (i.e., bike lane surface flush with adjoining surfaces):

One lane	3.3 ft (1 m)
Two lanes in same direction	6.4 ft (1.95 m)
Two lanes in two directions (not recommended)	8 ft (2.4 m)
Two lanes in multiuse	12 ft (3.7 m)

Minimum lane width with a lateral obstruction along the edge (parked cars, walls, barriers, or curbs): 5 ft (1.5 m)

Vertical clearance 8.2 ft (2.5 m)

Maximum grade 3 to 5%
For less than 200 ft 15%

Average speed 10 mph (16 kph)
Design speed 20 mph (32 kph) on level ground
25 mph (40 kph) on downhill

Horizontal curve is usually not a consideration, but:

To maintain 20-mph (32-kph) speed	100 ft (30 m)
To maintain 25-mph (40-kph) speed	155 ft (47 m)

Capacity 2000 bikes/h per lane (1.8 s and 7.5 m apart)

Two lanes, same direction	3000 to 5000
Four lanes, two directions	5000 to 10,000

Note: In China, considerably higher volumes are being achieved with riders experienced in high-density conditions.

The American Association of State and Highway Transportation Officials (AASHTO)* suggests:

Minimum width of:
3 ft for paved shoulders along rural roads
4 ft for lanes with no lateral obstructions

Recommended width of 5 ft

Lanes should be placed along the right-hand side of a one-way street.

* AASHTO, *Guide for the Development of Bicycle Facilities,* 1991.

experimentation and on-site exploration of various options. The capital investments are not large, and possible errors can be corrected later. Most important, a period of adjustment is needed for members of the community to modify their attitudes and habits and to do such simple things as purchase bicycles.

This can be illustrated by a series of events that occurred in New York City in 1980.[19] Mayor Ed Koch, having returned from a trip to China and being impressed by bike operations there, announced with some fanfare that he was implementing a progressive concept that would be good for the city. Several avenues in Midtown Manhattan were equipped with concrete and asphalt separators that carved out a series of dedicated lanes. It turned out that this was a hasty effort, with no adequate preparation of the public or alerting of possible riders and generating an immediate outcry by motorists, taxi drivers, and businesspeople who lost scarce circulation space and access. (They also lost opportunities to double-park.) The volume of cyclists, after the original novelty wore off, was rather low, and the experiment looked wasteful in the overburdened Midtown environment. Koch made another announcement after a few months: respecting the wishes of the population, the lanes and barriers would be removed. They were (except for Sixth Avenue, where painted lanes were kept), and the cause for cycling was set back at least a decade in New York. This is not the way to do experimentation: there was no overall planning, there were very limited education and promotion efforts, and the pilot project was much too short for potential users to become active participants and change their established travel habits. The Mayor, however, scored political points twice. The details of the case are more complicated than can be given in this summary, but the experience does illustrate the uncertainties and emotions that are often involved in bike programs.

Assuming that a decision is made—based on one set of reasons or another—to proceed with a bicycle program in a community, there are several options as to how the system can be structured.[20] There are, however, a number of characteristics that cyclists expect from all facilities:

- The pavement surface should be as smooth as possible, but not slippery when wet and not covered by loose material. Removal of debris and good street maintenance are important.

[19] See articles in the *New York Times,* July to November 1980, and *Time* magazine, November 24, 1980, p. 110.
[20] A number of states have prepared planning and design handbooks to assist communities and bike system proponents. Among those are California, Oregon, and New Jersey.

- The environment should be safe and secure in every respect, i.e., it should preclude crashes between bicycles or bicycles and other vehicles, minimize chances of criminal actions and vandalism, provide reasonable grades and widths, and so on.

- Movement should be as continuous as possible with few stops, because any interruption impairs the fluidity of motion and consumes excess energy.

- Because of the leisurely pace of the movement, riders can observe their surroundings and will appreciate visual quality, as well as amenities offering comfort and rest.

Bicycles in Mixed Traffic

The most basic approach is to regard bicycles as regular vehicles, which is a right they have anyway, and introduce them in the general traffic stream. The contribution of an organized program would be that bike presence on designated streets is encouraged and protected through a series of noncapital or minimal physical improvement programs. On streets that carry at most a few hundred cars in a single direction per hour, no special bike facilities or traffic treatments can be expected or are needed, nor can they be justified on a cost basis given the low user volumes. Motorists occasionally may lose a few seconds to accommodate cyclists who are equally entitled to use any public road space and actually enjoy some priority on designated streets. Good visibility across all parts of the circulation space by drivers and cyclists is to be provided, and proper traffic behavior by all parties is to be expected.

The program might have the following components:

- An extensive education program that prepares the general public—particularly motorists—for the acceptance of bicycle operations on streets and responsible behavior. Likewise, there has to be a parallel program for cyclists that instructs them in good and safe riding habits. There should be a convincing explanation of why the bike program is beneficial for the entire community, with a positive message that reaches all local residents, workers, and civic leaders.

- Enforcement of traffic regulations among cyclists in American communities has been uneven, to say the least. Police often do not take the mode particularly seriously, and cyclists, on the other hand, sometimes think that they can and must take shortcuts because nothing is being done for them anyway. With growing bike volumes, such situations, of course, cannot be tolerated, and a sense of responsibility needs to be established. Particularly dangerous practices by cyclists—as mentioned before—are disobedience of traffic signals, moving the wrong way on one-way streets or on the wrong side of a roadway, riding on the sidewalk, not signaling turns, and pushing through pedestrian clusters on crosswalks at high speeds.

• The designation of bike routes and the posting of the standard signs. This step by itself is not of great practical significance since it does not control much nor limit anything; there is nothing specific to be enforced. Indeed, in some instances this program has been an empty political action to show that something is being done and to placate cycling advocates. The signs do serve, however, to emphasize the overall intent and to alert everybody repeatedly that bicycles are likely to be operating on the designated street and can be expected on all other streets in the community (see Fig. 3.1).

• At points with high vehicular and bicycle traffic volumes, localized physical improvements can be provided. This may include special bike phases in the traffic signal cycle, reservoir spaces to accumulate bikes before turns or forward movement (they will filter to the head of a queue at a red signal anyway), improvements to visibility and lighting, and provision for safe passage under and through obstacles (underpasses, tunnels, bridges, railroad crossings, etc.).

There are two fundamental considerations regarding these programs:

1. All American communities have some cyclists who will persist in their activity whether anything is done to assist them or not. Therefore, the basic education programs and some spot improvements to remove dangerous situations should be mandatory in all places. Likewise, cyclists are subject to all traffic regulations, and therefore should expect to be ticketed for infringements.

2. In bicycle-oriented communities, even those with elaborate bike paths, much of the riding will take place on regular city streets connecting residences and establishments to

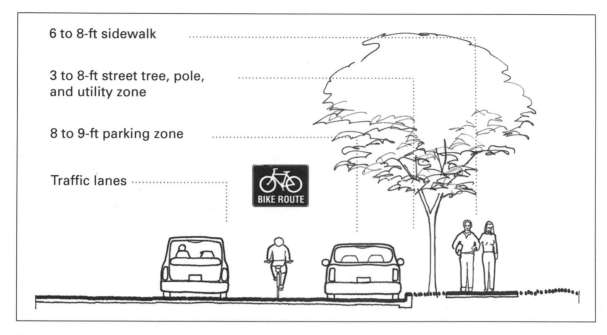

6 to 8-ft sidewalk

3 to 8-ft street tree, pole,
and utility zone

8 to 9-ft parking zone

Traffic lanes

BIKE ROUTE

Figure 3.1 Bicycle route on a city street.

any designated networks. Therefore, the basic nonphysical programs are again to be expected, and, indeed, regarded as a starting point for any other more far-reaching efforts.

Shared Lanes

One program that almost every bike system proponent supports is the designation and creation of lanes that are shared by both cyclists and motorists—at least in those instances where the flow volumes of either mode are not too high. Such facilities are also frequently rather difficult to implement because these lanes have to be wider than a regular traffic lane, and this additional width is difficult to find, short of rebuilding the roadway.

The concept of shared lanes supports mixed traffic operations, except that faster motor vehicles would overtake bicycles within the same lane, not intruding laterally into other lanes. Since cyclists would largely stay on the right side of the roadway, but a few feet away from parked cars (doors may be opening)[21] or high curbs (pedals may catch), there must be sufficient width for both

[21] The sudden opening of car doors in front of a moving bicycle is just about the most feared event by cyclists. To cope with it in complete safety, a 4-ft-wide buffer strip would have to be provided.

bike and automobile to move side-by-side as the car passes. This adds up to a width of at least 14 ft (4.3 m) or more with substantial traffic volumes, as shown in Table 3.1 and Fig. 3.2. (A regular street traffic lane would be 10 to 11 ft (3.0 to 3.4 m) wide or more.)

While wide shared lanes is a favored concept, it remains still mostly in the realm of theory because there has been very little if any implementation in American communities. Thus, they are not proven in practice, and there may be a potential problem: 14 ft (4.5 m) is wide enough for two cars to stand side by side (a standard automobile is 1.6 m wide) or even move together, which can easily happen with impatient motorists on badly congested streets.

In the early days of bike lane development, a classification system was devised that defined three possible types of dedicated and exclusive facilities:

•   *Class I.* Bike paths physically separated from vehicular roadways, following their own independent alignment.

•   *Class II.* Bike lanes separated from vehicular lanes by painted stripes or raised curbs.

•   *Class III.* Bike routes marked by signs, but not in exclusive bicycle use.

This classification is not used these days because each instance is different, and there are numerous complexities in structuring a network in any community. For the purposes of discussion, however, these distinctions among types of facilities are useful. They also give a historic continuity to earlier work. The arrangements discussed so far (mixed traffic and shared lanes) are the former Class III situations.

Table 3.1   Recommended Widths of Shared Lanes

	Vehicular traffic flow below 2000 units in 24 h		Vehicular traffic flow over 10,000 units in 24 h	
	Street with Curbside Parking	Street with No Parking	Street with Curbside Parking	Street with No Parking
Less than 30 mph (50 kph)	12 ft (3.6 m)	11 ft (3.3 m)	14 ft (4.2 m)	14 ft (4.2 m)
31–40 mph (50–65 kph)	14 ft (4.2 m)	14 ft (4.2 m)	14 ft (4.2 m)	Not advisable
41–50 mph (65–80 kph)	15 ft (4.5 m)	15 ft (4.5 m)	15 ft (4.5 m)	Not advisable

Source: New Jersey Department of Transportation, *Bicycle Compatible Roadways and Bikeways: Planning and Design Guidelines,* 1996, p. 7.

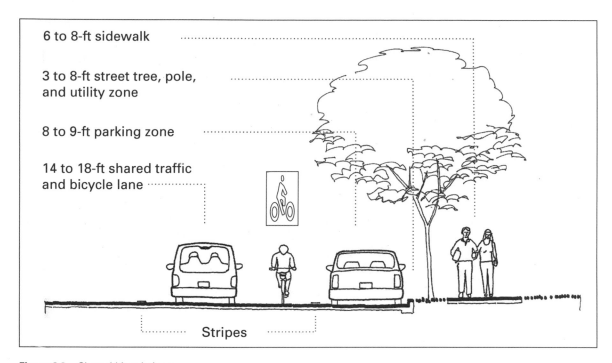

6 to 8-ft sidewalk

3 to 8-ft street tree, pole, and utility zone

8 to 9-ft parking zone

14 to 18-ft shared traffic and bicycle lane

Stripes

Figure 3.2 Shared bicycle lane.

The next series of actions outlined are devoted to the separation and dedication of lanes within the street rights-of-way for exclusive bicycle use (except for merging cars to make a right turn), i.e., former Class II facilities. The latter approach has become common practice, and the physical dimensions and configuration are reasonably set. Such arrangements have been tested through numerous applications in the United States and Europe.

Shoulders

In suburban and rural areas where there is no curbside parking and no curbs along the side, shoulders can be paved and used as bike lanes (moving in the same direction as motor traffic). The normal 8-ft (2.4-m) shoulder is quite adequate for this task, and the operative width can be less—as narrow as 3.3 ft (1 m) for streets with low traffic volumes.

In instances with a large vehicular traffic overload, it is possible again that motorists will regard the paved shoulder as a relief traffic lane. Thus, police supervision will be required, besides extensive stenciling of bike logos on the surface. Where roadway grades both up and down are steep (more than 6 percent), separate lanes for bikes become most advisable for the benefit of both motorized and human-powered traffic.

Striped Bike Lanes

An exclusive lane can be created by reserving at least a 3.3-ft (1-m)-wide strip along the right-hand side of a roadway by marking a line and stenciling the standard bike symbol at regular intervals (see Fig. 3.3). If there is a curb, which is a lateral obstruction that may be hit with a pedal, an additional width of 1 ft (0.3 m) should be provided; if the lane is inside of a parking lane, 2 ft (0.6 m) of additional width is appropriate to accommodate opening car doors (see Fig. 3.4).

The striped bike lane can also be placed outside a curbside parking lane, which is usually easier to implement (no parking spaces are lost), but brings the bike riders more into the motor vehicle space (see Fig. 3.5). In the old classification system, all these were Class II bikeways.

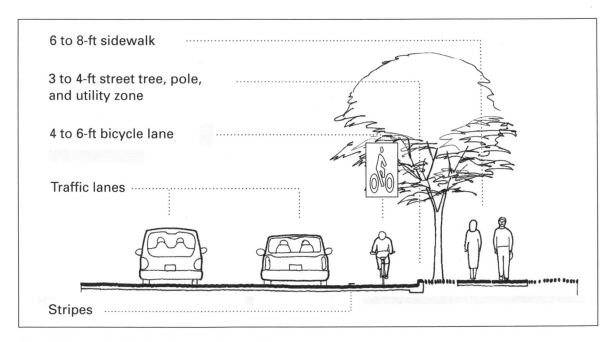

Figure 3.3   One-direction bicycle lane with no parking.

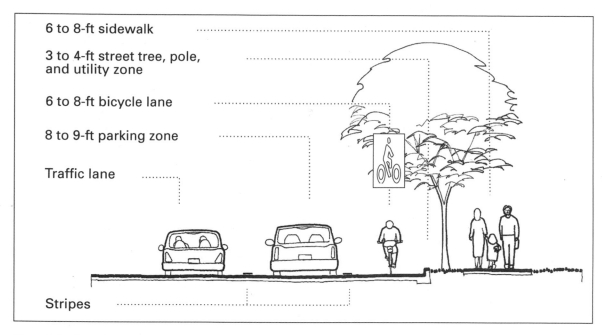

6 to 8-ft sidewalk

3 to 4-ft street tree, pole, and utility zone

6 to 8-ft bicycle lane

8 to 9-ft parking zone

Traffic lane

Stripes

Figure 3.4 One-direction bicycle lane inboard of parking lane.

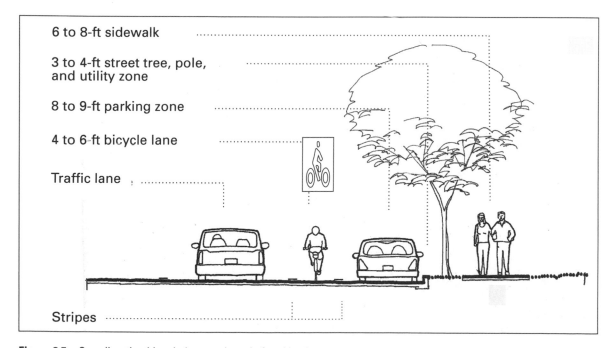

6 to 8-ft sidewalk

3 to 4-ft street tree, pole, and utility zone

8 to 9-ft parking zone

4 to 6-ft bicycle lane

Traffic lane

Stripes

Figure 3.5 One-direction bicycle lane outboard of parking lane.

Barrier Bike Lanes

To prevent motor vehicles from intruding into exclusive bike lanes, a common practice in the recent past was to separate on-street bike lanes using a low physical barrier such as a raised curb. This was frequently a series of precast concrete units 4 or 6 in (10 or 15 cm) high that could be bolted down along the lane's edge. This type of facility was included among Class II bikeways (physical separation but close proximity to motor traffic).

It has become apparent in the meantime that this response is counterproductive and even dangerous. It copes only with one type of bike–car conflict, which is among the least common occurrences—motor vehicles hitting bicycles from the rear and the side as they overtake them. The low barrier would probably not stop a fast-moving vehicle anyway. The barrier itself is a hazard because it may catch a pedal on the downstroke, or the rider and the machine may topple over into the adjoining traffic lane if the bike strikes a glancing blow against the barrier. Even protuberances only a few centimeters high, such as rumble strips and raised roadway reflectors, can cause problems for cyclists.

Sidewalk Paths

If sidewalks (and adjoining green strips) are wide enough, i.e., sufficient width is left for pedestrians, a bike lane can be carved out of this space. Care should be taken to ensure that there are buffer strips between the bicycles and pedestrians and bicycles and motor cars (even if they are parked). Networks of this kind are particularly common in West German cities.

Bike or Multiuse Trails[22]

The last, and presumably ideal, type of bikeway is a completely segregated and exclusively nonmotor trail (Class I). It will have its own right-of-way or run through public open spaces. Many such facilities have been created in the United States, but primarily with the recreational cyclist in mind. As discussed previously, they can also be used by com-

Curbside bike lane (Munich, Germany).

[22] New Jersey Department of Transportation, op. cit., Chap. 4.

muters or shoppers on bicycles, provided the alignment conforms to those origin and destination needs. The outlook for creating such facilities for bike transportation purposes in dense, built-up urban environments is not too promising.

The minimum width for such a two-way facility is usually given as 8 ft (2.4 m), while 10 ft (3 m) is listed as the preferred width. If the same paved trail is shared by walkers and skaters, at least 12 ft (3.7 m) will be necessary.

Segregated bike lane with barriers on Sixth Avenue in New York City.

The City of Davis, California, having much experience with bike systems, has the following guidelines[23] as to the employment of various control systems on any given street:

Bikes in mixed traffic	if	Average daily motor traffic (ADT) is below 2000 vehicles Normal car speed is below 30 mph (50 kph)
Designate on-street lanes	if	ADT is up to 8000 Speed is up to 35 mph (58 kph)
Create protected lanes	if	ADT is up to 14,000 Speed is up to 45 mph (75 kph)
Build separate paths	if	ADT is more than 14,000 Speed is more than 45 mph (75 kph)

Since there has been considerable experimentation with bike lanes, particularly in Europe, several possible refinements and modifications have been identified that can enhance the efficiency and safety of the system:

[23] See the city's home page on the Web at www.city.davis.ca.us.

- Extensive use of stenciled bike symbols on the pavement to stress use designation
- Eliminating all surface irregularities, including slight differences in elevations (such as manhole covers and settled pavement slabs)
- Employing colored pavement across the entire lane to stress further its separate character
- Ensuring nonskid (but not abrasive) surfaces throughout, possibly with special pavement materials
- Carrying lane designations across intersections, at least as dashed lines
- Providing head-of-the-queue reservoir space at intersections to give bikes a head start at signals
- Elevating the bike lane a few centimeters above street surface
- Bypassing bus stops on the inside to minimize weaving with heavy vehicles
- Marking bike lanes with a very low rumble strip or some other tactile markers, such as flush paving stones

The question of bicycle storage or parking permeates all discussion about systems that respond properly to the concerns of riders. Its importance is illustrated dramatically by views of the large spaces in the centers of Chinese cities that are packed with thousands of bikes awaiting their owners, having paid a small fee to the attendant. The possibilities can be summarized as the following:

- *Racks* outside movement space that allow tying down the machine in a secure fashion represent the minimum requirement. The problems of finding suitable space and managing these facilities have already been mentioned.
- *Lockers* offer weather protection and more security, precluding the need to carry heavy chains and secure auxiliary equipment and removable parts. Space requirements and capital expenditures will be higher than with simple racks.
- *Indoor parking* with attendant supervision is preferable, but becomes difficult to implement because convenient space for communal use has to be found near building entrances, or, if individual firms provide accommodation, the use of elevators

becomes an issue. Space reserved for bikes in regular automobile garages would be a constructive approach.

- *Special bike garages* would be the ultimate answer if the demand is high enough and permanent. Such facilities exist in Japan and a few other places, frequently accompanied by repair shops, sales rooms, and rental establishments.

Bicycle parking at a station of the Ottawa busway.

Possible Action Steps

In most communities current bike use for work and service trips is not very high, a major reason being that suitable facilities do not exist. On the other hand, bike paths and parking places are not there because there has been no appreciable demand for them. To get out of this dilemma and to get started, a critical mass needs to be established in terms of actual riders or at least in the form of sufficiently strong requests for action. It might be possible, for example, for a community to announce a major coordinated program, say, over a year, in the course of which the facilities are created and residents purchase machines and train themselves in their use. This is an idealized scenario and not likely to be a very practical approach in most instances.

Instead, it is possible to think of a gradual and incremental implementation program. Unlike, for example, rail systems that have to be built in their entirety to be of any use, most separate items of a bicycle program have utility in their own right.[24] A lane extending

Bicycle lockers at a station of the San Diego light rail line.

Parking lot for bicycles (New Delhi, India).

only for a few blocks or a special traffic signal will do some good by themselves and can be progressively incorporated into a larger network as it develops. An education and training effort would be productive almost anywhere at any time.

Any action program geared toward bike use for urban transportation should start with a survey of community attitudes toward bicycles and their use in general and in a given place specifically. The propensity toward cycling and likelihood of sufficient participation beyond recreational activity has to be gauged to establish feasibility. Various methods and procedures are available to do this, and it is not unfair to include promotional material in the overall effort. If the results appear favorable, and if support is assured by key civic, public, and business organizations as well as the local media, a strong mission statement by the local government would be appropriate. This would establish confidence that there is commitment for the program over an extended period and that facilities will be developed and regulations enforced. Members of the population would thus be encouraged to acquire bikes and gradually adjust their lifestyles to take advantage of the new transportation opportunity. Employers and property owners would be assured that any expenditures on their part would have a long-term utility.

The preparation of the physical plan is not a difficult task, utilizing the concepts outlined in this chapter. It is not, however, a task for rank amateurs because enough complications and subtleties will be present for the designs to benefit from experience, knowledge, and planning care toward an effective and safe sys-

[24] Chapters 3, 4, and 5 of the *National Bicycling and Walking Study* (FHWA, 1994) provide a detailed list of recommended actions that can improve the local cycling situation.

tem. There are many design details, not covered in this overview, that are crucial in achieving a well-functioning network. This includes in particular various crossing situations of paths, storage facilities, treatment of hazardous locations, accommodation of bus stops, etc.[25]

In addition, a critical dimension would be a close working relationship with each neighborhood and district included in the network. The residents are the ones who will be using the facilities at the local level, and they must feel an "ownership" of the plan. An implementation schedule is also called for, which may not be tied to specific future dates but rather the rate of growth in actual utilization.

Financing is always an issue, but in the case of bikes the costs would be scattered over the entire community: residents and workers purchasing their own rolling stock and many businesses and institutions providing parking and other support facilities. The public (municipal, in almost all instances) investments would be associated with the creation and maintenance of the lanes, trails, and control systems. Much of this can be done as a part of regular street and park development programs. In the case of bike systems, annual maintenance might be a more important issue than the original expenditures. Continuous training and enforcement of regulations are a part of the ongoing effort.

It should also be noted that, under recent national transportation legislation,[26] assistance from the federal government is available under certain conditions. Up to now, such help from national and state sources has been directed almost entirely to the creation of separate trail networks. It is to be hoped and expected that in the future such funds can also be used in retrofitting existing roadways for appropriate cycling operations.

[25] See New Jersey Department of Transportation, op. cit., or other state manuals.

[26] The *Transportation Equity Act for the 21st Century,* sections on enhancement programs.

Bikeways along the principal avenue of Xian, China.

Conclusion

The bicycle has been around for more than 100 years and is familiar to almost everybody. It is not an advanced technology mode, but it has never lost its utility. In some places bike use is advisable because it is economical and effective, in other places because it is healthful and resource-conserving. There is no debate that American communities should devote more attention to this form of transportation (not only as a recreational pastime) and take constructive steps to make routine use of bicycles attractive and safe. This is an easy concept to propose, but not so easy to implement. It will require in each place a dedicated program that creates the physical opportunities and brings a significant cohort of the local population into the ranks of bikers. It has been done here and there, but the challenge is to establish a viable nonmotorized system in parallel to the dominant automobile use, which will not be eliminated or supplanted, but perhaps tamed.

Bikes and pedestrians in central mall of Münster, Germany.

Bibliography

American Association of State and Highway Transportation Officials (AASHTO): *Guide for the Development of Bicycle Facilities,* Washington, DC, August 1991, 44 pp. A complete summary of the technical aspects and details, as seen by engineers.

Clarke, Andy and Linda Tracy: *Bicycle Safety-Related Research Synthesis,* Washington DC: U.S. Department of Transportation, Federal Highway Administration, 1995, FHWA-RD-94-062.

Federal Highway Administration: *National Bicycling and Walking Study—Transportation Choices for a Changing America* (Final Report), Washington, DC: U.S. Department of Transportation, 1994, FHWA-PD-94-023, 132 pp. Devoted largely to promotional programs, accompanied by 24 technical studies.

Forester, John: *Bicycle Transportation,* Cambridge MA: The MIT Press, 1994 (2d ed.), 346 pp. Comprehensive coverage, with extensive technical detail, submerged and colored by author's dislike of bike lanes.

Lowe, Marcia D.: *The Bicycle: Vehicle for a Small Planet,* Washington, DC: Worldwatch Paper 90, 1989, 62 pp. Summary of the arguments against the automobile and the search for bike applications.

McClintock, Hugh (editor): *The Bicycle and City Traffic: Principles and Practice,* London: Belhaven Press, 1992, 217 pp. Overall review of general issues, with extensive case studies from Europe.

New Jersey Department of Transportation (by the RBA Group, with the assistance of Lehr & Associates, Inc.): *Bicycle Compatible Roadways and Bikeways: Planning and Design Guidelines,* 1996, 62 pp. One of several complete bike system handbooks by a state DOT.

Perry, David B.: *Bike Cult: the Ultimate Guide to Human-Powered Vehicles,* New York: Four Walls Eight Windows, 1995, 570 pp. The encyclopedia of bicycles and cyclists, but nothing on bike lanes and paths.

Sign up for the Bike: Design Manual for a Cycle-friendly Infrastructure, the Netherlands: Centre for Research and Contract Standardization in Civil and Traffic Engineering, 1993, Record 10, 325 pp. The European approach, experience, and practice, with a precision and detail that can only be envied by Americans.

Wilkinson, W. C. III, A. Clarke, B. Epperson, and R. Knoblauch (Bicycle Federation of America & Center for Applied Research): *Selecting Roadway Design Treatments to Accommodate Bicycles,* Washington DC: U.S. Department of Transportation, Federal Highway Administration, 1994, FHWA-RD-92-073.

CHAPTER 4

Motorcycles and Scooters

Background

Riding motorcycles is a guys' thing. Women do not use them much, although some will sit on the back seat. There is no doubt that motorcycles are motor vehicles, but their role in regular transportation is hardly visible. In the early days, a motorcycle was the first step for a person looking to gain motorized mobility at an affordable cost. That is certainly no longer the case in the United States, and the machines respond to certain lifestyles, they confer an image, and they have much to do with recreation and social standing. The popular perception of the Hell's Angels and the dread of lawless biker gangs, which are parts of American folklore and are highlighted by the media at every opportunity, is difficult to change. These images color attitudes to a remarkable degree.

Nevertheless, motorcycles are a legitimate means of transportation, even within cities. Leaving aside the large biker groups on long cruises in the countryside over the weekend, there are difficulties today in defining responsible policies toward this mode and in identifying a proper place for them in the transportation spectrum. Those who wish to face a significant risk of accidents, weather conditions as they appear, and the ire of automobile drivers find the fluid ability to weave forward through clogged traffic a considerable advantage and even a thrill. (The mayor of Caracas, Venezuela, a notoriously congested city, occasionally does exactly that if it is essential that he reach an appointment on time.)

Like motorcycles, scooters originally were built and purchased as economical devices to extend the personal ability to overcome distances. They have faded very much from the American scene, but they still maintain their role in the developing world. They too are individual, two-wheeled, non-weather-protected vehicles for reasonably able-bodied persons, but they certainly do not carry any image comparable to that of motorcycles. A motorcycle tends to be intimidating; the scooter is faintly amusing among American urbanites. It is remarkable how influential two Hollywood movies were in establishing almost universal attitudes after World War II: the young Marlon Brando created a lasting association of motorcycles with menacing masculinity (*The Wild One,* 1954, by Laslo Benedek for Columbia), while Audrey Hepburn clinging to the back of Gregory Peck on a Vespa showed European sophistication as well as dashing urban mobility (*Roman Holiday,* 1953, by William Wyler for Paramount).

Some definitions pertinent to the discussion are as follows:

Motorcycle. A two-wheeled (inline) open motor vehicle, propelled by a two- or four-stroke internal combustion engine, with the driver straddling the machine. Has a seat in the back for another rider; a sidecar may be attached.

Scooter. A two-wheeled (inline) open motor vehicle, smaller and lighter and with smaller wheels than a motorcycle, with the principal differences that users can walk across the machine and a chairlike seat is provided.

Moped. A bicycle with a small auxiliary engine that assists but does not replace human power. (See Chap. 3.)

National statistics on motorcycles as a long-established vehicle type are generally available (e.g., Table 4.1), but it is difficult to gauge the breakdown between city use and operations outside on the open road. There is little doubt, however, that most of the mileage clocked by motorcycles is of the second kind, and that almost all scooter travel is confined to urban districts.

One-tenth of the national motorcycle fleet is found—not surprisingly—in California, followed by Ohio, Florida, Illinois, and

Table 4.1   Motorcycles and Automobiles in the United States

| Year | Motorcycles | | Passenger Cars | |
	Vehicles Registered (thousands)	Vehicle-Miles Traveled (millions)	Vehicles Registered (thousands)	Vehicle-Miles Traveled (millions)
1960	574	NA	61.671	587,000
1965	1,382	NA	75,258	723,000
1970	2,824	3,000	89,244	917,000
1975	4,924	5,600	106,706	1,034,000
1980	5,694	10,200	121,601	1,112,000
1985	5,444	9,100	127,885	1,247,000
1990	4,259	9,600	133,700	1,408,000
1995	3,897	9,800	128,387	1,438,000
1998	3,879	10,260	131,839	1,546,000

Source: Bureau of Transportation Statistics, U.S. Department of Transportation.

Pennsylvania. More than 65 magazines, some with a circulation of over 200,000, devoted to motorized bikes are currently available in the United States, testifying to the fact that there is a sizable group of people out there seriously connected to this piece of rather exotic hardware. While the total mileage accumulated by motorcycles has grown significantly in the last decade (back to early 1980s levels), the number of vehicles has not, and neither has their use for commuting inside urban communities.

Development History

The evolution of the motorcycle parallels that of the automobile as far as technology is concerned, except that in some instances motorcycles have been the pioneers because they are simpler devices and easier to experiment with. Thus, the machine that Pierre Michaux and Louis-Guillaume Perreaux built in 1868 by placing a small steam engine on a wooden velocipede was the first motorcycle and

Motorcycles parked in the Place de la Concorde, Paris.

the first self-propelled, mechanized personal transport. It was not very practical, but it did reach the frightening speed of 19 mph (31 kph).[1] Experiments with steam engines continued during the rest of the century, since the urge to develop a workable personal vehicle was overpowering, particularly in light of the successes in building effective mass transportation systems based on rail and powerful locomotives inside and between cities. The Smithsonian Institution in Washington, D.C., shows a Sylvester Roper steam bike of 1869.

The real ancestor of the modern motorcycle was the machine built by Gottlieb Daimler in 1885. In order to improve the early gasoline engine, Daimler placed it on a bicycle as a step toward an automobile. It was still a wooden "bone-crusher," but, except for the side stabilizing wheels, it had all the standard elements of a motorcycle and is easily recognizable as such. During the period before World War I, which was rich in various technical invention and development, many other similar devices appeared in France and Germany as well as in the United States. Some can still be seen in leading museums. The first production model, manufactured by Hildebrand & Wolfmüller in Germany (1892), was no longer just a modified bicycle but a distinct vehicle in its own right, with a step-through frame and a low-slung engine connected to the rear wheel by a rod. The prototype for the standard four-stroke motorcycle engine was built in 1895 by the French firm DeDion-Buton, with a battery and coil ignition.

As was the case with automobiles, this early period was characterized by the formation of clubs, the holding of races, and much experimentation—but only among a few dedicated devotees. Many small manufacturers tried various configurations and engine arrangements. Speeds of 100 mph were reached, and motorcycles entered World War I, thus undergoing strenuous testing under the worst possible conditions and doing generally quite well.

[1] Motorcycles have a large cadre of devoted admirers who have documented most thoroughly the history of the machine. The more accessible sources are the *Art of the Motorcycle* exhibition catalog for the June–September 1998 show at the Solomon R. Guggenheim Museum in New York (in electronic and print form), and F. Winkowski et al., *The First Century of the Motorcycle* (Smithmark, 1999). The technological aspects have been covered most thoroughly by C. F. Caunter, *Motor Cycles: A Technical History* (Science Museum, Her Majesty's Stationery Office, United Kingdom, 1982).

Motorcycles reached a state of maturity and established a regular market in the 1920s. American companies became very active in developing their own designs and entering the market with great vigor. The larger enterprises were Indian (1901) and Harley-Davidson (1902),[2] although the first brand-name manufacturer was the Metz Company (1898). Several other motorcycle companies (such as Cleveland, Henderson, Excelsior, and Wagner) were also active in the field, and the products represented a bewildering and ingenious array of technical concepts and details. Yet, all the vehicles operated more or less in the same way. At one time there were close to 300 firms building motorcycles in North America, but, over the years, the Harley-Davidson and Indian companies became the undisputed leaders. This was matched by activity in Germany (BMW and NSU), France (Clement, Rochet, and Peugeot), Great Britain (Triumph, Matchless, Norton, and Scott), Belgium (Fabrique Nationale), and Italy (Moto Guzzi).

As fascinating and as useful as motorcycles were, the much more comfortable and safer automobile, which became affordable very early, relegated them to a supporting and specialized role in the United States, even in the 1920s and 1930s. The technical development followed two distinct paths, which prevail even today: to build the fastest possible racing machines that have nothing to do with normal transportation, and to serve as large a "civilian" market as possible with vehicles that are reasonably efficient, safe, and easy to operate. The second component, in turn, can be subdivided further between machines intended to be used daily for transportation and those with a purely recreational purpose. The latter part is much larger than the first, and the market remains oriented toward young males.

The motorcycle had a very visible role in World War II, on both the Allied and the Axis sides. Indeed, many soldiers returning after the war who had used and seen motorcycles in action not only became owners of these machines as affordable transportation, but also organized themselves frequently into clubs (fraternities or gangs, according to one's perspective) that still color public attitudes and perceptions toward this mode. The vehicles of choice were the Indian Bob-Job and Chief and the large Harley-Davidson "hogs" and "choppers." This rebellious stance, epito-

[2] These two companies, as well as several others, have been covered by several full-size books and numerous articles.

mized by the motorcycle, stood in stark contrast to the desired calm and organized life of stable communities, with its own icon—the comfortable family sedan. In this dramatic context, the potential and applicability of motorcycles as a transportation mode became diluted, if not lost, in North America.

Another American film, *Easy Rider* (1969), described and encapsulated the national mood, with the starring roles clearly belonging to some motorcycles (not really Dennis Hopper and Jack Nicholson). As is frequently the case with motorcycles, they did not enter any cities and the action took place entirely on the road. During the same period, popular attention was captured by Evel Knievel with ever more spectacular and pointless (except for danger and excitement) stunts on the motorcycle.

By this time the Indian Motorcycle Company had faded from the scene, and Harley-Davidson worked hard to maintain its market position, particularly in face of competition from abroad. The British Triumph made a valiant attempt, but could not survive in the limited market; Ducati and NV Agusta of Italy were new and significant entries; and Honda from Japan became a major player in the racing arena as well as in the general use sector, as did Kawasaki, Suzuki, and Yamaha. The Japanese manufacturers were also most successful in establishing markets in developing countries where the lower price of a motorcycle as compared to a car was a significant factor. Honda in particular developed a large model line in the 1970s, including small vehicles for a general market (as well as off-road use) and very large units with extreme power capabilities.

The early 1980s marked a turning point in motorcycle presence in the United States. All-purpose urban use had declined, and since that time technological development for the American market has been oriented toward appearance and high performance. Nobody seems to build vehicles today specifi-

A contemporary power machine.

cally for normal city use (although such machines can be found among the many models available). The current superbikes look like jet fighter planes, and they are equally intimidating to the average city resident.

Motorcycles as a regular urban transportation mode continue to fade. In the decade from 1980 to 1990, for example, commuting to work by motorcycle in Manhattan dropped by 18 percent, while elsewhere in the region the decreases were well over 50 percent. On the other hand, the patterns in developing countries are quite different, particularly in those with vigorous economies. In Taipei, Taiwan, for example, there are still more motorized bikes than there are automobiles on the streets.[3] While the number of car registrations has increased 15-fold and that for motorcycles by 8.5 times in the last three decades, the ratio remains in favor of motorbikes:

Motorcycles for Each Car

1971	2.6
1981	1.9
1991	1.2
2000	1.4

The modern *scooter* is a deliberate invention to respond to specific transportation needs. It is cheap, easy to operate, attractive, light, and easy to store—in other words, it provides personal transportation for many in urban situations at a most affordable level. This was done in post-World War II Italy by Corradino D'Ascanio for the Piaggio Veicoli Europei company. Nevertheless, if the principal characteristic of a scooter is a step-through frame, there have been a number of earlier small vehicles that could be classified as scooters (for example, the 1894 Hildebrand & Wolfmüller machine and various "family bikes" of the 1920s and 1930s, which never caught on because of clumsiness or expense.)[4]

The new 1946 Italian design enclosed the engine to protect the rider from possible burns, provided a footrest and a shield in

---

[3] Taipei City Motor Vehicles Inspection Division.
[4] There are books covering the development of scooters, but much information can be found on the Web page of scootered.co.uk.

front, easy entry and exit from either side without having to climb over anything, a comfortable seat (accommodating women with skirts), simple operating controls, a low center of gravity, small wheels, and some enclosed space for baggage. The name *Vespa* (Wasp), given to the machine by the manufacturer, resonated well in the minds of the users and became a generic designation for all practical purposes. The very reasonable price made the vehicles an instant success in Europe as their first mass entry into personal motorized transportation.

The scooter was the ideal vehicle in countries that moved upward on the economic scale, and during the 1950s and 1960s it dominated the urban scene in European cities. Scooters appeared in the United States in measurable numbers too, but they never moved far beyond a novelty status and were popular only among a rather narrow population cohort—the more adventurous young urbanites. The vehicles found a large market niche in developing countries, including the use of the principal mechanical components as the base for various locally built for-hire transport devices (see Chap. 6, Paratransit).

Over the decades, a number of manufacturing firms in Italy, France, Great Britain, Germany, and Japan entered this industry, but, because of the limited and fluctuating demand, only a few have continued with production. The only American manufacturer of scooters—Cushman—attempted to equip paratroopers with light vehicles during WW II, but without any real success, and the company could not survive the uncertain market in American cities during the next decade.

Scooters remain in the urban transportation picture in high-density European cities, not because they are cheap, but because they can get through traffic jams and they are easy to park. They can also be seen in areas of lower prosperity and rural villages. Scooters are very rare in North America at this time, and do not even appear in official transportation statistics.

Scooter on a street in Shanghai.

Types of Motorcycles and Scooters

The mode under review here provides personal transportation to single users (rarely more than two) in individual vehicles under their own direct control, using public rights-of-way. The principal difference from automobiles is not only size, but also the lack of a container or shelter for the users. While all motorcycles and scooters are motor vehicles, the smaller and slower types may be barred from major arterials and highways. All motorcycle drivers have to be properly licensed; this is not always the case with scooters. Almost all states require that motorcycle riders wear helmets; some states also make them mandatory with scooter use. There are distinct types within this vehicle group.

Motorcycles

Since the 1950s, the all-purpose street vehicle has been replaced by a multitude of models in response to different specific purposes and interests. Some types are far removed from being a means of transportation; a few cannot even be registered for street use.

- *Traditional.* Vehicles in general use that have to be reasonably easy to operate by nonprofessional drivers, have adequate comfort and safety features, and are compatible with the urban environment. This is no longer the largest market segment.

- *Cruiser.* The most popular type, placing emphasis on appearance, style, and sound, purchased and used primarily for recreation and as an expression of individual personality.

- *Sportbike.* Vehicles built for racing, or at least to give that image, with ability to reach high speeds and good cornering. Not particularly comfortable to ride on.[5]

- *Touring bike.* Large vehicles with built-in comfort and convenience features for long trips—wind protection, communications systems, wide seats, luggage compartments, etc.

---

[5] Cognoscenti are also able to distinguish the following subtypes: road racers, dirt-trackers, superbikes, TT (Tourist Trophy) machines, dragsters, motocross bikes, hill climbers, trial bikes, endures, sidecar racers, ice racers, desert racers, and speedway bikes.

• *Dual-purpose bike.* Various combinations of features in search of markets.

The technical people have classified motorcycles from the very early days as to the amount of power that any model can develop, which characterizes the performance (speed, acceleration, carrying capacity) better than anything else. This is largely a function of the size of the engine, which, in turn, is measured by the volumetric cylinder capacity or displacement (expressed in cubic centimeters).[6] Such classification also allows races to be held among comparable machines at a fair level. The common classification ranges are 350 cc, 500 cc, 750 cc, and 1200 cc.

The first group encompasses mostly the general-purpose, lightweight machines (as low as 124 cc) that can be used effectively for city driving. The second and third groups are not particularly popular these days; at least they do not receive much attention. Most of the modern high-performance vehicles are now designed around a 1000-cc engine. There are cruisers that approach 1500 cc, which might be suitable for highway patrol work, but hard to visualize with any other good purpose. (The Honda Gold Wing is 1832 cc.)

Tricycles (Trikes)

These are stable three-wheeled vehicles, basically adding a passenger unit atop an axle with two wheels to the front half of a motorcycle. They are favored by the elderly and handicapped persons and are usually licensed as motorcycles.

Scooters

This category encompasses small and light vehicles for individuals and two persons for short trips. Besides building the standard open models, there have been repeated efforts to give more weather and safety protection to the rider and make the vehicles more comfortable. This includes full body enclosures, possibly utilizing three wheels, which very soon become miniature automobiles for two persons. Another development has been the

---

[6] Cubic inches can also be used, but nowadays the international standards use the metric system (cc). There are a number of other features and measures that distinguish engines and their performance, all important to specialists and devoted bikers.

recent introduction of a scooter (BMW C1) that is equipped with a "safety cell"—overhead curving bars to protect the rider in case the vehicle topples or turns over, thereby precluding the need for a helmet. It also has an ergonomic seat with a belt and shoulder loops.

Scooters are classified by engine size. They started in the 50- to 100-cc range, but, while they were very affordable machines, their performance did not particularly please anybody. At this time, the standard models by almost all manufacturers rely on 124-cc engines. Unavoidably, higher-performance models have been developed at about 250 cc, and there is at least one with 499 cc (Yamaha XP500) that is able to race most motorcycles.

Mopeds

(see Bicycles)

Unconventional Motorized Units

There continues to be an urge to develop the smallest possible vehicle that would allow people to move faster than walking pace with no exertion. Examples include rocket packs worn on the back and motorized roller skates. Many such devices have appeared, but invariably, at least so far, they have proven to be unreliable or dangerous to the rider and others in the vicinity. Motorized skateboards, however, have persisted and are occasionally seen on streets. Considerable athletic skill is required to ride them, and it is hard to imagine a more dangerous contraption in heavy city traffic.

Yet, there is a new development in this class—the Segway Human Transporter or "Ginger"[7]—which claims to be able to transport people in absolute safety at speeds up to 12 mph (19 kph) standing on a small platform. The device has two parallel wheels (about 15 in in diameter), and balance is maintained by an intricate system of gyroscopes that sense instantly any shifts in weight and make proper adjustments to the rotation of the wheels. There are no physical controls or brakes; forward motion, slowing down, or changes of direction are accomplished by the

---

[7] As reported in the *New York Times,* December 3, 2001; the *Urban Transportation Monitor,* March 18, 2002; and other newspapers and magazines. The official name is now *electric personal assistive mobility device*—a term with no chance of becoming a household word.

rider moving slightly, even instinctively, in the appropriate direction to send a signal to the gyroscopes. Power is provided by a rechargeable battery. This piece of equipment has to be taken seriously—even though experiences with "dream machines" have been abysmal in general—because the inventor has many very advanced and useful devices to his credit, prominent and highly respected sponsors support it, and some $100 million has been spent on development so far. The U.S. Postal Service and the Atlanta Police Department, among others, are considering full-scale utilization of this device by their personnel.

If it does prove itself workable, does Ginger represent a potential revolution in urban transportation by making everybody three times more mobile? A series of questions will have to be answered first. Will many people be able to acquire it? (The estimated cost is $3000.) What about surmounting curbs, ramps, potholes, and stairs? Since it cannot be easily carried (65 lb; 30 kg), can it always be moved around under its own power? Will it run on streets, risking serious collisions with automobiles? Or, will it be allowed to operate on sidewalks where impacts with pedestrians, given the mass and the speed of the machine and its rider, would be threatening? Can the extremely sensitive system of controls be made reliably immune to the shocks and bumps of everyday use? Will people accept it? (Of course, similar questions were also asked when steam locomotives, streetcars, bicycles, and automobiles first appeared.)

In the meantime the simple running-board scooters[8] have never left the scene. They do experience periods of popularity (minor fads) and deep obscurity. These scooters may be propelled forward by pushing against the ground with one foot, by simple mechanical pumping arrangements, or even by small engines. While they expedite forward motion above that achievable by walking, they require some skill in operation, the action does not look very dignified, and the devices become cumbersome accessories to be carried by the patrons when not in actual transportation use. Here again, they do not mix well with pedestrians and even less so with cars. It appears that these scooters have had much time and many chances to carve out a market niche for

---

[8] A running board supported by two small wheels, the front one being steerable by the handhold extending upward. Best known as a children's toy.

themselves, but this has not happened, beyond interior use in some airports and warehouses. This can only mean that they do not respond to a real transportation need, but, undoubtedly, will continue to be discovered and promoted with some regularity.[9]

Reasons to Support Motorcycles and Scooters

Minimal Space and Fuel Requirements

Both motorcycles and scooters are small vehicles, not extending much beyond the physical size of their users. Consumption of street and parking space is quite efficient and contained. Even the largest motorcycle does not require more than a 9 × 3-ft (2.7 × 0.9 m) area to be parked, as compared to a 20 × 9-ft (6 × 2.7 m) stall for an automobile—a sixfold saving.

While the fuel consumption of a passenger car is about 21 mpg, it is 50 mpg for motorcycles. Since no vehicle can operate with less than one person carried, and automobile occupancy in North America is usually 1.1 to 1.3 persons per car, the difference between the two modes in the efficiency of using space is considerable.

Personal Preference

Motorcycles are cherished by a sizable population component whose lifestyles frequently revolve around these machines. Since they are legitimate vehicles, there are no reasons to discriminate against them, as long as public safety is not breached. Biker organizations have lately asserted their rights with some vigor—for example, contesting the exclusion of their members from some eating and drinking places because of their unique and striking attire. This has little to do with transportation,

Motorcycle parking lot in Columbus Circle in New York City.

[9] The latest entry, shown at the 2002 Auto Show in Detroit, is the aluminum Scoot, to be powered by a fuel cell battery and selling for $500.

but much to do with the maintenance of just societies, even when the debate focuses on vehicles as the instruments of certain types of behavior.

Agility in Traffic

The great maneuverability and small size of these vehicles allow them to operate in spaces and situations that are not suitable for cars. Special lanes could be threaded through tight environments, but they do not exist. Motorcycles can, however, "split lanes" on regular roadways and move forward between cars. This practice is allowed in California (the state with the most motorcycles), but is outlawed in many others. It is dangerous, particularly if it happens in moving traffic, and it infuriates automobile drivers, leading to potential road rage.

Accessibility and Ease of Operation

As mechanical transportation equipment, motorcycles and scooters are relatively inexpensive and affordable to a much larger population cohort than automobiles. Operational skill requirements do not extend much beyond an ability to ride a bicycle (granted, not possible for everybody).

Reasons to Exercise Caution

A century of experience with motorcycles and half that time with scooters has generated good understanding of what the capabilities and drawbacks of these modes are.

Personal Danger

The high speed that can be reached by motorcycles, coupled with the lack of any protection for the users (save the helmet and possibly leather attire) and exacerbated frequently by the irresponsible traffic behavior of some drivers, has led to a dismal safety record. In 1998, motorcycles constituted less than 2 percent of the motor vehicle fleet in the United States and accounted for a mere 0.4 percent of vehicle-miles accumulated.[10] Yet, they were associated with or responsible for 5.5 percent of total traffic fatal-

---

[10] Data from the National Highway Traffic Safety Administration, U.S. Department of Transportation (National Center for Statistics and Analysis).

ities, and that figure rose to 5.9 percent in 1999. The fatality rate per 100 million miles traveled was 1.3 for people in automobiles and 23.4 on motorcycles. This means that a biker is 18 times more likely to meet an unpleasant end than a car occupant on American streets and highways. The ratios are comparable everywhere else (the estimates in Great Britain, for example, show dangers 15 times larger).

The situation is serious enough so that it needs emphasis, such as highlighting the record of motorcycle fatalities in the United States over a longer time period:

1960	790
1965	1650
1970	2280
1975	3189
1980	5144
1985	4564
1990	3244
1995	2227
1999	2472

Conditions today are not as tragic as they were in the early 1980s, but there has been no steady and reliable improvement in the last few years either, which should be expected given a drop in overall motorcycle use. Scooters do not appear in these statistics at all, which does not imply that they are much safer. The data are simply absorbed by other modes. If motorcycles were seen as regular means of transportation used by the general population, not just a few individuals by their own free choice, there would be a major public outcry across the country regarding the accident rates, possibly demands for banning the vehicles. The latter discrimination would not be acceptable, but there are causes for serious general concern in the safety sector.

Environmental Impacts

It has been argued (by motorcycle advocates) that the engines are small and that the total fleet is quite limited in size in any city, and, therefore, the machines should be exempt from all environmental controls. This is not a defensible suggestion since there is no particular reason to grant exceptions simply because of smallness in an overall unsatisfactory situation. The fact of the matter is that the engines used in motorcycles and scooters are not designed with pollution controls in mind, and the emission rates of identified pollutants are high.

A very serious urban environmental impact is noise generation by motorcycles. Some riders perceive loud operations as a necessary part of the biking experience, but the effects on the rest of the population can be extremely disturbing. Residents are quite sensitive to urban noise conditions, with such concerns sometime constituting 80 percent of quality of life telephone complaints.[11]

Traffic Mix

It can be argued that the speed and mode of operation, if not size, of motorcycles are compatible with those of cars and trucks, and, therefore, they all can mix on the same roadways. There will be frictions and localized competition for space, but, with proper behavior by every motorist and biker (which does not necessarily happen all the time) safe joint use of facilities is certainly possible.

Scooters, on the other hand, represent a different situation. They are small, slow, and fragile, and they do not mix well at all with regular vehicular traffic. They certainly cannot operate safely on sidewalks either, where conflict with pedestrians would be most disturbing and damaging. Scooters are somewhat akin to bicycles, and they could use bike lanes quite effectively. The conflict here is one of priorities and concept. If bicycle use is to be encouraged as a most desirable urban mode, then the intrusion by any kind of motorized device may be a problem.

Application Scenarios

The motorcycle mode is a dilemma for urban transportation planners. These machines undoubtedly will and should continue in their role as recreational vehicles, even as expressions of special lifestyles and subjects of a hobby, but such activities do not fit in well with regular urban operations in built-up districts. While motorcycles provide mobility, they do not solve many access problems—in fact, they tend to exacerbate them. Frictions and threats to safety will persist, but not at levels that would argue for

[11] As reported in the *New York Times,* January 15, 2002, referring to all noise incidents, not just those caused by motorcycles.

exclusion. Those who wish to use motorcycles for commuting and other regular purposes should be free to do so, but—needless to say—under full traffic control.

Scooters involve similar issues, and, even though their current presence in American cities is minimal, some attention has to be devoted to their role and impacts as a transportation mode. They are likely to be more visible in areas populated primarily by younger people (university towns) or in smaller communities where automobile volumes are not intimidating. If it could be argued that scooters will replace car purchases and use, they would be most welcome. If they are to be present in significant numbers, proper accommodations (such as parking spaces and some lanes) could be made, as has been done in some European cities. Short of all that, the situation remains uncertain, and appropriate policies and programs have to remain a local matter.

Components of Motorcycle and Scooter Operation

Vehicles

The rolling stock (Table 4.2) is completely in private ownership, with never any participation in public transportation. The only exception to the last statement is in some developing countries with severe resource constraints, where foreign-produced mechanical components of motorcycles and scooters are used to build indigenous for-hire vehicles (in taxi and paratransit service).

Safety Devices

Because the vehicle itself has no means to shield the driver in case of collisions, personal protection has to be carried when operation takes place in dangerous situations or when the regulations so require (or common sense prevails). The principal element here is the helmet, which has to be worn by all

Motorcycle as a base for a public vehicle-for-hire (Thailand).

Table 4.2   Selected Motorcycles and Scooters

Make and Model	Length, in (m)	Weight, lb (kg)	Type of Engine	Maximum Speed, mph (kph)	Fuel Consumption, mpg	Price, Current Dollars (with Great Variations)
Motorcycles						
Yamaha FZS600/S Fazer	88.9 (2.08)	417 (189)	599 cc, 4 cyl	138 (222)	43	7,600
BMW F650ST	85.8 (2.43)	421 (191)	652 cc, 1 cyl	109 (175)	42	7,200
Honda VFR800i	84 (2.43)	470 (213)	781 cc, V4 cyl	153 (248)	40	12,000
Ducati 996R	80.5 (2.05)	436 (198)	998 cc, V2 cyl	175 (282)	35	24,500
Harley-Davidson Night Train	95 (2.41)	630 (286)	1450 cc, V2 cyl	110 (177)	40	16,200
Scooters						
Peugeot Speedflight	68.1 (1.73)	209 (95)	100 cc	62 (100)	100	2,900
Piaggio Vespa T5 Classic	69.2 (1.76)	238 (108)	124 cc, 1 cyl	59 (95)	65	2,700
BMW Executive CI 200	81.7 (2.08)	400 (181)	176 cc, 1 cyl	70 (113)	88	5,200

Source: Bike magazine, December 2001, and manufacturer data.

motorcycle users in 20 states, but only by young riders in another 27.[12] Three have no helmet requirements at all (Colorado, Illinois, and Iowa). National statistics show that only about 65 percent of operators actually wear helmets, while 85 percent of passengers comply. Official government standards have been established for acceptable helmets. For racing and other forms of competition, gear includes leather full-body clothing, skid pads, steel-toed boots, and full-face helmets, which would not be very practical for regular city travel. The storage and safekeeping of the helmet when not in use is already a problem.

[12] Bureau of Transportation Statistics, U.S. Department of Transportation.

Space Use

Both motorcycles and scooters, of course, are only a fraction of the size of regular automobiles, yet they have to use the same movement space (traffic lanes) and storage (parking) accommodations. The incompatibilities in scale and resulting frictions have already been mentioned, but, given the expectable low volumes of bike traffic under any circumstances in American communities, this improvised situation will have to be accepted, provided that there is a clear understanding and respect for the rules of the road. This is a very complex field with specific details worked out (and largely enforced) in European cities, but it is doubtful that many people know exactly what the official requirements are on American roads and streets, and whether too many care. For example, under what circumstances is lane-splitting or filtering permitted, how much separation is to be maintained between different types of vehicles, when is overtaking allowed, what exact environmental safeguards are to be maintained, etc.? To find the answers, the traffic manuals of the different states would have to be studied in some detail.

The provision of parking spaces is also an unresolved issue. Even the largest motorcycle occupying a full parking stall represents a waste of space. If there are many motorcycles that regularly congregate at certain locations, special space allocations can be made for them—usually an 8-ft-deep area extending lengthwise in approximately 3-ft modules, depending on the number of vehicles expected. Such accommodations can be found frequently in core districts of the larger cities, at institutions, and near other places that generate motorcycle traffic. Parking garages can allocate such spaces as well. Security is a significant concern since the machines are vulnerable to vandalism and their owners usually feel very protective toward them.

Capacity and Cost Considerations

Since it cannot be expected that there will be large volumes of motorcycles or scooters at any time on any segment of the street network in American communities, it is not a matter of attempting to determine throughput capacities, but rather an issue of incorporating these machines into regular motor vehicle volumes without disruption of flow patterns or endangering any of the riders. Nor does this call for special control programs beyond a full observance and enforcement of standard traffic regulations.

The direct costs associated with motorcycle and scooter use fall just about entirely into the private sector. The riders will use public rights-of-way and other joint facilities, but no specific accommodations are called for, beyond possibly the designation of efficient parking spaces in some instances. Purchase prices of a good motorcycle start around $6000, but can go up to levels comparable to those for compact cars (leaving aside the extraordinary, customized units). Fuel consumption, of course, is quite low, but insurance costs, given the prevailing safety record, will be high.

Scooters can be acquired for less than $3000, but in this sector too there are opportunities for self-expression and status seeking through fancier models. It can be expected that, if regulations toward better environmental quality are tightened, unit prices may escalate.

Conclusion

It is not really possible, no matter how hard one may wish to try, to define and defend a significant role for motorcycles in regular urban transportation. They usually create more problems in cities and communities than they can solve. Yet, the fact that they do not always conform to general expectations is not a sufficient reason to consider exclusion or harsh restrictions. Motorcycles still are, and will continue for some time to be, legitimate means of transportation in places where a significant cohort of the population cannot afford a car but are able to acquire a motorbike. The argument that these people should rely on mass transportation does not help much, given the state of those systems in many places.

In American communities, there are two considerations. Motorcycles are utilitarian vehicles with a high capability for maneuverability and penetration. They are obviously essential for police work and delivery of messages and urgent shipments, and they can help in emergencies. We expect to see motorcycle escorts for heads of state and welcome their presence at other ceremonial functions. And then we have the sizable group of dedicated bikers whose entire lifestyle, not just weekend recreation, is focused on these machines. Not too many of us belong to this fraternity, but they certainly enliven the scene, and the rest of us can observe them from a distance with some trepidation, but also frequently with some envy. These bikers are entitled to be present, as long as they do not significantly constrain and endanger the rest of us.

Scooters represent a different situation. They never achieved the original promise in American communities of giving enhanced mobility to individuals without excessive costs and overload of traffic facilities. If an automobile is affordable, its speed and comfort overwhelm any urge to battle street traffic and the elements on an open seat. Yet, the question still remains open: is there a possible role for a small, simple, and agile device that extends individual mobility beyond the walking scale for people who are not likely to become users of regular bicycles? Perhaps in well organized, high-density districts, which we will continue to have in future communities.

Bibliography

Note: General transportation planning and traffic engineering references usually mention motorcycles in passing; scooters appear hardly at all.

Krens, Thomas, and Matthew Drutt (eds.): *The Art of the Motorcycle,* The Guggenheim Museum of Art/Abrams, 1998. Catalog of a major exhibition, with review essays.

Wilson, H.: *The Ultimate Motor-Cycle Book,* Dorling Kindersley, United Kingdom, 1993. An encyclopedic survey of the hardware, its use, and riders.

Automobiles

Background

Automobiles dominate the transportation picture today, both inside and outside cities. Cars are the blessing and the curse of American communities. They have given an unprecedented level of mobility to the larger part of this society (albeit not everybody), but they also threaten to choke our center cities, and they consume resources at a disproportionate rate. Mass transit accounts for only a few percentage points of all daily travel in the United States; it almost falls within the range of statistical error, and, thus, could theoretically be ignored in any general transportation discussion. Many Americans do exactly that, but that is not the attitude taken in this book. Nevertheless, we live in a country that runs on rubber tires, for better or for worse.

Commuting to work may no longer be the largest single trip purpose in the United States, but it is undoubtedly the most critical one. On any given day, as shown in Fig. 5.1, seven-eighths of these movements (87 percent) are accommodated by some type of individual motor vehicle—overwhelmingly cars carrying only one person. Among the large cities, only New York has more commuters on mass transit than in cars; and the younger a city is, the more closely its modal breakdown is likely to approach that of the national averages.

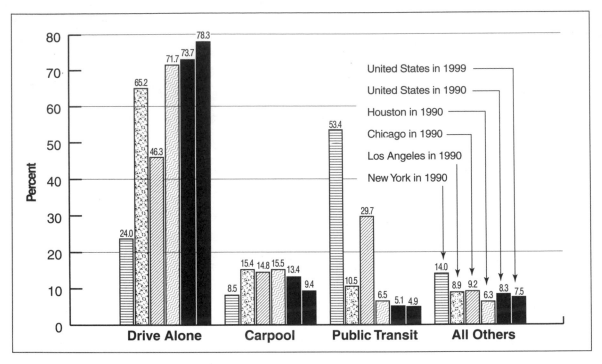

Figure 5.1 Commuting to work in the United States and its four largest cities. (*Sources:* U.S. Bureau of the Census for 1990; American Household Survey for 1999.)

The ubiquitous and persistent presence of automobiles in the daily lives of Americans is well reflected in contemporary literature, popular arts, and film. There are continuous references to family trips by car, driving to the shopping center, younger members of families trying to gain use of the car, and certainly commuting. The latter activity is frequently depicted with some irony and sarcasm since those travails are well known to everybody. Hardships and boring delays are expected, but they are tolerated or overcome with fatalism and resignation, sometimes with pride. The general impression is left that there really is no choice, and that normal life is dependent on the automobile, coping with traffic problems created by somebody else.

Automobile use percentages, of course, are substantially higher in the low-density rings of metropolitan areas than in the central cities, as well as for all family/social/recreational trips. Counting all daily travel, not just commuting, 2.15 percent of all person trips were on public transit in 1990 and 94.32 percent moved by automobile, van, or truck.[1] In the last decade (to 2000), the role of carpools across the country has decreased substantially, and single-occupancy automobile use has continued to grow.[2]

A mobility system based on motor vehicles consists of two very distinct parts:

1. The cars themselves, which are individually operated and privately owned, with the drivers expecting much freedom of movement and choice of paths

2. The network of streets and roads, which are publicly owned and maintained (usually by the local municipality), drawing upon government resources

The first element is not planned or managed in any overall sense, and the size, type, and composition of the fleet is not controlled by anything except market forces. There are countries where government has placed severe restrictions on the ownership of private automobiles (Singapore, some Caribbean Islands, China), but such policies would be unthinkable in North America. However, we do have traffic regulations, roadway entry restrictions, parking rules, tax levies, environmental controls, driver licensing, insurance requirements, and safety standards that together or separately can be effective in guiding the use of automobiles within urbanized areas toward a larger common good. Much of the discussion in this chapter will address those opportunities.

The second element—streets—constitutes public rights-of-way and is undoubtedly the most basic infrastructure system for any community. It is under the direct purview of the local government, as has been the historical pattern from the very beginning, even though in the United States and most other countries higher levels of government participate, at least with some financial assistance, definition of standards, and certainly overall safety concerns. Subdivision streets, which have been built by the thousands in the last half century, are paid for by land developers, but within municipal specifications, and they do become public streets after completion. All this calls for short-range and long-

[1] U.S. Department of Transportation, *Nationwide Personal Transportation Survey,* Federal Highway Administration, Washington DC, 1992, p. 34.
[2] Precise year 2000 figures were not yet available at the time this book went to press.

range planning, particularly recognizing the fact that rights-of-way, once established, are one of the most permanent features of any developed area. Regrettably, while land uses and transportation services have an obvious mutual dependency, they remain under different jurisdictions in American communities in the guidance of their development and control of their performance. Under strong home rule provisions, all decisions regarding land development rest with local municipal government; transportation systems need to be organized at regional levels, and responsibilities are usually assigned to metropolitan or state agencies.

The dialog between the various levels is not always as good as it should be. A component of a regional rail network may not get built because a separate municipality does not wish to have such access, even though generous federal assistance may be available and state agencies are supportive. A single-minded encouragement of major traffic generators (shopping centers, for example) in a specific locality, which will thereby strengthen exclusively its own municipal revenue base (real estate taxes), may dislocate overall traffic patterns and overwhelm parts of the regional road system.

Structuring and designing a street network with its many complex dimensions is a major responsibility, but falls outside the scope of this discussion. In reviewing the utility of the automobile as a transportation mode in this chapter, we will assume the street system as largely given, but by no means implying that it should not be upgraded in almost every community. Specifics on how to extend a street network, modify physical elements for higher capacity and better safety, or improve traffic control systems can be found in other references.[3] Nevertheless, the channels and the vehicles running along them constitute the total system that provides service at adequate or less-than-satisfactory levels. While street network design will not be discussed in its entirety in this chapter, there are many features and detailed elements that will be reviewed because they materially influence service capabilities.

---

[3] The most widely used reference materials on roadway design and motor traffic operations are generated and published by the American Association of State Highway and Transportation Officials (AASHTO), the Institute of Transportation Engineers (ITE), the Transportation Research Board (TRB), and the U.S. Department of Transportation, Federal Highway Administration (U.S. DOT, FHWA). Each of these organizations has lists of current publications and extensive Web sites. The key references are cited later in this chapter at the appropriate places.

There were 165.2 million household vehicles (not including trucks or buses) operating on the roads and streets of the United States in 1990. With a total population then of 239.4 million, it is clear that everybody could have gotten into a motor vehicle at the same time (on the front seat, or 1.4 persons per car) and all the automobiles could have driven off. While it appears on some occasions that exactly this has happened, it could not actually have occurred because there were only 163.0 million licensed drivers.[4] Is this surplus fleet a luxury, a necessity, or a habit?

Automobiles can be looked at from different perspectives, and car owners assume different attitudes at different levels toward them. First, cars are simple means of transportation that provide mobility with quite predictable costs of investment and operation. But they also offer enclosed private space that can serve many useful purposes, ranging from the carrying of purchases to being a boudoir. For a busy individual on a commuting trip, this may be the only instance during an entire day when quiet solitude prevails. (Being delayed in congestion does not generate road rage in everybody.) Next, cars certainly are visible indicators of the economic class and social standing of the owners, each model not only reflecting the amount of money that was spent for it, but also the taste and sensibilities of the owners. In some instances they are a deliberate expression of the self-defined image of the owner. Somebody who chooses a muscular pickup truck will not buy a Lexus limousine, and vice versa. Somebody who buys a zero-emission car today has to be a dedicated environmentalist. In many instances, the car is much more than a large household appliance; it is a pet of the family and may even carry a name. All this brings into play considerations of utility, economics, social standing, and psychology.[5]

[4] *Nationwide Personal Transportation Survey,* op. cit., p. 6.

[5] The role that automobiles have in American life is explored in D. L. Lewis and L. Goldstein (eds.), *The Automobile and American Culture* (University of Michigan Press, 1980, 423 pp.). Dozens of authors review the influence of the private car from different perspectives, ranging from its role in literature to being a decisive factor in rural life. Such books are R. Primeau, *Romance of the Road: The Literature of the American Highway* (Bowling Green State University, 1996, 170 pp.), P. Collett and P. Marsh, *Driving Passion: The Psychology of the Car* (Faber and Faber, 1986, 214 pp.), R. Flower and M. Wynn-James, *100 Years on the Road: A Social History of the Motor Car* (McGraw-Hill, 1981, 224 pp.), and many others. John Steinbeck, Jack Kerouac, Marshall McLuhan, and Tom Wolfe have had a lot to say about this industry and social preoccupation.

There are even more far-reaching consequences. The automobile may be regarded by each family member as the equivalent of an item of apparel or a pair of shoes. A type of vehicle suitable for each major purpose, ranging from commuting daily to the railroad station to attending a formal social function, may be desirable to have in one's closet (or garage). This situation is the case already among affluent families, and, if the purchase of vehicles is quite manageable within normal family budgets, the definition of a saturation point in car ownership (number of people per each vehicle, or vice versa) becomes an uncertain and receding target. (But surely not more than one per licensed driver—we thought!) It is no longer unusual for every adult family member to have his/her own automobile, plus perhaps one that is only safe for short trips to the train station, one for messy household chores with much cargo space, one for formal occasions, and one to go camping in.

Evidence shows that there are at least three generalizable kinds of motorists:

1. Those who regard the car as their castle and an extension of their private space, if not their personality. They will never give up the automobile, short of economic or regulatory coercion, which they will fight with any means possible. They are quite willing to wait out any traffic jams. They will also lobby vigorously for more roads and lower expenditures for transit.

2. Those who see the car as a transportation device—the most comfortable and convenient means available at any given time, or the lesser of various painful possibilities. They are candidates for switching to public and communal modes, but they will scrutinize comparative travel times, reliability of operations, need for multiple transfers and waiting, crowdedness, and personal vulnerability amidst many strangers.

3. Those who dislike driving for philosophical or practical reasons. They will use transit whenever it becomes available, and they will advocate the expansion of public systems. Regrettably, there are very few of them left behind a wheel because there were not too many of them to begin with and some gradually conform to the dominant patterns. Convenience does tend to erode conviction.

This article was originally published in the *New York Times,* Sunday, April 4, 1971.

The Car: Everyman's Island in the Urban Sea

Russell Baker

WASHINGTON—In the beginning a car was a sporting implement. Then came Henry Ford, turning it into a farm tool, and farmers discovered it could, in its spare time, take you to town. The car as a form of transportation device was born with that discovery.

For a while the car as shortest distance between two points flourished, but soon other uses began to overshadow it.

In the nineteen-thirties it began to double as a courting chamber.

In the nineteen-forties—the motel had by that time developed respectability and ended the automobile's role as trysting place—the automobile began to be a badge. Many persons believed that the car they parked in their garages (the cars were still not too obese to fit into garages, as they now are) was examined by the great world as a statement of a man's wealth, standing and personality.

With the nineteen-fifties, the manufacturers in Detroit decided that a car was really a sex symbol, which was very silly indeed. So silly, in fact, that it encouraged critics of the automobile to speak more forcefully.

Under the spew of criticism, we discovered that the car was a gasoline guzzler, a polluter, a death trap and a sinister tool of the slicksters who were out to kill cities in order to line the pockets of concrete, cement, asphalt and rubber czars. It is obvious that the car is no longer of much use as transportation, at least in cities, and yet we cannot give it up.

Why?

Why do we insist on nosing our gasoline guzzlers into those barely moving streams of traffic, there to sit, dumb starers into vacant space, as the engine consumes ever more expensive fuel?

The question is a puzzler. To come to grips with it, we will probably catch ourselves wandering out to our cars, turning the key, moving into traffic. A place where a man can think, traffic.

Think about it a moment as we sit here, frozen almost immobile in a sluggish river of cars. With the windows rolled up, we are a long way from the world. Ahead of us, perhaps, some fool half-mad with the need to perforate his ulcer curses the air and threatens the fenders of the harmless to gain a single car length, but for most of us these machines give peace, calm, serenity.

Here, and here alone of all the places we may be this day, our long thoughts cannot be interrupted by the telephone—unless, of course, we are among the minuscule minority who have insisted that life is juiceless without a jangler on the highway. If we tune out the constant-stream-of-news station and tune in the Mozart station on the radio, bad news cannot reach us.

We are out of touch with bill collectors, deliverers of coded punch cards, children who need coaching in the multiplication tables. If we have clearly foreseen the proper function of cars when we purchased, we can adjust the interior temperature to satisfy our whims. At the touch of a button, the seat may slide back, descend and tilt—all this simultaneously—allowing us to recline in much the same position we fancy the Emperor Tiberius assumed at dinner.

And there we sit, out of reach of the world, in traffic; reclining in the Roman-banquet position in the temperature of our choice; invulnerable to hateful telephonists; laved by Mozart or, if we choose, silence.

The car nowadays, in short, is everyman's green island in the sea. The rich can flee, in their need for peace, to islands in seas of water, where telephones can be made to never ring and television to never come romping head-on into you, shouting about tooth decay.

Most of us can't get to those wonderful islands in seas of water. There are too many of us nowadays. We have had to create artificial seas of traffic, and there we put our islands, those retreats from life's ugly press, the last places to which most of us can go and be absolutely alone.

Our islands have wheels. They emit foul smells. True, all true. We know that. We also know they are really not very good for getting us to work, which is what we tell ourselves we are buying them for. We have to lie to ourselves about things like this; it is the Puritan tradition.

We know that mass transportation is far more sensible for getting to work. But we also know, though we would not dare breathe it aloud, that there are no islands on subway lines. As the population continues its insane increase, we will eventually have to abandon the cars, of course, and take the bus, the subway, the trolley. It is probably inevitable.

That will be a great day for mass planners, but a bad one for most of us.

With a zillion people around us, green islands in seas of water will be priced at the size of the national debt, and there will be no room, even in the sea of an 18-lane freeway, for individually owned four-wheeled islands with custom-regulated temperature and seats that assume the Roman-banquet position at the touch of three buttons.

Where shall we go then? Bonkers, most likely.

These are not just idle musings that would merit attention only in a Sunday supplement article. It is the contention of this analysis that the use and the utility of the private automobile in contemporary American communities cannot be understood if it is assumed that it is solely a means of transportation. To be able to deal with it, its social and cultural value has to be a part of the examination. The desire for unconstrained and comfortable movement is embedded deeply in the human psyche, encompassing almost everybody—not just prosperous Americans.

Development History

People have never been entirely satisfied with their own natural abilities to overcome distances and to move from place to place. *Citius-altius-fortius* (faster-higher-stronger) is not just the Olympic slogan that comes into play every two years, it is a summation of basic human urges. It is applicable to urban transportation needs in all situations, with the further caveats that we also want to have free choice about when to travel and with whom, and that we wish to do all that in reasonable comfort and within an affordable range of costs.

Before there were automobiles and bicycles, there were horses, camels, and sedan chairs (but only for the rich and powerful). Just about every culture at any time in the history of civilizations has had legends and fairy tales that envision flying carpets and seven-league boots—magical devices that allow some people to move fast, freely, and far. Nobody has ever given away a flying carpet voluntarily.

Homer clearly had automobiles in mind when he described in the *Iliad*[6] the vehicles that Hephaestus had forged from three-legged cauldrons:

> He'd bolted golden wheels to the legs of each so all on their
> own speed, at a nod from him, they could roll to halls where
> the gods convene then roll home again—a marvel to behold.

As always where mechanical devices are concerned, Leonardo da Vinci thought of self-propelled vehicles in a systematic way first. Actual devices were built in the seventeenth and eighteenth

---

[6] R. Fagle's translation, published by Viking (1990), in Book 18 (The Shield of Achilles), lines 437–440.

centuries that utilized sails, windmills, air pumps, and clockworks to propel land vehicles. Some of them even worked, in a manner of speaking, but real possibilities in the practical utilization of nonanimal power only emerged with the beginning of the Industrial Revolution.

Early Experiments and Efforts

The history of self-propelled (i.e., motorized, individually controlled, over-the-road) vehicles can be traced back to the very early period when a mechanical power source—steam—became workable.[7] There is general consensus that the first automobile was a massive three-wheeled tractorlike vehicle with a huge boiler that was built in France by Nicolas-Joseph Cugnot in 1769. As an engineer of artillery, Cugnot was able to apply cannon-boring machines to the production of a reasonably precise steam cylinder. The vehicle ran at the stately speed of 2.5 mph[8] and could carry four passengers. By the end of the eighteenth century, various steam vehicles were running on the roads of France and Britain. This activity peaked in the 1830s, including scheduled services in a few cases. However, it also became apparent that the steam engines of the time frightened horses and people and were too cumbersome and too dangerous to operate in mixed traffic on the extremely poor road surfaces that then prevailed even in cities. Public opinion and legislation turned adamantly against over-the-road steam vehicles, and attention and resources were directed to the placement of steam engines on rails within their own rights-of-way. (The glorious history of railroading in cities is traced in other chapters of this book.)

Work with steam automobiles did not cease entirely, however, and there was a continuous chain of progressive improvements in the technology, resulting in a series of light but largely experimental vehicles in most of the industrializing countries. At the

---

[7] Among the many references tracing the history of the automobile, the following can be mentioned: The "Automobile" entry in the *Encyclopedia Britannica*, 1975, vol. 2, pp. 514–535; G. N. Georgano (ed.), *The New Encyclopedia of Motorcars, 1885 to the Present* (Dutton, 1982, 3d edition); New English Library, *History of the Motorcar*, 1971; and J. B. Rae, *American Automobile Manufacturers: The First Forty Years* (Chilton, 1959, 223 pp.). The most thorough history of automobiles is provided by J. J. Flink, *The Automobile Age* (MIT Press, 1988, 456 pp.).

[8] The same speed that characterizes congested street conditions in today's cities.

very end of the nineteenth century, when interest in self-propelled vehicles became a dominant preoccupation and other power sources had emerged, the steam vehicle again became a strong contender, with several successful models on the roads. The Stanley Steamer is still remembered in popular culture, and it was a very respectable engineering achievement; one set the land speed record of 127.66 mph (205.4 kph) in 1906.

Gasoline-powered cars, however, won out eventually because this fuel represented a more compact source of energy, but steamers have not been forgotten. There is assurance that, if we ever run out of petroleum-derived fuels (and we will), the individually controlled vehicle will not become extinct at all because other power sources are available. Steam engines, with only a little more engineering effort, can be made almost as effective as internal combustion motors, and other possibilities exist as well.

The experience with electric propulsion was somewhat similar. With the invention of the battery, leading to an operable device in 1881, its placement in a vehicle was an obvious next step. During

A Brief History of Automobile Technology

1769	Steam vehicle/tractor created by Nicholas-Joseph Cugnot in France.
1860	Stationary gas-fueled internal combustion engine developed by Jean Etienne Lenoir.
1876	Nikolaus August Otto builds a practical four-stroke gasoline engine.
1885	Gottlieb Daimler places a lightweight engine on a bicycle to create the first motorcycle.
	Karl Benz builds a three-wheeled motorcar.
1886	Gottlieb Daimler places a gas engine on a carriage.
1891	R. Panhard and E. Levassor create the standard car design with the engine in front.
1892	C. Jenatzy exceeds 60 mph (100 kph) in an electric car in France.
1894	The French Panhard becomes the first successfully marketed automobile.
1901	Ransom E. Olds introduces first mass-production assembly line.
1904	L. Rigolly exceeds 100 mph (161 kph) in France.
1906	The Rolls-Royce Silver Ghost first appears.

1908	Henry Ford starts production of the Model T on an assembly line.
1911	Cadillac introduces the electric starter.
1912	Henry Ford perfects the moving assembly line with a coordinated flow of parts.
1920	Duesenberg develops four-wheel hydraulic brakes.
1921	Lancia offers a unitary body with independent front suspension.
1927	H. Segrave reaches 203.8 mph (327.9 kph).
1928	Cadillac introduces an upgraded gearbox (synchromesh).
1934	Citroen offers front-wheel drive.
1938	The Volkswagen Beetle makes its appearance as the "people's car" and military vehicle.
1948	Michelin introduces the radial-ply tire; Goodyear offers the tubeless tire.
1950	Dunlop develops disc brakes.
1951	Buick and Chrysler offer power steering.
1954	Carl Bosch develops fuel injection.
1957	Felix Wankel builds a rotary engine.
1965	Publication of Ralph Nader's *Unsafe at Any Speed*.
1966	California introduces air pollution legislation directed at motor vehicles.
1972	Dunlop introduces self-sealing safety tires.
1979	Sam Barrett exceeds the speed of sound in a rocket-propelled surface vehicle.
1980	Audi mass-produces the first car with four-wheel drive.
1981	BMW introduces the onboard computer to monitor performance.
1982	Austin Rover offers the first "talking dashboard."
1987	Solar-powered vehicle travels 1864 mi (3000 km) in Australia.
1988	California passes stringent air quality controls aiming toward zero emissions.
1990	Fiat and Peugeot bring electric cars on the market.
1991	Satellite navigation systems are made available in Japan.
1992	Mazda and NEC offer collision avoidance and automated car control devices.
1995	Greenpeace produces an environment-friendly prototype (70 mpg).
1996	Daimler-Benz introduces a fuel cell car.

Sources: Webster's New World *Book of Facts,* IDG Books Worldwide, Inc., 1999, and other history of technology references.

the following decade, experiments took place in France and the United States, and the field expanded quickly. It was an electric vehicle that first reached 60 mph (100 kph) in 1899. In the early period of automobile development, battery-powered cars were fully competitive with the other types, and a number of manufacturers entered the field. However, sales peaked around 1912, as the limited speed and range of the electrics became a handicap and the other types of propulsion outpaced them. These operational problems due to an unsatisfactory power storage device are not solved yet, even though there has been considerable interest and pressure in the concluding decades of the twentieth century to reactivate the non-polluting electricals, at least as city cars.

The First Automobiles

The winner among the power choices, at least for the duration of the twentieth century, was the gasoline engine automobile, as everybody is well aware. Again, there were several inventors and engineers who had experimented with this type of engine during most of the nineteenth century, but the honor of being the fathers of the automobile belongs to two Germans: Carl Benz in 1885 and Gottlieb Daimler in 1886, who not only built working prototypes, but also carried their ideas to the level of fully operational models and actual production efforts. The other inventors fell short of such practical achievement.

The first spiderlike Benz tricycle, while it had a most unusual shape, showed many of the mechanical elements that became standard components of automobiles, ranging from differential gears to a carburetor. The car was exhibited and demonstrated at the Chicago World's Fair in 1893, and undoubtedly urged forward a great many American engineers who were active in this field during the following several decades. By 1898, Benz had a vigorous production effort under way.

Gottlieb Daimler participated in early experiments with stationary gasoline engines, until he successfully mounted one on a carriage in 1886. Unlike Benz, who envisioned the new technology creating special vehicles, Daimler's first program was to convert carriages to self-propelled units by selling the motors and other technical elements. Nevertheless, he soon developed self-contained models and was manufacturing vehicles for sale by 1890. (The Benz and Daimler firms merged in 1926, and started to produce cars under the Mercedes-Benz label.) The building of automobiles on a commercial scale was also soon established in France, Great Britain, Italy, and several other countries with technological capability. A market for cars, despite their early high price tag, emerged instantaneously at the beginning of the twentieth century. At first they were expensive toys—the equivalent of yachts—but the vehicles certainly fascinated everybody, even those whose financial means were far short of actual purchase.

The development of automobiles in the United States lagged in the early days,[9] as did actual production, but then—as could be expected—various entrepreneurs entered the field with considerable vigor. There had been any number of inventors and tinkerers who had probed various possibilities in self-propelled locomotion since the middle of the nineteenth century, and there were certainly significant successes with cars utilizing steam engines, but the first American gasoline-powered car appears to have been a vehicle built by J. W. Lambert and tested in 1891. Since this effort had no specific followup, the Duryea brothers are usually credited with being the automobile pioneers in the United States. They had a working model in 1893, established their reputation by winning races, and became early leaders in the sudden new industry. By 1898, there were more than 50 companies making cars, and the number grew to about 240 a decade later. Among them were such names as R. E. Olds and J. W. Packard, who made their mark on the American fleet some years later.

In 1890, more than 4000 automobiles were produced in the United States, of which 38 percent were electrics, 40 percent steamers, and 22 percent gasoline powered.[10] All were handmade, rather clumsy affairs, certainly expensive, but intriguing to most people. Only true enthusiasts with sufficient means could be motorists, much to the amusement and amazement of most everybody else. The vehicles had no real utilitarian function, and little thought was given at that time as to what the future implications might be. The real concentration on technical advancement and improvement was still to be found in Europe. The American car producers looked toward a mass market—unlike the European

---

[9] There was some confusion in the early days as to what name the new device should carry in English, until the French *automobile* was adopted, which did not please language purists since it is a mixture of Greek and Latin. Many other basic automobile terms have a French origin.

[10] *America's Highways 1776/1976: A History of the Federal-Aid Program* (U.S. Department of Transportation/FHWA, 1976, 553 pp.), p. 54.

manufacturers, who served the specialized luxury market with fancy machines.

Preautomobile Roads and Streets

The roads and streets of United States, with a few exceptions, were in a pitiful state to the very end of the nineteenth century.[11] The need for long-distance connections stretched to the limit the capability or the eagerness of any government entity to provide quality facilities. Some relief was offered since Colonial days by the many turnpikes—private roadways that collected tolls—but they too faded away with the establishment of railroad service. Roads outside cities were maintained reluctantly by local governments relying on property and poll taxes and statute labor. Loose sand during dry periods and deep mud during wet periods made any travel a laborious, slow, and expensive undertaking. Lack of access roads to farms was a particularly serious deficiency that hampered agricultural production and doomed families to isolation and a meager existence. State governments at that time did not participate in any road improvement programs.

In cities the situation was not much better. Almost all had adopted the standard gridiron plan, with only a few variations here and there, which suited urban functions and continuous growth patterns quite well. Some visual and functional relief to the uniform street networks was provided by the introduction of parkways. At that time, these facilities were seen as extensions of parks—civic design elements with a primarily recreational role. The inspiration was the boulevards of Europe, and many large American cities implemented one or more such roadways for strolling pedestrians and carriages. Parkways turned out to be also quite suitable for motor vehicles.

The street surfaces were another story. Industrialization trends placed heavy demands on access facilities, and railroads could not satisfy all movement needs, particularly at the local level. Much volume was carried between factories, warehouses, and transportation terminals by horse-drawn trucks and drays. They were heavy vehicles, with steel tires that ground to dust any sur-

---

[11] The history of roadway development in the United States is described most fully in *America's Highways: 1776/1976,* op. cit. Another reference is B. E. Seely, *Building the American Highway System: Engineers as Policy Makers* (Temple University Press, 1987, 315 pp.).

face covering. Cobblestone streets were expensive to build and maintain, and, while they provided reasonably good footing for horses, they did not offer a smooth ride for vehicles. Most streets in cities were not paved anyway, even though commuting needs started to emerge as cities increased in size and various special-purpose districts emerged. Complaints about noise and accumulating horse manure were quite common.

For industrial purposes there was little choice but to build heavy pavements of granite blocks or hard paving brick; other streets were fortunate to receive graded gravel surfaces with controlled drainage or macadam.[12] Asphalt surfacing became practical in the last decades of the nineteenth century.

The deplorable street situation in American communities was attacked and changed largely by the Good Roads movement within a rather short time period. Starting with the 1880s, more and more voices were heard identifying the poor state of roads and streets as an obstacle to economic development and an affront to human dignity. The most energetic impetus toward positive action, however, was generated by bicyclists, who were suddenly not only many in number as the bicycle craze swept the country, but also well organized and vocal in their demands. They wanted smooth asphalt surfaces, and various associations of "wheelmen" made themselves heard starting in the 1890s. The results were remarkable: even state governments were drawn into programs for pavement upgrading.

In 1907, there were about 47,000 mi (76,000 km) of streets in American cities with populations of 50,000 or more, but still only 44 percent of that length was improved.[13] Thus, of the total mileage, 16.3 percent had heavy-duty pavement, 13.3 percent had macadam surfaces, 8.9 percent had asphalt covering, and 5.4 percent had simple gravel surfaces. This spanned a range from engineered, watertight pavements to surfaces with no stability whatsoever. The condition of roads kept improving steadily over the decades, save during wars and periods of national economic dislocation.

---

[12] A type of pavement consisting of a thick layer of crushed stone with bituminous binder and covering course.

[13] Office of Federal Coordinator of U.S. Transportation, *Public Aids to Transportation,* vol. 4, 1940.

The Popularization of the Automobile

Henry Ford appeared on the scene in 1903, and soon revolutionized the market by placing the automobile within a reasonable price range for the average American family. He did not invent any technical elements himself, not even the assembly line; his genius is found in the ability to combine many innovative approaches and methods into a process and a product that was truly original in its new form. By taking advantage of new materials, Ford did design an improved vehicle that was lighter and smaller than the expensive models then on the road, more suitable for daily practical use. After a satisfactory prototype was developed following considerable design experimentation, the Model T (the Tin Lizzie) entered production in 1908. Ford's great achievement was the meticulous structuring of the moving assembly line (1913) and the organization of the flow of parts and semi-finished components with precise timing to be incorporated into the finished product. Ford succeeded in achieving his goal of creating "a cheap, versatile, and easy to maintain" vehicle—"a car for the great multitude"—and transforming the automobile from a luxury item into an affordable necessity.

In a few years, he expanded his production from not quite 15,000 vehicles per year in 1907 to 248,000 in 1914, and at the same time cut production costs. A Ford could be bought for $600 in 1917, which was not a trivial sum at that time, but certainly within the means of a family with a solid wage earner. The price came down to $290 in 1926, and a total of 15 million Model Ts were placed on the market between 1908 and 1927.

The following decades in North America to the beginning of World War II were marked by two general trends. One was the building of special cars at an extreme level of luxury, speed, and sometimes size. These are now called the classics, and include some names that are still in business (Rolls Royce, Mercedes-Benz) and some that left the scene but are certainly remembered (Pierce-Arrow, Bugatti). Their extraordinary purchase price and very high maintenance expense placed them in a class accessible to only a few. Their use was limited mostly to leisurely motoring in the countryside, visiting friends of equal standing, appearing in Hollywood movies, and racing. The results of this situation were that automobiles acquired great status, and practical technological improvements were tested under generous financial conditions.

This article was originally published in the *Yale Law Journal* in February 1908 as "The Status of the Automobile."

The Horseless Carriage Means Trouble

H. B. Brown

The invention of the automobile has introduced upon the public roads of the country a novel and not altogether welcome guest. Although barely ten years since it first made its appearance, it has already conquered an important position in the domain of travel. Indeed, its great power, speed and weight have made it a veritable king of the highway, before whom we are all invited to prostrate ourselves. Though admitted to the use of the roads, in common with other vehicles, certain restrictions have been found necessary to curb its masterful and dominating influence.

With the advent of the automobile the courts were confronted with the proposition that a self-propelled vehicle [would be] limited to no part of the highway, capable of the speed of an express train, and attended by a cloud of dust and smoke, and the emission of a noisome odor. Notwithstanding these objections, automobiles have doubtless done much to earn their popularity. They have brought suburban towns within easy access from the city; they do not run upon a fixed track, and have no monopoly of any part of the highway; they do not seriously interfere with its use by other vehicles, and afford a most convenient and expeditious method of traveling between cities and outlying villages or country seats. In the form of electric runabouts, doctor's coupés, express and delivery wagons, and other teaming, they are rapidly superseding vehicles drawn by horses. They have largely taken the place of traveling carriages with those who are desirous of speed, and are content with little more than a perfunctory view of the scenery, which, however, cannot be thoroughly "taken in" when running at a rate of over twelve miles an hour.

To those who occupy or drive them, they are undoubtedly a fascinating amusement. The speed of which they are capable intoxicates and bewilders the senses, and deadens them to the dangers which surround the machine, and by a sudden mishap may turn it in the twinkling of an eye into a terrible engine of destruction.

It is a fact too notorious to be ignored by the courts, that the excessive speed of automobiles costs the lives of many persons;

[1906]

THE HOL-TAN COMPANY, BROADWAY, Cor. 56th St., NEW·YORK

Price
$9000
any folding
top included

Model: **F I A T.**

Body: Optional.

Color: Optional.

Seating capacity: **Five, seven or nine persons.**

Total weight: 3000 pounds.

Wheel base: 114, 122 and 131 inches.

Wheel tread: 54 inches.

Tire dimensions, front: 910 x 90 m/m.

Tire dimensions, rear: 920 x 120 m/m.

Steering: **Wheel, with worm and sector.**

Brakes: **Internal expanding, water cooled.**

Gasoline capacity: 30 gallons.

Frame: **Pressed steel.**

Horse-power: 35.

Number of cylinders: **Four.**

Cylinders arranged: **Vertically, under hood; cast in pairs.**

Cooling: **Water; Daimler radiator.**

Ignition: **Make-and-break; low tension Simms-Bosch magneto.**

Drive: **Double side chain.**

Transmission: **Selective sliding gear.**

Speeds: **Four forward, one reverse.**

Style of top: Optional.

Descriptive catalogue sent upon application to the above-named company.

From the Handbook of Gasoline Automobiles, 1904-1906/Dover

and that scarcely a week, sometimes scarcely a day passes without chronicling from one to a dozen deaths occasioned by the reckless driving of these machines. Fortunately, the chauffeur and his guests are the usual sufferers, and in their misfortunes as lawbreakers, the general public do not much concern themselves. Our sympathies are rather reserved for the hapless farmer whose horses are frightened, or whose wagon is wrecked, for a failure or inability to comply instantly with the chauffeur's signal; or for the bystander who is run down and crushed by the enormous weight of these engines.

The automobile lacks one of the most attractive concomitants of pleasure driving in the companionship of the horse. This is a feature which may not be considered by those who are indifferent to him, but to those who recognize an instinctive sympathy, more easily felt than described, between man and certain of the lower animals, such as the horse, the dog and the donkey, the cold and heartless mechanism of the automobile furnishes a poor substitute. The automobile is doubtless a most useful vehicle, but one is not likely to lavish upon it the fond attention he bestows upon his horse or dog. A man may admire his own carriage, but his affections are reserved for the horse that draws it and the dog that follows it.

The future of the automobile depends principally upon the chauffeur and his sponsors. If he observes faithfully the speed laws of the various localities (and herein lies the main obstacle to his popularity) he may expect to be accorded such rights as his superior speed requires for the perfect operation of his machine; but if he persists in defying these laws, he must expect legislation more drastic than any yet attempted; for after all, those who do not use automobiles are still a large majority and control the legislatures. It has been proposed that special roads be constructed for automobiles, upon which ordinary vehicles shall be excluded, and to which the speed laws should not apply. This might be satisfactory to the general public, but probably not to the automobilists themselves.

How far the automobile is a mere whim of fashion, and how far it meets a real need of the community, time can alone determine. Judging from its rapidly increasing numbers, it seems to have made a place for itself in the hearts of the people. Whether it will take its rank as one of the favorite vehicles of pleasure and commerce, or supplant them all, we shall eventually know—but not now. The lesson of the bicycle, for years an absorbing amusement of the highest classes, now a harmless though useful vehicle for school boys and messengers, will not be lost upon us. The automobile has much to contend against in its offensive characteristics, and above all, in the arrogant disregard of the rights of others with which it is often driven; but new inventions may obviate some of these difficulties, and a few sharp lessons from the courts may inculcate more respect for the rights of others.

Whatever the outcome may be, every true admirer of the horse will pray that it may not be the extinction or dethronement of the noblest of all domestic animals.

(Reprinted by permission of the Yale Law Journal Company and the William S. Hein Company from the *Yale Law Journal*, vol. 17, pp. 221–231.)

The other trend, largely thanks to Henry Ford but also other manufacturers that had to remain competitive, was the gradual introduction of automobiles in the lives of most Americans. These new "appliances" met with great favor, and it is hard to identify anybody who tried having an automobile and then decided to give it up. Many of the early motorcars were used first by people in their regular profession or trade (doctors, salesmen), who found this new means of transportation most convenient and suitable. Many automobiles were also being acquired by families who saw in them a device to broaden their recreational and social opportunities. Farming families in particular could benefit from gaining connections to other activities beyond the homestead. Suburbanization and commuting to jobs by car was still a small component of city development and daily activity,[14] but there was a perception—actively promoted by the car manufacturers—that automobiles were a splendid means of escaping the difficulties and constraints of the city by taking recreational rides outside.

Community and Economic Impacts

Problems in American cities—particularly the repressively crowded and tragically unsanitary living conditions of the poorer classes—have been recognized since the onset of the Industrial Revolution. In the second half of the nineteenth century the advocated remedy was not sanitation and improvement of the neighborhoods, but rather the creation of parks as "urban lungs" to provide relief space for everybody. In the first half of the twentieth century, with the basic problem still festering, the attitude toward a solution was similar. The car was seen as a device to make urban life tolerable by allowing most people to temporarily escape to, and seek

Rally of antique cars in Connecticut.

[14] K. T. Jackson, *The Crabgrass Frontier* (Oxford University Press, 1985, 396 pp.). Suburban development until World War II was still largely tied to the railroad station.

recovery in, the countryside. In retrospect, this perhaps was so-cial engineering at its worst, but it placed the automobile on a shining pedestal as a constructive means toward upgraded urban livability. Nobody really expected that it would also cause a pro-found change in urban structure and daily habits of residents. Whatever motor vehicle congestion could be seen at some loca-tions, there was general relief that horse wagon crowding, steel-tire noise, and street pollution by horse excreta and carcasses were disappearing. The discovery of air quality problems was still decades in the future.

Over the first half of the twentieth century the automobile industry moved from hundreds of small manufacturers to a few large enterprises through mergers and acquisitions. This became a necessity because of the massive investments needed in the large-scale production facilities serving a mass market. Unlike in Europe, where many manufacturers largely served their national—thus limited—markets, the American car builders were able to tap into a very large territory populated by millions of people, most of whom were eager to become customers. This allowed and demanded production efficiencies and attention to general customer satisfaction. Besides Ford, General Motors (started in 1908) under the leadership of W. C. Durant emerged as a huge corporation by absorbing many smaller firms together with their brand-name models. Chrysler, formed on the base of the Maxwell Motor Company in 1925, became the other member of the Big Three. By the time of the Depression, only five inde-pendents were left with one quarter of the national market. The United States produced one-half of the total global motor vehicle volume, but sizable enterprises respected for the quality of their product also became established in Great Britain, France, Ger-many, and Italy.

As the automobile established its role in American society, there was also a parallel concern with the roadways on which the cars had to operate. The period up to World War II was charac-terized by an increasingly more extensive and supportive federal involvement in the construction and improvement of the national and local networks, starting from almost nothing in the early part of the century. It should not, however, be assumed that the sequence of ever-more-generous national legislation and programs was a prime cause of automobilization. Rather, these programs

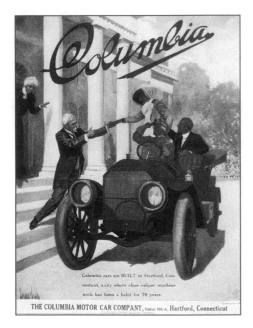

Columbia Motor Car Company ad.

were the consequences of a broad-based national demand for constructive action by the ever-growing motorist fraternity. These requests for action were fully supported in just about every community by merchants, newspaper publishers, and civic boosters.

The question still being debated today among students of modern America is not so much whether political pressure was applied for the building of roads (it certainly was, and still is), but the degree to which this was a true expression of grassroots demands by users or was materially pushed beyond immediate needs by the "highway lobby"—motor vehicle manufacturers, cement companies, construction firms, oil companies, steel suppliers, etc. The efforts can be seen as legitimate advocacy utilizing regular and established political procedures or as the promotion of self-serving interests inimical to the needs of the total society.

Early highway assistance legislation was generated largely by the national concern for "getting the farmers out of the mud," but in the course of years urban concerns also entered the programs. A detailed history of this sequence is not necessary here, but the general trend is of considerable interest as an indicator of attitudes and responses.[15] For example, the 1916 Federal Aid Road Act was the first piece of basic legislation that required each state to establish highway departments to receive federal funds; the 1921 act introduced the concept of a national roadway system (with a "primary" component of 200,000 mi) and set allocation levels of federal funds. There was no thought at that time about urban roadway needs. The Bureau of Public Roads (BPR), established in 1921, became a significant agency as the federal arm in all subsequent national work on arteries and roadways above the local level.

The automobile era in the United States started in the 1920s. The national fleet had more than quadrupled, and fatalities had

---

[15] A complete review of this period of national history, several times over, but always from a highway building perspective, is found not only in *American Highways: 1776/1976*, op. cit., but also in the several books on suburbanization.

already exceeded 30,000 by 1929.[16] A revenue stream from road user taxes (mostly on gasoline, starting in 1919) was turned on, political pressures from motorists became a real force, and intensive construction programs were initiated and carried through. Buses started to replace streetcars rather quickly, and motor trucks took over just about all short-haul freight movement. Construction standards had to be upgraded because the volume and weight of motor vehicles destroyed the old, lightly built roadway surfaces. This was the time when the motorcar started to become the principal element in a consumer-oriented society and a major factor in the change of lifestyles and social activity. The development of contemporary America is a most complex story, and the automobile and the motor truck are active instruments in a largely successful achievement. The result, as it stands today, has its faults and its gaps, but it is still the envy of most other countries.

Street congestion in cities became visible in the 1930s, and the first responses were to increase capacity by streamlining flows incrementally through one-way arrangements, traffic signals, lane markings, and rebuilt turning radii. String developments serving motorists started to appear on principal radials, and bypass roads emerged as a possible solution to internal congestion. Organized traffic studies also became necessary to document the developing situation (traffic counts in Chicago and Cleveland were the pioneering efforts), although it was not yet known how to use the data to full effect.

In the 1920s, the automobile became an indispensable device for farmers to break their isolation and for professional and business people who needed fast and reliable transportation throughout the working day to do their jobs effectively. It was not yet much of a commuter mode and not within full reach of lower-income families. The car became, first, a means to reach recreation places, undertake motoring vacations, and run family errands.[17]

---

[16] The events of the period when automobiles established their dominance in American communities (the 1920s and 1930s) are well covered by M. S. Foster in *From Streetcars to Superhighways* (Temple University Press, 1981, 246 pp.). An overview of the events leading to current urban patterns as shaped by transportation systems is provided by K. H. Schaeffer and E. Sclar in *Access for All: Transportation and Urban Growth* (Penguin Books, 1975, 182 pp.).

[17] J. J. Flink is a scholar of the social impacts of automobiles, and a summary of his observations is found in "The Family Car," *ITS Review,* May 1987.

Suburbanization was, however, abetted, even if work trips were still largely transit based.

The decade of the Depression was marked not so much by extensive new roadway construction (there were several significant examples, however, as noted later) as by a continuation of surface upgrading and widening of highways (particularly trying to eliminate deadly three-lane facilities). There was an urgent concern with safety issues. As a part of these efforts, it became necessary to define physical standards of roadway geometry and elements and to seek uniformity across the country. Makework projects were initiated as relief programs, and one such effort for white-collar workers was the further development and expansion of traffic studies. Landscaping programs along arterial roads were implemented.

The Beginning of Urban Highways and Expressways

This was also the time when highway design concepts became established and the first visions of major facilities for fast travel and commuting appeared in several places, going beyond the earlier formal recreational parkways. A significant pioneering effort was the Long Island Motor Parkway, built by William Vanderbilt as a grade-separated speedway where gentlemen motorists could really put their expensive toys through their paces after paying an entry fee. The facility was in operation from 1906 to 1911, and it showed what the physical channel requirements might be for this new, incredibly fast mode of travel.

The idea that highways would have to be quite different as the horse era faded away was initiated by the Bronx River Parkway in Westchester County (opened in 1923). The dual-carriage roadway, intended for leisure time motoring, was built as a part of a river cleanup project.[18] It was not exactly an urban facility, but the corridor was already quite developed. Grade-separated interchanges, variable-width right-of-way, no lateral access with entry/exit only at designated points, a flowing layout allowing smooth and rapid movement, extensive landscaping, and attention to architectural details marked this project. It was a dramatic departure from the drab, utilitarian highway designs heretofore. The motoring public loved it, the engineers were pleased with the

[18] S. Grava, "The Bronx River Parkway: A Case Study in Innovation," *New York Affairs,* no. 1, 1981, pp. 15–23.

operational features, land developers found it useful, and public officials and civic boosters approved.[19]

For all practical purposes a new form of traffic artery had been created, and it spawned the whole limited-access highway industry. The Bronx River Parkway was soon followed by other Westchester County parkways and similar facilities on Long Island. As demand and usage increased, the concept of a parkway became a convenient designation for highways excluding trucks, not a characterization of a roadway's role. The Bronx River Parkway has been "improved" many times since its origin to increase its traffic-carrying capability and can hardly be recognized today (except for a few downstream sections). The pattern was set, however, followed by other examples in other communities before the end of the 1930s, notably the Merritt Parkway in Connecticut (first half opened in 1938) and the Arroyo Seco Parkway between Los Angeles and Pasadena (opened in 1940).

Other locations and states under the pressures of automobile traffic took initiative on their own and started to implement major highways by relying on tolls or state-backed bonds for financing. The most visible and extensively developed example was the Pennsylvania Turnpike (opened in 1940), but work was done in Maine, New Jersey, and other states as well. This experience appeared to show that the American public not only demanded roadways, but was also quite willing to pay for them. It is said that German highway officials came to Westchester County and Pennsylvania to learn before they built their own *Autobahn* system (with a major military dimension) before World War II. After the war, American engineers, in

The first limited-access highway—the Bronx River Parkway.

---

[19] The largely grade-separated traverses and roadway loops of Central Park in New York City (built between 1858 and 1876) were conceptual models for many design elements later. The roadways were originally built for horse-drawn carriages.

turn, brought back some of the European experience when the U.S. Interstate system was started (more about that later).

The 1930s were also a period when major highway building started to take place inside cities. Much of it was assisted by the national work and economic recovery programs designed to cope with the Depression. The Henry Hudson Parkway in Manhattan, as well as the many highway projects undertaken by Robert Moses,[20] are examples of these efforts. Major programs were developed in Jersey City, Boston, Philadelphia, Chicago, and a number of other cities on a smaller scale. It was still possible to thread these roadways through the urban fabric without too much displacement. If the contemporary records and publications are searched diligently, some voices of concern about impacts can be found, but the general public attitude toward highway construction was still overwhelmingly favorable.

Federal highway legislation continued in the 1930s, with the 1934 Hayden-Cartwright Act making highway planning possible in an organized way and allocating significant resources for surveys and investigations. The 1938 Act was the first to pay attention to roadside improvements.

While the development of a national highway system outside cities is not within the scope of this book, it is necessary to make a quick reference to the beginnings of this effort in the late 1930s because eventually these roads affected cities and communities most extensively and sometimes severely. Talk about superhighways under federal control can be traced back to the 1920s, but the triggering document was a sketch drawn by President Franklin Roosevelt in 1937 that showed three routes spanning the country in the east-west direction and three north-south. The Bureau of Public Roads followed with a study immediately and recommended six transcontinental routes, placed quite precisely. It envisioned a 27,600-mi (44,400-km) network connecting all major centers within the United States having no at-grade intersections.

There was much concern with financing such an undertaking, but the method favored initially—toll roads—was soon deemed infeasible and inappropriate. It is important to note that there was no significant discussion whether such a system was needed

---

[20] At the peak of his career, Robert Moses held almost a dozen posts in New York state and city government that encompassed parks, highways, slum clearance, power, and other public works sectors.

in the first place; the concerns were solely how to build the network and how to allocate its costs. By 1940, it was an established fact that every American family that could afford a car had one or would soon acquire it; the assumption was that the fleet would be used on a daily basis.

The Transition Period of World War II

World War II military actions were characterized by much movement and dependence on motor vehicles, which was a lesson that most participants remembered. It was not just tanks and troop carriers, but also humble support vehicles to move supplies, maintain communications, and ferry the wounded and the commanders. The *Wehrmacht* bogged down in the mud of Russia and ran out of fuel;[21] the Allied forces rolled to victory, ably supported by the quickly expanded, versatile production capability of the established car manufacturers. Two vehicles with major roles in subsequent years emerged from the conflict: the German "people's car" (Volkswagen, 1938) and the U.S. Army jeep.

World War II was certainly an interruption of the trends in car manufacturing and highway building, but ultimately it made little difference in the United States because the recovery after the war was fast and the growth lines continued on the earlier trajectory. Car production jumped from 70,000 in 1945 to 3.5 million in 1947; registration moved from 30.6 million units in 1945 to 37.4 million in 1947. The roadways had certainly worn out during the war years under heavy use and minimal maintenance, and the demand for improvements was answered with urgency. The so-called Highway Needs Studies that were prepared on a regular basis showed a never-ending demand with no saturation in sight in terms of new construction of roadways.

To respond to the pent-up demand for cars, the Big Three quickly reestablished their production lines, but they were also joined by new manufacturers who saw great market opportunities for standard and special models. These were Kaiser-Frazer, Studebaker-Packard, American Motors, and a few other smaller firms. Some interesting and handsome vehicles were produced, but the new firms were not able to gauge consumer preferences accurately or to survive the competitive environment for very long.

---

[21] One of the most prominent slogans of wartime Germany, painted everywhere, was *"Rädern rollen für den Sieg!"* (Wheels roll for victory).

Another trend was the gradual emergence of strong automobile manufacturing and marketing capability abroad. While American manufacturers satisfied 76 percent of the global market in 1950, they had only 28 percent of the volume in 1970. European manufacturers were not particularly successful in convincing a large number of American buyers that small and fuel-efficient cars were desirable, but they were able to establish footholds on other continents. The Japanese firms, after an uncertain start, moved aggressively with reliable and affordable models into all markets, including the United States. They built production and assembly plants within countries that consumed a sufficient number of units, and by 1980 manufactured more vehicles than American firms.

The Highway Era

The Federal-Aid Highway Act of 1944 set the stage in preparation for peacetime. This landmark legislation extended federal aid to urban areas, defined primary and secondary systems, authorized significant sums for construction, established allowable assistance ratios for each type, allocated 1.5 percent of total project funds for planning, and formalized the concept of the National System of Interstate Highways. The last item received the most attention; a 40,000-mi (64,400-km) network was authorized that was defined segment by segment by the respective state highway departments within the next few years. The other noteworthy new situation was that after this point most federal highway aid went to urban areas.

Immediately after World War II, American communities were not yet fully motorized, but a huge latent demand had been generated that was filled most rapidly. In 1950, 41 percent of families did not own a car; by 1980, such families were less than 13 percent of the population. During this period, social patterns, courtship habits,[22] daily life, commuting behavior, and use of leisure time all changed through the almost universal means of mobility offered by the automobile. Even the shape of a standard house was altered.[23] It can also be said that the role of women in

---

[22] *Motor Trend* magazine reported in 1967 that 40 percent of all recent marriage proposals had been made in a car, presumably because of convenience and necessity.

[23] As analyzed by F. T. Kihlstedt in "The Automobile and the Transformation of the American House," in *The Automobile and American Culture,* op. cit., pp. 160–175, the front porch disappeared as a connector to the public life on the street, and the most prominent frontal feature became the garage door.

families and in the larger community was substantially expanded through the use of the automobile.[24] As the vehicles became not only more available (the second family car), but also much easier to handle (requiring no particular physical strength), the traditional housewife could reach a greater array of destinations every day and enter the full-time or part-time labor market. In many instances she also became the family chauffeur. As more women have become wage earners without shedding their traditional household roles, the overall use of the car has increased—as a device that can reach multiple destinations and serve many trip purposes during the day.

The traffic congestion problems in cities reached disturbing dimensions by the late 1940s in a number of places, and the only solution that appeared reasonable was not just to widen existing arteries, but to carve new large highways through the urban fabric. There was no patience to wait for full-scale federal assistance either, and many large cities started dramatic large projects on their own (by 1947). This includes the Cross-Bronx Expressway in New York City, the Northern Circumferential Highway in Boston, the Congress Street Expressway in Chicago, the J.C. Lodge and Edsel Ford Expressways in Detroit, and many more in almost all large American cities. These projects were accomplished by brute force, with practically no regard for local neighborhoods; they were placed where the motor traffic demands were the highest and right-of-way acquisition least expensive. In many cases that meant inner-city minority neighborhoods. From the point of view of responding to demand, these urban expressways were certainly successful, being filled by cars almost immediately after opening. This implementation practice with no land use planning to speak of set a pattern for the next several decades before the realization set in that the motor traffic demands may well be insatiable and that established communities deserve protection and preservation. Opposition to highways was local and seen officially as purely parochial, but the seeds of significant discontent had been sown.[25]

[24] See J. J. Flink, op. cit.

[25] Lewis Mumford, the consummate thinker about the urban condition of the twentieth century, praised highways and automobiles at the beginning for their ability to disperse loads while widening the range of opportunities (see his essays "The Fourth Migration," in *Survey Graphic,* May 1, 1925, p. 132; and "Townless Highways for the Motorist," in *Harpers Magazine,* August 1931, pp. 347–356). He changed his mind shortly after (see "The Highway and the City," *Architectural Record,* April 1958) and became one of the most articulate opponents of excessive motor vehicle presence.

Interstate highway interchange near Pomona, California.

Given all this construction activity and expenditure of massive funds, there was a reasonable expectation for supporting documentation of at least existing demand volumes that might have to be satisfied. This caused the development in the mid-1940s of origin and destination (O & D) surveys that were able to use sampling techniques to gauge volumes between all pairs of traffic zones, thus indicting quite clearly the accumulation of traffic flow in certain corridors. The availability of planning funds of 1.5 percent of total construction cost was instrumental in advancing traffic/transportation studies to a reasonably high level of sophistication within the next two decades. A key realization was that land uses and fixed activities generate the need for trips, which can be classified by purpose, mode, time of day, and type of traveler, according to the intensity and character of demand.

A predictive capability can also be developed by extrapolating (or deliberately planning) future situations. With the advent of computers in the 1960s, simulation models of transportation behavior over metropolitan areas relying on massive databases became possible. This is not the place to explain the workings of these quantified and carefully documented study procedures, except to note that they placed transportation planning on a rational foundation and today allow systematic investigation and study of transportation proposals not only for highways, but also for transit and other major modes. The means to develop these crucially useful procedures came substantially from the 1.5 percent funds as well, and it is apparent that advances in this field have slowed considerably since these research assistance programs were terminated. The seminal work with metropolitan area transportation simulation models was done in Detroit (DMATS), Chicago (CATS), Pittsburgh (PATS), and New York (Tri State Regional Planning Commission) in the early 1960s.

Approval of the interstate system ran into some political problems in the early 1950s: after all, it was to be the largest public works program undertaken by any government at any time in history.[26] It regained momentum when President Eisenhower, who as a young major had tried to move Army trucks across the country, assumed the stewardship for it and designated it as his grand plan. In 1954 the Clay Committee not only gave a positive endorsement, but also worked out some of the critical administrative and financing procedures. The bill was signed in 1956, together with the creation of the Highway Trust Fund, which assured a continuous, guaranteed stream of dedicated income for highway construction purposes. The first authorization was for $25 billion to the year 1969.

The National Interstate Defense Highway System (lately renamed the Eisenhower Interstate System) became a singular force that changed how the United States developed further, how manufacturing and communications were accomplished, and how Americans lived and conducted their activities. The effect was the equivalent to what the railroads did in the nineteenth century, and perhaps more. Continental mobility and access to all parts of the country were an unprecedented achievement that propelled economic activity and changed social patterns. There was also a profound effect on cities. (See Table 5.1.)

The original concepts called for linking up all urbanized communities with populations of 50,000 or more (which was basically

---

[26] An exhaustive review of the interstate program and its implications is found in Tom Lewis, *Divided Highways* (Penguin Books, 1997, 354 pp.).

Table 5.1 Inventory of Roadways in the United States, 1999

Urban

Principal arterials, interstates	13,343 mi
Principal arterials, other freeways and expressways	9,125 mi
Principal arterials, other	53,206 mi
Minor arterials	89,399 mi
Collectors	88,008 mi
Local streets and roads	592,978 mi
Total	846,059 mi (1,361,562 km)

Rural

The rural system is classified in approximately the same format, but the aggregate length is 3,071,181 mi (4,942,452 km).

Source: US DOT, Bureau of Transportation Statistics, 2001.

achieved), but also for establishing a system of belt and radial roadways in all the larger cities, penetrating the historical fabric (which ran into major difficulties). Each central business district was to be enclosed by a tight expressway ring to provide internal distribution and be the focus of radials connecting the suburbs to the center. Such a ring was fully achieved only in Rochester, New York. In all other places these rings encountered physical obstacles or adamant community opposition trying to protect the inner neighborhoods. The radials encountered similar problems, and there was a period when the designation of a corridor placed many vulnerable low-income districts under a cloud, with residents and businesses drifting away and creating swaths of wasteland pending Interstate acquisition. Also, while the radials were supposed to bring business and jobs to the center, it was forgotten that a road always leads in two directions, and the economic activity, as well as residents, had a means to move to the outside instead.

The outer belts—sometimes in several tiers—as well as the radials on the periphery of metropolitan areas encountered few obstacles because the land was still relatively open, and these networks were usually constructed in full. The highway development was immediately or contemporaneously accompanied by land development in the suburban rings. First came residences that still depended on the center for jobs and services, but they were soon followed by shopping centers, institutions, manufacturing plants, and office buildings.

The word *sprawl* has entered our daily language, and the descriptions and analyses of its characteristics have filled articles, papers, and books in a torrent during the last decade.[27] The problems of a highly inefficient built environment with much social isolation and segregation have been well documented. It is also clear that this is how the overwhelming majority of families wish to live—as long as they can reasonably bear the costs of acquiring a parcel of land with a house and are willing to accept long trips each day by every family member. Undoubtedly, the private auto-

---

[27] Such analyses include S. Hayward, "Legends of the Sprawl," *Policy Review,* September-October 1998; P. Gordon and H. W. Richardson, "Defending Suburban Sprawl," *Public Interest,* Spring 2000; B. Katz and A. Liu, "Moving Beyond Sprawl," *The Brookings Review,* Spring 2000; K. Kehde, *Smart Land Development* (LUFNET, 2000); and many others.

mobile (or two or three per family) is the active ingredient in the creation of this situation.

The millions of acres covered by single-family homes and other low-density development that today ring all sizable central cities exist because the federal government had favorable programs of mortgage security for home purchases and because regionwide road networks were sponsored by higher levels of government. But it can also be argued that these programs were not generated by external and independent forces; rather, they came from a broad-based popular demand that enabled the dispersal trends to happen. Families paid not only for their houses, but in almost all instances also for the local streets, built by subdividers as land was developed, connecting each property into the overall circulation network.

If it could be asserted—as some do—that all the costs associated with the car are paid for by their users through various operational and tax expenditures, then a simple observation of these patterns would suffice, and they would have to be accepted. If, on the other hand, there are costs thereby created that are imposed on the larger community or society without being fully borne by the users of automobiles themselves (such as impairment of air quality, depletion of petroleum resources, the operation of inefficient land use distribution, the excessive construction of roadways, the nonpayment of parking costs), then an inequitable situation is present. The nondrivers (largely central city residents) subsidize the lifestyle of the drivers (largely suburbanites). As will be discussed further in the section on cost considerations, evidence does indicate that the full costs of automobile use are not compensated for directly, and, therefore, a series of questions come to the fore. Should they be (because this is not done in all sectors)? Is this imposition of a cross-subsidy, whatever its exact amount may be, a surreptitious and dark cabal or has it developed this way along a historical course with the full acceptance of the electorate? If everybody knew exactly what the true costs and their distribution are, would we make adjustments? Would the majority accept a much higher federal tax on fuel (as Europeans do)?

Yes, the car made all these changes in urban life possible and created the complications. Whether the machine is the cause or the effect, and who owes what to whom, remains to be resolved—but not here and now.

Shaping the Modern Automobile

In terms of automobile technology, the 1950s and 1960s were a period of considerable advances in mechanical systems, making cars much easier to operate without the application of much strength or even great skill by drivers. Automatic transmission, power steering, and power brakes were introduced and quickly became standard features. There was experimentation with new types of motors, such as gas turbines and rotary engines, but practical applications were not achieved.

Another trend that was particularly visible in the 1950s was the attention devoted to the appearance and styling of the vehicles. This was driven to a large extent by the severe competition in the industry and the aim of salespersons to replace each family sedan every few years with a distinct new model.[28] The extremes were undoubtedly the monumental tail fins found on many models produced in the late 1950s. Since that time, more rationality in styling can be observed, one of the reasons being that each basic model change is a very expensive and time-consuming process for the manufacturer that can only be covered by a large production run. Today most models on the road, expensive or not, are hardly distinguishable from each other. (There are exceptions, and this may change.)

A major milestone in the development history of the automotive industry was the publication of Ralph Nader's book *Unsafe at Any Speed* in 1965,[29] which criticized severely the safety record of the motor vehicle fleet, finding particular fault with the manufacturers and sellers of automobiles. The accident statistics had, indeed, risen most alarmingly, moving toward a peak of more than 55,000 fatalities per year in the early 1970s. This issue resonated most strongly with the public and lawmakers, and generated a major national concern that found expression in the progressive introduction of a long series of safety features in vehicles, greater atten-

---

[28] The first influential book devoted to a scathing criticism of the automobile also appeared at this time—J. Keats, *The Insolent Chariots* (J.B. Lippincott Co., 1958, 234 pp.). It was widely read, and its title became a popular characterization of the machines (actually coined by L. Mumford). This was followed by other works in a similar vein—for example, D. Wallop's *What Has Four Wheels and Flies?* (W. W. Norton & Co., Inc., 1959, 192 pp.), a spoof of the automobile industry and dogs.

[29] R. Nader, *Unsafe at Any Speed: The Designed-in Dangers of the American Automobile* (Grossman, 1965, 298 pp.).

tion toward the qualifications and responsibilities of drivers, and the redesigning of roadways and equipping them with safety devices. These programs have been eminently successful. Today, despite much greater automobile use than three decades ago, the fatality count has been brought down to about 40,000—still a very large number, making car travel the most dangerous form of transport, but better relative to the statistics of most other countries.

Chrysler sedan of the 1950s—a vehicle of maximum size.

From time to time, it appears that American consumers may embrace small cars and purchase them in sufficient numbers to make a difference on the roadways. Such periods were the mid-1960s, when compacts and fastbacks enjoyed some popularity, and the mid-1970s, when the fuel crisis hit the industry. None of this has lasted, and consumer preferences have always steadily reasserted themselves toward as large a vehicle as is affordable. A similar attitude exists with respect to engine power, whether it is needed or not. The current vogue of sport utility vehicles (SUVs) is by no means an unusual and unexpected occurrence. Few politicians today would suggest or support any legislation against size.

There has been significant technical upgrading of the vehicle itself. It has been made not only safer but also much more reliable and dependable, and much easier to drive, with a number of forgiving features. Flat tires and boiling radiators are not really seen any more along roadsides. Better and lighter materials, stable distribution of weight, impact-resistant design, diagnostic performance-monitoring systems, and a number of other features are now standard elements of regular cars.

Pollution control quality remains a sore point. Today's car is undoubtedly greatly superior in those characteristics to what they were even a decade ago, but the American manufacturers have had to be dragged by threats and inducements to improve their product. A completely clean or "green" car is possible, but it will cost more (certainly at the beginning), and there is no longer a very persistent public demand for drastic measures (more about that later).

Reactions Against Cars and Highways

A watershed in the history of highway building in the United States was the series of organized actions in the late 1960s against specific expressway proposals in a number of cities. It was a time of social unrest in a country showing concerns with political, racial, and environmental issues, and the threatened disruption of neighborhoods by massive roadways was among these issues. Local community groups, assisted by some national organizations, decided to fight city hall, and they were successful. In some cases the opposition was exercised through established procedures; in others, such as Newark, New Jersey, highway proposals triggered riots that burned down large districts of those cities. Highway projects were stopped in Chicago, Boston, New York, New Orleans, San Francisco, Baltimore, and other cities.[30] The results were a better understanding of the implications of major public works projects, more responsive attitudes and programs toward compensation, and eventually the termination of new highway construction in established urban districts.

At the same time,[31] environmental concerns, particularly those related to motor vehicle–generated air pollution, entered the public consciousness and regulations were enacted at all levels of government to take corrective action. This affected the design of engines, fuel composition, and above all the unconstrained use of automobiles and trucks in urban settings. These problems are not solved yet, but significant progress toward more healthy and livable cities has been made. Control programs have become politically acceptable, within reason and provided that they are phased in gradually and that the reasons for them are understood. A global movement is afoot not to eliminate the automobile, but to

[30] These events are covered by a whole set of books and publications, some with an unabashed advocacy position. They include A. Q. Mowbray, *Road to Ruin,* (J.B. Lippincott Co., 1969, 240 pp.); H. Leavitt, *Superhighway—Superhoax,* (Doubleday & Company, Inc., 1970, 324 pp.); R. A. Buel, *Dead End: The Automobile in Mass Transportation,* (Prentice-Hall, 1972, 231 pp.); and J. H. Kay, *The Asphalt Nation: How the Automobile Took Over America, and How We Can Take It Back* (Crown Publishers, 1997, 418 pp.).

[31] The National Environment Protection Act (NEPA) was passed in 1969.

make it behave responsibly in civilized societies and well-organized communities.

Types of Automobiles and Roadways

The rubber tire mode consists of two basic parts: the rolling stock in private ownership and the infrastructure of streets owned and maintained largely by local governments, which form networks consisting of links between nodes (intersections). These three principal physical components of the overall system are briefly described in the following sections, recording information that is generously available in other reference works. (There is, of course, another component that makes the system work—drivers—but there is little that we can do about them in a planning context. Except for some instances of road rage, they appear to be becoming better and more responsible in the United States, judging from accident statistics.) As shown in Table 5.2, of the global fleet of more than two-thirds of a billion motor vehicles, 31 percent is in operation in the United States. (It was 58 percent in 1960.)

Automobiles

Of the thousands of models of automobiles that have been manufactured over a century and of the hundreds that are on roads worldwide today, certain classes can be identified (see Table 5.3). The issues that concern the presence and use of automobiles in communities have almost nothing to do with the features that dominate the design, production, and sale of these vehicles in the very large consumer market—top speed, acceleration characteristics, and styling. Leaving aside for the time being the questions of the need for private cars, the intensity of their usage, and occupancy rates, which will be reviewed later, the relevant characteristics of the vehicle itself are the following:

- *Size.* The Daimler-Chrysler Smart Car, shown in Table 5.3, as well as several other similar units now available, is exactly half the size of a regular sedan. Two of them can be parked in a normal parking slot; they could even perhaps be placed perpendicular to the curb. They can carry the loads normally required of a full-size car. Yet, the market for them

Table 5.2　Population and the Motor Vehicle Fleet

	United States	Car-Producing Countries of Western Europe*	Rest of the World	Total
Population (millions)				
1900	76.2	186.2	1387.6	1650
1920	106.0	193.4	1560.6	1860
1940	132.2	227.8	1940.0	2300
1960	179.3	237.3	2622.7	3039
1980	226.5	263.1	3967.1	4457
2000	281.4	278.4	5520.3	6080
Motor vehicles produced				
1900	4192	5,312[‡]		9505
1920	2.23 million	61,080[§]	0.22 million	2.38 million
1940[†]	2.51 million	1.09 million	0.42 million	4.02 million
1960	7.91 million	6.01 million	2.58 million	16.49 million
1980	8.81 million	10.74 million	19.77 million	38.51 million
2000	12.77 million	13.73 million	31.02 million	57.52 million
Registration of motor vehicles				
1900	8000	N/A	N/A	N/A
1920	9.24 million	N/A	N/A	10.94 million
1940	29.44 million	7.12 million	8.06 million	44.63 million
1960	73.77 million	22.91 million	30.20 million	126.97 million
1980	155.89 million	89.56 million	165.66 million	411.11 million
2000	209.51 million	150.22 million	322.15 million	681.87 million

Source: Data from R. A. Wilson, Transportation in America, Eno Foundation, 1999; Automobile Manufacturers Association; and various almanacs.

* Includes Belgium, France, Germany, Italy, Sweden, and the United Kingdom. These are the "traditional" producers; other countries have joined the industry in more recent decades.

[†] 1938 data are used here due to the disruptions caused by World War II.

[‡] France and Germany only.

[§] France and Italy only.

is extremely limited in North America. Admittedly, they are not particularly comfortable for long trips, but the hope persists that they can capture a role as an urban car and be a family utility vehicle.

- *Fuel Efficiency.* The average gasoline consumption rates have started to climb again as SUVs capture an increasingly larger share of the market. An easy remedy to adjust the purchasing habits of Americans would be substantial increases in

Table 5.3 Typical Characteristics of Automobiles

Class, Manu- facturer, Model	Seats	EPA City/ Highway mpg	Acceler- ation 0 to 60 mph, s	Length, in (cm)	Width, in (cm)	Wheel- base, in (cm)	Curb Weight, lb (kg)	Horse- power	Price Range
Miniatures Daimler Chrysler Smart Car	2	39/55	17.5	98 (250)	60 (152)	71 (180)	1587 (720)	55	$10,000
Small cars Chevrolet Prizm	5	29/37	10.7	175 (445)	67 (170)	97 (246)	2480 (1125)	125	$14,200–$16,200
Green cars Toyota Prius	5	52/45	12.6	170 (432)	67 (170)	100 (254)	2750 (1247)	70 and 40	$20,000
Family sedans Honda Accord	5	20/28	8.0	189 (480)	70 (178)	107 (272)	3295 (1495)	135–200	$15,500–$25,000
Upscale sedans Lexus ES300	5	19/26	8.4	190 (483)	71 (180)	105 (267)	3390 (1538)	210	$31,500
Large sedans Buick Park Avenue	6	18/27	7.5	207 (526)	75 (191)	114 (290)	3880 (1760)	205–240	$33,000–$37,600
Luxury sedans BMW 5-Series	5	18/26	8.0	189 (480)	71 (180)	111 (282)	3770 (1710)	184–394	$35,400–$70,000
Sports cars Mazda MX-5 Miata	2	25/29	9.0	155 (394)	66 (168)	89 (226)	2365 (1073)	142	$21,200–$25,700
Wagons Mercedes-Benz E320	7	20/26	8.4	190 (483)	71 (180)	112 (284)	3930 (1783)	221–340	$48,700–$51,500
SUVs GMC Yukon XL	9	14/16	8.6	219 (556)	79 (201)	130 (330)	5590 (2536)	285–320	$35,800–$40,300
Minivans Dodge Caravan	7	18/24	11.4	201 (511)	79 (201)	119 (302)	4210 (1910)	150–215	$19,200–$29,100
Small pickups Ford Ranger	5	17/29	11.2	202 (513)	70 (178)	126 (320)	3870 (1801)	135–207	$12,000–$20,600
Large pickups Toyota Tundra	6	14/17	9.4	218 (554)	75 (191)	128 (325)	4710 (2136)	190–245	$15,600–$23,400
4-wheel drives Subaru Forester	5	22/26	10.4	175 (445)	68 (173)	99 (251)	3225 (1463)	165	$20,300–$22,900

Sources: Consumer Reports, New Car Buying Guide 2001; manufacturers' Web pages.
Note: The examples in each class are rated "Recommended" by *Consumer Reports.*

fuel taxes. While the cost of petrol is frequently three times higher in most Western European countries, such a policy at the present time would receive no public support in the United States, and the political feasibility is close to zero. Nevertheless, there are vehicles that operate at very high fuel efficiency and are available today. Many depend on a diesel engine, which is not popular at all in the United States, and others have a hybrid power plant that is still in somewhat of an experimental stage. The Audi A2 hybrid diesel can maintain a driving performance of 78 mpg (33 kpl); the Volkswagen Lupo (not yet available in North America) can operate at 99 mpg (42 kpl); and the Honda Insight gasoline hybrid claims 68 mpg (29 kpl).[32] There are some other models, using the cheaper diesel fuel primarily, that reach efficiencies well in the 40 to 50 mpg range. It is thus clear that much better performances would be possible with available technology, even without reaching into the realm of exotic new power plants and fuels.

- *Safety Features.* Automobiles today are much safer than they used to be, but they are not foolproof and completely forgiving. It is hard to see how the many physical safety devices that are now routinely built into cars can be substantially improved. Further progress in reducing accident rates will have to concentrate on driver education and construction of safety elements along roadways, and await the full development of automated warning and collision avoidance devices under intelligent transportation system (ITS) efforts.

- *Control of Environmental Impacts.* Again, great strides have been made lately, but the total fleet is not at where it should be. Zero-emission vehicles are possible, and they exist, but the political arena and the marketplace are not yet ready to embrace them. Much is going on at this time in developing appropriate technology, but it is impossible to say whether the acceptable solutions will come from fuel cells, electric vehicles, modified or new fuels, or incremental upgrading of conventional hardware. Improvements and replacements will come, however, because a broad-based public concern

[32] As reported in the *New York Times,* May 27 and September 16, 2001.

Trivia Facts about Automobiles

The longest automobile ever built as production models was the 1938 Duesenberg SJ town car—246 in (6.25 m), which was also the heaviest standard car at 6400 lb (2903 kg), and the 1958–1960 Continental Mark III—229 in (5.82 m). Stretch limousines, seen frequently on our streets, reach 33 ft (10 m) and can carry 14 passengers in considerable comfort. The longest passenger vehicle/limousine apparently is the 72-ft (22 m)-long custom-made limousine with a hinge in the middle produced in California for an Arab sheik. A 100-ft (30.3 m) limousine with 26 wheels operates in show business out of Burbank.

The standard car with the quickest acceleration was the 1962 Chevrolet Impala SS 409, at 4.0 seconds for 0 to 60 mph. It was exceeded by a Ford RS200 Evolution in 1994, at 3.07 seconds. The highest speed reached by a standard production car is 240.3 mph (386.7 kph) by a McLaren F1 in 1998.

Among American cars, Checker, used mostly as a taxicab, had the longest production run of the same model, from 1956 to 1982. Worldwide, the Volkswagen Beetle has been manufactured since 1938. The Ford Model T was built from 1908 to 1927. The most popular car has been the Toyota Corolla—more than 24 million have been produced since 1966.

The car with the highest accumulated mileage is a 1966 Volvo P-1800S that has been driven for a total of 1,764,000 miles so far.

Sources: Various, including the *Guinness World Records*.

has been established. Even if individual motorists are not particularly eager to embrace new technology that will cost somewhat more, at least at the beginning, societal pressure does exist to take progressive steps.

The purchase and use of cars will continue in American communities because of the undeniable advantages cars offer to each separate user. The problems occur primarily at the communal level when many motorized units have to compete for the same limited circulation space and all the engines together impair the environment. If multiple car ownership by families expands, as can also be expected, it is likely that a range of vehicles for specific purposes will be available. Where people live in organized settlements close to each other, it will be most desirable, if not mandatory, that the vehicles be "green" and "urban," i.e., non-polluting, only large enough to accommodate a few people on regular short trips, and easy to park.

Streets and Highways

Streets and roadways have two principal purposes: to carry traffic and to provide access. A small neighborhood street, for example, may have a very light traffic load (a few dozen cars per day), but it has a vital role in allowing public access to each parcel not only for cars and pedestrians, but also for mail carriers, baby carriages, garbage trucks, moving vans, etc. The local street (or more specifically, the local right-of-way) must also accommodate access by services that are frequently buried beneath the pavement: water mains, telephone lines, sewer pipes, cable conduits, etc. A limited-access highway, on the other hand, will carry very large traffic loads but have no access along the sides at all. Any type of street between those extremes will have both purposes, but in different proportions.

There are various ways of classifying roadways, but the most useful format appears to be a recognition of the type of service that they provide and their defined role (see Table 5.4).[33]

1. *Access Streets.* These streets serve properties directly by tying them into larger networks, providing secure public access for vehicles, people, utility lines, and service links. They also connect parcels to larger roadways. The traffic that they carry should be local—i.e., associated with the contiguous residences and establishments—and should include as few through movements as possible. Capacity is not the issue; safety is. The traffic volumes should be very low. A physical design consideration is to allow heavy service vehicles (garbage trucks, moving vans, and certainly emergency vehicles) to enter and leave with reasonable dispatch. Pedestrians and children are likely to be present. Streets in downtown districts of cities, while they may be narrow and serve primarily as access facilities to adjoining properties, might constitute a separate subclass because the traffic loads are usually high and the service quality not always satisfactory.

[33] Among the many references that contain this information in some detail are American Association of State Highway and Transportation Officials, *A Policy on Geometric Design of Highways and Streets* (Washington, DC, 1994, 1006 pp.); Transportation Research Board, *Highway Capacity Manual 2000* (National Research Council, 2000); and C. H. Oglesby and R. G. Hicks, *Highway Engineering* (John Wiley & Sons, 1982, 844 pp.).

Table 5.4   Urban Roadway Characteristics

| Characteristic | Limited-Access Highways | | Restricted-Access Arterials | | Access Streets | |
	Express-ways	Park-ways	Major Arterials	Regular Arterials	Collec-tors	Local Streets
Usual number of lanes	3 and 3	2 and 2	3 and 3	2 and 2	2 and 2	1 and 1
Lane width	12 ft (3.65 m)	12 ft (3.65 m)	11 or 12 ft (3.35/3.65 m)	11 ft (3.35 m)	11 ft (3.35 m)	10 ft (3.00 m)
Type of median	Wide and variable	Wide and variable	Narrow	None	None	None
Desirable minimum right-of-way width	150 ft (46 m)	200 ft (61 m)	110 ft (34 m)	80 ft (24 m)	60 ft (18 m)	50 ft (15 m)
Interchanges/ intersections	All grade-separated*	All grade-separated	Some grade-separated	At grade	At grade	At grade
Separate left turn lanes	None	None	Yes	Desirable	None	None
Traffic controls	None	None	Signals throughout	Signals	Entry stop signs	None
Service roads	Possible	None	Desirable	Possible	NA	NA
Lateral treatment[†]	Barriers	Barriers	No driveways	Some driveways	Few driveways	Free driveways
Spacing between interchanges/ intersections	2 mi+ (3.2 km)	1 mi (1.6 km)	0.5 mi (0.8 km)	1200 ft (0.4 km)	600 ft (0.2 km)	NA
Curbside parking	None	None	None	Possible	Yes	Yes
Sidewalks	None	None	Possible	Yes	Yes	Yes
Speed limit	55 mph+ (88 kph+)	55 mph (88 kph)	45 mph (72 kph)	40 mph (64 kph)	35 mph (56 kph)	25 mph (40 kph)
Minimum horizontal curve radius[‡]	1600 ft (490 m)	1600 ft (490 m)	1000 ft (300 m)	600 ft (180 m)	400 ft (120 m)	200 ft (60 m)
Maximum grade[§]	3%	6%	6%	6%	8%	12%

Note: Since there is no official format for classifying roadways, different schemes are employed by different references. The scheme used in this table is not in conflict with those, except that it is based on a planning approach, stressing functional characteristics.
* Grade-separated facilities have crossing streets at two different levels requiring a bridge structure.
[†] Lateral treatment refers to controls or facilities placed along the outer edges of the vehicular roadway.
[‡] These are approximations only. The precise curvature limits depend on the selected design speed (which is not the same as the speed limit), superelevation, pavement characteristics, and other considerations.
[§] Grade or gradient is a measure of how much a roadway rises or falls in elevation along its length. It is expressed as a percentage: units of change vertically within 100 units horizontally.

2. *Arterials*. These are major city facilities providing the basic means of internal circulation for motor vehicles (and also most transit services) (see Fig. 5.2). The purpose is to move traffic expeditiously and safely. They give form to the urban pattern. Arterials should not divide neighborhoods and districts, but should serve as boundaries and connectors. Residences should not front directly on arterials, but business establishments might; the larger roadways can have service roads along the sides. Since these are urban components, the intersections will be largely at grade (i.e., streets at the same level with all movement paths intersecting each other); they should be, however, as few as possible and spaced at long intervals. Full traffic control devices (signals, signs, markings) are necessary throughout.

3. *Expressways*. The purpose of these facilities is to move large volumes of motor traffic with speed (and safety). They are of the limited-access type, i.e., they have grade-separated interchanges and no access from the sides (except at controlled nodes). They should have good landscaping, wide rights-of-way, attention to appearance and architectural quality, and screening from abutting land uses. Expressways range from facilities at "interstate standards" to true parkways. They should have no direct relationship to neighborhoods; the intent is to touch districts that generate major motor traffic demands, but not penetrate them. Almost all American cities have decided not to build expressways any more; some links have been removed (Portland, San Francisco), some have been buried (Boston, Seattle), and increasingly more miles are being screened by noise fences (acoustical barriers). Expressways represent the principal transportation infrastructure in suburban territories, and more links may be built in the outer parts of metropolitan areas. While some additions to the national network will be made between cities, there may be a reasonable consensus that we have close to enough limited-access highways across the country.

The basic structure, geometry, and elements of roadways have not changed much for a long time—indeed, ever since there has been wheeled traffic. The modifications and improvements have

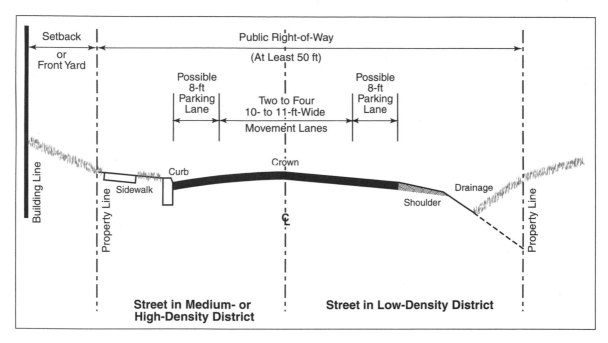

Figure 5.2 Cross-section of an urban street.

come and will continue to come in the details and in how the facilities are being used (and allowed to be used).

Intersections

Intersections are critical elements of any urban road network. They accommodate merging and turning movements, but are also where most delays occur, accidents tend to happen, and pedestrians battle for crossing space. The extent of traffic control devices and physical structuring of any intersection should be a function of the volume, type, and prevailing speed of traffic approaching and going through the node. The basic requirement is good visibility so that each motorist can see any other movements in front or on either side that may affect his or her actions. This means unobstructed views at eye level across all corners from which crossing vehicles or people may emerge.

The first step in providing specific controls are stop signs, either holding back the secondary movement or requiring that all approaching vehicles stop and look at the scene before proceeding across. Next are traffic signals, which can range from a single device at an intersection to an elaborate system of many elements that control specifically and separately each straight or turning movement. Frequently, the cycle[34] is 90 seconds long (40 cycles per hour), with 60 seconds of green allocated to the primary direction and 30 to the secondary. Cycles longer than 2 minutes are not unheard of, but they are not particularly advisable because motorists become impatient. The phases may be set permanently with possible manual overrides, but more likely devices will be used that change the timing according to the time of day (responding to changing traffic patterns). Right turn on red arrangements have become very common across the country, except in dense city districts where heavy pedestrian crossing takes place, and not only save time for motorists but also reduce air quality impacts from idling engines.

These days sensors are often embedded in one or all legs of an intersection to record and count the presence of vehicles. These devices provide information that is used automatically or manually to adjust the duration of the various phases in a cycle in real time (giving more green time to the heavier movement so that supply and demand is more balanced), replacing preset mechanical timing with demand-actuated response. These programs lead to intelligent transportation systems (discussed in more detail on subsequent pages).

Beyond traffic control elements or in conjunction with them, physical restructuring of intersections can be undertaken in those instances where the severity of flows warrants it. This encompasses strict channelization of lanes, designating and reserving space for right-hand and left-hand turning movements (see Fig. 5.3). The latter are particularly critical because they almost always involve potential conflict situations. Such turns are usually accommodated by parallel holding lanes at the intersection itself (frequently with a separate signal phase), or they can be siphoned off further upstream and kept clear of the principal street crossing.[35]

Grade-separated interchanges constitute a separate subarea of engineering design and are largely separate from transportation

---

[34] Cycle refers to the entire sequence of changes that a signal shows, returning to the same state as when the cycle started. A phase is each discrete step of the cycle.
[35] This is the rather rare "advance left turn" or "continuous flow" intersection (patented), which incorporates extensive arrangements of lanes and signals.

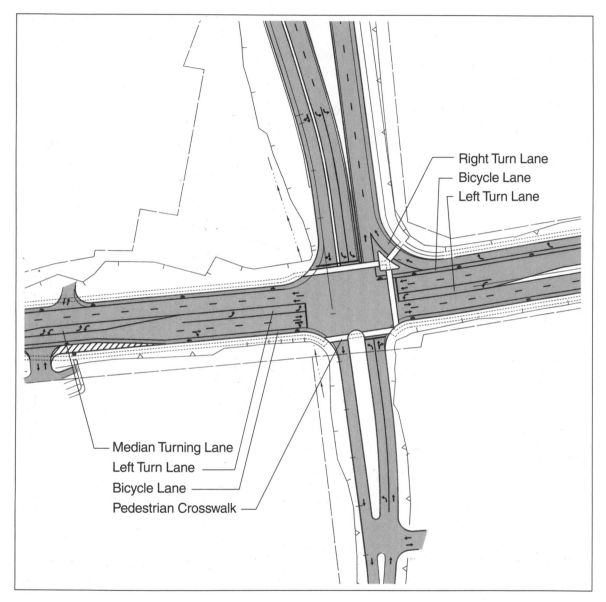

Right Turn Lane
Bicycle Lane
Left Turn Lane

Median Turning Lane
Left Turn Lane
Bicycle Lane
Pedestrian Crosswalk

Figure 5.3 Channelized, at-grade intersection of arterial streets. (*Source:* Rensselaer Co., New York, and Parsons Brinckerhoff.)

discussions related to communities. An extensive set of reference works is available,[36] keeping in mind that the construction of a grade-separated interchange in an urban environment is usually a major problem because of the space that it requires.[37]

Reasons to Support Automobiles

The popularity of individually controlled vehicles responds to the basic human impulse to act and move without constraints caused by others. Even if this is not always achievable due to limited resources or the lack of open movement space, the expectation or hope is always there. Americans have lived with these opportunities now for quite some time, and there is no mystery as to what the advantages of automobiles are. Yet, for the record, it is well to briefly list them since they fundamentally influence the choice of transportation modes in any contemporary setting.

Loosening of Geographic Constraints

Individuals and families owning automobiles are not constrained by the limited range of walking distances or the alignment of transit services. The car can reach any reasonable destination, as long as a road is available (and sometimes even if one is not). The radius of daily operations is extended manyfold, and a much richer array of possible destination points for any purpose is brought within accessible range. No rural location is completely remote; all suburban places are reachable from any other place most of the time. All job locations within a metropolitan area (save the very largest ones, where distances from one edge to another can exceed a hundred miles) are potentially available; all service nodes and entertainment/cultural centers are reachable. This is sprawl, and the automobile is in its natural environment.

Freedom from Schedules

Any travel for any purpose can be done at any time—at least theoretically—unconstrained by the schedules of any other transportation provider. It can be expected that most automobile

---

[36] See, for example, AASHTO, op. cit.
[37] A regular cloverleaf intersection may consume more than 30 acres of land. More elaborate modern interchanges may take up as much territory as an entire medieval walled city.

trips will consume less time than trips using any other mode because there is no waiting time, no delays due to transfers, and no stops to accommodate fellow travelers. Recreational trips to outside facilities are readily possible; extended vacation journeys are quite economical and governed only by the wishes of the participants. We all know that this does not happen all the time, but the periods of constraint and delay are mostly predictable and sometimes avoidable. When everything works, the automobile offers a nearly ideal state of mobility to those who have access to it, and this, in turn, gives accessibility to desired destination points.

Privacy

The car is an exclusive and private capsule that requires no sharing of space with strangers or coming in close contact with them. Privacy and security are valued benefits that people cherish since they allow each person to do things without considering the presence or preferences of others—listening to music, adjusting the thermostat, or smoking (yes, even a cigar!). In a crowded city, privacy is a significant boon.

Status

In a society that tends to make most features uniform, there is some value in being able to express one's individuality or level of achievement with a tangible expression in the form of a major possession. (See the section on the status of automobiles.)

Reasons to Exercise Caution

The problems associated with automobiles have been reviewed at great length in numerous publications.[38] It is almost a separate major branch of journalism and book publishing in the United States, and nothing has escaped the various investigators. The material has appeared in the popular media and in penetrating philosophical and technical analyses. The latter, however, are

---

[38] Among others, J. H. Kay, *Asphalt Nation: How the Automobile Took Over America, and How We Can Take It Back* (Crown Publishers, 1997, 418 pp.); J. H. Kunstler, *The Geography of Nowhere* (Simon & Schuster, 1993, 303 pp.); S. B. Goddard, *Getting There: The Epic Struggle between Road and Rail in the American Century* (BasicBooks, 1994, 351 pp.).

read primarily by a self-selected group of advocates and academicians who are already "converted." The mass market is continuously inundated by material that accepts the automobile as a matter of course, explores the advantages of new models, and describes automobile-based lifestyles as normal and desirable, without thinking too much about the impacts.

The fact that Americans know what societal and environmental consequences are being generated by motor vehicles does not mean that members of the public will change their behavior voluntarily if their own convenience and preferred lifestyle are affected. (Such recognition, however, may make corrective and compulsory programs politically acceptable, even if grudgingly.)

Excluded Population

Not everybody has access to a car, can drive a vehicle, or may wish to do so. The reasons may be age, physical capability, financial means, or personal preference. The simple conclusion is that automobiles cannot be the sole means of mobility in any community because such a state would strand a sizable component of the total population. In neighborhoods with older and poorer populations, as well as in districts with severe space crowding, this condition is particularly critical.

If the elderly, people with temporary disabilities, pregnant women, mothers with small children, persons with shopping bags or suitcases, people with cognition and orientation difficulties, and similarly constrained or encumbered potential travelers are included, this cohort in any community at any given time may reach 25 percent.[39] On the other hand, while there were 1.42 licensed drivers for each household vehicle in the United States in 1969, this ratio had dropped to 0.99 in 1990 (163 million drivers and more than 165 million cars and vans).[40]

Congestion and Space Consumption

The popularity of the car is a major drawback if a number of motorists decide to use their vehicles at the same time to go to the same destination or simply utilize the same roadways. Since a basic

[39] As estimated by the International Association of Public Transport (*Union Internationale des Transports Publics*, UITP), June 2001.
[40] U.S. DOT, *National Personal Transportation Survey* (FHWA, 1992), p. 6.

characteristic of cities is the concentration of specialized activities in defined centers, the convergence of traffic is inevitable. If the density of payload in the traffic stream is low, any available movement space will be quickly exhausted.

In a number of instances these conditions have reached intolerable levels, making some city districts unworkable. It is a major reason for outmigration of business and service activities seeking locations where employees, visitors, and freight deliverers can reach the premises with some dispatch. The much-maligned sprawl development has the significant virtue of dispersing trip origin and destination points over a large space. Nevertheless, suburban traffic congestion is now a common occurrence, particularly on arterials that lead to limited-access highways.

The situation with respect to urban congestion continues to be monitored, and at this time good documentation of conditions is available. For example, it has been estimated that in Los Angeles—the most congested urban area in the nation—the average delay per year (1999) for each person was 56 hours, which translates as an individual loss of $1000 for every man, woman, and child.[41] The other areas within the group of the 10 most congested were San Francisco–Oakland, California; the Washington, DC, area (including parts of Maryland and Virginia); Chicago–Northwestern Indiana; Seattle–Everett, Washington; Boston; Atlanta; San Diego; San Bernardino–Riverside, California; and Portland, Oregon–Vancouver, Washington. Prominent within this group are the growing cities of the West Coast, with annual per capita delays ranging from 34 to 53 hours. The least congested urban area was Corpus Christi, Texas (7 hours), followed by

[41] D. Schrank and T. Lomax, *The 2001 Urban Mobility Report,* Texas Transportation Institute, May 2001. The estimate of loss is based largely on an assumed value of time and wasted fuel.

High-density traffic on 34th Street in New York City (level of service F).

Typical dispersed roadway scene in suburbia (Route 22 in New York State).

Buffalo–Niagara Falls, New York; Bakersfield, California; Albany–Schenectady, New York; Rochester, New York; Pittsburgh, Pennsylvania; Kansas City, Missouri-Kansas; Spokane, Washington; Colorado Springs, Colorado; and Oklahoma City, Oklahoma (ranging from 6 to 24 hours). It is sobering to note that the places under less traffic pressure include a number of cities with economic difficulties.

The overall finding of the study was that congestion has increased substantially across the country in the last two decades, as measured by all indicators. The national congestion bill in 1999 was $78 billion (lost time and excess fuel), and sufficient relief could not have been provided by building thousands of new lane-miles.

From time to time, the American Automobile Association publishes a list of the chronically worst traffic locations in the United States. The selection has no scientific base except continuous reports assembled from local clubs. The "winners" are invariably nodes inside cities where several interstates come together, where interstates meet major local arteries, or where "mixing bowl" configurations exist (in Boston, Chicago, Dallas, Houston, Los Angeles, Minneapolis, New Orleans, New York, Seattle, and Washington, DC).[42]

Air Pollution

Internal combustion and diesel engines have received much attention ever since they were identified as major contributors to the unhealthy state of cities. Extensive investigations have resulted in a specific description of the problem and a definition of appropriate standards, followed by legislation to provide controls and improvement. The attention focuses on seven pollutants: ozone

---

[42] As reported in *The Urban Transportation Monitor,* September 29, 2000. See also the November 26, 1999 issue.

Space Needs of the Automobile

The normal dimensions of a parking space are 9 × 20 ft (2.7 × 6.1 m). Thus, each automobile, even when stationary, consumes 180 ft² (16.7 m²) of surface, and more because an empty space should await a vehicle at the end of each trip.

When the vehicle is moving on a highway with 12-ft (3.7-m) lanes, safe spacing has to be maintained between units, which may be, say, 3 carlengths. Then the space consumption is 12 × 80 ft = 960 ft² (89 m²). Since the average occupancy of commuting automobiles is 1.1 persons, the per capita surface requirement is 873 ft² (81 m²).

For the sake of comparison, the 50 patrons carried by a bus on the same highway will account for only 38 ft² (3.5 m²) each. If the car carries 3.5 people, as is frequently the case on recreation-based trips, and the large bus is loaded with only 20 riders, the respective per capita space consumption is 274 and 96 ft² (25.5 and 8.9 m²)—still a significant difference.

Rail transit can achieve an even higher density of useful service, but too many different parameters enter into the calculation to offer a clear comparison with roadway-based services.

(O_3), carbon monoxide (CO), nitrous oxides (NO_x), sulfur dioxide (SO_2), lead (Pb), volatile organic compounds (VOCs), and particulate matter (particularly microscopic particles, PM_{10}).

Remedial programs, extending now over some 30 years, have achieved significant results, primarily due to improvements in the engines of vehicles. These efforts started with positive crankcase ventilation (burning blow-by gases), followed by engine modifications to improve exhaust characteristics, sealing of gas tanks against evaporation, and adding catalytic converters (blowing hot exhaust gases over porous material). As old cars have been gradually replaced by cleaner models, and other programs have taken effect, the quality of air in American cities has improved measurably. However, a quarter of Americans still live in areas where at least one of the main pollutants remains at an unhealthy level. Six of the largest metropolitan areas are failing to meet the standards for three pollutants or more (New York City, Los Angeles, Chicago, Phoenix, Pittsburgh, and El Paso). Denver has the distinction of being the first large city that has moved from a most severe state of pollution to full compliance with the Clean Air Act.[43] As can be

---

[43] US Environmental Protection Agency data as reported in the *New York Times*, September 1, 2001.

seen in Table 5.5, significant progress has been made in the last decade in improving air quality across the country, but a zero emission level is still far away.

The evidence is quite clear that it would be possible to eliminate air pollution from mobile sources by adopting new engine technology, switching to cleaner fuels, or substantially upgrading existing engines. This would involve somewhat higher costs of manufacturing and operating vehicles, certainly at the beginning, but there appears to be no urgency or political will to take the next step at this time. The danger is that increasing automobile use may erase the gains that have been made so far.

Ultimately, the answer lies most likely in the full-scale adoption of alternative power sources and new types of engines. This encompasses at least electric motors, gas turbines, fuel cells, and steam engines—all of which are under intensive development and current testing. There is not much point in speculating at this time as to which of these possibilities will be the most appropriate choice, except to urge these efforts forward and await the engineering results. The hybrid vehicle (batteries and electric motor, recharged by a small gasoline or natural gas engine running at a steady, efficient rate) is the leading candidate today, with many such cars and buses on the streets already.

Table 5.5   Vehicle Emission Rates, g/mi

Emissions	Gasoline-Powered Vehicles		Diesel-Powered Vehicles	
	1990	2000	1990	2000
Light-duty vehicles				
Hydrocarbons	3.08	2.16	0.73	0.63
Carbon monoxide	24.68	19.28	1.68	1.57
Nitrogen oxide	1.81	1.38	1.65	1.33
Heavy-duty vehicles				
Hydrocarbons	11.89	5.32	3.30	2.22
Carbon monoxide	131.19	48.67	13.71	11.53
Nitrogen oxide	6.49	4.72	21.05	11.24

Source: Estimates by US Environmental Protection Agency.

Noise

A significant livability problem in cities is the level of noise generated by motor vehicles, particularly if they are heavy, are not well maintained, or have defective mufflers. Even the triggering of automobile burglar alarms can become a nuisance contributing to the overall problem. Noise analyses, therefore, are one of the standard elements included in environmental impact statements of any major project. Corrective actions include not only the improvement of the power plant, but also containment and screening devices, ranging from massive acoustical walls and sound barriers to double-glazed windows for residences near traffic arteries.

Accidents

Heavy vehicles operated at high speeds by basically amateurs (sometimes under an impaired state of alertness) can represent a lethal combination. There are, often enough, property damage, personal injury, and fatalities. Ever since the first automobile-caused death in 1889, when an alighting streetcar passenger was run over in New York City, the toll in the United States mounted alarmingly, reaching a peak of 56,378 fatalities in 1972. Since that time there has been a very encouraging trend toward significant reduction (39,000 deaths in 2000). Fatalities will never be eliminated entirely (among other causes, there always has been a component of suicides), but further steps certainly can be taken. Recognizing that accidents can be caused by inadequacies of the roadway or failure of the vehicle, or can be mostly the fault of the driver, the latter represents the area where programs hold the greatest promise for results.[44] In the 1970s, accidents associated with alcohol use represented more than half of the total; today that number is below 50 percent. It is well to keep in mind that the safety record in the United States on a per-mile basis is quite good, if not the best in the world. In many other countries the volumes of injuries on roads and streets have tragic dimensions.

The most frequently used measure of overall traffic safety is the number of fatalities per year per 100 million vehicle-miles

---

[44] A useful summary of recommended safety programs is *The Traffic Safety Toolbox* (Institute of Transportation Engineers, 1999, 2d ed., 301 pp.)

traveled. This indicator currently in the United States is 1.7, down from 20+ in the early 1920s.[45] As an international comparison, another indicator is the number of fatalities per year for each 10,000 vehicles. In the United States, it was 2.2 in 1996, while the worst record was held by Ethiopia with 151.

Depletion of Petroleum Resources

Gasoline is distilled from oil that has been generated through slow natural processes over millions of years in underground strata. No matter how many new discoveries are still being made, the overall supply is finite. Since oil is also the base for a multitude of vital petrochemical products, it can be argued that the time has already been reached when this material has to be conserved for better purposes than burning up. The United States is increasingly concerned, if not yet about the total supply, then certainly about dependency on foreign sources and its ability to tap them. Dramatic steps have been taken to assure that this currently crucial flow will continue—ranging from the building of massive reserve storage facilities to going to war. Eventually, and perhaps soon, a switch to other fuels and transportable energy sources will have to take place. There are no insurmountable obstacles to doing that, except higher costs, which may be a temporary condition anyway.

Disposal Problem

Every new vehicle will wear out, and, while some used cars may be exported to other countries, the removal of wrecks from the environment remains a challenge. Scrap and junkyards do their job, but they too can become eyesores unless they are carefully controlled. Every year 10 to 13 million motor vehicles have to be scrapped in the United States.[46] During periods when scrap metal has a limited market, hulks tend to be abandoned along streets and roadways.

The suggestion has been made repeatedly that each unit should be so designed that its eventual separation by type of material can be accomplished easily and that each new purchase be accompanied by a deposit for disposal that becomes available at the end of the vehicle's useful life.

---

[45] Data from National Highway Traffic Safety Administration.
[46] U.S. DOT, Bureau of Transportation Statistics, 2001.

Economic Inequity

While the purchase price of a second-hand automobile can be quite low, the acquisition of individual vehicles with all the associated operating costs is still beyond the means of many households. There are poor neighborhoods in almost every American city where such conditions prevail—particularly where female-headed households with dependent children are seen frequently. Since the structure of contemporary cities and the distribution of activities largely reflect the assumption that most people will travel by car, those without automobiles are effectively barred from full participation in normal urban life.

Dominance of the Transportation Field

The overwhelming presence and popularity of private automobiles in American communities have created a distorted inventory of transportation capability. The severe reduction in transit use during the last half-century has brought a virtual disappearance of public service in many places, at least far below what could be considered a responsive level.

Separation and Isolation

The current patterns in American community development are characterized by low-density single-family residences that accommodate in separate neighborhoods people of the same social/economic/ethnic types. Despite many reasoned analyses that this is a tendency leading to a segregated society, the trend persists, and the automobile has made it possible. Since most nonwork activities concentrate on the immediate family and the dwelling, participation in public efforts (attending meetings and community gatherings, taking part in communal efforts, and even voting) declines, and a growing disinterest is experienced by members of the public. The public realm is losing its central place in our daily lives.[47] The private car then becomes not only the general instrument toward such a state, but the actual embodiment of segregation. The sealed capsule containing usually one person limits all contacts, except for the radio, cell phone, and passive observation of the street scene.

[47] This argument is advanced most strongly by R. D. Putnam in *Bowling Alone: The Collapse and Revival of American Community* (Simon & Schuster, 2000, 541 pp.).

This article was originally published in the *New York Times,* Sunday, February 19, 1989.

From Bad To Worse: Angelenos' Traffic

Robert Reinhold

LOS ANGELES, Feb. 18—It has never been easy, but this has been one of the worst months in memory for the harried freeway drivers of Los Angeles, where the price of living near warm palm-fringed beaches, lovely mountains and deserts is that one can hardly get there.

Just Friday, a tanker caused a colossal nine-hour traffic jam by jackknifing on the San Diego Freeway just before the evening rush. That was worse than when another truck accidentally dumped 64,000 pounds of manure on the Foothill Freeway on Tuesday evening and when 1,000 head of sheep made a mess of things at the Las Virgenes Road exit of the Ventura Freeway the other day. And if all that were not enough, Angelenos last week faced what so many of them moved here to get away from: snow.

To be sure, traffic woes are not new here. But the consensus is they have become worse in the last year with Southern California's relentless population growth. Traffic has become the great equalizer, the common denominator of all Angelenos.

Traffic Controls People's Lives

Traffic is Topic A, at the dinner table at home, in the elevator at work. Occasionally some sensational news will briefly seize local attention, like the separation of Jane Fonda and Tom Hayden or the $22 million jury award to Rock Hudson's lover. But it is traffic that controls people's lives and is the source of jokes and daily tales of horror and heroism. Friends and neighbors trade stories and secret shortcuts.

"You just have to reorient your whole way of living and doing business," says Alison Grabell, a former Foreign Service officer who moved here six years ago from Washington. "It's mad, almost chaos, just overwhelming."

But Ms. Grabell has made the ultimate adaptation; she works at home, commuting from bedroom to study.

Few can do that, though, and Angelenos have devised elaborate adaptations. People time their breakfasts to enter the freeway at just the right minute, knowing a short delay can double their commuting time.

Coffee and Toast in Car

Jennifer Rodes, a graduate student and French tutor, can be seen in her Toyota Tercel on the Santa Monica Freeway, with coffee and toast, preparing her lessons in the front seat when traffic slows, sometimes changing clothes in the car.

Hope J. Boonshaft-Lewis, who does public relations, says she finally "broke down" and bought a cellular phone, which she often uses to cancel appointments she cannot make because of traffic.

Anyone driving from downtown Los Angeles to Orange County, about 40 miles to the south, is best advised to bring a snack and a thermos of water.

For all that, when they are clear, the freeways of Southern California are marvelous for getting around, knitting together a vast area into one metropolis. It just takes a little ingenuity, and luck.

Knowing the Traffic Patterns

"I have made friends with the freeways," says Lynn Tuite, who commutes about six miles from Pasadena to the University of Southern California south of downtown. The trip can take from 15 minutes to an hour, depending on the traffic and time of day. "Certain lanes move faster than others and I know where they are now. I do a lot of lane jumping. I just make up my mind it's going to take an hour and a half to get someplace that ordinarily takes a half hour."

One man, a college teacher, uses the Santa Monica Freeway for a 12-mile commute from Westwood on the West Side of Los Angeles to downtown. He knows the traffic patterns as well as he knows his wife. "I know that if I leave anytime before 7 A.M., it takes just 20 minutes," he says. "If I leave after 7:10, it takes 35 minutes." He brushes his teeth and does his dental flossing in his Ford Mustang convertible. One day a car of smiling young women honked and they waved; when they passed he read their bumper sticker: "Dental hygienists do it better."

One reason the traffic is so bad is that there is little public transportation. Another is that rising housing prices have forced thousands to live on the edge of the Mojave Desert, or deep in the "Inland Empire" near Riverside and San Bernardino, forcing commutes to Los Angeles and Orange County, where the jobs are, of 50 or 60 miles each way.

Ellen Bendell lives in Lancaster in the once-barren Antelope Valley north of Los Angeles. She must get up at 4 A.M. for the 62-mile commute to her job in Burbank near downtown Los Angeles. "I leave when it's dark, and I get home when it's dark," she says. "I don't remember what my house looks like."

All of this has spurred renewed efforts to find alternatives. A subway is under construction downtown. Officials from throughout the region are considering methods to move jobs closer to where people live. Gov. George Duekmejian held a meeting on traffic in Sacramento on Feb. 8. And last Wednesday the Transportation Committee of the Los Angeles City Council gave tentative approval to Mayor Tom Bradley's proposal to limit heavy trucks on city streets during peak hours. Also, the city is offering to pay up to $5,000 per vehicle to companies that buy vans for employee van pools, and the council is considering a plan to compel all large employers to pay $15 a month to subsidize their workers' bus passes.

No one is more sensitive to commuting problems than William E. Bicker, the Mayor's transportation aide, who is the target of what he calls "every conceivable Buck Rogers transit scheme." He gets letters from many former New Yorkers who live here saying the solution is a subway system like New York's. The elderly suggest a return to the streetcars that used to operate until the tracks were torn up 25 years ago.

One man offered a scheme that would limit rush hour to commercial vehicles, cars with two or more occupants and single-passenger vehicles whose owners pay a $2,500 annual fee for the privilege of driving alone. Another man sent in drawings for an upside down monorail that would hang from cables, move at 300 miles an hour and carry 100 passengers in each car.

But driving is such an ingrained way of life here, that few seem optimistic about improvement. "Every year it gets worse and worse," says Arthur Groman, a lawyer who lives in Beverly Hills. "But my strong feeling is that the 'I' principle will prevail and people won't cooperate. Angelenos are so married to their autos they will not ride the subway. They cannot understand they may have to park and walk three blocks or be at the mercy of someone else's driving."

Status of Automobiles

The production of motor vehicles is the largest single manufacturing enterprise in the United States. General Motors is the largest corporation of any kind in the world, and most major carmakers have plants here to serve the huge North American market. In terms of employment and gross national product, this industry with all its associated services represents about a quarter of all business. It is an operation concentrated in large units, with regional assembly plants and an elaborate system of distribution and service establishments in every community. In 1970 there were 34,000 franchised dealers, 28,000 used car sales places, more than 200,000 gasoline stations, and 100,000 enterprises in car repair.[48] In some extreme instances, any programs to curtail the presence and use of automobiles in communities have been interpreted as damaging to the national economy.

Automobiles are an intrinsic part of American life. Daily operations revolve around the use of this means of mobility, and every new development and construction project has to recognize its presence.

A diversity of models is available that should respond to every budget and taste. Since differences in styling are not too pronounced today, major attention is devoted in the market to reliability and affordable comfort features. The streamlining of the distribution and sales systems ensures quick consumer satisfaction, and the next steps are likely to be procedures that allow purchasers to state their individual preferences in appointments and features, which are sent back to the assembly lines for quick delivery of the exact desired type of vehicle.

Safety devices including front and end units that collapse, absorbing impact energy; automatic seat belts and air bags; headrests; a strong passenger compartment; antilock brakes; and a number of other elements are now standard on most models. Modern cars are easy to drive and are somewhat forgiving of driver errors.

Computer-based control and performance-monitoring systems enhance efficiency and assist with maintenance tasks. On-board navigation systems are available and may become common features as well. Intelligent regionwide information and guidance

---

[48] *America's Highways,* op. cit.

systems that appear to be around the corner should expedite way-finding on most trips, even though they will not be able to solve congestion and overload problems. All this makes automobile use safer, more user friendly, and attractive.

The opportunity to observe the car-buying habits of the American public over half a century leads to a few broad conclusions, not necessarily inspiring ones. The principal finding is that big cars are much preferred over small units. While everybody knows that large vehicles consume much space and fuel, are rarely filled to capacity, and impose constraints on others, buyers tend to ignore these considerations. There have been periods of fuel shortages and economic downturn when smaller automobiles do better in the showrooms, but the practice does not last. It is always back to the largest vehicles that can be afforded, with scant regard for societal responsibility by individual consumers. If asked, the answers are that one's purchase does not make a difference within the huge general fleet, that safety is a major personal concern (a large car will crush the other guy), that you never know how much groceries and equipment will have to be carried, and that there really isn't a fuel shortage. The recent popularity of SUVs, which are actually trucks, illustrates this contention. It can be argued that if fuel costs were higher, these habits and the orientation toward larger vehicles would be modified. Perhaps so—assuming that the prevailing political attitudes could be somehow overcome—but such increases would have to be very large to result in measurable changes in buying and usage patterns. There have been repeated instances where gradual and marginal increases in bridge and highway tolls have made no difference at all in volumes because the increases have been below levels that would be significant enough to register with the customers and affect their behavior.

Curiously—because of their ubiquitous presence in American communities—having an automobile confers no special status on the owner. This is certainly not the case in just about all other countries, where car ownership is a major step in establishing and maintaining one's status in society. In the United States, a useful gambit to open a conversation with a stranger may still be to ask, "What kind of car do you drive?" but only if the answer is "None" will significant interest be generated. Assuming that the respondent is not very poor, he or she must be then some sort of nonconformist or eccentric, and it would be wise to find out soon whether it is safe to continue the conversation.[49] Indeed, there is a bit of reverse snobbism associated with nonownership, implying that a superior lifestyle is being practiced (supported by an occasional rental, since it is not possible to avoid cars entirely).

A Lamborghini at the curb in Los Angeles.

It is, of course, possible to score points of social standing via a personal vehicle, but significant and deliberate steps have to be taken by entering the luxury car or special accessory market. This market has existed from the very beginning, and it remains strong by catering to financial abilities in tiers of classes that are well understood by everybody and can lead to stratospheric levels. A regular Cadillac on the street will not turn any heads, but a yellow Lamborghini will stop everybody in their tracks.

Application Scenarios

Automobiles and street networks are all around us every step of the way in our daily lives, and we now have communities where, practically speaking, this is the only mobility system in place. Not an ideal situation by any means, but one that works (with significant frictions and deficiencies) and is preferred by the overwhelming majority of the public. Information on how to accommodate this mode—how to design highways, lay out local streets, provide parking facilities, regulate traffic flows, and incorporate physical safety features—is readily available in technical publications and need not be repeated. Neither will the strategies discussed here include the possible rebuilding of city districts to accommodate the automobile or the procedures for building new completely car-oriented communities or districts.

[49] The author is somewhat of an expert in this field, having proudly survived without owning a car for decades as a Manhattan resident, but succumbing eventually due to family and second-home obligations. When in Rome, one cannot not do as the Romans do forever.

The task here is to go one step further and to search for and evaluate possibilities for more responsible and effective approaches of dealing with the opportunities and the existing problems. The goal is not to eliminate the automobile and restrict it universally, but to make it behave effectively and responsively under a diverse set of conditions.[50] Surface traffic conditions can become worse yet; if present trends continue unabated, real mobility crises in many communities are to be expected. As useful and as attractive as the car is, livability and the ability to function in many urban environments are at stake if some corrective actions are not taken.

The problems and possible remedial actions are not unknown; they have been identified many times over. The challenge is to evaluate their suitability and consequences, and to generate the political will and public acceptance needed for implementation. This will be done next, searching for applicability in specific instances for different types of districts, times, and purposes.

At this time, the remedial and ameliorative possible actions have been grouped into the two following classes.[51]

Travel Demand Management (TDM)

TDM measures encompass all those possible actions that would achieve greater efficiency in the use of travel services and facilities (supply) by adjusting or minimizing the demand for automobile operations.[52] Since it has been obvious for some time in

[50] Several analyses have been published that outline various strategies to minimize dependency on automobiles, particularly by addressing land use planning possibilities. These include: R. Ewing, *Transportation & Land Use Innovations* (APA Planners Press, 1997, 106 pp.); D. Carlson, *At Road's End: Transportation and Land Use Choices for Communities* (Island Press, 1995, 168 pp.); K. Alvord, *Divorce Your Car!* (New Society Publishers, 2000, 305 pp.); *Evaluating the Role of the Automobile: A Municipal Strategy* (City of Toronto, 1991, 191 pp.); and R. T. Dunphy, *Moving Beyond Gridlock: Traffic & Development* (Urban Land Institute, 1996, 100 pp.).

[51] Regulations in the United States under the federal Clean Air Act Amendments of 1990 (CAAA) require that all transportation studies include an examination of Travel Demand Management/Transportation System Management measures before any capital-intensive projects are considered.

[52] A useful summary is found in *Implementing Effective Travel Demand Management Measures: Inventory of Measures and Synthesis of Experience,* prepared by Comsis Corporation for the U.S. Department of Transportation, September 1993, DOT-T-94-02. *Transportation Research Records* frequently publishes special issues on TDM/TSM or associated topics. See, for example, #1346 (1991), #1360 (1992), #1394 (1992), and others. Another reference is E. Ferguson, *Transportation Demand Management* (Planning Advisory Service 477, 1998, 68 pp.).

American cities that the growing trend and demand for increased mobility and expanding travel needs cannot be satisfied by continuing infrastructure expansion over an extended period in most communities, solutions and relief could be sought in the rational use of available facilities or in the expansion of more effective high-density modes. Mobility and accessibility have to be maintained, but not necessarily by the same traditional means. Programs can be developed that attempt to change demands on the surface roadway system or modify modal choices by changing user behavior. The intent is to lessen the total loads on roadways or to shift travelers from cars to modes that can perform with greater overall efficiency.

1. Improved alternatives to the single-occupant vehicle (SOV):
   - Transit improvements (see Chaps. 8 through 15)
   - Carpooling
   - Vanpooling (see Chap. 6)
   - Pedestrian and bicycle facilities (see Chaps. 2 and 3)
2. Incentives and disincentives:
   - Employer support measures (see Chap. 6)
   - Preferential high-occupancy vehicle (HOV) treatments (see Chap. 9)
   - Ride-sharing incentives (see Chap. 6)
   - Parking supply and price management
   - Tolls and congestion pricing; user charges
3. Alternative work arrangements:
   - Variable work hours; alternative work schedules
   - Telecommuting; work-at-home options

Transportation System Management (TSM)

TSM programs strive to adjust existing roadway networks and elements to improve their capacity and facilitate traffic flow without incurring major capital investments. This assumes that the demand may remain approximately the same, but that a higher level of performance and better safety can be extracted from the infrastructure already in place. There are dozens of TSM methods and programs, including both physical features and operational approaches that constitute the arsenal of traffic engineers toward the improvement of flow conditions. Thus, they are regular procedures[53] that largely attempt to deal with the symptoms of the problem—to expedite traffic flow. They are undoubtedly necessary, but most do not fall under the category of approaches discussed in this book.

1. Expediting traffic flow:
   - Improved signage to improve safety and cut unnecessary travel
   - Pavement markings to guide movements and enhance safety
   - Coordinated traffic signals to achieve continuity in movement
   - Channelization of traffic lanes to control flow
   - Left and right turn lanes and traffic signals to expedite movements
   - Keeping lanes open at intersections (daylighting) to increase processing ability
   - Intersection widening and streamlining to remove friction points
   - Computer-based traffic control to expedite all operations

2. Monitoring and metering:
   - Ramp metering signals to avoid overloads on vital facilities
   - Surveillance systems to monitor traffic conditions (particularly on highways)

3. Giving attention to public services:
   - Bus priority signals to expedite high-density services
   - Bus turn-out bays to remove blockage of lanes
   - Control of taxi operations; provision of taxi stands to minimize cruising

4. Controlling parking:
   - Strict enforcement of parking regulations to minimize entries

[53] J. L. Pline (ed), *Traffic Engineering Handbook* (Institute of Transportation Engineers/Prentice Hall, 1999, 704 pp.), and other traffic engineering reference works.

- Parking permits for local residents/workers to preserve livability of districts

5. Adjusting the use of the network:
   - Reversible traffic lanes to balance supply and demand by time of day
   - One-way streets to increase aggregate throughput

6. Providing responsive management of operations:
   - Deployment of traffic police to discourage irresponsible behavior
   - Incident management programs to clear obstacles quickly

7. Upgrading safety features:
   - Rumble strips along pavement edges or approaching stops to warn motorists
   - Motorist information systems to avoid unnecessary travel
   - Public education to foster responsible behavior

8. Restricting automobile use or entry:
   - "No drive" days (by selective indicators, such as license plate numbers)
   - Auto-free zones; auto-restricted zones to allow important districts to operate

Most of the items on the TSM list (not intended to be exhaustive) are quite obvious and are purely traffic engineering considerations. A few, however, are of more fundamental importance and are discussed on the following pages, primarily under ITSs and vehicle restriction programs.

Components of a Potentially Reformed Physical System

This section consists of a series of reviews of various programs at different levels of economic/social/functional/institutional/political feasibility that could be adopted by communities to reform the current state of affairs of almost unbridled and somewhat irrational use of the automobile. In effect, it is a menu from which choices can be made. Each program has some features that would change prevailing habits and upset some cohort of the population.

Yet, most of them can be found in one place or another (Europe, mostly), and, as the situation tightens up more in North America, some may have to become mandatory.

Sharing of Vehicles

The fundamental fact of automobile use in the United States is an average vehicle occupancy rate of 1.6 persons (1.1 for work trips, 2.1 for social-recreational).[54] Since almost all cars have five or more seats, this means that two-thirds of the actual operating transportation supply remains empty. Large vehicles consume space and resources carrying very little. To put it another way, each year 1.4 trillion household vehicle-miles are generated nationally, resulting in at least 7 trillion[55] seat-miles, of which only one-third is actually used. These are numbers beyond the comprehension of anybody except federal budget experts, but, since the average vehicle trip is 9 miles long and the total population is 281 million, each U.S. resident could take 2000 trips per year (or 6 each day) without affecting the existing loads on the roadways at all (but not necessarily reaching the desired destination).

Clearly, this is an example given only for the sake of its dramatic image; it is not a workable scenario. However, it does point to the fact that, if there is a transportation crisis, it is not one of total supply, but rather of its useful distribution. It has been long thought that there might be some reasonable ways to tap into this capacity, i.e., to increase the average load factor.

CARPOOLS

A group of travelers, usually commuters to and from work, can make arrangements on a daily basis to share a vehicle. The prerequisites, of course, are that their origin and destination points are approximately the same, that they travel at the same time on the same schedule, and that they are willing to share a small container with each other every day. This is a program that all levels of government have promoted for several decades because the

---

[54] The data in this section refer to 1990, as assembled and published in the *Nationwide Personal Transportation Survey* by the U.S. Department of Transportation/Federal Highway Administration, 1992, FHWA-PL-92-027.
[55] In American usage, a number with 12 zeros.

potential reduction of vehicle loads on streets is quite obvious. Unfortunately, while 19.7 percent of all commuters relied on carpools in 1980, usage had dropped to 13.4 percent in 1990, and it went down further to 9.4 percent in 1999,[56] with only about one-fifth of these patrons in cars with three or more persons.

Carpools can be formed by individuals privately (i.e., people who know each other and live and work in the same districts), or public agencies can act as brokers to assemble compatible groups or at least provide a means of information exchange. Arrangements can also be made by employers for their own workers. (These possibilities are reviewed further in Chap. 6). Sometimes special programs are instituted by transportation management associations (TMAs), which are local organizations encouraged by federal transportation agencies to act within communities and districts assisting employers and commuters to expedite daily travel. The scope of their activity may encompass any program that would rationalize transport operations within their area, ranging from the running of buses to the distribution of maps; it can certainly include assistance toward carpool operations.

The members of the carpool can alternate in driving their own vehicles, thus having no money change hands; or they can rely on one of the members and his or her car, for which appropriate compensation would be paid. It is important to note that under extant regulations vehicle owners may not sell rides to any members of the public who may be standing along the roadside. That would be the offering a commercial service on a public right-of-way, which requires a franchise from the government. Carpools with regular participants do not fall in that category. The gain to each member is that the commuting costs in a shared vehicle will be substantially lower than for each using a separate vehicle. At the same time, the carpool does provide more privacy, perhaps good companions, more comfort, and greater responsiveness than regular public transit services.

It appears that people will use carpools when they have to, but it is not the preferred transportation choice regarding personal convenience. The constraints are that there cannot be much vari-

---

[56] U.S. Census data; the source of the 1999 estimate is the U.S. Department of Housing and Urban Development, American Housing Survey. (The official 2000 U.S. Census figure was not available at the time this book went to press.)

ation in terminal points or schedule from day to day for any rider, unless all the members agree. Difficulties are encountered if somebody has to run personal errands, needs a vehicle in the middle of the day, or has to work late or reach other points at the beginning or end of the day. Emergencies cannot be readily responded to. Freedom of mobility is significantly curtailed, and that is a quality that Americans value highly. It also happens that the compatibility of a group, confined to a tight space each day for extended periods, may unravel. (Should the windows be open or closed?)

To make carpools work well, there is a set of conditions that should be reasonably satisfied:

- The work destinations have to be clustered, preferably in the same building for the same employer. Large firms or government agencies are promising venues. Likewise, the homes should be in the same neighborhood, which is likely only if the employment place is large and many colleagues reside in the same neighborhood.

- The jobs should have very regular schedules, with overtime work and trips to other places rare occurrences.

- The trip itself should be reasonably long, because otherwise the time needed to assemble and distribute the riders becomes an excessive proportion of the entire twice-daily operation.

- Carpools merit every preferential treatment on the street system, as is already the case with the use of special HOV lanes and sometimes conveniently located designated parking spaces.

- If carpool members are not assembled along a route by the driver, a marshalling place is needed where individual cars can be parked for the day, transfers from buses can be made, and standing space is available. If this is an improvised action by each carpool, local frictions may result by preempting parking spaces, double parking/standing, etc. A better solution is to have designated and reserved places with the appropriate layout. Edges of large shopping center parking lots, for example, provide such opportunities since they are not likely to be needed by shoppers during regular working days. TMAs can act constructively in this field.

- It is necessary to have reliable backup systems in place so that personal emergencies can be accommodated at any time and late workers can get home. This can usually be provided by local taxi companies, but other arrangements are possible as well. There are a number of instances where the employer or an organization ensures such service without excessive costs to the user (guaranteed ride home).

- Since carpools can only operate successfully in regular and repetitive situations, the only other conceivable suitable situations, besides commuting to work, might be connections to large institutions with a regular clientele (such as universities and medical centers).

Carpools are a rather intimate and personal approach to operate communal transportation. They require a great deal of social network support and compatibility among members, unless they occur on a completely casual basis, with users entering without prearrangement and not saying anything to each other (see the section on organized hitchhiking).

There is a significant near-future possibility that carpools may gain a new lease on life, despite the current discouraging trends. With enhanced communications systems (such as the Web and e-mail) and more powerful and individually accessible computers, it might be possible to maintain information systems that assemble information on trip requests and match that with service availability even on a daily basis. That would represent a dynamic system with quick responses in the search for appropriate rides, surpassing the current rather inflexible set of arrangements. This is not a simple task because the trip needs of participants change from time to time, and keeping the information current takes some effort. The necessary data have to include at least the following for each potential participant:

- Name, address, and telephone number
- Origin and destination locations
- Daily schedule
- Personal preferences (nonsmoker, coffee drinker, etc.)

A variation of the carpool concept is the *vanpool*. This, however, is not only a matter of utilizing a larger vehicle, and features of a public service are included. Therefore, this submode will be discussed further under Paratransit.

COOPERATIVE CAR OWNERSHIP

Of the 168 hours in a week, most family sedans are used a few hours per day at best; they may be quite busy on weekends or sit completely idle. If the occupancy rates of automobiles are low, their hours of actually being in useful motion are even lower. It is, therefore, possible to envision a system under which small groups of individuals or families on a block or within a neighborhood own and use a few cars jointly. When needed, they draw a vehicle from this pool by prearrangement, as units are available. A cooperative organization would seem to be appropriate, but a limited-profit commercial enterprise would also be workable.

Under this scenario, the total fleet located in any given area would be reduced, thereby achieving greater overall efficiency and reducing parking needs. The total vehicle-miles generated may not necessarily be lower, but they might be because there would be an implied encouragement to use public transit more. The greatest benefits and best feasibility would be found in high-density urban districts.

There are, however, serious practical problems facing this concept:

- Many people regard their automobile as a personal item, sometimes an extension of their personality. (Would you lend your overcoat to somebody else?) To jump in a car on impulse would not be possible.

- Since trip demands tend to concentrate at certain times, a vehicle may not always be available when desired. This will be seen as a serious constraint on individual mobility by some. If a federated system of local organizations were to be established or a large rental firm were to be in charge, some shifting of units geographically according to demand may be possible.

- A management, record keeping, and cost allocation system would be necessary. This may be a personal chore assumed by somebody, dependent on mutual trust, or the operations would have to be placed on a business-like basis. Somebody at least has to keep the keys. Repairs and maintenance have to be arranged for. Compatibility among the participants would be a significant element toward successful operations.

- Parking spaces would have to be created or designated so that a vehicle can be readily found, used, and dropped off.

It would have to be returned to the proper place at the time promised in a good and clean state.

All this does not bode well for eager acceptance of cooperative car ownership in American communities. Yet, the concept has merit, and has worked in many instances. There are several hundred such organizations that operate in almost 500 cities in Switzerland, Germany, Austria, the Netherlands, Great Britain, Denmark, Sweden, Norway, Italy, and probably in quite a number of other places with less visibility. In the United States, the examples are fewer, but they do exist—in Seattle; Chicago; Portland; Boulder, Colorado; and Riverside and the Bay Area in California.

ORGANIZED HITCHHIKING

The low occupancy rates of automobiles mentioned earlier suggest another (theoretical) approach. Fill those empty seats! (This was actually a popular slogan during World War II in the United States.) The extreme form of this scenario would be binding legislation that every car in motion carry a sign indicating its destination and that every potential traveler standing along the roadside have the right to flag down any car going his or her way and get a ride. Some money would have to change hands.[57]

Before cries of outrage are heard, let us state immediately that this is not an acceptable concept in a free country for any number of reasons. However, therein lies a glimmer of a solution, and it has been practiced under emergency conditions with voluntary arrangements even in the United States—not only in wartime but also during the subway strike in New York City. Such a system would also resemble very much the spontaneously self-generated jitney operations in many cities (see Chap. 6); there would be serious problems with keeping such operations under civilized control. Not the least of the concerns would be the possible impairment of the already fragile public transit services. Just as in all other instances where strangers enter somebody's private space (regular hitchhiking, for example), security issues loom very large. Some legislation and local regulations would have to be amended; administrative and supervision systems would have to be implemented. Above all, confidence of prospective riders and the reliability of the service providers would have to be established.

[57] See F. Spielberg and P. Shapiro, "Slugs and Bodysnatchers," *TR News,* May–June 2001, pp. 20–23.

Recognizing that this concept in its full form has almost no feasibility at this time in North American communities, it serves nonetheless to highlight the point that much transport capacity actually exists. It is not always used rationally, and, if mobility truly breaks down due to surface congestion, relief measures can be found, even if they are draconian.

To some extent, such practices can be observed at a few places where extensive HOV lane systems are in operation (Washington, DC; San Francisco; and Houston). This is casual carpooling, or "slugging" in the vernacular, involving people who wait along curbs at strategic locations to provide enough riders so that otherwise single-occupancy cars can enter HOV lanes. The interesting feature is that no money changes hands—the hitchhiker gets a free ride, and the motorist can use the fast preferential roadways.

These improvised and eminently logical responses to transportation demand under specially created conditions do offer hope that nonconventional responses are possible and may point toward solutions that have so far escaped the attention of official service providers or even are contrary to their established practices.

STATION CARS

A program that has received considerable attention, is frequently encountered in Europe, and has been implemented on a pilot project basis in a few places in the United States is the so-called station car concept.[58] The idea is that there would be a fleet of vehicles associated with a rail station that could be picked up by commuters to do errands at the destination end, or, at the home end, to drive home, be kept overnight, and then be returned to the station next morning. Obviously, a management system has to accompany this rolling stock inventory, but it is not a particularly difficult task with the use of magnetic cards and special keys. It is similar to a communal bicycle system, as long as simple regulations are in place as to who may use the vehicles, what insurance and maintenance responsibilities exist, and how fees are to be collected. The concept might also be applicable to other instances where repetitive access patterns exist—campuses, business parks, airports, etc.

Again, the benefits to the community would be that the total number of vehicles would be reduced, families might not have to purchase a second or third car, and the vehicles themselves would

[58] There is even a National Station Car Association with its own Web page.

be environment- and people-friendly. Electrically powered two-seaters would be particularly suited for this purpose.

European "invalid's car" that may be a model for a station or urban car.

A number of rail stations of the Bay Area Rapid Transit District (BART) have implemented station car programs (CarLink) utilizing small electric Ford vehicles (the Think model) under the management of the Hertz car rental company, or relying on Honda natural gas vehicles. Participants have to subscribe to the program, distinguishing between home-side and work-side users. Similar programs are in operation serving districts in Seattle (Flexcar), Boston (Zipcar), and Portland, Oregon (Car Sharing).[59] Other cities, including Chicago and Atlanta, are considering station cars as well.

The Washington Metropolitan Area Transit Authority (WMATA) is in the process of initiating an extensive communal car-sharing program at its suburban Metrorail stations geared to customers who arrive at these stations by rail and need convenient means to reach local destinations to conduct business and then return to the station. It would be managed and operated by a commercial car rental or system management company and take advantage of Metro-owned parking. Customers would enroll on a subscription basis, allowing reservation of vehicles even for short periods and at hourly rates below usual car rental tariffs.

Parking Management

Every car has to be placed somewhere at the end of each journey, and, if the parking supply is restricted or limited, the incentive to undertake a trip in the first place may be curtailed.[60] This concept

---

[59] *The Urban Transportation Monitor,* May 25, 2001, p. 3.
[60] The planning and design of parking facilities are covered by most standard traffic reference books, and there are special publications. Among the latter, a useful handbook is M. C. Childs, *Parking Spaces* (McGraw-Hill, 1999, 289 pp.).

has been employed for some time either by implementing strict obstacles to the creation of parking spaces in any district or by imposing high fees. Paris; Vienna; London; Singapore; Stockholm; Tokyo; Boston; Portland, Oregon, and quite a few other places have had extensive parking space management (i.e., restriction) programs in place for years.

This is certainly a negative measure, albeit an effective one, opposed by most motorists and business entrepreneurs who believe that their customers will only come if parking is available. This debate has been going on for decades. At one end of the scale is Midtown Manhattan, where there are no regular (legal) curbside spaces at all but no parking shortage exists for those who are willing to pay the tariffs charged by commercial garages ($15 for the first hour and $29 for 10 hours, including a tax, plus surcharges for better locations). At the other end are suburban shopping centers with huge free lots that get completely filled only on the Saturday before Christmas and the unrestricted free spaces provided for all employees of firms and federal agencies at suburban locations.

Each community has to make a hard choice. The means to implement restrictive policies are readily available by passing strict zoning regulations that bar (or strictly limit) the construction of any parking spaces associated with various types of buildings and by not issuing permits for commercial garages (as is done in London). The contrary program is the building of municipal garages with low fees or subsidies by local merchants in the hope of attracting business (as is done in many American cities).

There is always the issue that demand management through high charges will be seen as discriminatory against the less prosperous members of the community. Any actions toward restricted parking opportunities should be accompanied by programs that enhance public transit services, thereby maintaining reasonable means of accessibility. Qualified low-income patrons who need to use commercial parking spaces can be assisted through direct subsidies, special coupons, or tax credits.

A variation of the same theme is the reservation of curbside spaces in a neighborhood for the use of local residents only, identified by a sticker on the windshield. This generates some legal concerns because a street is a public right-of-way, but the interpretation that neighborhood residents have a priority claim on such spaces does hold. Somewhat similar actions are the awarding of permits to "privileged" parkers—government officials, policemen, judges, the working press, doctors, and others who merit ready access to their destination points in the public interest, assuming that they exercise this privilege only when on duty.

Traffic Calming

In the 1970s, a number of communities in Western Europe became concerned about the excessive presence of motorcars on neighborhood streets and their threats to walking residents and playing children. Out of this emerged the concept of *woonerven* in the Netherlands and *Verkehrsberuhigung* in Germany. Communities in Denmark quickly embraced this approach as well.[61] The German term became translated literally as "traffic calming" in English. This unusual label caused some merriment in the early days, but nowadays the designation is accepted and fairly well recognized among the general public. The program is in direct opposition to what traditional traffic experts have always advocated—expediting traffic flow by removing all obstacles.

The idea of traffic calming is not to bar or eliminate the automobile, but to make it behave responsibly when it is operated in places where adults are at home and children are around.[62] Basically, it means slowing down the car to a walking pace, making the drivers always cognizant that they are moving on streets where pedestrians enjoy a distinct priority. Safety against accidents, injury, and even the perception of possible harm is the basic aim.

This can sometimes be achieved by prominent signs and warnings at the entry to traffic-calmed districts. The standard signs are a schematic pictogram of houses and children and a 30-kph (18.6 mph) speed limit warning. These work in some societies, but are

---

[61] The Danish term is *Trafiksanering;* the French use *moderation de la circulation routière.*

[62] By this time, much literature is available in English: Road Directorate, Denmark, Ministry of Transport, *An Improved Traffic Environment: A Catalogue of Ideas* (Report 106, 1993, 172 pp.); J. A. Yuvan, *Toward Progressive Traffic Management in New York City,* 1996, unpublished Master's thesis, Columbia University; S. Grava, "Traffic Calming—Can It be Done in America?" *Transportation Quarterly,* October 1993, pp. 483–505; County Surveyors Society et al., *Traffic Calming in Practice* (Landor Publishing Ltd., London, United Kingdom, 1994, 199 pp.); and R. Ewing, *Traffic Calming: State of the Practice* (Institute of Transportation Engineers, 1999, 244 pp.).

Traffic-calmed local street in Copenhagen.

not a sufficient guarantee of proper behavior by everybody all the time. For example, stop signs placed frequently along a street would be regarded by some drivers as arbitrary obstacles for no good functional reason and are likely to be ignored. Therefore, physical elements are introduced under traffic calming programs to make it simply impossible for any motorcar to move too fast and with lack of attention to the surroundings. These devices, which can be used singly or most likely in various combinations, include the following:

1. *Speed bumps* (see Fig. 5.4*a*), also known as humps or sleeping policemen in other countries, are low horizontal barriers across pavements that create no disturbance if a car crosses them at low speed, but result in a significant shock if the velocity is high. The shape of the bump can be designed to limit speed to any predetermined level. They are quite common today and are found even in garages as prefabricated units. They have to be placed at intervals along any calmed street to preclude speeding up after crossing the first one.

2. *Raised platforms* (see Fig. 5.4*b*) are similar in function to speed bumps, except that they extend along the length of the pavement. They are particularly useful for indicating places where pedestrian crossings are to be anticipated. Any bump or platform has to allow surface drainage along gutters and not impede bicycles.

3. *Full barriers* (see Fig. 5.4*c*) can be placed across streets or diagonally across intersections to modify a local gridiron network so that shortcuts by through-movement vehicles are eliminated or made most cumbersome. In effect, a sim-

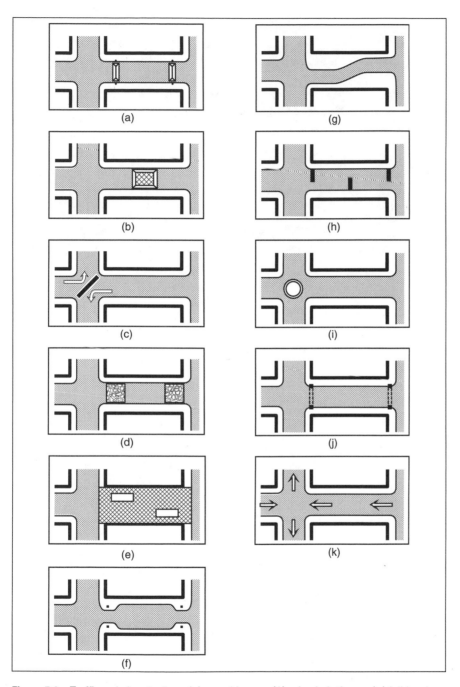

Figure 5.4 Traffic calming devices: (*a*) speed bumps, (*b*) raised platforms, (*c*) full barriers, (*d*) special pavement textures, (*e*) elimination of curbs and sidewalks, (*f*) narrowed lanes, (*g*) staggered alignments, (*h*) chicanes, (*i*) traffic circles, (*j*) gateways, and (*k*) street reversals.

Traffic barriers that are effective in Münster, Germany.

ple maze can be created that represents no problems for local residents but would certainly discourage outsiders. In some countries a solid white line is enough to stop cars; actual fences, landscape elements, bollards, and high curbs represent a stronger statement. Such mid-block closures will create two cul-de-sacs (dead-end streets) for cars, but allow pedestrians and bicycles to move freely. Diagonal barriers at intersections result in loops within the street grid. However, all these barriers and their locations should be designed so that quick access by emergency vehicles is not impeded.

4. *Special pavement textures* (see Fig. 5.4*d*) serve to alert drivers that they are not on a motor traffic street. Paving blocks, brick, and even cobblestones are such indicators of changed character in place of smooth blacktop.

5. *Elimination of curbs and sidewalks* (see Fig. 5.4*e*) is a signal that the entire width of the street is in joint use by pedestrians, bicycles, motor vehicles, and baby carriages. Landscaping and resting places can be placed along the alignment, and automobiles and service vehicles may enter but only at a slow pace.

6. *Narrowed lanes* (see Fig. 5.4*f*) and localized constrictions (pinch points) are devices that slow down drivers instinctively, particularly if heavy vertical elements are placed directly along the side. These may be posts, containers for plants, or walls. A narrow wall or pedestrian island can be placed along the center line of a street, taking away some of the width of adjoining lanes. These constructions threaten to scrape sides of automobiles if care is not exercised in driving, which sometimes actually happens, but is regarded

as a reasonable price to pay for calm driving.

7. *Staggered alignments* (see Fig. 5.4*g*) are deliberate distortions of straight movement paths by adjusting the usual lane configuration (if space is available). Any change in direction, particularly if such locations are visibly highlighted, will slow down cars.

Semiclosed street in Brooklyn with improved recreation space.

8. *Chicanes* (see Fig. 5.4*h*) are a variation of the previous strategy in which protruding physical elements are placed on alternate sides of the roadway, forcing cars into a zigzag (or slalom) pattern and thereby precluding any fast movement.

9. *Traffic circles* (see Fig. 5.4*i*), also known as rotaries or roundabouts, with a small diameter are also effective in slowing down traffic and making drivers cautious. They also generate focal points for the local circulation system that can be neatly landscaped.

10. *Gateways* (see Fig. 5.4*j*) are an old tradition in residential districts (in St. Louis, for example), and even if they are only symbolic with no actual closed gates, they serve as unmistakable indicators that a special district is being entered and that appropriate motoring behavior is expected.

11. *Street reversals* (see Fig. 5.4*k*) are changes in movement direction on one-way channels at relatively short intervals, thereby precluding any fast through movement. Such systems applied fully would only be comprehensible to local residents with repeated experience.

Starting with neighborhoods in Berkeley, California, the 1970s, where traffic calming concepts were first introduced in the United States and battles between local residents and through-motorists were fought in the streets and in the courts (as could be expected), the idea has caught on quite well in a number of American communities. Projects have been initiated, mostly as local grassroots efforts, and frequently specific demands for government assistance are being made. Since traffic calming is just the opposite of what traditional traffic engineering has tried to accomplish, there is not yet complete acceptance of the concept in all municipal traffic and transportation departments.

For traffic calming programs to succeed, a number of concerns have to be satisfied:

1. The local neighborhood must not only accept reorganization of its street network passively, but there should be active local support that leads to broad-based compliance and even surveillance by the residents. It is not too difficult for a few irresponsible drivers to act contrary to the expectations; a positive proactive attitude by the entire community helps in enforcement.

2. It is most advisable to improve the flow on surrounding and bordering arterials because additional loads will be placed on them. This will also minimize the temptation for through-motorists to seek interior shortcuts.

3. The requirements of emergency vehicle access have to be respected. Frequently, for example, local fire departments may oppose traffic calming programs because the physical elements may impede direct movements. Drivers of such vehicles should be well familiar with the local layouts. On the other hand, as a sad commentary on our current urban situation, some police departments have utilized traffic calming measures to reduce opportunities for fleeing drug dealers and impede accessibility for their customers.

4. The various elements and pavement textures should receive design attention because they become very visible parts of each neighborhood. Quality and appearance do count if the overall livability of the community is to be enhanced.

"No Drive" Restrictions

It is possible for municipalities to pass a selective regulation that bars the use of automobiles from certain areas during specified times by defined characteristics. For example, all blue or two-door cars may be excluded from the central business district during the working day. While this would be rather easy to police because any violators can be easily spotted, it would be a rather arbitrary system.[63] Most frequently this is done by digits on the license plate. For example, cars whose plates end with an even number might not enter on Mondays, Wednesdays, and Fridays, while those with an odd last digit would be excluded on Tuesdays, Thursdays, and Saturdays.

This program, needless to say, is a harsh one, and can only be regarded as a desperation measure when street congestion becomes completely unmanageable. It imposes severe restrictions on commuters and shoppers who have to seek alternate modes, but only on certain days. Service providers will experience significant fluctuations in demand that will be difficult to accommodate. Businesses are likely to suffer because the program is an admission that normal operations are no longer possible in the affected district.

It is also likely that the public will regard such a program as arbitrary, excessive, and undeserving of support. Civic disobedience becomes a sport. In the city of Lagos, Nigeria, where congestion within the constrained geography had created something close to an absolute standstill, this scheme was implemented a few years ago, and the results were both ineffectual and amusing. Many people did not necessarily obtain a second license plate, but it was found that those who could afford an automobile to begin with could also afford a second one. Thus, the total fleet increased substantially, and the daily volumes did not drop much. In addition, every day after 6 P.M., when open entry resumed, most of the cars in the city took joyfully to the streets to celebrate the evening.

Automobile-Free and Automobile-Restricted Zones

Beginning in the 1950s, cities started to exclude cars from designated areas to retain and recapture the viability of business dis-

---

[63] For a time, hatchback cars, station wagons, and jeeps were barred from a central avenue in Beijing. (As reported in the *New York Times,* April 17, 1999.)

tricts that were being overwhelmed by motor vehicle traffic. In Europe this involved many if not most of the core districts still maintaining their tight medieval street patterns, where volumes of motor vehicles simply could not operate. In North America the effort was propelled by a desire to introduce in downtowns some of the features of suburban shopping malls, which were siphoning off business activity at a growing rate. These actions were revolutionary programs that represented a new way of managing and structuring city districts. As such, they have received much attention and numerous analyses. There is a voluminous set of documents recording the experience.[64]

In Europe, automobile-free zones (AFZs) are doing well and are established parts of dense city cores. Experience shows that successful programs have to encompass many features, including effective penetration by transit services, good access for pedestrians and bicycles, and attractive spaces and walkways inside. Vehicular access roads have to be carefully structured and peripheral distribution arteries are essential; parking facilities have to be conveniently located, but on the periphery. Controlled entry by service vehicles (usually off-hours) and emergency vehicles has to be provided.

In North America, the dominant type of AFZs are downtown pedestrian malls, with the pioneering example being instituted in Kalamazoo, Michigan, in 1959. The national experience with dozens of examples has not been entirely successful. Indeed, at this time many of these projects are being eliminated, reverting back to open car access since the original expectations of recapturing business strength have usually not been fulfilled. The most visible example of such a sequence of events is State Street in Chicago, which was transformed into a mall in 1979 at considerable expense and then eliminated in 1996. It is becoming clear that localized and limited car exclusion programs are not sufficient to materially affect the massive dispersal trends across metropolitan areas.

Nevertheless, there are successful projects in the United States that show that good results can be achieved in appropriate situa-

[64] Among these references, the following can be mentioned: *Revitalizing Downtown* (National Trust for Historic Preservation, 1991, 127 pp.); *Main Street Success Stories* (National Trust for Historic Preservation, 1997, 191 pp.); and K. Halpern, *Downtown USA: Urban Design in Nine American Cities* (Whitney Library of Design, 1978, 256 pp.).

tions with coordinated programs. This includes the core areas of Boston and of San Juan, Puerto Rico, where not too many other choices exist. Short pedestrianized blocks in the midst of very high-density development are also logical and workable and are found in the cores of just about all the largest cities. There are also examples in smaller places, mostly associated not so much with intensive retail activity as with entertainment and cultural enclaves. Transit malls (streets that allow only buses and other public service vehicles) have emerged as the AFZ form with the best promise of positive results in North America. This includes the Nicollet Mall in Minneapolis, Cherry Street in Philadelphia, Fulton Street in Brooklyn, and facilities in Portland, Oregon.

Automobile-restricted zones (ARZs) fall in the same family of efforts to curtail the use of automobiles. They are publicly more acceptable because the controls are not as extensive as with AFZs, but the results are not as far-reaching. These programs allow the entry of motor vehicles, but on a selected and time-controlled basis. The most common form of this program is to establish gates that can be activated by magnetic cards, allowing in only vehicles that belong to local residents and businesses. Taxis may enter or may have to pay a fee; service vehicles are admitted for a limited time period; other vehicles may be excluded entirely or permitted with the payment of a relatively high charge for a fixed period. Such programs have been in place for years now in Norway (Trondheim, Oslo, Bergen), but they are also found in many other places in Europe—practically every city center that still retains its original medieval street pattern.[65]

The best-known and oldest project of this type is the area licensing scheme of Singapore that requires all vehicles entering the central business district to pay a fee and carry proof of such transaction. At this time the gates and payments are managed electronically. In the case of Singapore, where people are used to strict government controls, the system has been successful and has worked well. In American communities, there are political problems and concerns about the restriction of free movement. In Europe, many such projects exist and show that the systems provide much flexibility in managing traffic behavior and operations.

---

[65] The author has an apartment (and the use of a family car) in Riga, Latvia, within the medieval core of the city—an ARZ with electronic gates that aspires to be an AFZ, except that anybody who is willing to pay a $9 fee may drive in.

Control gates to Old Town of Riga, Latvia, an ARZ.

Coupled with parking controls, there are possibilities of closing the district entirely during special events or opening the gates to free entry when demand is low (at night, for example).

An interesting and effective variation on these themes has been the program in Göteborg (Gothenburg), Sweden, with a system of subdividing the central district by lines (barriers) that can be crossed by public vehicles but not private automobiles. Vehicles may enter and exit any of the five cells from the peripheral ring road, but they cannot get very far inside or cross the district. The charges for parking in garages decrease with distance from the very center. Thus, anybody can make a reasoned selection balancing convenience versus cost. The city of Bremen in Germany has also established a similar system with movement constraints.

Intelligent Transportation Systems

The sector of the entire transportation field where the development of new concepts and advanced procedures is most intensive today is the multitude of programs listed under the label of intelligent transportation systems (ITSs, originally called intelligent highway and vehicle systems). In all instances advantage is taken of computer data processing capabilities and the power of electronic means of communication. The basic aims are to provide good and immediate information to vehicle operators that will allow them to make more effective decisions before and during a trip and to manage traffic flows through a real-time ability to react to overall demand situations. While these procedures cannot create additional capacity for the physical networks in an absolute sense, they should be able to substantially manage the utilization of the available capacity toward maximum aggregate effectiveness. This is the first real opportunity to proactively operate the roadway systems and constructively influence driver actions, which so far have been governed by a multitude of separate and uncoordinated individual decisions on the public rights-of-way. In effect, advantage can be taken of whatever open movement space is available, and drivers can begin to have an overview of the total operational landscape within which they find themselves. Rational decisions could also be made at any given time not to make a trip at all or to seek another mode. Safety can be enhanced, and instances of accidents and other constrictions can be identified quickly.

All this does not come for free. Monitoring devices have to be placed at many locations on the network, elaborate communications connections are needed, central or interlinked computer systems have to be organized, vehicles and homes have to be equipped with information receivers, and individual navigation systems have to be deployed. Taken together, these arrangements represent the next infrastructure layer that most likely will cover our communities and enter private places. Such systems are being implemented now since the basic engineering has been accomplished and operational readiness exists. Undoubtedly, as years go by, upgrading and modification will be repeatedly necessary. At this time, the various elements of ITSs are classified as follows:

- *Advanced traffic management systems* (ATMSs), which encompass devices that can monitor traffic conditions on streets, analyze the received information, and, based on previously established patterns and current information, control traffic signals and other elements that guide traffic behavior on streets and roadways.

- *Advanced traveler information systems* (ATISs), which, using data from ATMSs or other sources, provide information to consumers that is available continuously in homes, in vehicles, or at workplaces through various means. The latter may be direct links with display devices, the telephone, the Web or e-mail, or radio and TV announcements.

- *Advanced vehicle control systems* (AVCSs), which become components of the rolling stock and encompass lateral control (steering), longitudinal control (acceleration and braking, maintaining safe intervals between vehicles), and collision avoidance. The ultimate development of this concept may be the automated highway requiring no manual driving of vehicles, which already exists in the form of several test tracks. This system was predicted in the GM Futurama exhibit at the 1939 World's Fair, and it may be in place a century later.

- *Electronic toll collection* (ETC), which is a separate subject outlined in the next section.

- *Commercial vehicle operations* (CVOs), which encompass procedures that are intended to expedite the movement of freight and make operations safer. These are not necessarily a part of this discussion.

- *Advanced public transportation systems* (APTSs) represent material for chapters on public transit (Chaps. 11 through 15).

- *Advanced rural transportation systems* (ARTSs) are outside the scope of this urban discussion.

Tolls, Congestion Pricing, and User Charges

For many years, starting with the 1950s, only a few voices[66] argued that the use of roadways is not an unrestricted right, and that economic control measures should be employed when the supply of space becomes scarce (i.e., drivers should pay for the use of roadways when the demand is so high that movement is impaired for everybody). This concept has now become an acceptable topic for discussion beyond the tolls on bridges and tunnels and specifically constructed turnpikes and toll roads. Requiring entry fees at district boundaries—such as the area licensing scheme in Singapore—or imposing user charges on regular city streets are programs that would still not have political support in North America. Yet, it is difficult to see what else could be done in many urban areas where the traffic overloads have reached crisis proportions and the trends continue. Since building more highway lanes into and within those districts is a remedy that could cripple the patient, the only choice left is for activities to move out—an option that has continuously been exercised by individual enterprises for some time now and that will ultimately bring destruction of traditional city patterns.

If such charges were to be collected at toll booths and documented with paper tickets, the traffic tie-ups would be monumental and idling cars would seriously affect urban air quality. Today,

---

[66] Principally William Vickery, economics professor at Columbia University and later Nobel laureate.

however, electronic devices make the system easily workable, although, admittedly, costs are involved. The engineering has been done, and the necessary elements are available on the market. One technical scenario would be to place sensors or monitors at many or key locations on the street network that would record the presence (or passage) of any vehicle, which would have to carry transponders (simple electronic devices that respond by identifying the vehicle). The charges could be proportional to the demand for the circulation use of any section of the street network: pay nothing or very little in the middle of the night and experience very high charges during peak periods that would discourage all traffic except that portion that absolutely must be there. At the end of the month, each car owner would receive a bill, similar to what we get from the power company, water supply agency, or cable TV provider.

Obviously, there are any number of implementation problems, but they concern operational details and appear to be solvable. How do you deal with cars that do not carry a transponder or are from out of town? How do you accommodate low-income people who must drive? How do you inform users what the rates will be at any given time on any specific roadway? How do you give priority to essential vehicles? Should small "green" vehicles get a break? There is much system design work to be done, but the principal task is to gain public acceptance of this concept, even if it has to be presented as the last reasonable measure to preserve the viability of high-density areas.

Halfway programs may not be effective or acceptable. For example, for many years suggestions have been made to place tolls on all the bridges and tunnels entering Manhattan. This has generated vocal opposition, with some justification, as being discriminatory against residents of the other boroughs. A partial response might be to establish a series of cordon lines throughout the city, but that may encounter problems as well. Eventually, a regionwide, if not national, system appears to be indicated.

Modification of Work Patterns

There is a family of programs that do not address the use of automobiles directly, but attempt to minimize peak hour transportation demands, thus affecting street loads nevertheless. They have been tried with some success and are in effect in numerous instances around the world, including the United States. They encompass the following:

- *Staggered work hours*. Different firms and institutions located in the same district can set their workdays to start at different hours. The arrival and departure loads of employees will thus be dispersed over a longer period, shaving and spreading the extreme peaks. There are some problems of maintaining communications among workers, but we face that anyway because of time zones across the country and in global operations. A partial response would be to have uniform hours for each industry (for example, financial firms, design consulting offices, universities, advertising agencies, etc.).

- *Flex time or alternate work schedules*. A further step, quite common today, is to allow each employee to set his or her own hours, as long as the total required working time is accumulated and there is a core period when everybody is present and reachable. This has the additional advantage of accommodating individual schedule needs (for example, taking children to school, doing regular errands, etc.).

- *Working at home/telecommuting*. With the ubiquitous presence of personal computers and universal communications systems, many employees can be productive without sitting at their office desks. More and more are doing just that, often with the encouragement of their management, and the easing of traffic loads can already be observed. There are firms today where one-fifth or more of the employees are not in the office on any given day; and when employees do come in, they have to make a reservation for a work space. If somebody works only one day per week at home, the commuting load is reduced by 20 percent. A variation on this theme is the establishment of satellite business centers at scattered locations open to anybody, which are reachable by short local trips and can provide all office services communally (for a fee) that may not always be available at home.

Most transportation specialists regard such programs with some favor, primarily because they do not require investments in infrastructure or modification of established patterns. Yet, it is not likely that the reduction will be more than 10 or 15 percent (which would be a major relief nevertheless if applied to rush hour conditions). Some effects may already be visible. For example, commuter traffic in the 5 to 6 A.M. hour on the Hudson River crossings reaching Manhattan increased by 38 percent from 1996 to 2000, and dropped slightly between 7 and 9 A.M.[67] There may be any number of factors in play for this phenomenon, but the rush period is undoubtedly spreading out.

Education and Information

It can be argued that people will behave responsibly if they are fully aware of the problems that excessive automobile use creates for themselves and their communities—provided that this does not curtail their mobility, does not entail significant additional costs, and above all does not reduce their personal convenience appreciably. A prerequisite, of course, would be the availability of effective alternate transportation modes besides the private car.

All that, based on rationality and good citizenship, may not be sufficient either. What is probably also needed is a shared community spirit, peer pressure to do the right thing, and public visibility as to who participates and who does not. This is not an easy task, with no real examples to show in North America, save for some incremental and short-lived efforts here and there. It can be done, however, as seen in a number of places in Europe where bicycle and transit use is the norm, even though automobiles are accessible to most urban travelers. One example is the inspiring, but rare, program instituted in South Perth, Western Australia.[68] Under the TravelSmart program, which is basically a marketing tool for nonautomotive transportation, households sign up and receive detailed information on travel options. This encompasses electronic ticket availability, reliable data on schedules and service operations, and unabashed promotion of healthy and environmentally friendly modes. Estimates indicate 61 percent greater use of bicycles and 35 percent more walking among the program participants, as well as a 14 percent decrease in single-occupant automobile trips and 17 percent increase in public transportation ridership.

The key to this success may be an extensive system of self-monitoring of travel behavior by the participants, communal record keeping, and distribution of the results to the public showing significant progress. There are motivation and recruitment surveys and debriefings of members. In other words, it is a pur-

[67] Port Authority of New York and New Jersey data.
[68] As reported in the *Urban Transportation Monitor,* July 6, 2001.

"Full strength in No. 3 turret!" shouted the Commander. "Full strength in No. 3 turret!" The crew, bending to their various tasks in the huge, hurtling eight-engined Navy hydroplane, looked at each other and grinned. "The Old Man'll get us through," they said to one another. "The Old Man ain't afraid of Hell!" . . .

"Not so fast! You're driving too fast!" said Mrs. Mitty. "What are you driving so fast for?"

"Hmm?" said Walter Mitty. He looked at his wife, in the seat beside him, with shocked astonishment. She seemed grossly unfamiliar, like a strange woman who had yelled at him in a crowd. "You were up to fifty-five," she said. "You know I don't like to go more than forty. You were up to fifty-five." Walter Mitty drove on toward Waterbury in silence, the roaring of the SN202 through the worst storm in twenty years of Navy flying fading in the remote, intimate airways of his mind. "You're tensed up again," said Mrs. Mitty. "It's one of your days. I wish you'd let Dr. Renshaw look you over."

Walter Mitty stopped the car in front of the building where his wife went to have her hair done. "Remember to get those overshoes while I'm having my hair done," she said. "I don't need overshoes," said Mitty. She put her mirror back into her bag. "We've been all through that," she said, getting out of the car. "You're not a young man any longer." He raced the engine a little. "Why don't you wear your gloves? Have you lost your gloves?" Walter Mitty reached in a pocket and brought out the gloves. He put them on, but after she had turned and gone into the building and he had driven on to a red light, he took them off again. "Pick it up, brother!" snapped a cop as the light changed, and Mitty hastily pulled on his gloves and lurched ahead. He drove around the streets aimlessly for a time, and then he drove past the hospital on his way to the parking lot.

James Thurber, "The Secret Life of Walter Mitty," from the book *My World—and Welcome to It.* (Copyright © 1942 by James Thurber. Copyright © renewed 1970 by Helen Thurber and Rosemary A. Thurber. Reprinted by arrangement with Rosemary A. Thurber and the Barbara Hogenson Agency, Inc. All rights reserved.)

posefully managed approach, with a coherent organization in place (involving its own budget), not something left to casual voluntarism that may not have a lasting impact.

Capacity Considerations

Individual motor vehicles as a transportation mode can only provide low-density service, even in the best of circumstances—certainly as compared to most public transit operations. Automobiles are large physical units requiring considerable space, as discussed earlier, and they require buffer zones because they are operated by drivers with different skill levels.

When limited-access highways were first built at a large scale, it was determined that the *base capacity* of a lane (with no stops or obstructions along the way) was 1200 to 1500 vehicles per hour.[69] Over the years this "official" figure has moved upward, and today it stands at 2000 to 2200.[70] Theoretically, a lane should not carry more than 2200 to 2400 vehicles per hour under safe conditions,[71] since such volumes would be difficult to maintain in a steady state over an entire hour. Furthermore, there are always physical or operational impediments that reduce the base capacity to what is called *practical safe capacity*. This includes a series of conditions that make drivers react instinctively to constraints and reduce driving speeds:

- Lanes narrower than 12 ft
- Lateral clearance of less than 6 ft (vertical elements of any kind)
- Frequent access points
- Steep grades above 3 or 5 percent
- Sharp horizontal curves
- More than 5 percent heavy vehicles in urban areas

There are tables and equations that allow in each given case a precise calculation of the real capacity by applying reduction factors.[72] The results will most likely be less than 2000 vehicles per hour per lane. In spite of all this, survey data show repeatedly that

---

[69] Actually, the units used in precise traffic studies are *passenger car equivalents* (PCEs), recognizing that many vehicles are larger than automobiles and create greater disturbances in the traffic stream. For example, a regional or express bus may count as 2 PCEs, a local bus 3 PCEs (it weaves in and out of lanes frequently), and a tractor-trailer rig as 6. On expressways the differences are not as pronounced; in difficult terrain and in crowded areas the large vehicles have a most significant impact as compared to passenger cars.

[70] It is difficult to explain this change in a basic parameter—except that it reflects contemporary driving skills and abilities of motorists, the quality of vehicles and equipment, and the evolving definition by traffic experts as to what constitutes safe and acceptable conditions on the road.

[71] Proper spacing between units according to speed, cars with no mechanical defects, driven by skilled and always alert drivers under perfect weather conditions.

[72] See Chaps. 12 and 15 in the *Highway Capacity Manual 2000* (Transportation Research Board, Washington, DC, 2000), or any standard traffic engineering reference book.

actual hourly traffic volumes on many major highways routinely exceed 2000 vehicles per lane. The apparent record is 2650 per hour per lane on I-66 in Fairfax, Virginia. How can this be?

The first answer is that loading under such conditions has gone beyond the safety threshold and the highway operates in what chemists would call a supersaturated state. Any disturbance in the stream will cause a breakdown of the flow and may result in a chain of accidents. The other answer is that traffic engineering is not an exact science, despite what some specialists would like to maintain, and that traffic behavior has much elasticity and internal flexibility. This general observation applies to all further discussion about traffic engineering elements. The general patterns are certainly repetitive and predictable, but the specific numbers will always be different. The causes of traffic are numerous and complicated; they are subject to numerous forces, including human factors, all of which should place any apparently precise calculations under suspicion. Actual traffic flows will not be the same from hour to hour; they will differ for the same hour from week to week. There is an implicit agreement, however, that we will all accept traffic analyses as showing accurate results and act in accordance with them, even if this may only be an apparent reality. So be it, but experienced judgment and comprehensive estimates based on an understanding of overall patterns should carry weight, i.e., a "reasonableness" check is always advisable.

These observations also apply to one of the benchmarks of traffic analysis—the existence of regular cycles in flow behavior. These are:

- The hourly variations over a day, with traffic volumes building up before 9 A.M., dropping off and having perhaps a mild midday peak, and building up again to an afternoon peak between 4 and 6 P.M.

- The weekly cycle, with average volumes varying by day of the week and being very low or very high in certain places on the weekend (depending on their activities).

- The seasonal cycle, which is less pronounced today than it used to be, reflecting largely the inclination of motorists to use cars regardless of weather conditions.

With all this in mind, the capacity concerns and traffic management approaches regarding city streets are considerably more difficult than the simple highway cases because many more factors are in play.[73] To begin with, capacities are governed not by the characteristics of street segments but rather the flow-processing ability of intersections. These are the nodes in the system that have to accommodate flows from different directions, following conflicting paths, and may have deliberate gateway controls (traffic signals, stop signs, selective flow restrictions). The capacity of an intersection in each direction is not only a matter of the number of lanes and what percentage of time they have a green signal (or, with unsignalized intersections, how quickly the crossing can be made). Additional factors to consider are the time lost in deceleration as the intersection is approached, the reaction time of drivers when signals change and the start-up time, the flow-processing ability of the intersection upstream, the volumes of right turns and the right-turn-on-red situation, the presence of pedestrians, and the maneuvers of buses and trucks. A major determinant of the capacity of any urban intersection is the left turn situation—how large is this volume, does it conflict with the opposing flow, is it permitted, and are separate lanes and signal phases provided?

This is a complex situation, but (computer-based) methods are certainly available that allow the calculation of capacities for any type of situation, utilizing extensive field survey data or relying on reasonable assumptions. As can be expected, the hourly capacities of urban street lanes vary widely—from 300 or 400 vehicles to about 1000. Frequently, arterial lanes will accommodate volumes in the 600- to 800-vehicle range. On the other hand, a minor street may only be able to allow 100 or 200 cars to enter or cross a major street with heavy volume and few gaps in the stream.

The capacities of streets in terms of people, of course, is the product of vehicular capacity multiplied by occupancy rates. The latter ranges from 5 (theoretically only) to 1.1 (for commuting), with everything else in between (for example, cars traveling to football games will carry an average of 3 persons, those going to baseball games 2.5).

Conditions on roadways are usually described by two measures:

1. *Volume-to-capacity ratios* (*V/Cs*), which contrast the actual volume of vehicles observed or counted on any street segment (or intersection) with the throughput capacity of the

[73] See Chap. 10 and others in the *Highway Capacity Manual 2000*.

Illustration of Capacity Estimates

The following procedure (known as the Creighton-Hamburg method) is intended as an illustration only to explain some of the considerations in capacity analyses. It was developed in the early days and used widely in urban situations; it has been surpassed today by the more precise and elaborate procedures of advanced traffic engineering. Yet, it does provide reasonable estimates, if used with care and if the expectations are rational. The procedure outlined here applies only to normal intersection configurations with simple red-green traffic signals.

Capacity of a signalized intersection per hour = number of effective lanes × number of green phases in an hour × number of cars processed by a lane per green phase × correction factors

Effective movement lanes.   Lanes in actual operation. For example, if the curbside lane is partially impeded by parked or standing vehicles, it would count as a fraction depending on its processing capability at the intersection (estimated by judgment).

Number of green phases in a cycle.   The number of times the signal turns green during an hour. For example, if the cycle is 90 seconds long, there will be 40 instances when the gates open.

Number of cars processed per cycle per lane.   The time (in seconds) for each green phase divided by the headway as they move across gives their total number. The time interval between them is about 2 to 3 seconds. The total time available, however, has to be corrected by the time it takes for the first driver in the queue to react and reach normal speed (about 3 to 4 seconds)—start-up lost time. Thus, a 60-second phase can accommodate 22 vehicles per lane [(60 − 3.3) / 2.5].

Correction Factors—Various conditions that create constraints on the operations of an intersection:

- Pedestrian crossing and taxi friction—reduction factor from 0.90 to 1.00, depending on intensity

- Presence of buses—reduction factor from 0.80 to 0.97, particularly with local buses (assuming that the number does not exceed one per minute)

- Right and left turns—from 0.74 to 1.00 (with a 12 percent turning component) depending on the total number of lanes in operation

- Peak hour conditions—from 0.88 to 0.92, recognizing the general state of stress during those periods

Figure 5.5 shows an example of a capacity calculation for the northbound lanes.

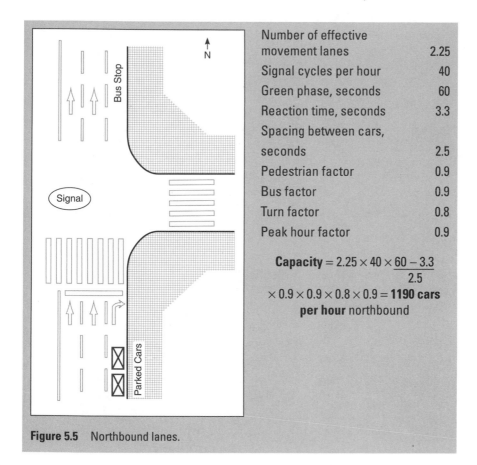

Number of effective movement lanes	2.25
Signal cycles per hour	40
Green phase, seconds	60
Reaction time, seconds	3.3
Spacing between cars, seconds	2.5
Pedestrian factor	0.9
Bus factor	0.9
Turn factor	0.8
Peak hour factor	0.9

$$\text{Capacity} = 2.25 \times 40 \times \frac{60 - 3.3}{2.5}$$
$$\times\, 0.9 \times 0.9 \times 0.8 \times 0.9 = \textbf{1190 cars}$$
per hour northbound

Figure 5.5 Northbound lanes.

same facility. Expressed as simple ratios, any measure higher than about 0.8 indicates congestion, a condition that approaches saturation. The number can actually exceed 1.0 in the field, pointing to an overload situation in a fragile state and with extensive flow stoppages.

2. *Level of service* (*LOS*), which is a characterization of prevailing conditions. Precise definitions and procedures exist, gauging the extent to which actual speeds approach free flow conditions and the seconds of delays experienced by motorists. LOS A denotes situations with no movement constraints generated by other vehicles, and LOS F is overall breakdown with stop-and-go movements at best. In high-density urban districts, LOS C becomes the best reasonably attainable objective. (See Fig. 5.6.)

Figure 5.6 Highway levels of service. (*Source:* ITE, Traffic Engineering Handbook, p. 127.)

Cost Considerations

The use of automobiles as a mode of regular transportation is an expensive proposition, but one that most people are willing to accept and pay for. It is estimated that having an automobile costs the average American family $4500 to $6500 each year.[74] (It was about $3000 in 1977.) Opponents of automobile use argue that the total costs to the society are much higher than those experienced by motorists themselves; even proponents will concede that there are (some) noncompensated for public costs.

The costs to the car owners consist of a variable component (i.e., dependent on the amount of miles driven) that includes gasoline and oil, maintenance, and tires, and a fixed component that includes insurance, license and registration, depreciation, and finance charges. The remarkable thing is that, as shown in Table 5.6, while most costs have steadily moved upward, the price of fuel in the United States has dropped during some periods in current and even in inflated dollars. In 1975, the cost of fuel and oil constituted more than one-third of the total variable costs in operating an average automobile; in the late 1990s, it was barely 13 to 14 percent. The pending petroleum crisis apparently has not made any difference so far. Any appreciable increase in the price of fuel has been regarded as a national calamity, calling for immediate corrective measures. If gasoline costs constitute about 15 percent of the per-mile expense of operating a car, and the federal fuel tax is 12 percent of the retail price, the tax represents only about 2 percent of the total cost. Given the fact that gasoline prices in European countries are likely to be two or three times as high as those in the United States, the level of the federal tax does not appear to merit much concern, and substantial increases would not make much difference at the pump.

The other half of the cost situation associated with automobiles as a transportation mode is the price of the road network—its construction and maintenance. These costs, except for the original construction of local streets in subdivisions built by private developers, are the full responsibility of government. Most of the work is done by municipalities and counties, albeit assistance for major

---

[74] As estimated by using AAA information on per-mile costs, with total annual mileage from 10,000 to 15,000. The Surface Transportation Policy Project estimates that Houston residents pay $9722 each year to drive a car, or 21 percent of household expenditures, far more than for housing.

Table 5.6 Costs of Owning and Operating an Automobile in the United States (in Current Dollars)

	Costs per Mile						Average Purchase Price, New*
	Variable Costs				Fixed Costs	Total	
	Gasoline and Oil	Mainte-nance	Tires	Subtotal			
1975	$0.048	$0.010	$0.007	$0.065	$0.079	$0.144	$4,950
1980	0.059	0.011	0.006	0.076	0.136	0.212	7,878
1985	0 056	0.012	0.007	0.075	0.158	0.232	11,902
1990	0.054	0.021	0.009	0.084	0.246	0.330	15,364
1995	0.058	0.026	0.012	0.096	0.316	0.412	18,957
1999	0.069	0.036	0.017	0.122	0.369	0.491	20,679 (1998)

Source: American Automobile Manufacturers Association.
Note: Costs are calculated for a standard American-made vehicle, driven the approximate national average of 15,000 mi/yr.
* U.S. Department of Commerce, Bureau of Economic Analysis, 1999.

facilities is available from higher levels of government through various assistance programs as they have evolved through the decades. Even with the current slowing down of new highway construction in the country, these budget items remain high, particularly as facilities age and become worn, thereby requiring continuous and extensive repair and upgrading.

General cost estimates cannot be presented as particularly useful guides since every case is different. Major variations are always found in right-of-way acquisition. If land has to be purchased for new alignments or even marginal strips have to be taken for street widening, the costs can be extraordinarily high; if reconstruction takes place within existing rights-of-way, these costs would be zero.

Recent (1990) estimates, i.e., national averages,[75] show that the cost of major reconstruction of highways in built-up areas, including the building of additional lanes and adding to the right-of-way width, exceeds $2.3 million per lane mile; reconstruction with wider lanes is not less than $1.4 million per lane mile; and complete replacement of pavement is about $1 million per lane mile. Any less extensive repair and maintenance will be proportionally lower, of course, down to perhaps a few thousand dollars per mile to fill in potholes.

[75] Jack Faucett Associates, *The Highway Economic Requirements System Technical Report,* U.S. Department of Transportation, FHWA, July 1991.

New construction does not necessarily cost much more than complete rebuilding. A reasonable approximation at this time for a two-lane street (with curbs and sidewalks and normal subgrade preparation, but not including right-of-way costs, any utilities, and major earth moving), would start at $1.7 million per mile.[76] For larger highways, a reasonably detailed specification of the physical standards is required before any reputable estimator will venture a projected number.

The debate about the costs generated by automobiles and roadways as a transportation system does not end here. For decades there has been a vigorous discussion as to what elements should be included in these accounts and who bears the cost or is responsible for various components.[77] Highway proponents insist that this activity generates sufficient public revenues through various taxes and payments to build and maintain the roads and operate its control systems; opponents point to a long list of secondary costs and impacts that are not compensated for by the users and fall back on the society at large.

There is sufficient evidence, however, to indicate that gasoline taxes and user charges cover less than two-thirds of all the tangible costs if, besides the construction and maintenance of roadways, highway patrols, traffic management, emergency response, and police investigations are also included. The rest is covered by general revenues from various agency budgets. The analysis becomes even less certain if it is recognized that parking provided free to employees and customers is also an expense (although it is presumably passed back to users in general indirectly); accident costs (including pain and suffering) that are not compensated by insurance and constitute losses to the society in any case; impairment of health, loss of productivity, and destruction of materials due to automobile-caused pollution; and congestion delays resulting in lost time and personal stress.[78] The

[76] Estimates by engineering design firms.

[77] Among a great many publications and analyses, the following are useful: J. J. MacKenzie et al., *The Going Rate: What It Really Costs to Drive* (World Resources Institute, 1992, 32 pp.) and K.T. Analytics, Inc. & the Victoria Transport Policy Institute, *Review of Costs of Driving Studies,* for the Metropolitan Washington Council of Governments, 1997.

[78] The Port Authority of New York and New Jersey estimates that annual congestion losses in its region may amount to $9 billion, encompassing largely the value of lost time and wasted fuel.

list can continue with the preemption of land that could be used for a productive purpose, the exhaustion of petroleum resources that may be needed as vital raw materials, the need to maintain military strength that can protect fuel sources, and even the acceleration of global warming with all the consequences that this entails.

If all this is included, the per-mile driving cost can be easily doubled, and the argument can be made that the larger society absorbs these costs and thereby subsidizes automobile use. This is not the place to resolve the argument, particularly since the answer depends on where the boundaries of responsibility are drawn. Two observations can be made, however. If the overwhelming part of the society is the beneficiary of this assistance, then they are simply shifting their own resources from one budget class to another, and they have the right to do that. And, if a rigorous cost-benefit analysis were to be demanded, the American public would easily see sufficient benefits in this situation to justify the costs. It only remains to be hoped that each member of the society is aware of the prevailing resource distributions and all the associated implications.

Land Development Effects

Automobiles started to dominate land development in American communities well before World War II, and they changed the metropolitan structure and the activity and density patterns entirely in the second half of the twentieth century. Motor vehicles may not have been the cause of these revolutionary events, but they certainly were the powerful means toward the results. Under their influence, metropolitan areas are now different in size and shape, and urban operations have changed for almost everybody. The old city districts built in a compact configuration are impacted by this transportation mode as well and often have difficulties adjusting to the new requirements. Some central districts have lost much of their former vitality and significance, to a large extent because they cannot be effectively serviced by motorcars, and other transportation means have not been provided for one reason or another. City life is quite different at the beginning of the twenty-first century than it was in the first half of the twentieth century.

Cities have never been like this before, where unprecedented distances are routinely covered by millions of workers and residents every day and houses are close enough to each other that residents can see their neighbors, but apart enough not to hear them. Even districts built during the previous eras experience the consequences of having to accommodate volumes of large machines.

The automobile made all this possible. It allowed the basic human urge to possess and control large private space (a house and a piece of land) to be fulfilled to varying degrees of satisfaction, while still being able to maintain contacts with all the services and work opportunities that characterize an urban area. It is a wasteful way to build cities, since compact development would minimize the expense of providing all services, but it is a model that most people prefer, and which—in the case of America—they can afford. Opinion and attitude surveys document well that this is exactly what most people want, and they claim to lead happy lives in this environment. They are the envy of the rest of the world, although with many caveats and observations about the profligacy and selfishness of it all. Some evidence is emerging, however, that there is dissatisfaction here and there with the automobile's dominance of daily lives and some willingness to consider alternatives. That would be the foundation on which to build a more responsible attitude and acceptance of management programs leading to efficient transportation systems.

The most striking feature of automobile access is that practically any location can be reached. This means the traditional concentration of destinations, which gave character and life to cities the way the older generation understands them, is now a detriment rather than an asset. To place any activity at almost any location within the surrounding urban field is workable, provided that a road network is in place. In terms of traffic operations, the effect is that trips are distributed in various directions because origin and destination points are scattered over space. Finding places where the urban scene can be enjoyed by long walks is becoming difficult.

It can be suggested, however, that manufacturing and distribution activities, which by and large have left central cities, benefit materially from peripheral locations. They can have as much space as they need for efficient operations, they are not constrained by congestion (usually), and motor trucks not only allow full service of their local markets, but also can reach, if necessary, any destination on the continent via the national highway system. Offices do not have to be in physical proximity to each other in all cases because electronic communications are effective, and their employees can reach their desks utilizing their own means of transport (mostly). A definitive analysis of these conditions has not yet been made, but there is enough evidence to suggest that the American economy works in high gear because of the freedom in locating enterprises and establishments.

The tight cities as they have been built for centuries, largely because transportation and communications were constrained, have been broken apart, and they will never be the same again. Even the changes brought by railroads, streetcars, and metro services, which expanded the radius of effective operations but kept activities tied to corridors and station locations, have been submerged by the unconstrained access capability of the individual motor vehicle. Any path will do, and, therefore, any piece of territory can be reached and developed. If there is demand for better roadways—a request that is usually shared by everybody locally—they will be provided. Thus, all metropolitan areas in the United States are covered by a well-developed network of streets and highways, extending to their outer reaches. There is continuous griping about the quality and loading of these facilities, but that is largely an indication of the amount of attention that this system enjoys and the perception of its vital role in the daily lives of the residents. There is the political will and the resources to keep the system in very good condition.

If the consequences and the resulting patterns at the metropolitan level are clear and unambiguous, the local land use–motor transportation relationship is not so easy to define. If all locations are about equally good and accessible, then a prediction of what will happen at any specific site is uncertain. At least, given that there are many more acceptable locations than there is demand for, some will be used and others not. The last part is, therefore, capricious and bothersome to planners who would like to act with some confidence and specificity. Consumer service establishments tend to cluster at highway interchanges, but not all of them. Retail, ranging from used car lots to megastores of household goods, will line major arteries, but not every mile of all of them. Regional shopping centers select locations accessible by several highways, but there cannot be one at every potential such location. Edge cities (clusters of office buildings and associated services) seek roadway access as well, but there will be only a few such developments in any metropolitan area, and real locational constraints apparently do not exist.

Possible Action Programs

Edge city development along a major highway (Tysons Corner, Virginia).

The basic question facing planners and urban managers today (somewhat akin to Hamlet's quandary) is whether to accept the conditions as described on the previous pages as "natural" and unavoidable, whether to oppose them vigorously for reasons of efficiency and the maintenance of the traditional city form, or whether to seek controls and modifications within an overwhelmingly popular urban environment that would make it workable, equitable, and efficient.

The first option—to do nothing—is not acceptable, not only for conceptual and professional reasons, but also because serious problems exist, as has been pointed out in numerous analyses as well as in this discussion. The principal problems are:

- A part of the population remains without adequate mobility.

- The automobile service system works with great inefficiency (leaving aside for the time being the inefficiency of the overall land use pattern).

The second option is to fight the oppressive presence of motor vehicles—a philosophically attractive proposition. It could be seen as an inspired crusade on a white horse in shining armor to slay the dragon that is gobbling up the livability of cities as we have seen them and experienced them from a gentler past. While this would hardly be the attitude of the overwhelming majority of the American population, serious people have mounted serious campaigns to do exactly that.[79] That attitude is not taken here for several reasons:

- It is a war that cannot be won because of the overwhelming power of the other side (admittedly, not a very inspiring admission).

- Every period in the history of cities has been characterized by its own urban form, surpassing and adding to, if not replacing, the previous one. Neither the medieval city nor the streetcar city can be built today, and they should not. Preserving the good elements from the past is vital, but they should be adapted to contemporary needs and not remain only as museum artifacts.

- The public has made its desires known repeatedly, consistently, and unmistakably. This is not just an American phenomenon, but one that is found anywhere else where economic and political opportunities exist. If participatory democracy is to be practiced and if the voice of the people is to be respected, then there are limits to the extent of "corrective" programs. Blind acceptance of trends is certainly not advocated, and leadership has to be exercised, but there are thresholds beyond which actions may become unacceptable. Where is the boundary between effective programs and futile social engineering?

That leaves the third option—the search for workable, relevant, and effective means to cope with the automobile and define its proper place. The task is thus to establish situations and implement programs whereby the negative features of automobile use are minimized. The positive features will take care of themselves. These opportunities, outlined in the section on the components of the physical system, either attempt to increase the density of vehicle loading or control vehicle use under various scenarios. All of these programs should be accompanied by the constructive actions reviewed in Chaps. 6 and 7, supported as

---

[79] The most complete and coherent recent statement of that kind is the book by Professor Vukan Vuchic, whose name has appeared repeatedly in the footnotes of this book, *Transportation for Livable Cities* (Rutgers, 1999, 352 pp.). The other side of the story—that automobiles are useful and essential but should be properly managed—is presented by B. Bruce-Briggs, *The War Against the Automobile* (E.P. Dutton, 1975, 244 pp.) and J. A. Dunn Jr., *Driving Forces* (The Brookings Institution, 1999, 230 pp.).

well by mass transit services as much as possible. Which specific program—or more likely combination of programs—is suitable for any given community or situation has to remain a local decision, assuming that the systems can operate within regional frameworks of coordinated actions.

Conclusion

If 90 percent of urban travelers opt for the same mode under almost all circumstances, that is no longer just a landslide choice or an overwhelming preference.

The new town of Reston, Virginia, with a hierarchically organized road system.

It is a well-nigh complete dominance. But not quite.

The current attitudes in American communities toward automobile use are neatly and convincingly illustrated by a *Chicago Tribune*/WGN-TV survey of Chicago's suburban residents.[80] The percentage of respondents who regard traffic congestion as a "major problem" or "somewhat of a problem" increased substantially from previous levels to 68 percent. The most favored solution (by 51 percent) is more road construction; 35 percent also support more investment in public transit, but only with the expectation that *others* will use it, thereby freeing highway space. Only 15 percent of men and 9 percent of women drivers commute to downtown. Eighty percent drive alone, and about half insist that they would never use public transportation, even if financial incentives were available. About two-thirds of the respondents like the possible options of working at home or having flexible hours.

Obviously, the automobile mode is a powerful force in today's communities, and it offers great convenience and a high degree of mobility to those who are able to operate these vehicles and have them readily available. The advantages and attractiveness of individually controlled vehicles are so great that they will never be

[80] As reported in the *Urban Transportation Monitor,* September 14, 2001, p. 4.

Typical urban highway view, with the skyline ahead (Los Angeles).

abandoned, short of a complete personal or national economic disaster. If they kill too many people, internal and external safety devices will be applied; if they poison the air, clean engines will be developed; if we run out of petroleum-based fuel, other types of power will be utilized. If they overcrowd our cities and make them unworkable, engineering solutions are not sufficient, as we have seen after several decades of intensive highway building. Neither are extreme approaches, such as banning automobiles entirely or abandoning dense urban districts and starting again in open fields, of any use. The car can be tamed, as many effective but not universally applied examples show, and it will have to be tamed.

Car culture is now an intrinsic part of American life. The industry itself is a major component of the gross national product, and efficient productivity is a key element in national economic well-being. The automobile has changed social customs and living habits for most of the population within the last half-century—even courting practices and sexual mores are different than they used to be. Overall mobility has never been so good. In the perception of most Americans, this is a positive situation that allows the enjoyment of a preferred lifestyle; it is the envy of most of the rest of the world.

But all is not fine. Significant concerns persist about waste of resources, an unhealthful envi-

Display of a wrecked car in Riyadh, Saudi Arabia, as a warning.

Automobiles in Kuwait, the United Arab Emirates, and Saudi Arabia

The United States does not represent the ultimate example of car culture. That dubious distinction belongs to Kuwait, the United Arab Emirates, and Saudi Arabia today. The native-born population of these countries has access to wealth far beyond the means of the average American, and much of it is devoted to the acquisition of motor vehicles. In the Emirates, there is one car for every 1.1 citizens of the country[81] (not including the expatriate workers). Gasoline and oil prices are extremely low. Recreation choices are few, and the most popular activities are visiting friends and relatives, going to the shopping mall, or simply driving on the well-developed system of highways, frequently at amazing speeds. Given the harsh climate, all buildings are air-conditioned to a subarctic level. All of these movements are accomplished in an equally air-conditioned car. The only variation may be extended picnics in the desert, also reached by automobile. The dominant and universally demanded housing type is one-family residences completely separated from each other and fully enclosed and screened for maximum privacy. Self-contained family life is the guiding principle that can also be readily extended and accommodated by a large car.

This situation is the visible environment; there are gaps in the picture. Women, for example, are not allowed to drive in Saudi Arabia, and can travel anywhere only if accompanied by a close male relative or husband. The very large expatriate worker population (mostly east and south Asians) remains outside the "normal" operations. They have their own accommodations or are live-in servants, and few of them own an automobile. Some rely on rather sketchy bus services; most are transported by vehicles (vans and minibuses) belonging to their employers. This constitutes extensive private paratransit systems in all cities and employment centers.

For example, the new town of Jubail in Saudi Arabia, built in conjunction with the massive petroleum-based industrial complex, is well planned and generously equipped with services. It started to operate a complete public bus service, but experienced extremely low ridership. The Saudis use their automobiles, the industrial workers are ferried between their compounds and plants by vans, and there are not enough women and children riders to fill the special compartments for them in the rear of public vehicles. The only testimony to these good intentions remaining today is the many attractive but idle bus shelters along all principal streets.

The prevailing transport systems in Kuwait and Saudi Arabia do the job, but with certain shortcomings. It can only be suggested that any American who is an uncompromising advocate of a complete car-based mobility system spend some time in these places to experience what the results and implications may be.

[81] Compared to 1.6 in the United States.

ronment, sprawling development, increased segregation and separation, erosion of a sense of community, and immobilized cohorts of the population. There are challenges for the near and the distant future.

Bibliography

American Association of State Highway and Transportation Officials (AASHTO): *A Policy on Geometric Design of Highways and Streets,* Washington, DC, 1994, 1006 pp. The "bible" on roadway design and all physical elements, used around the world.

Downs, Anthony: *Stuck in Traffic: Coping with Peak-Hour Traffic Congestion,* The Brookings Institution, 1992, 210 pp. A readable and comprehensive statement about the contemporary issues associated with motor vehicles.

Dunn, James A., Jr.: *Driving Forces: The Automobile, Its Enemies and the Politics of Mobility,* Brookings Institution Press, 1998, 230 pp. One of the few reasoned defenses of the automobile, stressing its benefits and outlining some corrective measures.

Pline, James L. (ed.): *Traffic Engineering Handbook,* Institute of Transportation Engineers/Prentice Hall, 1999 (5th ed.), 720 pp. The most widely used comprehensive reference on all procedures and devices employed in traffic management.

Transportation Research Board: *Highway Capacity Manual 2000,* National Research Council, Washington, DC, 2000. Not only technical details on capacity and levels of service calculations, but also a full review of vehicle behavior.

U.S. Department of Transportation, Federal Highway Administration: *America's Highways: 1776/1976: A History of the Federal-Aid Program,* 1976, 553 pp. Detailed history of roadway building, motor vehicle development, and federal assistance programs in the United States.

Vuchic, Vukan: *Transportation for Livable Cities,* Center for Urban Policy Research, Rutgers, NJ, 1999, 352 pp. Complete review of the problems caused by automobiles and the argument toward severe restrictions.

Paratransit

Background

Mobility will not be denied to people who need transportation services if they have some resources to spend or if the society in which they live recognizes its obligations. A way will be found everywhere, except in places in the most desperate economic state, to move urban residents, even if the systems have to be improvised and generated through local entrepreneurship.

If city buses do not or cannot reach all districts, or if the service is very sparse, neighbors will start running their own cars along obvious routes and offer rides to others (for an affordable fee). If low-wage, but essential, employees cannot reach job places by themselves, employers will have to pick them up with their own vehicles. If nondrivers who live in American low-density suburbs need to go anywhere, they should be able to call for a community-based service that costs less than a regular taxi. If affluent commuters are not pleased with the quality of regular transit, besides opting for the private car, they may pay premium fares for special services. If many people have to reach a specific node repeatedly (an airport, shopping center, or major institution), but they wish to do that with greater comfort and speed than a bus can provide, some sort of shuttle service will be initiated sooner or later.

All this is *paratransit*—a service that is not quite full public transit and that has some of the convenience features of private automobile operations. It is most often smaller in scale than real transit, utilizing smaller vehicles, and it can be legal or illegal as defined by local rules and regulations.

In many respects there is nothing much new about paratransit. Names such as *shuttle service, minibus, jitney,* and *downtown circulator* describe operations that are quite well known and have been around for some time. *Paratransit, demand-responsive service,* and *dial-a-ride* operations, however, are of recent coinage within the last three decades. They represent deliberate efforts to fill gaps in the transportation spectrum, and today they are official modes under various government and private programs in this country. The etymology of terms identifying paratransit services in cities of the developing world is a fascinating subject in its own right, and examples are given later in this chapter.

Systematic analysis of paratransit is difficult. The general structure can be outlined easily enough, but what happens in any given community is a most complicated matter. It is not only a question of trying to hit a moving target, it is also a target extremely fuzzy around the edges.[1] Services are initiated and dropped, the levels of operations ebb and flow, the same vehicle may be used in different ways on the same day, and government attitudes and regulations evolve and are modified repeatedly. True demand-responsive services may appear and operate illegally and thus not even be visible in the official records.

Yet, while the actual number of passengers carried by paratransit may be quite small in any community, it is a vital auxiliary service in an automobile-oriented society. Indeed, if current trends continue, it will merit major attention everywhere in North America as the indispensable mode to secure mobility for all residents (i.e., nondrivers). In 1990, this activity still carried less than 1 percent of the total unlinked trips (79 million out of 8526 million).[2]

---

[1] The basic reference today is Robert Cervero's book *Paratransit in America: Redefining Mass Transportation* (Praeger, 1997, 281 pp.), which is as complete a documentation of the paratransit situation in the mid-1990s as could be expected. However things change continuously and rapidly.

[2] In 1998, demand-response modes accommodated 55 million unlinked trips, out of a total of 8746 million (American Public Transportation Association data).

Table 6.1 shows the official statistics for the total passenger-carrying motor vehicle fleet in the United States.[3]

Development History

Since paratransit-like services often appear spontaneously and may last for limited periods, it is quite possible that there have been operations in the past that assisted local residents and workers with mobility services and then disappeared without a trace. Paratransit, however, is only a meaningful concept if there is real transit in operation, and where an unmet need becomes painfully apparent and official agencies have to step in, or where potential service providers possess underutilized vehicles that can be put in communal use effectively.

Minibus operating as jitney (*dolmush*) in Istanbul.

This happened in the United States around 1914 and earlier, when a number of people who had purchased automobiles (mostly Model Ts) faced economic hardships during an economic recession and had no full-time need for their vehicles. These peo-

[3] U.S. Department of Transportation 1995 data.

Table 6.1   1990 Distribution of U.S. Motor Vehicle Fleet

Demand-responsive vehicles (vans)	16,471*
Paratransit vehicles (mostly private)	68,000
Taxicabs	32,000
Intercity buses	19,491
Public buses	59,500
School buses	380,000
Regular automobiles	143,500,000

* The Bureau of Transportation Statistics records 32,899 "demand response" vehicles in 1998. This number, however, is not directly comparable to the 1990 data since paratransit vehicles and taxicabs are not listed separately in the more recent table.

Definitions of Terms

animal- and human-powered vehicles (carriages, rickshaws, and pedicabs) Historic private and for-hire means of transportation in all cities. Now replaced by motor vehicles. Horses create sanitation problems; human-powered vehicles (even pedicabs) are considered to be socially unacceptable and to present a poor civic image as regular transportation modes, and are largely outlawed in such use. They exist in some remote cities and villages around the world, and they are frequently present as a recreational service for tourists and sightseers in specific locations (Honolulu, New Orleans, and New York City, for example). (Not reviewed further as a transportation mode.)

carpools Private vehicles used by groups of people with approximately the same origin and destination points and the same personal schedules. Sharing of expenses is at their own discretion. (See Chap. 5.)

car services Transportation enterprises, usually with a local base, that provide motor vehicle service for hire. The regular pattern is to offer taxi service upon telephone request, but jitney and feeder service on set routes may also be provided, as well as charter service. In some inner-city neighborhoods, the scope of operations may be stretched beyond authorized limits. (See Chap. 7.)

circulators Same as **shuttles,** except operating within a district, frequently as a loop.

demand-responsive services Vehicles that pick up rider(s) at any point or at designated nodes, when summoned, and carry them to a selected destination. No schedule, no fixed route; with or without payment of fare. (Taxis fall in this general category, but the designation is usually reserved for shared-ride vans operated at the community level.)

dial-a-ride Same as **demand-responsive services,** except that the vehicle is called by telephone, frequently a day in advance. (A taxi is expected to arrive immediately.)

gypsy cabs/bandits/poachers Vehicles operated illegally in a jitney or taxi mode. Have no operational authority or license of any kind (or significantly exceed allowed service limits), probably are underinsured, and drivers may not be properly qualified. All financial transactions are by cash, generating no records or tax revenues. If a dial-a-ride or jitney vehicle, for example, picks up street hails when it is authorized only to accommodate prearranged rides, it operates in a gypsy mode, although it may be otherwise legally licensed.

jitneys Same as **public transit,** except that the vehicle is usually small, and the number of vehicles on the route at any given time is expected to be sufficient to accommodate the demand. No schedule, but a fixed route (with deviations possible) and a fixed uniform fare. Entry/exit is on individual signal (street hail).

paratransit Vehicles in communal service (unlike a private car), but without all the traditional public transit features (unlike bus or rail transit). Entry is on payment of a fare, by showing a pass, or available only to a preselected group of patrons (may be free). Travelers can usually summon a service vehicle, which will take riders to different places, when needed. The U.S. Department of Transportation definition from the 1970s is as follows: "Those forms of urban passenger transportation which are available to the public, are distinct from conventional transit (scheduled bus and rail) and can operate over the highway and street systems."

private individual transport (automobile, motorcycle, bicycle) A vehicle operated for the private purpose of the owner and/or driver. No schedule, no fixed route, no direct tariff. (See Chaps. 3 and 5.)

public transit (bus, rail, and others) Vehicles accessible to the public upon payment of a fare or showing a pass. Operate on schedule and along designated, fixed routes. (See Chaps. 8 through 15.)

shared taxis Same as **taxis**, except carrying several parties with different but proximate origin and destination points. (See Chap. 7.)

shuttles or feeders Vehicles operated on specific routes that are frequently short and simple, with at least one strong terminal. May or may not have a schedule; various entry controls may be present or be entirely absent.

subscription buses Semipublic paratransit vehicles with fixed routes and schedules, but accessible only to patrons who sign up and pay a monthly tariff. Such operations, if successful, most often have been taken over by public agencies in the United States. (See Chap. 8.)

taxis Vehicles accessible to the public for hire to carry one party between points of its choice. No schedule, no route; tariff is based on distance or duration of trip or by zone. "One party" is one or more persons traveling together with the same origin and destination. (See Chap. 7.)

vanpools Similar to **carpool**, except that a larger vehicle is used. Service may be fully or partially sponsored from outside the user group. No public entry.

ple are credited with inventing the first American *jitneys*.[4] They went out on the streets in significant numbers—particularly in Los Angeles—followed transit lines, and offered rides to waiting passengers. Since the fare was reasonable[5] and the service quick and agile, they did not lack customers. These patterns were

[4] This story is a part of American folklore, repeated many times. A complete review and examination of the national situation is provided by R. D. Eckert and G. W. Hilton, "The Jitneys," *The Journal of Law and Economics,* October 1972, pp. 293–325. An earlier article was A. Saltzman and R. J. Solomon, "Jitney Operations in the United States," *Transit Planning and Research,* Highway Research Record No. 449, 1972, pp. 63–70.

[5] The same 5¢ as on streetcars at that time. The nickel coin was called a "jitney" in contemporary slang, hence the name of the transport service.

repeated in other cities, and the situation became threatening to the public transit companies, which were not in a strong financial state anyway. They had political clout, however, and over the succeeding years they were able to convince municipal governments to outlaw jitney operations.

Some services lingered longer than others, but eventually anti-jitney legislation made a clean sweep of the country—almost.[6] Some of the more successful jitney operations became regular municipal bus routes or feeders to streetcars; most were simply eliminated, even with some loss of convenience to local patrons. An interesting exception was Hudson County, New Jersey, a place with a checkered history in its public administration style, where private operations prevailed until they were absorbed by the regional transportation agency in the mid-1970s. St. Louis had a jitney service until 1965.

Another well-known legal survivor was the Mission Street jitney in San Francisco, which faded away only in the 1970s and 1980s. Service was provided by largely part-time owner-drivers of vans, feeding the downtown area. The most famous American jitney, still going strong, is the operation in Atlantic City, New Jersey. It has several well-serviced routes, basically parallel to the boardwalk, and operates as an integral part of the city's transportation system. It is managed by an association (actually, a brotherhood) that ensures service quality, assigns shifts to owner-drivers, negotiates fares with the public regulators, provides repair and maintenance facilities, makes bulk purchases, and generally protects the interests of its members. Special vehicles are used (converted bread and milk delivery vans) that allow passengers to board and move around inside easily. Reduced fares, supported by the municipal government, are available to seniors and students.

Another major episode in the history of self-generated trans-

Jitney vans in Atlantic City.

[6] The magazine *Jitney Bus* changed its name to *Motor Bus* in 1915.

portation services in the United States started in the late 1960s. This was the spontaneous emergence of local operations in many districts of the larger cities, almost entirely confined to low-income and ethnic neighborhoods, as the devastating effects of segregation and property abandonment became dramatically visible in the latter third of the twentieth century. Local entrepreneurs, owners of regular cars, started to ferry the residents of their own communities in significant numbers to mass transit stations and medical and shopping centers, and made themselves available for trips with special purposes and destinations. To tell the truth, gypsy cab operations had maintained a sporadic underground existence all along in the intervening years in many inner-city districts. Jim Crow laws, making it difficult for African Americans to use regular buses, frequently fostered the emergence of a parallel jitney service. Sometimes efforts were made to suppress them, sometimes not.

These operations, which reached considerable volumes in the 1970s, happened without the support, authorization, or even knowledge of local government. They represent a dynamic situation that changes constantly by adjusting to demand and the severity of police enforcement of rules and regulations. To keep track of all the events and service changes, to document exactly the scope of activity in any given place, and even to count the vehicles in play were and are well-nigh impossible tasks.

Frequently, the triggering event for starting these services was a dislocation in regular transit services (such as a bus and subway strike), but their lasting persistence needs more explanation. One of these is that, since such informal services are largely found in Latino districts of American cities, this is a matter of transferring the experience from the home countries, where private operators provide the bulk of transit services, to the new place. The riders feel comfortable with the service, they know the driver, who is from the same neighbor-

Self-generated feeder service to a subway station in Brooklyn.

hood, they speak the same language and local news can be exchanged in a social setting, and there is some satisfaction in thumbing one's nose at the establishment. Fares are the same as or even lower than on regular buses, and the vehicles move faster. Money is kept in the neighborhood; local jobs are generated, including repair and maintenance of the fleet.

The other theory is that the emergence of such services is simply a reaction to the real or perceived inadequacies of conventional transit that is available locally. There is the charge that poorer districts are neglected by regular service providers, and, therefore, deficiencies are corrected at the grass roots level, whenever possible. No handbook is needed to invent or structure a local jitney operation. When transportation is needed, some inadequacies, even serious ones, of the local operations can be tolerated by the patrons—such as fast and reckless driving, underinsurance, no job security or benefits for the drivers, and the stigma of illegality. As could be expected, the larger centers of such activity were found in Miami, Pittsburgh, Chicago, and New York, but similar car services have also been reported in San Francisco; San Jose; San Diego; Baltimore; Boston; Omaha, Nebraska; Chattanooga, Tennessee; and the list goes on. They still exist today.

There is no question that the illegal neighborhood operations, which sometimes extend to services over long distances, are a thorn in the side of regular transit agencies. The basic charges are that dangerous street traffic conditions are created, passengers are endangered, and—above all—ridership is siphoned off from already hard-pressed public operations. The latter effect is characterized as "skimming the cream." On the other hand, some transportation planners not associated with the official agencies suggest that, if each trip on public transit requires a subsidy, then it is really a matter of "slicing the deficit."[7] The wisest policy would appear to be to tame the efforts and take advantage of the private energies and capabilities, to recognize the operations as useful auxiliary options and incorporate them into the overall systems—to "shave the peaks," at least.

[7] S. Savas, S. Grava, and R. Sparrow, *The Private Sector in Public Transportation in New York City*, U.S. Department of Transportation, 1991; and S. Grava, J. Gaber, and N. Milder, *Private Auxiliary Transport at Jamaica Center*, Columbia University, 1989.

Nevertheless, the situation in a number of places where regulations are openly flaunted or ignored is not acceptable; it is not just an inelegant state. If the local cars serve a purpose, then constructive action in terms of legalization and utilization would be indicated. One can analyze examples in cities of the developing world where such services are the norm, and considerable experience has been gained regarding private, self-generated transport services. Or, one can look at cities in Western Europe where they do not exist because regular transit operates at high quality levels and people tend to respect rules. Or, one can examine countries of the former socialist bloc where, at least during the transition phase, such spontaneous activities flourish extensively. An ambivalent attitude still prevails in American cities regarding the inner-city car services and jitneys. They should not be there, but they provide a useful mobility service to the local population. If they are noted at all in official documents, passing references are made to "subsidiary," "auxiliary," "neighborhood," or "adjunct" services. Occasional police blitzes ticketing illegal jitneys have not resolved the issues.

The 1960s was also a period when it became apparent that all was not well with transportation in the fast-growing low-density suburban areas. The car did not serve everybody, but suburban congestion started to threaten. Something had to be done, and the obvious response was to establish automobile-like services that could be summoned when necessary to carry the nondrivers to their destinations. People had to be enticed to use vehicles jointly. Government and social service agencies had to step in. The prototype already existed—the taxi companies in every community that could be contacted by telephone since there was no reason for them to cruise the streets when very few walkers could be seen. Thus, dial-a-ride services were born, and they became a part of the modal inventory. At the same time, an overall designation—*paratransit*—was coined, principally because government agencies presumably could not cope with any type of operation that did not have an identifying designation.[8]

First, there were research efforts in the late 1960s to explore the options and structure operational procedures—as if new systems had to be invented. Soon after, the federal government,

---

[8] The International Taxicab and Livery Association changed its name only in 2001 to the Taxicab, Limousine and Paratransit Association.

through its Urban Mass Transportation Administration (UMTA), now the Federal Transit Administration (FTA) within the U.S. Department of Transportation, instituted demonstration programs to test the new concept and to identify factors that would create problems or show opportunities for success. The first projects were organized during 1970 to 1972 as taxicab-based dial-a-ride services in Davenport, Iowa; Hicksville, New York; Little Rock, Arkansas; Lowell, Massachusetts; Madison, Wisconsin; Merced, California; and Richland, Washington, and as bus- or van-based services in Haddonfield, New Jersey; Ann Arbor, Michigan; Batavia, New York; and Columbia, Maryland. At the same time, explorations were also made in Bay Ridges, Ontario, and Regina, Saskatchewan. These were all small or suburban, car-dominated communities. Inner cities received attention as well in the context of local Model Cities programs in Columbus, Ohio; Detroit; Buffalo; Ft. Walton Beach, Florida; and Toledo. A few years later, the demonstration efforts were extended to Knoxville, Tennessee (subscription van services), Rochester, New York (demand-responsive small buses), Naugatuck Valley, Connecticut (elderly and handicapped), Albuquerque, New Mexico, and Portland, Oregon (elderly and handicapped in midsize cities), Chicago (elderly and handicapped in a large city), Cleveland (demand-responsive), Cranston, Rhode Island (dial-a-ride), and Danville, Illinois (taxi tickets).

All of these projects received extensive federal assistance, augmented in many instances by state and private contributions. Eventually, there were some hundred demonstrations across the country. There have been many official and scholarly follow-up studies of these cases, with Ann Arbor, Batavia, Haddonfield, and Westport, Connecticut, receiving the greatest attention. The findings formed a clear pattern.

The major conclusion was an acknowledgement that in areas that are too sparsely set-

Van service (Minny Bus) in low-density communities of Connecticut.

tled to support regular scheduled transit service, there is no other viable choice to provide reasonable mobility to all the nondrivers who have no family members with a car to accommodate them at all times. A market does, indeed, exist, but the potential patrons remain sensitive to the quality of service. They would rather take the car, if at all possible.

It was also found that in districts with bus and rail service, which by law has to be able to carry even severely handicapped people, the concept of *mainstreaming*—adjusting all facilities and vehicles so that all people with disabilities can use regular services, rather than providing them with special "equal but separate" services—is not always the best response. Out of these considerations two types of dial-a-ride operations have evolved: those that are open to the general public and for which a fare has to be paid, and those that serve a preauthorized clientele, usually the elderly and the handicapped, with specialized equipment, personnel, and involvement.

The other finding regarding the early services was that the goal of grouping trips so that a given vehicle could follow the most effective route by picking up and dropping off passengers as it moves along was not easy to achieve. The human dispatchers were not quite able to arrive at the most efficient schedule and routing at all times, and the computers of the day were not yet particularly fast and smart. Patrons did not like waiting for a pickup, having to follow a circuitous path for the convenience of others, and sharing a small vehicle with strangers. Resistance by established transit agencies and labor unions was quite apparent as well.

The major problem, however, with the demonstration programs was costs. The acquisition of the vehicles was the easy part. There were significant expenses associated with the daily processing of requests and establishing reliable record-keeping systems; labor input in everything was considerable. In some instances, the cost per ride approached the local taxi fare for a comparable trip. Significant permanent subsidies were needed in all instances, and in a number of communities the service was stopped once the federal demonstration grant ran out. At least in one place (San Jose County), great success killed the operation— the local government could no longer keep pace with subsidizing an increasing number of trips.

Lessons were learned, however, and a national sense of social responsibility has been established. There probably is no reasonably settled American community today that is completely devoid of some paratransit service—at least there are those offered by social service agencies, religious and charitable organizations, or medical centers. This does not necessarily mean that the available service is as quick and convenient as the patrons would like it to be, but it is basically there at an affordable level or free of charge. It has also become an increasingly common practice for regional transportation agencies to include paratransit as a component of their total inventory of operations. In no small measure these changes in policies have been caused by the increasingly more stringent mandates under the Americans with Disabilities Act (1990) that asserts that everybody has a right to mobility. The new operations may be not only special services with special vehicles for defined groups (most often the elderly and handicapped), but they may also be modified transit services in low density corridors. These programs are a welcome recognition of the public's responsibility toward mobility for all residents, and may even include elements of cross-subsidy (accepting the fact that some operations will run a larger deficit than others) within comprehensive budgets.

Types of Paratransit

The great variety of forms that paratransit can take makes it necessary to attempt classification structures that allow some sensible discussion of their capabilities and suitability in application. Table 6.2 shows one such typology, advanced by Robert Cervero,[9] which is based on the public versus private distinction among services.

While the classification structure in Table 6.2 covers the field very well, paratransit can also be examined from the perspective of what kind of service is being provided. The result would be the typology chart in Table 6.3.

Another possible way to classify paratransit services is according to the type of operation in place (see Table 6.4).

A specific type of paratransit, which is encompassed in Tables 6.2 through 6.4 but should be highlighted because it is encountered frequently, consists of services provided by universities with large or scattered campuses. Usually vans or minibuses are used

[9] R. Cervero, *op. cit.,* page 15.

Table 6.2 Paratransit Classification by Public/Private Nature of Service

Paratransit Service	Service Types	Service Configuration*	Primary Markets
Commercial services			
Shared-ride taxis	On demand, hail requests	Many-to-many	Downtown, airports, train stations
Dial-a-ride			
Specialized	On demand, phone requests	Many-to-many	Elderly, handi-capped, poor
Airport shuttles	On demand, phone and hail requests	Few-to-one	Air travelers
Jitneys			
Circulators	Regular route, fixed stops	Loop; one-to-one	Employees, low income, specialized
Transit feeders	Regular route, hail requests	Many-to-one	Employees, low income
Areawide	Semi-regular route, hail requests	Many-to-many	Low income, recent immigrants
Commuter vans	Prearranged, scheduled	Few-to-one	Commuters
Employer- and developer-sponsored services			
Shuttles	Prearranged, regular route	Loop; often one-to-one	Commuters, students
Vanpools	Prearranged, scheduled	Many-to-one	Commuters
Buspools	Prearranged, scheduled	Few-to-one	Commuters

* In this case referring to the number of trip origins and trip destinations that are being serviced by each vehicle.

that connect academic buildings, dormitories, and parking lots on fixed schedules. Fares may be collected or the service may be supported by the institution's general funds or parking lot fees. These operations are present not only at universities with large, sprawling campuses, but also in urban settings with the institutional buildings and mass transit nodes at various locations (Berkeley has the Humphrey GoBart and Columbia the Lamont Shuttle).[10]

[10] L. Davis, "Transit Agencies in Demand at Local Universities," *Metro*, April 2002, pp. 20–25.

Table 6.3   Paratransit Classification by Type of Service

Type of Paratransit Service	Service Configuration	Descriptive Name and Service Provider
Community-based services		
Demand-responsive	One origin to one destination	Taxi, by private enterprise
	Many origins to one destination	Dial-a-ride, by public agency, institution, or private firm
Street hail	Along a route	Jitney, by private enterprise (could be a public agency)
Commuting services at city or regional level		
Private, prearranged	Few origins to one destination	Employer vanpool with hired driver or with designated self-driver
Commercial, prearranged	Few origins to one or few destinations	Van service, by private enterprise
Public, prearranged	Few origins to few destinations	Van service, by public agency
Service to special nodes (airports, terminals, shopping centers, institutions)		
Node to node	One origin to one destination	Shuttle*
	Along a route, to one destination	Jitney*

* These shuttles or jitneys may be operated by private enterprise or public agency; entry may be on payment of fare, on showing a pass, or free to designated customers.

Another paratransit type well known to business people is the very common airport shuttle that is associated with just about every airport in the United States providing commercial air service. These operations have been most successful. They are perceived to be not much inferior to taxis, and they certainly are much cheaper. They are clearly more convenient than regular transit, even in the few places that have a direct rail connection to the regional airport. Major commercial centers and hotels are connected with reasonably frequent schedules. These days large commercial enterprises are the principal operators, often active in a number of cities (such as the Supershuttle).

Table 6.4   Paratransit Classification by Type of Operation

Fixed service

Fixed routes, basically identical to conventional transit, except that the vehicles are smaller and run at shorter headways. Jitneys fall in this category.

Demand-responsive to origin and destination points

Exclusive ride for one party
No different than taxi service, except that there may be a public subsidy for eligible patrons.

Shared ride
Clustered origin and destination points for several parties. This is the most common type of paratransit, but the class includes also carpools and vanpools.

Group ride
Accommodation of a number of patrons with the same origin and destination points; basically a short-term charter.

Variations in service

Route deviations
Service with side trips to accommodate individual patrons. This may be done at the expense of increasing the trip duration, or be permitted only if arrival schedules can be maintained.

Limited stops
Passengers are picked up and discharged only at preselected and fixed points.

All of the paratransit services included in Tables 6.2 through 6.4 (except for the specific van operations) are independent of the type of vehicle used. For example, while most jitneys historically have been passenger sedans, there is no reason why a bus cannot be operated in such a way (as it sometimes is). Increasingly, however, passenger vans with a standard or a modified configuration are being employed for most paratransit operations.

Shuttle van in Reston, Virginia.

Reasons to Support Paratransit

Paratransit has emerged as a legitimate urban transportation mode due to necessity. It fills a niche in car-dominated societies, and it responds to very specific needs.

Communal Transportation

Paratransit, by assembling at least several travelers in the same vehicle, improves the total performance of transportation systems that would be otherwise completely overwhelmed by single-occupancy automobiles. This saves space, conserves fuel, reduces air-quality impact, and gives more choices in individual mobility. Even persons who usually rely on cars may from time to time opt for a good jitney or shuttle. The availability of a responsive demand-activated public service may preclude the purchase of a family's third or perhaps even second automobile. More transportation choices are offered.

Paratransit may very well be the only remaining communal transportation mode that is still workable in very low-density districts. Indeed, there is no density limit because some form of emergency and nonemergency access is needed in rural areas as well. Paratransit has a considerable secondary benefit in that its presence alone can be a reason for conventional transit to improve its efficiency and responsiveness. Some competition is always healthy, and transit is not really a natural monopoly (as is, for example, water and power supply). Transportation services relying on individual motor vehicles do not benefit particularly from efficiencies of scale (as rail transit does), and disaggregated systems can work quite well. Paratransit has the special advantage, due to its small-scale operations, of being closer to its customers.

Mobility for All

The immediate reason why communities have to have paratransit services is to accommodate all those members of society who do not, cannot, or do not wish to drive. This includes not only the elderly and the handicapped, but also the young, people who have sprained an ankle, motorists whose cars are being repaired, and any number of other permanent or temporary nondrivers. The better forms of paratransit, if they are properly managed, can respond very well to general demands for mobility, as well as to immediate trip requests.

Service Quality

In almost all instances, unless reference is made to some dilapidated sedans operated as neighborhood jitneys, paratransit should be able to offer a service at least one comfort level higher than conventional transit. This is due to the smaller vehicles used, the frequency of service (or strict adherence to schedules), the quality of the vehicles themselves, and the relationship of the driver to passengers. Patrons' sensitivity to comfort, convenience, privacy, and security is becoming a dominant factor in the choice of transportation modes in North America.

Agility

Because the vehicles carry few passengers and make stops only on demand, any comparable trip duration will be less than on regularly scheduled transit. With private services, where the driver's income depends on the number of passengers accommodated, the movement is even quicker.

Flexibility

The same basic vehicles can be utilized in different ways. On any given day, if appropriate dispatching arrangements are in place, a van can be used as a jitney or commuter unit on a fixed route, while responding to dialed-in requests during the off hours. On weekends, it may be chartered by groups making special trips. Service patterns can be easily adjusted as needs change. Operations can be scrapped without too much of a loss by placing the vehicles on the used car market.

Community Spirit

Paratransit, because of its size, is an intimate transportation mode. If it is used routinely, the passengers and drivers get to know each other, which—beyond its basic social dimension—also fosters mutual trust and a sense of security. It is a service that belongs to, and is associated with, neighborhoods. This is an important consideration in ethnic low-income districts, but even the Hampton Jitney (started in 1974), which is not a jitney at all but rather a subscription bus that carries prosperous New Yorkers from Manhattan to the prestigious recreational area, has a social aspect with considerable cachet.

Job Creation

While labor input is usually a serious cost consideration in North America, there are plenty of people in poorer areas who need jobs, even if such employment does not offer high wages. Frequently, driving a vehicle is the entry toward gainful employment that may be sufficient to support a family and can lead to better opportunities. The required skills are not particularly high, and startup barriers are not onerous. Entire cottage industries in vehicle adaptation and maintenance at the neighborhood level are possible. Management skills in running services can also be acquired. All these considerations in job creation are particularly significant in developing countries.

Ease of Implementation

The start of a paratransit service does not require large initial lumpy investments. Off-the-shelf vehicles can be acquired, and some support equipment is needed; the right-of-way is already present; the personnel need rather limited training. The major issue in starting a service is likely to be securing operating authority—reviewing or possibly asking for adjustments regarding local rules and regulations of entering in public service, reaching some compromises with the local transit agency (unless the agency is the implementor itself), and considering the attitudes of organized labor.

Reasons to Exercise Caution

Experience shows that the various types of paratransit operations are associated with larger or smaller problems, which deserve attention if the full potential of this mode is to be achieved.

Cost Considerations

Regular paratransit service in North America is expensive, without doubt. Each small vehicle carries only a few passengers yet requires a driver, as well as support staff. Under normal conditions, private jitneys cannot compete with subsidized buses along the same route on the basis of ticket price. But such confrontations are not decided on a cost basis; the matter rests usually on quality of service and responsiveness.

Semilegal and illegal car services frequently do not charge the same fare as bus and rail transit, but presumably only because their drivers work at the minimum wage level (or lower) and receive no fringe benefits. Needless to say, they are not union members. It has been argued that many of the paratransit operations can be very cost-effective because they consist of one-man businesses with minimal overhead expenses. While this is true, it does not account for the need to have some overall managerial system. Nevertheless, neighborhood jitney services have the considerable distinction that they do not draw any resources from public coffers.

Use of Motor Vehicles

Automobiles and vans are employed, and, therefore, all the street congestion, safety, and air quality concerns associated with large street vehicles and not-so-clean engines are present. Accident rates can be very high if driver behavior is not well controlled; emissions of pollutants can be most excessive if engines are not well maintained; fuel consumption with aggressive driving under urban stress will be above norms. All these impacts are usually particularly severe with self-generated operations in the realm of questionable legality.

Labor Issues

As has already been mentioned in the "Job Creation" section, the paratransit mode is associated with very high labor input. If the operating and support personnel receive normal wages and all benefits, the resulting costs will be high, especially when considered in light of the inherent low productivity of small vehicles. On the other hand, with marginal, self-generated inner-city services, the workers remain at the edges of economic survival.

Section 13(c) of the Urban Mass Transportation Act of 1964 requires that any proposed service that involves federal funds in any way must obtain acceptance from local unions, ensuring that existing jobs will not be adversely affected.[11]

Driver Behavior

In those instances where the driver's income depends on the number of fares that are collected, there is a natural inclination to hustle and cut corners. This becomes very apparent on the

---

[11] "Capital assistance can only be provided if fair and equitable arrangements protecting individual employees against worsening their positions with respect to their employment are assured."

streets, with opportunistic disregard for traffic regulations and aggressive seeking of passengers. The business firms may extend the same short-cut approach to insurance coverage and vehicle quality and safety. All this, of course, does not apply to the many paratransit services that operate legally and correctly.

Profit Motivation

While issues of profit maximization are encountered with all privately offered services, they are of particular concern in the paratransit sector because such arrangements are more common with this mode. There simply is no natural incentive for an operator to run serrvices where the income does not cover costs, or to do it during low-demand hours. This is an obvious reason why overall controls have to be exercised so that steady service is available where and when it is needed. Internal cross-subsidy arrangements may be necessary, whereby the prosperous routes contribute resources to the weaker ones.

Institutional Issues

Evidence shows that the history of paratransit-type operations in the United States has been full of strife, as services have frequently emerged spontaneously and intruded onto the turf of conventional (i.e., official) transit. As a matter of fact, it is an embarrassing story encompassing outright criminality in some instances and purposeful refusal in many cases by municipal governments to face the issues or even acknowledge the existence of such operations. There is little doubt that government has to exert reasonable controls because a communal service is involved, and, if left to its own devices, cutthroat competition will ensue.

It can be suggested that the frequently unhappy situation has been caused by a rigid attitude that only full-scale conventional transit should be operated, regardless of profound changes in urban patterns, densities, and user needs during the last half-century. There is also the significant factor, as is usually the case, that regulations are structured to protect established interests. This refers not only to the conventional transit agencies, but to all other service providers and participants who have managed to secure a foothold in any activity. Fortunately, the rigid and inflexible attitudes are changing, compromises are being accepted, and various cooperative accommodations are being made.

Social Status of Paratransit

Paratransit is perceived in different ways in different places, always according to the role that it plays locally and the history of how the service came to be implemented. Attitudes are hardly ever neutral, ranging from an envious admiration of high-priced exclusivity on some services to a belief that only the marginal members of a community would use illegal service. There is also a wide array of paratransit operations in the United States that are specifically geared to the needs of special groups, such as the elderly and handicapped, or that are provided by specific institutions (religious, educational, medical) for their own clients or by employers for their own workers. These services may be operated at a higher or lower level of quality, but they are exempt from the status evaluation debate, being regarded as necessary and appropriate in any community for the tasks that they perform. They serve a somewhat captive ridership.

The interesting issues are associated with services that are available to the public, with the riders making their own choices. At the top of the list are various commuter van operations that provide a comfortable and rather direct ride for regular patrons, usually from prosperous neighborhoods, at premium fares. These services are equal to and perhaps exceed the attractiveness of suburban commuter rail operations. Variations exist, as for example the special taxi service from the East Side of Manhattan to offices downtown. An exemption from taxi regulations was granted allowing cabs to queue up on York Avenue near 72nd Street, being filled on a first-come-first-served basis by customers paying a flat fare. The vehicles enter FDR Drive directly for a quick trip to Lower Manhattan. The service is almost as good as taking a regular taxi, but it costs less.[12]

At the other end of the spectrum are the improvised and frequently illegal jitney operations that flourish in some low-income areas. They are patronized only by the residents of those neighborhoods, who are well aware of the risks that they may be facing and of the lawless status of the operations.

In the cities of the developing world, the class distinctions tend to be even more pronounced. Services are associated with specific neighborhoods, and the lines are rarely crossed. In many

---

[12] These are shared taxis, but they actually operate in the jitney mode.

Marshrut taxis in Moscow operating as jitneys or shuttles.

places, middle-class riders will never enter a public bus, but their own jitneys offer an acceptable alternative with satisfactory comfort, security, and status. Upper-class travelers, in turn, will shun the jitneys. The same patterns can be observed currently in Eastern Europe, with *marshrut* taxis (usually line-haul vans), operating at somewhat higher fares than buses and streetcars, being very popular—less crowded and much faster.

While jitney operations of this type of improvised and auxiliary nature do not exist in the well-organized communities of Western Europe, there are paratransit services that accommodate the elderly and handicapped, and special services with their own clientele at the luxury level can be found here and there.

Self-Generated Jitney Services in Developing Countries*

If, on any given day, all the users of public and communal transportation systems around the world were to be counted, the results would quite likely show that the largest group depend on privately operated, small-scale services generated at the local neighborhood or corridor level—in other words, paratransit, the way we understand the term in the industrialized countries. These travelers are not aware that they are doing anything unusual, and the teeming metropolitan areas as well as the remote villages of Asia, Africa, and Latin America are unthinkable without their *jeepneys, matatus,* or *publicos.* There are literally thousands of jitney and minibus systems in vigorous operation everywhere (except Western Europe and some cities of North America). At first glance, they all appear to be different, but this is primarily because of the hardware used—ranging from homemade bicycle rickshaws to sleek production-line minibuses. The institutional and managerial structures and basic operations are quite similar, which allows a general discussion and comparative descriptions. In a nutshell, private individuals acquire the highest-technology vehicle that is reasonable under the circumstances and respond to the mobility demands of their neighbors at tariffs that the customers can afford, but that are high enough to provide a living for the service provider.

These services tend not only to be self-generated, they also exist in a Darwinian environment (survival of the fittest) and, therefore, those that are visible on the streets are by definition successful. They probably have found their market niche by trial and error, but undoubtedly they have managed to achieve responsiveness to needs and flexibility in how they provide the service.

The principal positive characteristics of these services are the following:

- The operations grow out of the grass-roots service needs at the local level. There are no overall comprehensive plans; there is often little control by government. These conditions change as the services become more visible; in some instances they become officially recognized as base transit operations and a public management structure is superimposed.

- The vehicles are the most affordable devices that the community can muster. They may start with bicycle rickshaws or may be manufactured in local workshops as passenger compartments atop imported scooter engines and mechanical parts; local carpenters may build a passenger box that can be added to a pickup truck; or fully operational minibuses may be acquired. The type of hardware does not affect the basic format of jitney operations. Changes to the equipment can be made rather quickly if service levels demand it.

- The service is rather personal in nature because the vehicles tend to be small and they operate at the local level. Usually, the owners, drivers, and mechanics are members of the same community (except when service is offered to middle-class neighborhoods). Personal relationships and loyalties prevail between service users and service providers, and are a significant factor in continuity.

- The services are certainly labor intensive. Sometimes besides the driver there may even be a conductor (a small boy) who acts as a barker and fare collector. The generation and availability of such semiskilled jobs, including repair and maintenance work, are most often exactly what the community needs. Resources are kept within the neighborhood.

- The operators are in a position to respond to changing service patterns. They can make adjustments on an hourly basis or they can permanently change service routing without any real costs or difficulties. The drivers always know where service is needed, and they will pick up fares when and where they appear.

- Since small vehicles are usually used, a high frequency of service can be maintained. These vehicles may be the only motorized units that can penetrate the narrow and twisting lanes of many neighborhoods.

- The operations are undoubtedly incubators of technical and managerial skills for the participants, leading from doing simple tasks to the ability to tackle more complicated business challenges. In some cases local workshops producing or completing vehicles are the start of industrial activity.

- Large investments are not needed since fleets can be built incrementally, and the equipment can be progressively upgraded. No new streets or lanes need to be built (although that would certainly be most desirable in almost all instances). No research or development is ever undertaken; hardware is bought off the shelf as finished vehicles or sets of key components around which locally adapted vehicles can be produced. No public assistance or subsidy is given or expected.

All this is possible because individuals work hard to gain the largest possible income to support their families, to do better economically, and perhaps build a larger business. The industry tends to attract the more aggressive and energetic people from the local communities, who can enter the service with rather modest initial investment. There is powerful motivation, based on self-interest, to respond to the needs of the customers.

Lest this review become a one-sided paean in praise of unbridled private enterprise in public transportation, the negative features of these operations have to be outlined as well. While there are many variations from place to place, the following problems appear with some frequency:

- Service will be provided within a given corridor or even during a certain period only if it is profitable to do so in that instance. Low-density and very poor neighborhoods may remain without service, and operations may not be consistent.

- The high competitiveness of the activity frequently keeps the net income of the drivers very low, barely at survival levels. (It also keeps the fares down.)

- Situations are encountered where strong and sometimes even criminal-based organizations (warlords and gang leaders) have been able to establish monopoly situations, with all the abuses that accompany such conditions. Government agencies may not be strong enough or they may be preoccupied with other challenges, leaving operations alone that work somehow.

- The scramble for fares and the need to have a high turnover of riders results in aggressive driving that tends to ignore most traffic regulations and shows little consideration of other users of streets and roadways. Proper insurance is often a luxury.

- While in many instances public agencies maintain a hands-off policy, there is a growing number of cases, as cities reach higher states of development, in which governments actively discourage or even combat these operations. This is usually done in the name of upgrading the civic image of the city, since the improvised and self-generated services tend to give an impression of being primitive and threatening to middle-class attitudes. They do not look "modern."

- The quality of equipment maintenance and the presence of rider safety and comfort features may be seriously deficient within the freewheeling and constrained business environment.

- The regularity and reliability of service can be seriously affected by the individual behavior of drivers or monetary inducements by patrons.

These paratransit services, so dominant in cities of the developing world, are caught between two powerful forces. On one hand, there is the ever growing need for mobility services in expanding urban areas, where the spontaneous creation of communal transit is a welcome and frequently indispensable aid in maintaining livability. On the other hand, many internal problems are created that are difficult to accept in organized societies. It is quite clear that the negative features of rampant and rapacious capitalism and incipient criminality should receive corrective attention by public bodies. It is not equally clear how to structure and control these services so that everybody in need of urban transportation will receive it. Yet, it should be possible to take advantage of the energy and local resources characteristic of these operations to structure critically needed mobility systems for communities with minimal investment and great flexibility.

* This material is largely abstracted from the articles and papers published by the author, starting with the jeepneys of Manila in 1972. R. Cervero makes a strong point in *Paratransit,* op.cit., that the experience in developing countries should be a direct model for services in the United States. That is a good thought, and much is happening informally already, but the context and expectations are too different to assume an easy transferability of practices appropriate in the developing world to most American communities.

Application Scenarios

To start, it can be asserted that every American community should have mobility services that accommodate people who have serious difficulties using regular transit, particularly if that service is skimpy (or rarely available). This applies to high-density and low-density districts, with full awareness that the Americans with Disabilities Act (ADA, 1990) mandates that all public services be fully accessible. It is not only a mat-

Elaborately decorated jeepneys on the street in Manila.

Bicycle rickshaws in a rural town in Thailand.

ter that some people have severe disabilities that make entering bus and rail vehicles close to impossible even in the best of circumstances, it is a question also of responding to individual desires for comfort and convenience. There is every reason to equip conventional transportation facilities and the rolling stock with devices that assist entry/exit and to train the elderly and handicapped in their use (as a number of agencies are doing) to increase the number of people on regular transit. Dallas (DART), which operates vans and sedans in paratransit service as a part of its total scope of activity finds, for example, that it costs $35 for each such trip, while the expense (not the fare) on regular transit is only $2 to carry a passenger.

Programs to accommodate people with mobility impairments appear not to be enough by themselves in a prosperous society, but they are the starting point for comprehensive efforts. Undoubtedly, special paratransit services are expensive on a per-unit basis, but that is a cost that we should bear. While institutional and charitable organizations can share the load, the overall responsibility still rests with local government toward all its own citizens.

Tap taps on a major street in Port-au-Prince, Haiti.

The issues are more complicated with respect to general patronage, i.e., ensuring mobility in all instances, in all districts, for all residents and workers. There is no question that regular bus and rail transit should be the preferred modes, but it has to be recognized that a sprawling built environment now characterizes more territory of metropolitan areas than compact city districts. High-volume, capital-intensive transit becomes unworkable in these situations, and is certainly not cost effective, leading to a search for alternative transportation systems. There is nothing feasible on the horizon today, beyond the private car and some bicycles, except paratransit. Until the day when American communities are rebuilt in a true city configuration with intensive corridors and medium-high densities throughout (which would be welcomed by urbanists and

Special vehicles for service in Dakar, Senegal.

transportation planners, but may be a long time in coming), the choices are extremely limited.

Within this context, two situations can be distinguished:

1. Districts with overall low density that cannot even support a reasonably responsive bus network

2. Districts where transit with conventional vehicles is lightly loaded and operates with huge deficits

In the first instance, dial-a-ride services appear to be an appropriate response being able to fill the service gap, albeit at rather high costs. Such operations could be run by public agencies or private entrepreneurs, and there are some technological and managerial features that can make the services more workable than has been the case with the earlier experiments. The second instance may involve a rethinking of fleet compositions, the use of suitable vehicles, and modifications to operational practices, leading to some reforms in the established practices of transit providers—in other words, replacing large buses with smaller demand-responsive vehicles on appropriate routes. These possibilities are outlined and highlighted on the following pages of this chapter, but first a few illustrative cases can be reviewed.[13]

[13] The aforementioned *Paratransit* by Robert Cervero looks at just about every example that has existed in the United States and summarizes the experience up to the mid-1990s. Those cases will not be repeated here.

Names of Various Paratransit Services in Developing Countries*

Bicycle Rickshaws and Pedicabs

Trishaw	Hong Kong
Ojek; Becak (betjak)	Indonesia
Beca roda tiga; Lancha; Trishaw	Malaysia
Samlor	Thailand

Light Motor Vehicles Based on Scooters or Motorcycles

Tricycle rickshaw	Pakistan
Bajaj; Minicar; Mebea/bingo; Helicak	Indonesia
Tricycle	Philippines
Four-seater; Minitaxi	India
Samlor; Tuk tuk	Thailand
Honda-om; Lambro	Vietnam

Jitneys, Vans, or Converted Pickup Trucks

Opelet; Bemo; Mikrolet	Indonesia
Minibas	Malaysia
Auto calesa; Jeepney	Philippines
Rot song teow; Silor	Tailand
Auto rickshaw	Pakistan
Cyclopus	Vietnam
Pak pais; Public light bus	Hong Kong
Pesaro; Libere	Mexico
Publico	the Caribbean Islands
Route taxi	Trinidad
Dolmush	Turkey
Sherut	Israel
Carro por puesto	Venezuela
Lotacão	Brazil
Colectivo	Latin America
Taxi colectivo	Chile, Colombia
Service	Middle East
Yek toman	Iran
Arabea ogra	Egypt
Bakassi	Sudan
Mammy wagon	Africa
Gbaka; Taxi baggage	Ivory Coast
Mutato	Kenya
Louage	North Africa
Kia kia; Bolecaja	Nigeria
Fula fula	Congo
Tap tap	Haiti

* This is not a definitive listing; it is based on the author's own experience and a scanning of the literature. There must be quite a number of other local terms and nicknames out there. Some of these systems are no longer in operation.

Puerto Rico

The conventional bus service of Puerto Rico is sketchy, confined to San Juan only, and not very popular; the real mobility is provided by a very extensive network of *publicos* (jitney vans) that cover the entire island and concentrate on urban corridors. The system is endorsed, assisted, and regulated by the government, but otherwise it is a privately operated endeavor. Owner-drivers provide the service, they can largely decide themselves when they wish to work, and they are organized in route associations that manage schedule assignments and provide various cooperative services, such as vehicle purchases. They also represent the industry before the public and lobby for its interests. The Public Service Commission (PSC) of the commonwealth is the principal agency in charge of the system. The PSC examines the qualifications of operators for public service and does "needs and convenience" studies for specific new route requests, with public hearings. The Department of Transportation and Public Works registers and inspects vehicles and locates terminals and stops on principal highways. Municipal governments control the location of stops on city streets and off-street terminals.

There are intercity routes, local routes that serve neighborhood centers and transfer nodes, *lineas* that provide door-to-door delivery upon telephone requests, and some other service variations. The vans leave a terminal for a specific destination from a queue, departing when a full load is assembled. A total of about 15,000 vehicles on some 900 distinct routes have been in operation. A major government effort, involving considerable public resources, has been the building of terminals that also include parking garages. They are usually well-designed facilities, with proper accommodations for passengers and vehicles, becoming activity centers in their own right. Now that a heavy rail system is

Paratransit (Caguas) terminal and garage in Puerto Rico (Bayamon).

under construction in San Juan (the *Tren Urbano*), the *publico* network is expected to be modified as feeder service to those stations. The *publicos* have become the base transportation infrastructure for the island and are well liked and accepted by their customers for their responsiveness and convenience.

Des Moines, Iowa

Des Moines can be expected to encounter as much difficulty as any place in the United States in keeping public transit going.[14] ("Transit in Iowa is two farmers in a pickup.") The regional transportation agency, however, under its Central Iowa RideShare Program with a staff of three, has achieved considerable success in gaining the assistance of major employers in the downtown district in organizing vanpools. The interesting feature is that the 60 vehicles are acquired by the agency utilizing federal assistance, and then assigned to recruited drivers who each accommodate 13 coworkers. The driver pays nothing, the fares from the passengers cover operating expenses, and maintenance is contracted out by the agency to a commercial firm. Many of the routes are rather long, making vanpools a particularly attractive choice for commuters.

The State of Georgia

Georgia, among others, has instituted a brokerage system to accommodate elderly and handicapped persons with individual responses in each case.[15] This is classified as *non emergency transportation* (NET), whereby the local Medicaid office contracts with a broker to administer and coordinate operations. In this case the firm is LogistiCare, which has a network of its own transportation enterprises with vehicles and drivers, but screens each request for a ride according to the applicant's eligibility and medical state, seeks to determine the most effective mode for each trip among many potential providers (which may be regular transit), assigns and informs the rider (and issues passes, if necessary), and finally reimburses the specific service provider according to a predetermined fee schedule. The results after several years have been most encouraging, with the total patronage

[14] J. Duffy, "Des Moines Rising," *Mass Transit,* June 2001, pp. 8–18.
[15] See "Brokerage System Brings Efficiencies to Paratransit in Georgia," *Mass Transit,* June 2001, pp. 30–32.

growing (even on public transit) and costs being kept in check because the cheapest and most appropriate service is found in each instance. It also shows that effective paratransit operations are not so much a matter of hardware as of responsive overall management, targeting specific needs and opportunities. There is no reason why brokerage arrangements cannot also encompass other communal programs (such as welfare-to-work trips, delivery of meals, and even paid-for urban travel).

Palm Beach County, Florida

Palm Beach County on the other hand, has decided to assume the responsibility for paratransit directly.[16] This is a completely automobile-dominated territory, deliberately sprawling, with many large and scattered gated retirement villages that are not within walking or bicycle distance of anything and not welcoming to conventional buses. It is, however, an affluent county, and the transit agency (the Palm Beach County Surface Transportation Department, or Palm Tran) provides 120 paratransit vehicles, which exceeds the size of the bus fleet (106 units). All service is contracted out, responding to telephone requests, and a sizable portion of the county's annual budget goes to paratransit support, i.e., to subsidize the deficit operations. There are plans to build a county garage for the vehicles, thus carrying further the concept of government-owned but privately operated service.

Components of Paratransit Systems

Vehicles

Any vehicle can be used for paratransit service, ranging from bicycle rickshaws to full-size buses. Some, however, are more useful than others. Desirable specifications would include the following items:

- Enough seats to provide some efficiency, but not too many so as not to impede agility (10 to 15 might be a good range for general-purpose applications).

- Large doors that expedite boarding and alighting by passengers.

[16] See "Florida is *the* Paratransit State," *Mass Transit,* June 2001, pp. 21–28.

- Sufficient headroom and aisle width to allow convenient movement inside.

- Use of standard production models, if at all possible, to minimize purchase costs.

- Good maneuverability in city traffic, built-in safety features, good fuel efficiency.

- Arrangements so that the driver can collect fares (if any) quickly.

- Environmental protection features.

Indeed, such vehicles are available from a number of car and truck manufacturers—either passenger vans (possibly with a raised roof "bubble top") or minibuses that allow passengers to walk inside (such as the vehicles used by car rental agencies at airports). The reason why buses are rarely suitable for paratransit service is that the many passengers that they can carry will require many long stops to board and disembark riders, even on short trips, thus slowing down operations and impairing one of the principal benefit of this form of transportation—quickness. Passenger cars, of course, have a low payload ratio and are rather inconvenient to get in and out of, especially from the middle of the seat.

Vehicles used to serve elderly and handicapped persons should have the added feature of being able to accommodate wheelchairs

Van with wheelchair lift.

and plenty of handholds; floors as low as possible are also highly desirable. UMTA initiated a research and development project in 1975 with the aim of designing a vehicle specifically suitable for the patrons with impaired personal mobility.[17] Demonstrations were done in the early 1980s, and the findings have been useful in building or adapting suitable vehicles since that time.

[17] By McFarland and Minicars, Inc. of California.

Drivers

Since drivers of paratransit vehicles come in closer contact with their customers than on any other transportation mode, their personal attitudes and behavior are of importance. The technical qualifications are not extensive since small vehicles are in operation, and a regular chauffeur's license is quite adequate. This fact allows part-time staff to be engaged. In many instances, the operators are also the owners of the units. Problems are usually encountered only with the neighborhood self-generated jitney services where less-than-adequate driver qualifications are sometimes encountered.

Stops, Terminals, and Garages

These cause the least problems with paratransit services, although some complications may have to be faced. Stops are simple affairs along street curbs, usually without shelters because of the low frequency of service or lack of precise schedules. There may be questions as to whether regular bus stops can be used jointly and whether arrangements should be made to keep the stopped vehicles out of traffic streams (reserved curbside space within a parking lane, for example) and how to police proper use. Terminals usually consist of turnaround arrangements and reservoir space to accommodate a queue of the waiting vehicles. There may be exceptions, however, in instances where jitneys represent the base service for a community (see the preceding Puerto Rico case). Overnight storage of vehicles is a minor issue, with the units returning to the operators' homes or being parked in regular garages or on the street.

Communications Systems

For dial-a-ride operations, easy communications between customers and operators are crucial. Requests have to be taken, trips scheduled, and on-time pickups have to be completed as promised. For systems of any size, this can become a task of some complexity. It can be expected that today's computers and those that will be available in the near future will expedite these matters and bring efficiencies and reliabilities that earlier systems were not able to achieve. All this represents investment in hardware, software, and skilled personnel.

Ownership and Management Structures

The trend today is toward the incorporation of paratransit operations within the array of services provided by comprehensive transit agencies, recognizing the need to ensure mobility for all residents of a region. Paratransit services can also be contracted by engaging private business enterprises that specialize in this field. Some of these firms are of considerable size, with operations in a number of locations.

Jitneys, however, traditionally have been auxiliary feeder services started and operated by individuals or self-arranged groups. At various times and in various places, the organization takes the form of a loose association of owner-drivers who establish their own area or corridor of operations, a cooperative, or a formal business firm with a complete administrative structure. In all cases, the principal tasks of the organization are to protect its operations from intruders (poachers), to ensure that customers are reasonably satisfied with the service, to maintain discipline and competence among members, and to represent the enterprise to the outside—particularly the local government. The actual daily management of a jitney service primarily consists of the assignment of sufficient vehicles on any given route to carry the demand, hour by hour, in a continuous round-trip stream.

The duty of the local government (for operations contained within a single political jurisdiction) is to ensure that a safe and responsive service is offered to the public, which encompasses the issuance of an operations permit (franchise or license), checking of driver qualifications and insurance coverage, certifying and inspecting vehicles, guarding against excessively high fares, processing possible complaints from riders, ensuring that corridors with low demand receive service as well (cross-subsidies within the organization may be necessary), and stopping any unauthorized operators and violators of traffic regulations.

The principal responsibility for regulating paratransit services rests with municipal governments, encompassing the tasks just outlined, but state agencies may also have a say, usually through public service commissions or their equivalents. Gray areas exist because of the sometimes unconventional or pioneering nature of proposed or emerging operations; therefore, legal vigilance is called for in all instances, since no two places are likely to have the same set of rules.[18] A major distinction is whether a service is *private,* with no fares and available only to a preselected group of patrons, or *public,* with unrestricted entry by patrons. Frequently, it also matters whether street hails are acceptable or all rides are to be prearranged, with stated origin and destination points. (Historically, the classification was among common, contract, and private carriers.)

Capacity and Cost Considerations

The capacity of paratransit service is a flexible concept—whatever it takes to satisfy the potential ridership within the parameters of the established service. This means that a sufficient number of vehicles and drivers have to be available from period to period to be able to respond to all service requests (within reason, insofar as some trips may have flexibility in their timing and can be accommodated during slack periods). However, the general service criteria should respect the assumption that paratransit is a base service for most customers who have few alternatives, and, therefore, that not providing sufficient capacity or reliability in trip timing will impair the viability and principal purpose of paratransit. Perhaps there will be another jitney coming down the street if the first one is missed, but all other paratransit services are not in a good position to offer backup vehicles.

A capacity and reliability issue may appear in connection with prevailing traffic congestion on streets because paratransit vehicles have to use regular surface facilities. This, however, is a general problem. It can be argued that paratransit responds to a public purpose, and, if a choice has to be made, paratransit vehicles should receive priority on the street network. Close access to origin and destination points is particularly critical for elderly and handicapped patrons.

The remarkable thing about paratransit costs is that they span a wide spectrum in terms of operation and maintenance expenses. Neighborhood jitneys and car services, being a form of local transit, are able to operate without any subsidies whatsoever, while some dial-a-ride services generate unit costs that approach those of individual taxis. The explanation is apparently to be found

---

[18] These issues are discussed at some length in part 2 of R. Cervero's book, op. cit.

almost entirely in drivers' salaries and fringe benefits. Services that operate in low-income neighborhoods have to maintain fare levels that their customers can afford. This results in very low compensation for drivers who have to work long and hard hours to eke out a living with almost no job security. This could be regarded as an unacceptable situation, except that there usually is no lack of willing participants.

It is well to note again that paratransit involves very little capital investment. That is represented primarily by the rolling stock, purchasing used vehicles that start at a few thousand dollars or new ones from about $20,000 and up. (Units equipped with special features and devices, being in effect custom-made, can command a high purchase price.) Other equipment, such as computers and office furnishings, is of a routine nature. Larger operations may have their own repair and fueling facilities; in most instances these tasks will be accomplished through normal commercial establishments. The vehicles operate on public streets, and no charges are levied for the use of the surface traffic lanes (except through normal taxes).

Possible Action Programs

Just about all paratransit services that we are aware of have emerged or been instituted in situations where the demand has been quite apparent. It is not a question of doing elaborate feasibility and planning studies (although they may be useful if the effort is to develop comprehensive transportation systems), but a matter of filling obvious gaps and deficiencies. The specific requests most often come from underserviced corridors, districts, and potential user groups. It can be anticipated that these first steps in identifying needs will prevail, and there is nothing wrong with that, except that some constructive foresight from overall planners should be expected.

After defining the need, the principal question is the matter of identifying the agency that should be responsible for initiating and providing the service: private or public, already established organizations or newly created ones, as components of comprehensive systems or as freestanding local endeavors. What should be the level of subsidy (if any), and where should the resources come from? It is well to note that the federal government recognizes the significance of paratransit operations, and that assistance in well-defined instances is available.[19] It should also be recognized that experimentation and trial programs, in lieu of elaborate and detailed implementation studies, may be quite appropriate. The startup costs and the wrap-up expenses are not large, and, given enough time to break in a program and gauge ridership, actual demonstrations will give conclusive data on whether to continue, expand, or stop.

The operational plans for feeder/jitney systems are simple—it is a question of providing enough capacity on a route to carry the load without discouraging patronage, and responsive vigilance and flexibility are indicated. The other critical matter is a serious effort to maintain at least base service, even during periods of low demand and on key corridors with limited patronage.

For dial-a-ride services, the critical operational task in achieving reasonable efficiency is to arrange the schedule and routing of any vehicle to accommodate jointly as many rides as possible. Advanced information-handling procedures relying on suitable computer programs are indicated. The other promising effort to streamline and expedite operations would be to forgo efforts to provide door-to-door service for patrons, but instead move them only between well-defined nodes in a rather dense network.[20] Instead of attempting to serve a practically infinite number of origin and destination points in a community, limit the entry/exit opportunities to a still large but manageable number. Call-in requests would be accommodated between well-identified fixed neighborhood locations that are always within walking distance of the surrounding residences and establishments. They might be associated with city bus stops, have shelters, and provide communications portals.[21]

Suggestions have been made from time to time by various investigators,[22] referring principally to the experience in develop-

[19] These programs change continuously, and a review of the opportunities at any given time is always advisable.

[20] This concept has been explored and advanced by the National Transit Institute at Rutgers University. See various materials by A. Nelessen and L. Howe.

[21] See "Integrated ITS Technology and Flexible Route Bus Rapid Transit in Central Florida," *Urban Transportation Monitor,* July 21, 2000, p. 28.

[22] This is a principal policy proposal by R. Cervero, and it was also advocated by this author in earlier research. At this time, now that more experience is available and the general public attitude has become more understanding, the views of S. Rosenbloom would appear to be more supportable, as expressed in a review article, "The Paratransit Alternative," *APA Journal,* Spring 1998, pp. 242–244.

Self-organized commuter van service at the Port Authority Bus Terminal, New York City.

ing countries, that if substantially all current restrictions on transit service provision were to be removed, private enterprise would quickly respond to the challenge and accommodate the demand. Jitneys and commuter vans would provide service wherever it is needed and do it efficiently with minimal costs. While this remains an intriguing concept, it may not quite apply to American cities. Experience shows that models from the Third World work in districts that still operate at Third World levels. Regrettably, there are still quite a few of those in the United States. In better organized communities, however, greater responsibility toward patrons and operators has to be expected than that which is characteristically attained by private entrepreneurs. The principal issues are providing reliable service in places and under conditions that do not generate a clear profit and ensuring that the workforce is adequately compensated.

If those two conditions can be satisfied, there is every reason to explore possibilities in relying on private enterprise and individual energies to provide badly needed mobility in all communities and to foster competition.

Conclusion

It is apparent at this stage in the development of American cities that there are two areas where paratransit is critically needed to fill serious gaps in an otherwise rather rich mobility situation. The first of these are the very low-density districts where any conventional transit system would generate extraordinary and unacceptable expenses. The alternate option, besides the private car, appears to be limited to dial-a-ride. It will be costly on a per-unit basis, and much attention will have to be devoted to the task of keeping expenses at a reasonable level. There is a basic need to offer public transportation service to all those who not only cannot drive a personal car, but also those who choose not to do so. The expectation would have to be that the service can be made attractive enough so that its user base expands, which should constitute the eventual clientele for real community transit.

The other essential service would address social policy needs and be available only to those who are unable to enter and use regular services (or find it rather difficult to do so), even normal public-access paratransit services. This would encompass the elderly and handicapped, who would be screened and placed on a list of eligible patrons. These local transport operations would actually fall within the purview of social service programs, but they might be functionally best integrated with other transit services. There is not much point in charging fares for this paratransit service, except perhaps as a symbolic contribution, since extensive public support will unavoidably be required.

Beyond the two necessities, there are also three areas where paratransit may be potentially useful within the entire array of a community's mobility system. The first of these is the use of vehicles smaller than regular buses in a flexible or demand-responsive mode to feed travelers from lower-density districts toward mass transit stations/stops or activity centers. This service has to be seen as a full partner in an integrated network expediting the movement of patrons within a seamless system.

The other opportunity, which has proven itself in a number of instances, is to institute premium services at premium fares. This may not represent a politically attractive concept in an egalitarian society, but we have classes within many service areas anyway, including seats in theaters and airplanes, and this form of paratransit may be the best means to keep at least a portion of urban travelers away from automobile use in all instances. The comfort and semiprivacy that commuter vans can provide are powerful inducements toward at least some form of communal transport.

The final item on this list of paratransit opportunities is localized service on specific corridors or within intensive districts as jitney shuttles, accommodating short trips with a high turnover. Variable schedules may be in effect, and off-hours charter service may be available. This is a boundary area, entering the realm of regular transit, except with a stress on flexibility and responsiveness, which should be the principal characteristics of any paratransit service.

What this country needs today, and what it will need increasingly more tomorrow, is good paratransit. It works, it fills an important niche, and we will just have to pay for it.

Bibliography

Cervero, Robert: *Paratransit in America: Redefining Mass Transportation,* Praeger, 1997, 281 pp. A thorough review of all services in operation during the 1990s, as well as earlier ones. A survey of examples abroad, and an advocacy statement.

Kirby, Ronald F., et al.: *Para-Transit: Neglected Options for Urban Mobility,* The Urban Institute, 1975, 319 pp. An early summary and overview of the emerging mode.

Transportation Research Board: *Transportation Research Records,* National Academy of Sciences. This series of publications frequently includes papers on paratransit services—for example, Number 1433 (1994) and Number 1571 (1997).

Taxis

Background

The general perception may persist that taxis[1] provide a premium service to the affluent members of our society. This is true in the larger cities, as far as it goes, but it is not the whole story. For-hire services very much have to be counted among public transportation modes in any community today. A dressed-up couple going out for a night on the town will use a taxi (and they certainly should use it coming home); so will a businessperson going from the regional airport to a hotel or meeting place. But a middle-class urban resident who does not own a car, will use a taxi from time to time, as will a mother on welfare who wants to bring home groceries once a week from a store that offers better buys than the neighborhood establishment. Everybody needs a taxi occasionally for emergencies and as a backup for the personal car and regular modes of transit, as a personalized public carrier. In smaller places with limited public services, taxis or local car services are the backup means of mobility for just about everybody at one time or another.

---

[1] The name *taxi* comes from the French *taximetre*, referring to a charge—i.e., the meter that registers the fare.

Thus, in any community, taxis are an indispensable public transportation mode. They provide service of last resort, of reliable emergency response, of maximum convenience, and of luxury and indulgence if and when people can afford it. Yet, given the extremely high parking charges in some downtown areas, hiring a taxi may be cheaper than using a personal car for some trips. The taxi is also a public service that has never received—nor has expected—any public assistance except the use of public rights-of-way, for which it does not pay directly. There is, however, some indirect assistance since many social service programs (Medicaid, health insurance, and assistance for older citizens) reimburse their clients for transportation expenses, including the use of taxis when appropriate.

Taxi operations represent an industry that employs many people—today most frequently as a first step onto the more stable lower rungs of the economic ladder in American cities. Driving a taxi is a favorite occupation of recent immigrants, more so today than ever. It is a hard job with long hours, and it does not pay very well, but it is certainly a few notches above a menial service job or even factory employment.

In 2001, there were 180,000 officially registered taxi vehicles across the country, serving 1.8 billion riders each year. The cabs are licensed by and based in some 3500 communities, but service can be summoned from any location—even from remote places, if the customer is willing to pay the charge. Trips tend to be local (usually not much more than 3 mi), longer trips tending to be connections to regional airports.[2] Occasionally, the newspapers will report that a patron has taken a cab from one of the Eastern Seaboard cities to Florida (for a negotiated fee).

Taxi lineup at the Port Authority Bus Terminal in New York City.

[2] The average taxi trip in New York City is 2.6 mi (4.2 km). Short trips in Manhattan and long ones to the airports dominate.

The Taxis of New York

The taxi fleet of New York City, the largest in the country by far, with its distinctive cabs which eventually were all painted yellow, consisted of exactly 11,787 units for many decades. Ever since the local Haas Act was passed in 1937, that number has been one of the strongest constants in American urban affairs.* The total number of medallions carried by cabs was frozen, initially to cope with oversupply of vehicles and later to protect the investment that—not surprisingly—kept increasing in value because the demand for service was growing. From the original price of $10, the cost of a medallion has escalated at an average rate of 18 percent per year, reaching well above $200,000 today. The population of the city has not grown very much during this period; the demand has, however, with a change in lifestyles and the expansion of a prime global business center.

In the last few decades, the medallion has become more than just a license for transportation; it is also an instrument of investment for people who have no intention of driving a cab, but can lease vehicles out on a daily basis for a very good, risk-free income, reaping the benefits of a rate of appreciation not matched by any stock or bond. (There have been significant fluctuations in the last few years, however.) Those who purchase a medallion as an entry into their own independent business make a financial commitment in the vicinity of $250,000, including the purchase of the vehicle itself. After a down payment of about $35,000, the rest is borrowed through well-established channels. Of course, this loan has to be repaid, with interest. By working very hard for long hours in city traffic, the average owner-driver is reported to gain a net annual income of $20,000 to $30,000. The wisdom of doing this remains a mystery—particularly because the regulations insist that any given vehicle can be operated for no more than five years (thus, sufficient funds have to be accumulated to purchase the *next* cab), and there are no paid vacations or sick leave. That leaves the satisfaction of being an independent entrepreneur and of expecting to resell the operating permit later at a much higher price. So far, the medallion has been the New York taxi driver's retirement fund.

It is therefore easy to see why everybody in this industry has opposed most adamantly any public action that might erode the value of their investment. It took many years and much political energy to allow the city to sell an additional 400 medallions in 1996, but that number apparently has not made much difference either in the availability of for-hire vehicles on the streets or in the market value of the medallion. (Values actually rose during the auction.)

* The local press has followed the fortunes and events of this sector quite thoroughly. There have also been analytical articles and planning studies, particularly in association with various proposed modifications, which have appeared with some regularity. The NYC Taxi and Limousine Commission has prepared a series of informative reports on the various components of the for-hire activity.

The traditionally yellow New York cabs serve about 200 million passengers each year; they generate annual revenues of approximately $1.4 billion (1999).[†] After all overhead and fixed costs are subtracted, not much is left for the 40,000 individuals who carry hack licenses.

The legendary New York cabbie, as portrayed in countless Hollywood movies and immediately recognizable as a cultural type, retired some time ago. There are no more middle-aged wisecracking, cigar-chomping Jews or Italians behind the wheel on the streets of Gotham. Cabbies today are just about all recent immigrants from Asia, Africa, or the former Soviet Union. Applicants for hack licenses may speak any of at least 60 languages. Their English is frequently a problem, but this is no obstacle to receiving a permit to serve the public. The shortage of drivers is apparently serious enough to allow them to learn the city's geography by trial and error.

(The character created by Robert De Niro in the most famous movie about New York cabbies—Martin Scorsese's *Taxi Driver* [1976]—is not supposed to be typical. Unfortunately, another strong image has emerged.)

A significant peculiarity of the New York system was the distinction between fleet medallions (58 percent) and individual medallions (42 percent). Over the years, in response to changing demand and external economic forces, many adjustments have been made within the framework of the original legislation. For example, a large fleet was not the preferred form of organization from the 1970s to the mid-1980s, so two-person fleets were invented. These had the legal appearance of a business firm, but allowed owner-drivers to operate with considerable personal freedom. The corporate structure was a screen for stockholders, protecting them from personal liability and large insurance claims; injured passengers found it difficult to collect compensation. Brokerage arrangements under which licensed drivers "horse-hire" cabs are widespread. The driver rents a cab with a medallion by the 12-hour shift or by the week from a garage that is responsible for vehicle maintenance. The driver pays $100 up front and also pays for the gas consumed, which means that the fares go in the driver's pocket only after a number of hours have lapsed. The driver will net about $100 for a long day's work.[‡]

(The popular ABC TV series *Taxi,* with Judd Hirsch, Danny DeVito, and Andy Kaufman [1978 to 1983], was set in a fleet garage that somehow managed to stay in business despite all the problems that swirled in and around it.)

All this is controlled by the Taxi and Limousine Commission, a separate municipal agency. The commissioners are appointed by the mayor and the city

[†] NYC Taxi and Limousine Commission, *Taxicab Fact Book,* 1994 (the latest edition; an update is being prepared).
[‡] For full details of the New York City taxi system, see three articles by B. Schaller and G. Gilbert in *Transportation Quarterly:* "Factors of Production in a Regulated Industry," Fall 1995, pp. 81–91; "Villain or Bogeyman? New York's Taxi Medallion System," Winter 1996, pp. 91–103; and "Fixing New York City Taxi Service," Spring 1996, pp. 85–96.

council. The commission's responsibilities encompass setting tariffs (usually after pleas and documentation by the industry and public hearings), testing and qualifying drivers, maintaining discipline and proper behavior among drivers, and certifying vehicles for public service. In the past decades, much effort has also been expended to include under the commission's umbrella the other sector of New York's for-hire services that has emerged strongly in response to various needs and demands—public livery vehicles that may sometimes operate in a jitney or gypsy mode. Indeed, this sector holds many more vehicles than does the yellow fleet. (Perhaps 45,000 officially licensed units, plus many more unlicensed vehicles, which are difficult to count.)

This situation basically developed because the number of medallion cabs was not enough to serve the entire city, and the official fleet retreated to Manhattan's central business district and the airports, venues that provide the best returns. Since the 1960s, yellow cab drivers have rather consistently refused to go to the low-income districts for fear of crime, and have generally refused to pick up anyone who might wish to go there. This means people of color. This practice is illegal, but it persists.

The grassroots response has been the blossoming of neighborhood-based car services, originally licensed by the state as *public livery* operations. These are allowed to accommodate patrons only by prearrangement or telephone request; they usually rely on radio-dispatched vehicles. Given the overwhelming demand and the absence of yellow cabs, they do much more than that—including operation in full taxi mode, cruising the streets and soliciting street hails, and even running as jitneys. (See Chap. 6, Paratransit.)

To cut a long and complex story short, at this time a truce prevails between the two sectors, which by and large respect well-established turf boundaries and do not encroach in each other's sphere of operations. The two sides have accommodated themselves to the implicit conditions, even though the nonyellows attempt to work in Midtown Manhattan from time to time. It is an unsatisfactory and inelegant situation, but it works. There is little inclination by anybody to undertake drastic corrective action or to think about basic reforms. The immediate concern is the control of gypsies and poachers, who enter the scene without being licensed in any way. They are despised as damaging competition by everybody in the business, particularly by those one notch above them in legal standing—but not always by people who need transportation.

But that is not all. In 1982, to increase the number of cabs cruising the streets, existing radio-dispatch systems run by some yellow cab operators were transferred to a new class of for-hire vehicles—*black cars* (not necessarily always painted black), which were to provide a higher level of service by prearrangement. Black car services offered better vehicles, driven by chauffeurs dressed in suits, for those who were willing to pay a somewhat higher fare. These operations, together with limousine service, have become the modes of choice for New York businesspeople. Large firms frequently have standing arrangements with black car associations to keep vehicles waiting near their offices so that personal transportation is always available to their employees or visitors. The operators are owner-drivers, properly licensed and franchised, working through organizations that provide radio contacts and other managerial services. As an illustration of the fluidity of the for-hire transportation business, it can be mentioned that black car and limousine drivers are known to solicit street hails during slack periods.

As a result of all these efforts, the city enjoys reasonably good personalized public transport service—except sometimes on Friday afternoons, and when it rains. The principal message that seems to emerge from the New York example is that mobility services will be created if the demand is significant and if customers have expendable funds. Regulations and controls may catch up—or they may continue to lag behind, which makes no crucial difference in terms of actual operations on the street.

Development History

The operation of vehicles for hire as an organized activity can be traced back several centuries, and taxi-type operations are, indeed, the oldest form of transportation available to the public. It would be possible to start this review with Charon, who ferried the souls of the dead across the river Styx in Greek mythology, but that association should not be highlighted in a discussion of urban mobility services. Reference could be made to the many known instances in ancient cities in which boatmen carried passengers across waterways for a fee, sometimes under franchise.[3] There must have been countless unrecorded informal instances of people paying a fee to be carried by animals or vehicles owned by others.

In the seventeenth century, major European cities—particularly Paris and London—became large enough and had a vigorous enough business and social life to need transportation assistance beyond walking.[4] The logical response was to place horse-drawn

[3] The right or license granted by government to an individual or group to market (sell) a company's goods or services in a particular territory.

[4] There are not too many reviews of the history of taxis. The most complete examination is found in G. Gilbert and R. E. Samuels, *The Taxicab* (University of North Carolina Press, 1982), the only book so far that covers the entire industry. See also G. N. Georgano, "Historical Survey of the Taxicab," *The Taxi Project* (Museum of Modern Art, 1976), pp. 109–129, which provides a detailed history of the evolution of the vehicle.

carriages at strategic locations and make them available for hire by the trip. These were known as *hackneys*. They were socially acceptable, and after about the 1630s everybody who could afford the fares used them in large cities. English literature, particularly that covering urban life in London and New York, is full of references to hackney cabs, in works by authors ranging from Samuel Pepys to Arthur Conan Doyle. This transportation mode continued for two centuries virtually unchanged, except for the form of the vehicle. A remarkable array of carriages and carts was tried and used during this long period, developing certain criteria that are still largely applicable today. The vehicle had to be as light as possible to minimize consumption of propulsion power, but sturdy enough to last under the strains of city driving and poor pavement. The comfort of paying passengers was always a major concern, which encompassed ease of entry and exit (particularly for ladies with voluminous skirts and gentlemen with swords and top hats). Patrons had to be shielded from inclement weather, the aft end of the horse, and too close a proximity to the coachman. The best method of collecting the fare and the determination of an equitable amount were explored continuously.

Certain operational procedures and elements were established from the beginning. There were designated private and public hack stands or *standings;* the vehicles sometimes cruised or *played for hire.* The drivers became members of a well-recognized trade with some social status and a colorful reputation. *Cabbies* were assistants to the coachmen and took care of the horses. Some operators were owner-drivers, but associations and companies were the dominant business format, based on *livery* stables that hired out (i.e., leased) horses and equipment. Hackneys caused traffic congestion at some locations, cheating on fares was a steady complaint from the clientele, and there were requests from time to time to improve dilapidated and dirty carriages. Yet, business was generally good, and there are recorded instances of favoritism and attempts to corner the market in various cities.

Regulations became necessary as the business expanded and freewheeling competition ensued. Hackneys blocked the movement of carriages carrying members of the nobility, and broke up the already inadequate street surfaces. Government bodies tried various forms of rules to address various issues. Frequently, these involved a fixed maximum number of operating licenses, which needed continuous adjustment. Attempts were made to regulate fares, and license fees were imposed. Eventually, in England in the early nineteenth century, the rules were extended to the permissible length of the workday, standards of behavior regarding passengers, inspection of horses and coaches, and other concerns of effectiveness, safety, and quality of service. The license, in the form of a metal plate—a medallion—had to be displayed on the equipment.

Sedan chairs also appeared, but they did not last very long; *fiacres* and other types of coaches were popular vehicles during some periods in different cities. While the term *hack* has remained the generic designation, there are other terms derived from historical coach types that are still familiar today.[5] For example, *cabriolets*—light two-seaters with two wheels and one horse—appeared at the end of the eighteenth century; this soon became shortened to *cab*. Another effort in the search for a better vehicle was the *hansom cab*—a low-slung two-wheeled carriage with the driver sitting in the rear on a high perch—which became the dominant form of horse-drawn, for-hire vehicle until mechanical engines arrived.

All the efforts previously mentioned refer to individual and exclusive service, not mass transportation. The growth of cities reached a point in the early nineteenth century at which larger volumes of users of modest means had to be carried in regular patterns, and public transportation as a new and ever-expanding branch of transport service had to be developed.[6] (See Chap. 11, Streetcars and Light Rail Transit.) Personal for-hire services continued, of course.

Motorized taxis came a decade later than streetcars, in parallel with automobiles. As a matter of fact, the earliest applications of self-propelled vehicles were in the for-hire sector. In the last few years of the nineteenth century, electric cabs (the Electrobat) were introduced in Philadelphia, New York, and Paris in significant numbers. The horse in front of a hansom cab was removed, and a very heavy battery was placed under the driver's seat. This peculiar device did not survive for technical reasons, and horse hackneys kept expanding their presence in cities for the next

---

[5] One of them—the *victoria* (open carriage)—survives today in some cities serving the tourist trade.

[6] The short-lived attempt by Blaise Pascal to operate a *fixed-route* hackney service in Paris in 1661 is often mentioned as the first example of public transit.

decade. However, the internal combustion engine very soon reached a reasonable level of workability, and the gasoline automobile became the vehicle of choice for the next 100 years.

A new term—*taxicab*—was coined, and it spread throughout the for-hire mode. This occurred in New York in 1907, when cabs in one of the first fleets were equipped with a device to measure distance traveled—the *taxi-metre*. This was much liked by the riding public, since it minimized the likelihood of overcharges and arguments with the driver over the fare to be paid. The vehicles kept changing their shape during the following decades, but the operational and organizational structures had been set some time before, and they continued. There were, however, repeated adjustments and adaptations in different places in response to contemporary needs.

In the years before and after World War I in the United States, taxis quickly replaced hackneys, and the industry was characterized by numerous small enterprises, including owner-drivers. The service capabilities expanded in cities during the 1920s, but there was a definite shift toward large fleets and multifaceted business enterprises. This was a characteristic of the times, and taxi tycoons emerged, several of whom also owned transit companies and car manufacturing plants. John D. Hertz, for example, had fleets in several cities; among his pioneering innovations were attractive fares, telephone dispatching, reserved stands, and driver-training and service-quality procedures, as well as the discovery that yellow is the most suitable color for vehicles because of its high visibility. He also started the Yellow Cab Manufacturing Company, which developed and built a specific taxi vehicle. Later he expanded into other transportation sectors, before he found his real business arena in the rental field.

The issue of building a taxi vehicle suitable for North America runs through this history, and it remains unsolved today.

Standard passenger sedan used as a taxi in Istanbul.

The most successful enterprise was the Checker Cab Manufacturing Company, started in 1922, which set the industry standard for many years. A number of regular automobile manufacturers entered the market at various times with their own models, but none of them lasted—the excuse usually being that the market was not large enough to maintain a profitable production line. Checker went out of business in 1982, and no satisfactory replacement has emerged.

The Depression brought many drastic changes in most areas of American urban life and operations, including the taxi industry. Many people lost their jobs but still owned cars, and, since most cities had no entry restrictions at that time, they went out on the streets to make a living. Because the number of potential riders who could afford taxi service also dropped significantly, the oversupply of capacity generated a series of unfortunate results. The desperate cutthroat competition sometimes depressed fares below survival levels, and drivers everywhere engaged in illegal and unfair practices. In many places, the situation was chaotic and criminal. There was a public outcry; the industry and its members gained a deplorable reputation, and government had to step in with a much more severe approach than had been the case previously. The experience seemed to show that a service to the public cannot be relied on to work properly by itself in a presumably free market.

The new regulations adopted in many cities addressed entry controls, with the extreme case being New York, which set the maximum number of permits at 11,787;[7] fixed and published fares, which attempted to ensure a reasonable return and income; administrative responsibility, which guaranteed insurance coverage and financial ability to stay in business; quality and safety of vehicles; and driver qualifications, which also encompassed character references. Stricter policing practices were instituted. In this process, the taxi mode was transformed from a private business activity into a public transportation operation, carrying responsibilities that are not too different from those of a public utility. Nevertheless, it did not receive any public assistance or funding, and it still does not today.

World War II interrupted the production of civilian vehicles, but the cabs then in use were apparently sturdy enough to last

[7] As discussed earlier, it was increased only recently to 13,595.

during this period, and it was shown that various programs of shared taxi use (to save fuel) were possible and workable. After the war, there was an interlude when returning soldiers and sailors entered the presumed remunerative taxi business in great numbers, and the illegal "vet cabs" threatened to dislocate established practices. They were, however, absorbed rather soon by the booming economy, albeit after some difficult and complex administrative efforts.

A major change in taxi operating practices was made possible by the deployment of two-way radio communications systems. Cruising and waiting at hack stands could be largely replaced by quick and efficient response to individual service requests by telephone. As American cities became increasingly more dispersed through suburbanization, which reached intensive levels in the 1950s, localized car service companies with a telephone switchboard and a radio set became the normal model of for-hire service in these communities.

The change in inner-city districts was not technological but rather institutional. As is discussed in Chap. 6, with increased segregation and the abandonment of low-income neighborhoods by regular taxis, home-grown and sometimes semilegal car services filled the gap. These too were and are very much local enterprises, often organized along ethnic lines and providing mobility to the entrepreneurs' own neighbors.[8]

Otherwise, the latter half of the twentieth century did not see much physical change or innovation in the personal for-hire urban transportation mode. There have been, however, adjustments to organizational structures and shifts in types of service and their relative volumes. Since the 1970s, the taxi industry has been particularly concerned about federal (and local) assistance programs that encourage public transit expansion and use. This can be seen as a loss of potential taxi users, and there has been very little in the way of cooperative programs that would bring the taxi mode into the larger systems. Representatives of the industry claim that requests for fare increases to keep pace with inflation have met with much resistance. One result is that the take-home pay of drivers has dropped to a very low level. The services operate

[8] They are sometimes referred to as *vernacular cabs*. See P. T. Suzuki, "Vernacular Cabs: Jitneys and Gypsies in Five Cities," *Transportation Research,* 1985, vol. 19A, no. 4, pp. 337–347.

mostly with a very local base, employing drivers who leave for a better job at the earliest opportunity.

The number and size of very large taxi firms has declined in the last decade, but there may be a modification of that trend under way. In small and midsize cities, it is common to find a dominant company providing service; centralization of operations can also be observed in large cities. Several large business enterprises, including investors from Europe, have recently acquired taxi fleets in American cities.

An interesting event was an exhibition arranged by the Museum of Modern Art in New York in 1976 that displayed five different full-scale prototype vehicles in the search for a suitable urban taxicab.[9] The show provided a tantalizing glimpse of what could be, but there was no manufacturing follow-up.

Generally speaking, the taxi mode has established and maintained its service niche within the entire spectrum of mobility services available to communities, and reasonable stability prevails—but only as long as there is a supply of recent immigrants in the larger cities who are willing to work as low-paid drivers. Some modifications in service formats and entry into the paratransit field have been made, but only around the edges.

Types of Taxi Service

To the public, all taxi service looks about the same: a prospective rider or a few people going together summon a vacant vehicle, board, and tell the driver where they want to go. The cab will take the quickest route, or the patrons can instruct the driver as to which path they wish to take. When they arrive, they pay what the meter registers and usually give a tip. (In the large cities, the driver hopes for a 20 percent tip, but does

Cabs at a metro station in Mexico City.

[9] E. Ambasz, *The Taxi Project: Realistic Solutions for Today* (Museum of Modern Art, 1976).

not always get that much.) There are some variations, but this is the classic scenario of a *one-party exclusive ride*. In many places, this is the only form of operation allowed by regulations. If some members of the original group are dropped off along the way and the last rider pays the fare, it is still a one-party ride, although variations beyond this soon strain the nuances of strict legality.

The average occupancy of a cab throughout the United States is 1.4 passengers (plus driver), with very few exceptions. This is not a very high figure, suggesting that the exclusive ride is the preferred format.

The other possibility is a *shared ride*—several parties (strangers to each other) use the same vehicle on a continuous trip with different origin and destination points. A few cities allow this (e.g., Washington, D.C.), usually only if the first party agrees; most do not. There may be exceptions—for example, during high-demand hours, at nodes of concentration (airports), or during emergency periods (e.g., snow storms or transit strikes). The shared ride has the obvious advantage of gaining greater productivity and lessening street traffic loads; the disadvantages are that the length of the trip for any given rider is likely to be longer and the asset of privacy is lost. Many taxi patrons on most trips regard these conditions as unacceptable. There is also the problem of how the tariff is to be determined and distributed among the riders in a shared cab. The driver is certainly entitled to a higher fare than would be the return for an equivalent one-party ride, but not excessively so.

The other way to distinguish among taxi operations is according to the process of soliciting the vehicle:

- *Street hails,* whereby any vacant vehicle may be flagged down at any place at any time by anybody.

- *Hack stands,* which accommodate waiting cabs; patrons may enter them only at those locations. This scenario has the aim of preventing wasteful empty cruising, but it is practically unenforceable.

- *Prearrangement,* usually by telephone (dial-a-ride); patrons request a ride by specifying origin and destination points and time. This service, with radio-dispatched vehicles, is now available in all American communities; in many it is the only option.

The most significant classification form of the taxi mode is by ownership and organizational arrangements, even if they are not always visible to the public.

- *Independents (owner-drivers).* These are one-person business firms, with an individual owning both the vehicle and the license to operate. The owner (plus perhaps a relative on other shifts) does the driving, collects all the fares, and is responsible for all operational and legal aspects. This is the classical form of for-hire transportation, except that it does not entirely work under contemporary conditions—only in large cities where street hails are adequate to provide sufficient ridership and income. Otherwise, radio linkages and other joint auxiliary services (such as advertising, lobbying, bulk purchases, cooperative maintenance facilities, etc.) are almost indispensable. Frequently, therefore, independents maintain affiliations with a radio base.

- *Fleets.* These are business firms that own multiple vehicles and operating licenses, and usually can do all routine maintenance in house. In almost all cases (except in New York), the company will provide radio-dispatch service. The drivers may be *employees* who receive a commission and carry all fringe benefits. This is no longer the preferred format by the management side because of the associated costs. It is more likely that the vehicles will be leased by the shift to *lessee-drivers* who carry personal hack licenses. They are independent contractors in the eye of the law; they pay a rental charge and buy their own fuel. Except for insurance coverage, the firm has no financial or other responsibilities toward the drivers; the drivers are completely on their own to do the best that they can in making a living.

- *Minifleets.* These were two-person business arrangements (particularly common in New York City) that provided the legal protection of a corporation, but allowed each member to act independently as an owner-driver. They do not exist anymore.

- *Associations or cooperatives.* These are organizations of car services formed for the specific purpose of operational convenience—to take advantage of radio dispatching, to make bulk purchases, to pool repair and maintenance tasks, and to protect the interests of the participants. Sometimes the difference from fleets is only in the ownership of the vehicles; they may be enterprises that only provide telephone and radio connections to their paying members. Most often they take the form of a cooperative; they may have the characteristics of a brotherhood in ethnic neighborhoods.

Reasons to Support Taxi Service

- *Fully personalized service.* The taxi mode is more responsive to individual travel needs than any other option—more so than a private automobile because somebody else does the driving and there are no worries about parking. It operates at a high premium level, which is important to those who seek exclusivity and privacy; it is quite reliable and fast in emergency situations; it can serve lower-income residents if trips are infrequent and have key purposes (or if special services exist or supportive arrangements for payment are made).

- *Substitute for private automobiles.* Some urban residents may find that the occasional use of taxis in dense environments is a more cost effective practice than owning, maintaining, and garaging a car. In many instances, for-hire services are reasonable backup choices precluding multiple car ownership.

- *Availability and flexibility.* All American communities at this time are assured of taxi or car service that can be summoned by telephone with quick response times, or service can be ordered in advance. Any destination is accessible, even remote ones if the fare is acceptable.

- *Luggage accommodation.* Taxis are effective in carrying hand luggage; arrangements can be made for special shipments or larger loads.

- *Technology and skills level.* The vehicles are ordinary automobiles and can be repaired and maintained by regular mechanics. Drivers' qualifications need not exceed the level of an ordinary chauffeur's license. It is normally expected that taxi drivers, due to experience, will be the persons most knowledgeable about local geography and will be able to find any location within their territory. Unfortunately, with the short tenure of many drivers, this is no longer always the case.

Reasons to Exercise Caution

- *High tariffs.* All for-hire services are labor intensive; in North America, these costs can be high. This is particularly true if the drivers receive a normal salary or income with all fringe benefits.

- *Motorized traffic issues.* For all practical purposes, regular automobiles are used; therefore, taxis suffer from and often contribute significantly to street congestion, pollution loads, and accident occurrence.

- *Operators' behavior.* If the income of the drivers depends on commissions or the amount of fares paid, there are incentives to drive and solicit riders aggressively. This can lead to a series of unacceptable practices, ranging from minor traffic infractions to the endangerment of patrons. Extensive policing may be required in such instances. Many cases of discrimination and refusal to provide service have been recorded.

Taxi stand in Paris equipped with telephone and information display.

- *Operational friction.* Unlike regular automobiles, taxis stop frequently to pick up and discharge passengers. Since this does not always happen in designated spaces or along the curb, traffic flows are often impeded, and dangerous situations may be created.

- *Inadequate vehicles.* Unlike many European cities, American communities allow regular production models (only slightly modified) to be used in for-hire service. While this is the most economical approach and serves the driver well, passenger spaces tend to be constrained, and getting in and out of the vehicle can be difficult.

- *Low income and job security.* In its current form in the United States, the taxi industry tends to provide a meager economic return to its workforce. It constitutes a low rung on the job ladder in many places. Organizational arrangements frequently leave workers to their own devices, with practically no benefits or job security.

Social Status of Taxis

Taxis retain their position at the top of the list in terms of social prestige, exceeded only by chauffeur-driven sedans and limousines—which, in turn, can be classified as to whether they are leased or owned by the user. The superior standing of cabs continues, even though there have been times, usually during periods of economic hardship, when the behavior of many drivers has deteriorated into fare gouging and unsafe driving. These situations tend to be corrected, as do recurring instances of inferior vehicle maintenance and sanitation. The fares remain high, constituting a screen toward exclusivity and selectivity. There does remain the question of whether all patrons can enter and exit the production automobiles that are used as cabs in American communities with a semblance of dignity.

However, there is a different attitude toward the neighborhood-generated and sometimes semilegal car services that operate in the taxi mode. These are generally seen as utilitarian, necessary services with no particular higher or lower social standing. Everybody has to use them, at least from time to time, because other mobility choices are scarce. Some people, but not all, will draw a line against using for-hire vehicles that are not fully certified and authorized for service.

Application Scenarios

Every community has to have, and does have, individual transportation services available for hire that provide fast, direct, and exclusive service. The major question is what else the taxi mode can do. The possibilities include joint use to increase the productivity of the vehicles, lessen street traffic loads, and utilize the available fleet in some way to assist mass transportation. Of course, taxis are able to go beyond the fixed limits of transit service, and thus are able to extend transport coverage and accessibility to territories that otherwise would be left entirely to private cars. (These possibilities are discussed further in Chap. 6 and later in this chapter under "Possible Action Programs.")

An associated issue, always present with modes that command high fares, is affordability by potential patrons with lower incomes, who may be dependent on for-hire services when some special or recurring mobility needs have to be satisfied. At the other end of the scale, the Tipsy Taxi program of Aspen, Colorado, supported by community donations, provides a free ride home for anybody who should not be driving.

Components of Taxi Operations

The physical parts of taxi systems are quite simple and uniform across the country; the institutional arrangements—not a principal area of inquiry of this discussion—are most complex, and they vary from place to place.

Vehicles

Over the years, any number of models have been tried, always in parallel with contemporary developments in automobile technology. In this process, the specifications for an appropriate vehicle have been distilled, and they encompass the following:

- Sufficient space inside for at least three or four riders, with reasonable headroom
- Ease of entry and exit, without the need for passengers to double over; floors as low as possible
- Secure and easily accessible space for luggage
- Adequate and comfortable driver's compartment that can be screened off for passenger privacy and driver's safety
- High visibility and signals to indicate availability
- Superior maneuverability of the vehicle and ability to make tight turns, thus maintaining agility in tight urban settings
- Nonpolluting and fuel-conserving power plant

The vehicle that comes closest to satisfying these norms is the legendary London cab. Two models by Austin and LTI Carbodies (1958 and 1987) have been manufactured over the years, but even they are encountering problems today in the market. They are expensive as compared to equivalent regular cars, and large-scale production is not warranted by demand. Their best features are high doors for ease of entry (they are able to accommodate top hats), good luggage racks, and the ability to turn around on a regular London street. The vehicles were tried in New York, but they generated intensive opposition by American car dealers, and they were not certified for use on environmental and safety grounds. New York drivers did not particularly like them either, because the driver's seat was not comfortable. A pity.

The last real American taxi vehicle was the Checker cab, now seen only in motor museums, which was discontinued because of escalating production costs to serve a limited market. Imports of various European cabs have not fared well either, because of maintenance and spare parts complications. Thus, the fleets consist of regular automobiles from major U.S. production lines that are "hacked up"—equipped with heavier chassis, more durable brakes and shock absorbers (similar to police cars), signal devices, and the required livery (exterior color and markings). There is an expectation that some minivans and sport utility vehicles could be adapted for use as more suitable cabs than the usual options available today.

A normal cab will last about 3 to 4 years in service, but probably only 18 months in heavy city use under multiple shifts. Frequent inspections (perhaps three times a year) are necessary to maintain environmental and service quality. Some cities place a limit on how long a vehicle may remain in public service (5 to 10 years).

Possible Specifications for a Taxi Vehicle

Capacity	4 to 5 passengers (plus driver)
Length	175 in (445 cm) or less
Width	75 in (191 cm)
Height	72 in (183 cm) or more
Wheelbase	90 in (229 cm) or less
Turning radius	17 ft (520 cm) or less
Height of floor above pavement	11 in (28 cm) or less
Door opening	48 in (122 cm) high × 36 in (91 cm) wide
Headroom inside	56 in (142 cm) or more
Power source	Electric batteries, natural gas, steam, or any other zero-emission engine
Wheelchair accommodation	Desirable

Note: No such vehicle exists today.

Cities in Europe and on other continents frequently utilize small vehicles as minicabs—either for the entire fleet or as a second component in addition to regular taxis. This practice has never caught on in the United States.

Stands versus Cruising

Hack stands at key locations in high-demand districts are useful components of city taxi systems. They should be equipped with light signals, visible over several blocks, that show the availability of cabs, and they should have telephone booths to call for vehicles when none are waiting. Their principal advantage is that stands placed at frequent intervals should minimize wasteful empty cruising by cabs searching for fares. Drivers tend to engage in this practice even if the chances of being hired are equally good at fixed locations. Usually, 40 percent or more of the vehicle miles accumulated during a shift carry no paying passengers.

Another way to look at cruising is to regard the situation in high-demand districts where patrons expect to be picked up as soon as they reach a curb and raise their hand, instead of walking to a hack stand. Cabs then have to cruise to be effective. Indeed, the relative amount of cruising miles then becomes a measure of taxi service availability.[10]

Garages

Garages for overnight storage of vehicles, cleaning, and minor repairs are a necessary component of the system. They should be located as close as possible to service areas and should also accommodate owner-drivers.

Communications Systems

Except in high-density districts, taxi operations usually rely on two-way radio communication systems, locally controlled by a fleet or an association. The use of radio bands comes under the purview of the Federal Communications Commission. It can be expected that with further development of communications and information-handling technologies, more efficient and responsive service capabilities will emerge. At the present time the possibilities appear almost unlimited, most probably leading to arrangements whereby any prospective rider will be able to tap into a

[10] B. Schaller, "Taxi and Transportation Policy in New York," *New York Transportation Journal,* Summer/Fall 2000, p. 20.

comprehensive service database. (See the discussion of ITS potentials under "Possible Action Programs" later in this chapter.)

Drivers

Cab drivers come in much closer contact with their customers than do the operators of any other transportation mode, and they are directly responsible for passenger safety. Thus, not only their qualifications and skills but also their personal behavior and customer relations are a matter of public interest. They are usually licensed by a municipal agency (perhaps the police, the transportation or consumer affairs department, or a separate commission), which reviews not only their technical and medical competence but also their reliability and social responsibility.

In the United States today, at least in the larger cities, driving a taxi is, unfortunately, no longer regarded by most participants as a lifelong profession. The income is quite low, the hours are long, and personal dangers exist. The driver always sits with the back turned to potential assailants, and, while the take will seldom be large, there will be some cash, and robbery will appear to be temptingly easy to many criminals. A number of these incidents will escalate into murder. The partitions separating the driver from the passenger compartment provide significant protection for the driver,[11] but they are not foolproof, and they constrain the passenger space. There have been years when 40 or more cabbies (including gypsy drivers) have lost their lives at the wheel in New York.

All this severely limits the attractiveness of employment, and taxi drivers today tend to be recent immigrants and others who see this job as a temporary step before establishing a real career. The reason so few women work as drivers in the industry is presumably their concern for their personal safety. This overall situation is most unsatisfactory.

Fares and Their Collection

The choices in fare arrangements are:

- Flat fare
- Zone fare
- Metered fare

---

[11] Data from Baltimore show dramatic decreases in crimes affecting drivers after shields have been installed.

Through the window I hear, with a strange sort of pity,
the relentless, heart-breaking sounds of the city.
I'm part of its energy, power, and pain.
Tomorrow I'll do it all over again!
And again and again, 'til I'm shot in the head,
get killed in a crash, or just plain drop dead.
Potter's Field, Randalls Island, not known for its beauty.
On my stone let it say: Prince of Cabbies—OFF DUTY!
        Poem by Cabbie Prince (anon.)

Source: Published in the *New York Times,* November 15, 1991. The Hack Poet Society credits the poem to Peter Borovec (d. 1995).

The flat fare is seen very rarely—except on airport trips—and could be applied only within strict service district limits. The zone fare is also not common anymore (Washington, D.C. has it); it consists of specified increments when zone boundaries are crossed. The most equitable way to determine the fee to be paid is by measuring the distance (and time) that a trip consumes. The meters are quite reliable and tamperproof, and they give riders a comforting sense that no unpleasant incidents will occur at the end of the trip. The usual format (required by regulations) is an initial "flag-drop" charge, plus a set amount for each fraction of a mile traveled, plus a per-minute charge for time when the vehicle is not moving. (In New York in 2002, the flag-drop was $2.00, plus $0.30 for each one-fifth of a mile, plus $0.20 for each minute of stopping or slow driving, plus a $0.50 surcharge after 8 P.M.). In some places, at the discretion of the local regulators, there may be a special charge for luggage, for additional passengers, for going beyond a set service area, etc. The fare structure is often adjusted to generate an adequate supply of vehicles during periods of otherwise unsatisfied demand.

It is a curious fact that, while short trips are more remunerative because of the initial flag drop, drivers prefer long runs (such as to airports) because of the little effort needed to keep the meter clicking.

Ownership and Operational Arrangements

As has been previously outlined, there is great variety in the organizational structures of the various enterprises that constitute the taxi industry—ranging from single individuals owning large fleets to loose cooperatives of owner-drivers. They tend to be locally based, i.e., to remain close to the patrons and the regulators. Increasingly, the business enterprises provide the institutional framework only, leaving the operational responsibility to the drivers. What holds the activity together is primarily the communications system that receives requests for service from prospective customers and passes it on to the drivers in the field, who can then pick up the fares.

The classic London cab.

There are, however, other tasks that are best accomplished jointly or cooperatively. These include bulk purchase of supplies and fuel, basic repair and maintenance, provision of garage space, and—above all—protection of the interests of the participants, which encompasses securing adequate fare levels, deterring intruders and poachers, ensuring fair treatment by the enforcers of rules, etc. The taxi industry, both owners and labor, tends to be well organized, and it sometimes speaks with a single voice. As such, it is an effective local political force in most cities.

Capacity and Cost Considerations

The service capability of any given taxi system is usually measured by the number of licenses (or vehicles in operation) per 1000 people within any given community. This is not a perfect indicator—vehicle hours on the street during various periods would be better—but it is readily obtainable. The actual numbers may range from 0.2 to 1.2 in regular communities, which is a wide span.[12] The numbers clearly should reflect the propensity for using for-hire vehicles in any specific place, which should correlate with the level of business activity (bringing in outside visitors with no personal cars) and the availability and accessibility of

[12] G. Gilbert, op. cit., p. 108 ff.

public transit services. Among large cities, Washington, D.C., leads, with more than 11 per 1000 people (free entry into the business), but it is also high in Boston and New Orleans (about 2.5), and usually remains above 1.0 elsewhere. The New York case illustrates the uncertainties associated with this measure: if only medallion cabs are considered, the rate is about 1.5; if all livery vehicles, black cars, and other service cabs are included, the measure is more than 7.

However, the national database on these factors is not yet fully developed, and too many complex factors are in play to attain reliable ratios and to define norms. For example, the cabs with licenses issued by any one political jurisdiction are not confined in their operations to those municipal boundaries. The mode has sufficient internal flexibility to fill the service needs without too much of a time lag, even if strict entry controls are supposed to be in place (see the earlier description of experience in New York).

The principal criterion in setting taxi tariffs is to ensure a reasonable return on investment (vehicle, license, and infrastructure costs) and a fair income for the work force—provided that the charges are acceptable to a sufficient number of patrons. This always has been a dynamic situation, bouncing between high fares that are no longer affordable to a large segment of the population and low fares that no longer provide a decent living standard for the drivers. Attempts have been made to correlate the fare on an average taxi trip to the ticket price for an equivalent trip on public transit. If the average taxi trip is 2 or 3 mi, the ratio is expected to be 4:1 or 6:1. The comparison is not very satisfactory, because taxi charges are based on distance and transit charges are usually not; a for-hire vehicle may carry several passengers for the same fare; etc.

Possible Action Programs

Ensuring that a responsive, responsible, and reliable taxi service is available in any community, at any time, for any person, is the core task of planning and structuring a local system. There are a number of discrete actions that can be taken to expedite and improve for-hire service.

- *Hack stands.* Taxi stations at key locations to be used by all for-hire services, providing nodes of operation and passenger access.

- *Restrictions on cruising.* Attention by police to reduce waste in operations.

- *Preferential use lanes and access.* Permission for all public service vehicles (for example, taxis carrying fares) to utilize facilities that give preference to multioccupancy vehicles; possible creation of exclusive lanes and spaces for taxis where the volume warrants such allocations (for example, loading bays at stations).

- *Integrated dispatch and control systems.* Expansion and improvement toward comprehensive intelligent transportation systems wherein taxis constitute a significant component.

- *Use of an appropriate vehicle.* Encouragement (and certification) of vehicles with superior passenger convenience features.

- *Vehicle standards.* Surveillance of the fleet regarding the age, condition, and safety features of each individual unit in service.

- *Environmental protection and fuel conservation.* Because cabs are replaced at a rapid rate, and they are under strict surveillance, pioneering programs in this sector toward a "green" urban fleet could be effective.

- *Shared rides.* Authorization of arrangements, in appropriate situations, that achieve greater vehicle productivity and reduce costs to users, leading to paratransit concepts.

- *Coordination with transit.* Recognition of the taxi mode as a form of public transportation, with direct-access arrangements and coordinated schedules. (Cooperation between the public and private sectors in transportation.)

- *Special night service and backup to transit.* Expansion of the coordination concept, seeking opportunities to respond to service needs with an effective and appropriate mode. There are many opportunities in this area—for example, using taxis for the guaranteed ride home that is a part of many carpool and vanpool programs for participants who may have to work late.

- *User-side subsidies.* Given the fact that many low-income residents depend on taxi service from time to time, expansion of public assistance programs is warranted in many instances (such as the issuance of transportation coupons or sale of half-price vouchers to selected users).

- *Driver qualifications*.   Programs that would improve and maintain the operating and communications skills of drivers and their expertise in local geography are frequently welcome. (In London, for example, prospective drivers sometimes spend 2 years on a bicycle to acquire "the Knowledge"—the ability to locate instantly any street in the large service area.)

In recent years, some interesting intelligent transportation system (ITS) developments have emerged in the taxi industry.[13] These efforts rely on advanced electronic communications and management systems and promise more reliable and safer service to the patrons, although perhaps accompanied by some increase in cost. In Australia and the United Kingdom, systems are being set up that accept taxi bookings via the Internet and keep a record of the individual preferences of repeat customers. The use of the global positioning system (GPS) is likely to become a routine process, enabling instantaneous and responsive management of fleets. Singapore appears to be leading the field, with an ability to locate the nearest vacant cab immediately after a service request has been received, and dispatch it through clear digital (not voice) instructions. In Hong Kong, similar procedures would be elaborated further by written instructions and confirmations to overcome potential language problems, with fare payment by credit card.

Another dimension of ITS procedures is that they would enable dignified communication between drivers and persons with sensory or cognitive difficulties.

Use of shared cabs at York Avenue in Manhattan.

[13] See *Proceedings of 6th World Congress on ITS,* ITS America, Toronto, November 1999.

Conclusion

The taxi mode is well established in American communities, and it serves as a premium service, emergency service, and backup service for nonmotorists. It is available, and it works in almost all instances. Steady vigilance, however, is needed to maintain service coverage and quality, which may deteriorate if left entirely to private enterprise and market forces. It has to be emphasized again that the taxi mode continues to operate as a publicly available service without any external subsidies. The principal issue today is the ensurance of an adequate and secure income for the workforce, which should upgrade the quality of service. Much more could be done to integrate individual for-hire operations into the total array of urban transportation systems.

Bibliography

Ambasz, Emilio: *The Taxi Project: Realistic Solutions for Today,* Museum of Modern Art, New York, 1976, 160 pp. Exhibition catalog and background essays on prototype taxicab vehicles.

Gilbert, Gorman, and Robert E. Samuels: *The Taxicab: An Urban Transportation Survivor,* University of North Carolina Press, 1982, 200 pp. The only book with comprehensive coverage of the taxi mode; an advocacy view of the industry and its components.

Schaller, Bruce: Articles in *Transportation Quarterly* and reports issued by Schaller Consulting documenting just about every aspect of the New York City taxi business.

Transport Research Laboratory: "Taxi and Paratransit Transport Update (1997–2000)," *Current Topics in Transport,* United Kingdom, 2000, 29 pp. Abstracts of articles, reports, and papers in English on taxis, paratransit, minibuses, and jitneys. There are also earlier compendia, such as the one published in 1997.

University of North Carolina at Chapel Hill (various authors): *Taxicab Operating Characteristics, Cooperative Forms of Organization, Taxicab Regulations,* series of reports prepared for the Urban Mass Transportation Administration/U.S. Department of Transportation, distributed through the Technology Sharing Program, 1982 and 1983. Research reports describing and analyzing the taxi industry nationwide, based on surveys and interviews of service providers and regulators.

Webster, Arthur L., et al.: *The Role of Taxicabs in Urban Transportation,* U.S. Department of Transportation, Office of Transportation Policy Analysis, December 1974. A summary of the state of the industry at that time.

CHAPTER 8

Buses

Background

Buses are without question the workhorses of the transit world. There are a great many places where they are the only public service mode offered; to the best of the author's knowledge, no city that has transit operates without a bus component. Leaving aside private cars, all indicators—passengers carried,[1] vehicle kilometers accumulated, size of fleet (see Fig. 8.1), accidents recorded, pollution caused, workers employed, or whatever else—show the dominance of buses among all transit modes, in this country as well as anywhere else around the world.

Yet, their presence causes no excitement; there is no shred of glamour in their role on the urban stage. They are taken for granted, and, particularly in American communities, they are seen as the mode for the less-prominent part of society. If a bus service is substantially improved, which usually means adding more vehicles on routes, the photo opportunities for local politicians and celebrities are limited. Monies are spent in a quiet and routine way; no major public works efforts (except for largely invisible garages and yards, and occasionally a terminal) are involved; there is not much ribbon cutting. The service tends to

---

[1] See the statistics at the start of Chap. 11, Streetcars and Light Rail Transit.

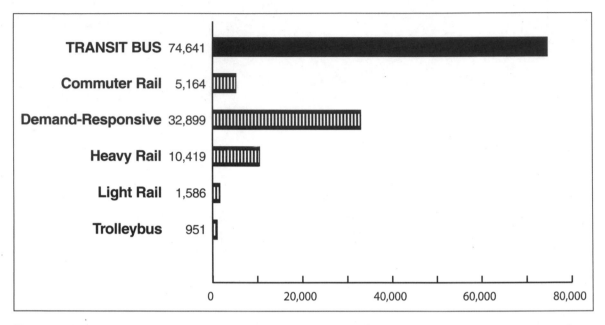

Figure 8.1   Number of public transit vehicles in the United States, 2000. (*Source:* American Public Transportation Association.)

be slow, and the vehicles are frequently not particularly comfortable. They are regarded as *common,* both in the sense of "occurring frequently" and of "lacking special status."[2]

Nevertheless, buses provide the base service in most places, they can carry considerable passenger loads, and the service can be significantly expedited if proper attention is paid. No advanced engineering or special skills are required to run them, they are economical, and, indeed, they may be the real mode for the future, given the current development trends in American cities.

To ride the bus is to experience the city. If one does not become frustrated by street congestion and confused fellow passengers who delay departures at bus stops, a seat at a window gives the best, somewhat elevated, view of local urban life, the ongoing street theater, and the changing streetscape with its many attractions and features. This does not have to be a sightseeing bus with canned commentary and a pricey ticket, either; any regular bus along a busy route (and a good map) will do.

[2] *Webster's New Collegiate Dictionary.*

As of this writing, there are 2262 agencies in the United States that operate transit buses.[3] The total active fleet is almost 75,000 units, with some 11,000 more in Canada. This is by far a much larger number of vehicles than those counted under any other public transportation mode. In addition, but not considered in this discussion of urban transit, there are about 4000 private operating enterprises in North America that own 44,000 commercial motorcoaches for service between cities, as well for charter, tour, and other non–common carrier operations. The school bus fleet, which is a separate industry in itself, represents an incredible inventory of some 350,000 units of rolling stock.

At the global scale, there are probably 8000 to 10,000 communities and cities that provide organized bus transit.[4] The larger places have other modes as well, but the bulk of these cities offers buses as their sole public means of mobility.[5] Some sources suggest that urban areas below 200,000 population are likely to be able to operate only buses for transit purposes, not any rail-based modes;[6] for American communities, with their low densities, this boundary line may go up to at least 1 million.

While buses are ubiquitous in just about every city, they have not left much of a mark on contemporary American folklore and popular culture. There is Ralph Kramden, the blustering but lovable bus driver, suffering through endless domestic calamities on *The Honeymooners,* but not much else. (This TV show aired on CBS from 1955 to 1971, but we will see reruns forever.) Ralph, however, was never actually shown in a bus.

The only Hollywood movie with a bus in a leading role was *Speed* (20th Century Fox, 1994), in which the vehicle could not be stopped (!) because of an attached bomb. So, it naturally had to barrel through Los Angeles, overcoming incredible obstacles. Otherwise, buses have had numerous supporting roles in movies as part of the urban background, but principally to give some variety to car chases and the tailing of bad guys. The use of a bus

[3] Data from the American Public Transportation Association (APTA), *Public Transportation Fact Book 2000.*

[4] Data assembled by the *Union International des Transports Publics.*

[5] While bus service may be provided by private firms in a number of instances, if entry is open to all customers, they are classified as public transportation services.

[6] As reported by G. A. Giannopoulos, *Bus Planning and Operation in Urban Areas* (Avebury, 1989), p. 10.

now and then becomes an exotic departure from the routine formula of these standardized episodes. Movies that show the destruction of cities by natural forces or monsters almost invariably include a scene in which a bus filled with passengers is demolished. The audience usually reacts with glee.

Buses have figured quite prominently in recent American political and social history. Before desegregation, separate bus services (not just separate sections in vehicles) for black and white patrons existed in quite a few places. One of the defining moments in the civil rights struggle was Rosa Parks's refusal in 1955 to give up her seat on a Montgomery, Alabama, bus to a white patron. The busing of school children to schools outside their neighborhoods to achieve some racial balance has been a hotly debated subject for several decades. The use of buses by predominantly low-income and nonwhite patrons in most cities remains a critical issue today. It is not something we wish to talk about, but the point is clear that, at the base level, urban transportation is not just an operational or technical concern. Issues of social policy and equity come to the fore most vividly. This fundamental service allows daily life to take place for many who would otherwise be immobilized, and it reflects society's expectations and attitudes.

While everybody knows what a bus is, a few precise definitions might be in order to give structure to our discussion. Buses come in various sizes, and they are used for different purposes. In most instances, a specific model of vehicle is intended for a specific type of operation, but it does not always have to be so. For example, a standard full-size bus can be run as a jitney, and a minibus can carry regional commuters.

A *bus,* as a vehicle, is a large over-the-street unit accommodating many riders, individually driven (controlled and steered), almost always utilizing a diesel engine and rubber tires (at least so far).

When this type of vehicle is operated on a public right-of-way (street or highway) in mixed traffic, along a fixed route and on a set schedule, admitting all who wish to enter, but usually upon the payment of a fare, it is a *public transportation* mode or *bus transit.*

Only bus transit on streets is covered in this chapter, not intercity service and transportation provided by institutions and enterprises for their own employees or customers. Buses can be given various priority treatments, they can be placed on exclusive rights-of-way, and they can even be automatically guided. Then they are classified as *bus rapid transit,* which is discussed in the next chapter. Demand-responsive or *paratransit* operations are also reviewed separately;[7] they too use bus-type vehicles, albeit of a smaller size. (None of these are rail-based, of course. Rail-based transit modes are covered in several other chapters.)

Standard (advanced design) American city bus (Grumman 870) in Honolulu.

Development History

If a *bus* is defined as a self-propelled, individually controlled vehicle in public service, then the first examples were the horse-drawn stagecoaches that started to operate inside cities when cities became too large for walking trips alone. Paris had a number of routes as early as the seventeenth century (*carosses a cinq sols*). The real start of such service, however, can be traced back to the 1820s, when several cities in France developed public systems and even coined a name for them—the *omnibus*. London and New York followed suit almost immediately. G. Shillibeer was a major developer of systems in Great Britain and a significant transit innovator. The Broadway Accommodation left its mark in Gotham's history, and by the late 1830s more than 100 units pulled by horses were on the streets of New York.[8]

The idea spread to a number of large cities in Europe and North America, since the ever-longer distances between places in the growing cities called for mobility assistance. Omnibus drivers acquired a reputation from the start as colorful urban characters, engaged in fierce competition—a reputation that has not been lost among taxi, van, and bus drivers even today.

---

[7] These are operations that do not follow a published schedule, and frequently not even a fixed route.
[8] E. G. Brown and M. Walker, *Gotham* (Oxford University Press, 1999), p. 565.

Very soon after omnibus service was initiated, the next step in transit development was taken—placing horse cars on rails. Thus, technological evolution split into two branches—trackless buses and rail-based streetcars—and the two modes existed in parallel for the rest of the century and beyond.

The search for suitable mechanical power started almost immediately as well, and the first effort was the large steam-driven stage-coach developed in 1830 by Goldworthy Gurney in Great Britain. It and other efforts were not practical, and horse omnibuses continued, a few well into the twentieth century. Their ability to follow any path within the already available street network was always a strong feature, even though the ride on cobblestone or poorly paved streets was not particularly comfortable. Further experimentation with different power sources, however—whether horses, steam, compressed air, wound-up springs, pulled cables, or electricity—addressed rail-based vehicles (i.e., streetcars), not bus-type vehicles. The latter needed a mobile and compact power plant, which was beyond the capability of nineteenth-century technology. For more than half a century, the desire to upgrade the mechanical efficiency of rails attracted the principal attention of inventors and engineers. Flexibility in operations apparently was not seen as a strong enough benefit to generate intellectual or technical investment; thus, there were no real advances in horse-drawn omnibuses during this period.

The *Compagne General des Omnibus de Londres,* established in 1851, became a principal agency not only in providing ever-expanding service in the capital of the British Empire, but also in taking leadership in the technical development of rolling stock and other equipment.

The true motorized bus is a transport mode of the twentieth century. While the internal combustion engine was invented in 1859 (and in 1878) and the diesel engine in 1892, "automobiles," or self-contained motor vehicles, emerged as practical machines only at the beginning of the new century (see Chap. 5, Automobiles). This was a period rich with various ingenious transportation devices that advanced the mechanical arts in general but usually represented dead-end efforts in a practical sense. Between 1895 and 1905, several omnibuses that experimented with the crude engines of that time were built in Germany. They attempted to attract urban passengers, but were not workable in general use. Petrol-driven motorbuses appeared in 1905 both in London and Paris.

When the automobile had established a foothold in North America, it did not take long to think of larger vehicles with space for many passengers as a further development of the motor car. Streetcars provided a model, and, since they were perceived as having a number of flaws, they were seen as candidates for replacement. The first practical buslike vehicles were basically motor trucks, with a boxy passenger compartment placed on a truck chassis.[9] Such improvised vehicles are still produced locally and used by the hundreds in developing countries around the world.

Early efforts (before World War I) were reasonably successful in Great Britain. The easing of legal restrictions against motor vehicles, the use of gasoline engines, and the development of various engineering features resulted in workable buses that started to provide regular service in London and elsewhere. Buses also appeared in the New World, and the first bus line was opened in New York City in 1905, when the Fifth Avenue Coach Company replaced all its horse omnibuses with motorized double-deckers. Similar operations soon emerged elsewhere as well, notably in Cleveland (1912), St. Louis (1916), and Chicago (1917). St. Louis is credited with establishing the first transit agency that operated buses exclusively—the Division of Parks and Recreation, Municipal Auto Bus Service. The Chicago vehicles were built by the Yellow Motor Coach Company, which was soon acquired by a minor but growing automobile manufacturer with the grand name of General Motors.

The jitney episode in American cities (1914 to 1916; see

[9] Among the various sources that have reviewed the history of bus transit development in the United States, the following can be mentioned: *Public Transportation Fact Book: APTA 2000,* American Public Transportation Association (51st ed., 2000); "Trucks and Buses," *Encyclopaedia Britannica* (1975, 18:721), Vukan Vuchic, *Urban Public Transportation* (Prentice-Hall, 1981); Gray & Hoel, *Public Transportation* (Prentice-Hall, 1992); and Brian J. Cudahy, *A Century of Service* (American Public Transit Association, 1982).

Private buses providing public service in Buenos Aires, Argentina.

Chap. 6, Paratransit) also left its mark on bus development. Not only passenger sedans were used for this service; there was an advantage in employing larger vehicles, and jitney buses appeared on the streets in various shapes. After the prevailing chaotic situation was constrained and somewhat regularized, starting in about 1920, some of these operations became regular bus lines, particularly as the still-dominant streetcar companies converted some of their lower-volume lines to bus service. Legal jitneys remained in operation in only a few places.

It is said—mostly by Americans—that the first true bus appeared on the scene in 1920. It was the Fageol Safety Coach, specifically designed for the purpose of carrying passengers as conveniently as possible. This claim of being a pioneer can be debated, because European buses of that time already had some of the same advanced features; nevertheless, the Fageol brothers made a breakthrough in assembling a coherent and useful vehicle. The frame was lowered, and the wheelbase was lengthened. Most important, passenger boarding and exiting was made easier with better-designed doors and steps. A few years later, the same company developed the integral frame, in which the sides and the roof are parts of the structural system, and moved the engine inside the shell. It is also interesting to note that at about the same time, other manufacturers (Mack and Yellow Coach) developed a propulsion system in which a gasoline engine drove a generator, which fed electrical power to motors near each wheel. (This concept has recently been rediscovered, and it is presented to us today as the new and promising hybrid vehicle.) By the late 1920s, the design of city buses and long-distance motorcoaches diverged—they became different vehicle types, responding to their own needs.

The early transit bus models created much interest since they showed that a viable alternative to streetcars was available (see Chap. 11, Streetcars and Light Rail Transit). By that time, trams were widely criticized as being responsible for growing traffic congestion and safety problems in American cities. They almost always ran along the middle lanes of practically all major streets, and many accidents occurred as riders had to exit into traffic lanes or cross them to reach the vehicles. The quality of service and level of maintenance was frequently lower than substandard. The public was ready and eager for a new and "modern" replacement. The fledgling bus mode not only seemed to offer operational agility and flexibility; it also presented an advanced image, particularly when new models in the 1930s placed the engine in the rear and gave the vehicle a streamlined envelope.

A period of profound transformation of mobility systems in American cities started in 1923 when several smaller cities (Bay City, Michigan; Everett, Washington; and Newburgh, New York) replaced all their streetcars with buses. The trend accelerated, with larger cities following suit. San Antonio was an early example in 1933.

A highly controversial episode in the history of city service management, much discussed in the literature,[10] occurred in the mid-1930s when large bus manufacturers (General Motors primarily) embarked on a deliberate program to assume control of streetcar companies and to rapidly introduce their own replacement vehicles in public service. Many see this in retrospect as an evil, self-serving monopolistic practice that irreparably damaged the quality of transit service in American cities; others regard it as a normal competitive business program to establish and capture a market. There is no doubt that the public at the time not only tolerated but applauded this transition. By 1940, the volume of bus riders exceeded that of streetcars nationally.

Meanwhile, in Great Britain, the London Passenger Transport board took over all transit operations in that world city in 1933. Under its auspices the double-decker became the dominant vehicle type, providing service throughout the city in support of the Underground network, and achieving the status of a permanent icon in London. The basic model was in operation from 1939 to 1979, when it was replaced by the Leyland Titan.

In the years from the 1920s to the 1940s, not too many changes in the shape of the vehicle or its form of operation were observable in America. The typical bus was rather cramped, with only about 30 seats. There were, however, significant technological improvements. The most important of these developments was the adoption of the diesel engine as the standard power source for all buses, after much experimentation, in the late 1930s. The two-stroke diesel engine saved fuel and was efficient and reliable, very

---

[10] A summary of the events, presenting both sides, is provided in several articles published in *Transportation Quarterly*, Summer 1997 and Winter 1998 (vol. 51, no. 3 and vol. 52, no. 1). In addition, there is a large volume of articles and testimony examining the issues, much of it rather emotional.

powerful, and quite easy to maintain. Hydraulic transmissions were made workable at about the same time. There was also steady growth in the size of the passenger compartment in response to demand; even an articulated unit was attempted in the 1930s.

After World War II, European manufacturers took an active lead in supplying their own as well as global markets. A large variety of models was developed, and a series of technical improvements were made, such as air suspension, quieter motors, and articulated units (in the late 1950s). Perhaps even more important was the attention paid to passenger amenities—larger windows, wider doors, better seating arrangements, etc. Eventually, the low-floor bus was also developed in Europe.

In the United States, the major long-range trend was the progressive abandonment of streetcar service, with much but not all of the work being taken over by buses. This turnover was basically completed by the 1970s, and while there is a respectful resurgence of light rail transit operations today, buses hold on to first place with overwhelming numbers.

Progress in upgrading rolling stock in North America during this period was not equally strong, even though the fleet needed replacement after the wear and tear of the war years. While there were several dozen active bus manufacturers in the late 1940s, in actual practice, just one of them (the General Motors Corporation) commanded more than 90 percent of the national market. Its 50-seat model was the standard of the industry. The other manufacturers—ACF Brill, Mack, Twin Coach, and a host of small enterprises—produced the same type of vehicle, which had not changed much from the prewar models and had become outmoded. Only Twin Coach came out with a new design, which primarily concentrated on automobile-like styling. Being protected by the Buy America clause,[11] there was no great initiative for very large car manufacturers to engage in expensive development programs to serve what they perceived to be a rather small market. Just as in the aircraft industry, it was expected that the federal government should make this investment.

In a certain sense, the boost for post–World War II efforts to develop progressively more attractive vehicles was the 1954 Sceni-

[11] The Buy America clause—Title 49 of the Code of Federal Regulations—requires that at least 51 percent of each vehicle's value be created in the United States, and also that final assembly be performed here.

cruiser. This was a vehicle designed for very long trips—transcontinental, not urban. It boasted two levels, with various passenger amenities and comfort features; it was a massive vehicle with 10 tires, two engines, and dramatic appearance. The Scenicruiser was inspirational, but also too expensive for regular low-fare use.

From the late 1950s onward, there have been periodic attempts to redesign and improve the entire vehicle—i.e., develop a new model for general urban use. The first such major effort addressed almost solely the external appearance of the vehicle, and the result was promoted as the "New Look" model, identifiable by its large slanted windows. It was produced from 1959 to 1978 by American and Canadian companies.

Vehicle research and development efforts have consistently been concerned with lower weight, better engines, smoother ride quality, greater passenger comfort, and a more contemporary appearance. Nobody has expected any revolutionary departures in the configuration of the vehicle. Improvements to the vehicle have been made, step by step, but it also has to be noted that just about every new model has encountered serious difficulties when first placed on the streets, requiring considerable redesign and replacement of components. For example, air bag suspension and air conditioning were introduced in the 1950s. There were significant difficulties with the latter, which extended over at least a decade until a robust enough unit was developed. The contemporary press was not kind in discussing all these initial problems, leaving in its wake many red faces and bruised reputations. Eventually, however, the hardware has worked.

The New Look bus is a case in point. Within the first year of operation, significant structural problems became apparent, which required stripping down the vehicles and rebuilding them with a strengthened frame. After these modifications, the units performed well, and we look back on them as the old reliables.

All through the decades leading to the 1960s, bus service in American communities had been operated under private or municipal auspices, relying on local resources. The Housing Act of 1961 signaled a new era under which the federal government (as well as state governments) started to assume increasingly greater responsibility for local public mobility. At first these efforts were selected bus demonstration projects that supported metropolitan and city-scale operations. In the mid-1960s, such experiments took place in Baltimore; St. Louis; Chesapeake, Virginia; and Peoria and Decatur, Illinois. They were reasonably successful, but even then it was apparent that riders on these routes were primarily diverted from other transit services; there were not too many customers who had abandoned their cars.

There were also demonstration projects to link poverty areas to workplaces (Long Island, New York, and central Los Angeles). In any case, a pattern became established that generated a sequence of federal legislation—starting with the Mass Transportation Act of 1964 and leading to the current Transportation Equity Act for the 21st Century (TEA21)—with a growing and more organized financial participation by the federal government in the provision of transit services, including buses. Principal attention was devoted to capital investment; construction and acquisition programs were eligible for up to 80 percent funding through federal grants.

While it is not the intent of this book to review the institutional arrangements under which transit systems operate, it is necessary to note that a consolidation movement took place in most large U.S. cities beginning in the late 1960s and early 1970s. Individual municipal and private systems had a difficult time maintaining services, and they were not able to respond to demands that now had assumed a regional scale because urban development had extended across all local political boundaries. The response was to establish regional authorities or metropolitan transit districts, thereby promising comprehensive system development and management. While the American agencies did not quite achieve the seamless service integration that characterizes their West European counterparts (metropolitan federations of operators), the results were definitely positive. There did remain a lingering question of whether massive central organizations can always cope effectively with purely local services, such as buses, but the trends were unavoidable, and the regional agencies represent the institutional structures within which most transit services have to be managed today.

A major effort to develop a contemporary American urban transit bus came in 1970 when the U.S. Department of Transportation initiated the *Transbus* program.[12] This was the first

[12] The events associated with this program were national news from 1970 to 1980. A summary and evaluation of the experience are provided in a 1979 follow-up, *Report by the Transbus Procurement Requirements Review Panel for the National Research Council.*

time that the federal government assumed responsibility for developing a transit vehicle, primarily because by that time operating agencies qualified for federal subsidy for the purchase of bus fleets. The aim was not only to develop a better and more attractive vehicle, but also to encourage competition among manufacturers. Contracts were awarded to General Motors Truck and Coach Division; Rohr Industries, which acquired the Flxible (*sic*) Corporation; and AM General Corporation. The spec-

Interior of Transbus—never placed in service.

ifications called for a bus that would be able to operate effectively under all urban and suburban conditions, have advanced nonpolluting engines, reach a top speed of 70 mph (110 kph), provide full wheelchair access, have cantilevered seats for easy floor cleaning, and be built with what was then considered a low floor—17 in (43 cm) above street surface. (The regular bus at that time had a floor height of 34 to 35 in [86 to 89 cm].)

The Transbus saga became a comedy of errors and a sequence of misdirected efforts. As has happened in other instances, the bureaucrats in Washington demanded a device that would do all things for all conceivable users, incorporating a wide range of new and untested features. While prototype vehicles, which did have a striking appearance, were produced and tested, they were found to be too heavy, and they guzzled too much fuel. They also had fewer seats than the regular buses then on the streets, and probably cost twice as much. The worst stumbling block was the floor height. Even when the requirement was raised to 22 in (56 cm)[13]—while user groups complained all along that the climb was too high to begin with—manufacturers insisted that such vehicles could not be made workable. It was assumed that the axles had to run across the width of the vehicle, thus mandating small-diameter wheels, which would result in excessive tire wear and difficulties in braking.

---

[13] It was later further modified to 24 in (61 cm).

To cut a long and sorry story short, when bids for making Transbuses for a consortium of cities (Philadelphia, Miami, and Los Angeles) were to be opened in 1979, there was no response from any manufacturer. There was a shakeout in the American bus industry along the way as well. Flxible was purchased by the Grumman Company; AM General left the bus business; and European manufacturers (such as M.A.N.[14] and Neoplan) started to probe the American market, while American firms were not able to secure any international customers.

In the meantime, however, GM had hedged its bets and developed its own new model—the RTS II (rapid transit series, 2 axles). So had Grumman Flxible with its Model 870. The U.S. Department of Transportation and all cities in need of new vehicles had to accept the fact that the shopping list was now limited to these two items and some exotic and unconventional units. The new models were dubbed *advanced design buses* (ADBs), and the federal specifications were rewritten to fit the interim product. The new buses appeared on the market in 1978. The debate still continues over whether national agencies should maintain uniform standards for public-service vehicles across the country or whether cities know best what they need and wish to purchase.

The ADBs did incorporate many improvements, but the floors were at 29 to 32 in (74 to 81 cm), and the price broke through the $100,000 line. The GM RTSs and the Grumman 870s are what we still mostly see on the streets of American communities in the early twenty-first century. The former are distinguishable by their rounded edges and corners; the latter are sometimes called *Darth Vader buses* because of their prominent forehead, which carries destination information.

The initial difficulties with ADBs were numerous.[15] The doors did not always operate properly, brake linings and tires wore out too quickly, air conditioning units were fragile, and there were

[14] *Maschinenfabriken Augsburg-Nürnberg.*

[15] During the 1970s and 1980s, as the Transbus and ADBs were designed and introduced in service, they received continuous attention in national media. Hundreds of articles in magazines and newspapers appeared during this time, almost on a daily basis. A summary is provided by David Young, "A World of Buses: Their Problems and Possibilities," *Mass Transit,* December 1980, pp. 6–9, 52–56.

transmission problems. The most serious issues were the heavy weight, which caused excessive fuel consumption, and structural cracks in the frame. The latter problem was particularly serious, which led New York's MTA to abandon and sell its 837 Grumman Flxible 870s.[16] The debate over whether the designs and construction quality were faulty, agency maintenance practices were inadequate, or broken street surfaces shook the vehicles apart continued for some time—until proper repairs, the addition of strengthening elements, and some redesign of vehicles made them acceptable.

As the Transbus events unfolded, the first articulated buses— European models (M.A.N., Icarus, Volvo, and Neoplan)—entered regular urban service in the late 1970s. They were found to be most suitable for high-demand routes because of their large size. After a probing start in Seattle, Chicago, and Los Angeles, by the end of the decade 11 U.S. cities had them in regular operation. Today they are seen in many communities, and they have become a regular component of agency fleets.

Since the middle of the twentieth century, we have gone through periods when fanciful technological improvements have been seen as answers to bus service shortcomings.[17] Undoubtedly, these attitudes have been generated by rather spectacular accomplishments in aerospace, weapons systems, and automated guideway transit. These expectations have rarely been satisfied within the rough environment of the urban street, but hope persists. One such example is the need to accommodate handicapped persons on public transit. It is not easy to bring a wheelchair onto a bus, and lifts (first introduced in San Diego in 1944) have been made workable, but at significant cost and with service delays. A good answer is probably to be found in a completely different vehicle.

In the late 1980s, American bus manufacturers started losing ground as European corporations, primarily German, started to

[16] These events are fully recorded in the *New York Times* from 1981 to 1986. Some criticized this divesting action as too harsh and precipitous; others praised the uncompromising concern for public safety.

[17] See U.S. Department of Transportation/Urban Mass Transportation Administration report *Bus Guidance Technology: A Review of Current Developments,* (UMTA-IT-06-0247-84-1), December 1983.

establish plants in the United States and penetrate the national market. At that time, vehicles made by the General Motors Corporation still constituted more than a third of the fleet, and Flxible accounted for a quarter.[18] The rest was made up of vehicles from Neoplan USA Corporation, AM General Corporation, and M.A.N. Truck and Bus Corporation.

As of 2001, there were 41 bus and chassis manufacturers in the United States,[19] but just 7 of them produced more than 90 percent of the vehicles placed in service. In the approximate order of sales volumes, which changes from year to year, the big seven are: New Flyer, Nova BUS, Gillig, Orion, North American Bus, Neoplan, and Motor Coach Industries.

The next American bus model, currently under development, will be the *Advanced Technology Transit Bus* (ATTB), a federally financed project executed through the Los Angeles County Metropolitan Transportation Authority, with the participation of the Northrop Grumman Corporation, a major aerospace and weapons producer. The project was started in 1992, and the design effort was expected to rely on new materials technology borrowed from the aircraft industry (stealth bomber), advanced operating components from military vehicles (electronic suspension system for tanks), and nonpolluting power plants using a range of alternate fuels. The specs call for a series of user-friendly, human comfort features, as well as a strong structure and modular design to allow easy replacement of parts. Much is expected from reducing the weight of the vehicle to about 70 percent of what a comparable conventional bus would weigh, perhaps even doubling its useful life to 25 years. Prototypes are being tested, and there is hope that the Transbus experience will not be repeated.

The 1970s and 1980s saw a disturbing trend in bus service in American communities—a serious decline in ridership, particularly in small and midsize cities. This situation overshadows anything else that can be said about the historical development of the bus transit mode. It was basically caused by the expansion of low-density, sprawling districts and the single-minded popularity of the private car, with continued growth of the national automobile fleet and its daily use by a prosperous society. Bus operators did not or could not combat these events, and service quality, gener-

[18] Gray and Hoel, op.cit., p. 150.
[19] *Metro Magazine 2001 Fact Book* issue, Fall 2000 (vol. 96, no. 8).

ally speaking, deteriorated as well. If ridership is lost, the natural tendency of an agency is to reduce the underutilized service, which further impairs the attractiveness and responsiveness of the operations.

There was a period when state and federal governments expanded their operational assistance programs to keep systems operating, but those days passed as well. Subsidies to maintain day-to-day operations have waned, and agencies have had to subsist largely on local resources. The nadir was reached in the early 1990s, with bus service sinking to disaster levels in some places.

In the last few years, there has been a noticeable recovery—though not exactly a resurgence yet—of bus ridership in most communities across the country. It cannot be said that these encouraging events are solely due to new progressive and effective programs by transit agencies; there may be a groundswell in the general rediscovery of transit and a ceiling in private car use. This remains a reason for hope, but not for complacency. It suggests that substantial upgrading of bus service is not only necessary but may, indeed, bring positive results.

A major area of controversy, which has not yet reached resolution, is the question of air pollution caused by city buses. The long-standing but erroneous assumption that diesel exhaust is not harmful has been reversed, and the occasional black cloud from the tailpipe of a poorly maintained bus reminds the public to be concerned. The first response was to switch to compressed or liquified natural gas, which burns pollution free but requires a series of adjustments to vehicles and fueling systems. Such vehicles are being tested in a number of places, and they appear to be ready for regular service at slightly higher costs. They are identifiable by their large rooftop fuel containers and the prominent markings on the sides of the vehicle highlighting the environmental sensibilities of the operating agency (see cover).

Fuel cell power plants (the direct chemical generation of electrical energy from hydrogen, with water vapor as the only emission) may be the engine of the future, but so far development problems persist. Chicago started actual testing in 1997, but cost and performance issues are still to be resolved.

The supporters of diesel technology have mounted a counterattack by insisting that the fuel and the engine can be significantly improved to reach acceptable air quality standards, thus causing no disruptions to established practices. In the meantime, hybrid buses have appeared as well; they utilize electric motors backed up by small internal combustion engines and batteries. The general concept is that the engine runs at a constant efficient rate, feeding batteries, which, in turn, power the separate electrical motors and provide for extra surge requirements. While Orion, the Gillig Corporation, and New Flyer have established an apparent lead in this field, almost all other manufacturers are at least experimenting with the new technology. European manufacturers have been in this field for some time. At this time the hybrid bus appears to be the leading choice among agencies experimenting with environmentally friendly technology, and it may become the standard model for city service. We will see in the next few years.

Last, but not least, there is the emergence on this continent of true low-floor vehicles, and their gradual introduction in more and more communities. Following French and German examples, the first American effort was the shuttle service on the Denver Mall, started in 1980, using an airport service vehicle with floors 11.8 in (30 cm) above street surface. This was an immediate success, particularly judging from user responses. The first two operating agencies in the United States to purchase standard-size low-floor buses for regular service were in Champaign-Urbana, Illinois, and Ann Arbor, Michigan, both university towns with a progressive attitude and patterns of surge loading by students.

The list of low-floor applications continues to grow. For example, in 1995 the Metropolitan Atlanta Regional Transportation Authority (MARTA) purchased 51 such buses from New Flyer Industries that are of conventional size (with 39 seats), but have floors at the 14-in level, which can be dropped another 3 in by the kneeling feature.

As will be pointed out repeatedly in this chapter, the use of low-floor buses is not only becoming more common in American communities, it may also be a harbinger of a complete changeover in the fleets.[20] Because of its user-friendly features, from the perspective of the passengers, it is the only way to go.

The conclusion from the historical overview would have to be that the bus mode in America is about to cross a significant but not revolutionary threshold—the use of vehicles that are environmentally acceptable, easy to enter, and comfortable for riders. For

[20] A series of columns by George M. Smerk in *Bus Ride* magazine since 1994 suggest such a path.

a long time the vehicle and the way it is operated had remained basically unchanged, with only a slow and gradual upgrading taking place. Since the mid-1990s, however, new fuels and engines, low floors, and various passenger comfort features have become active ingredients in the industry. Together with operational reforms, discussed in the next chapter, there are reasonable expectations that the bus mode may become not only a strong but also an attractive contender in the urban transit field. Statistically, buses are the leaders already; they do not generate much loyalty and affection from their patrons. There really is no reason for American riders to accept inferior vehicles and procedures when transit users in other advanced countries have enjoyed better service for years.

But let us keep the current exuberance associated with revolutionary technical improvements within a realistic range. Such advances will be most welcome, but a bus will still be a bus—a basic transit vehicle. The quality of service will still be judged largely by how frequently it runs and how comfortable a ride it provides. People will continue to fret if they have to stand in the rain for many minutes, they will complain about rude bus drivers, they will sue if they trip and dislocate an ankle, and they will battle for a seat—before they praise the beauty of advanced engineering.

Types of Buses and Bus Operations

All buses look largely alike—as we all know—and they have basically the same configuration: a large passenger compartment where riders can move standing up, engine in the rear, doors along one side (the right side in countries with traffic on the right side of the street), and a single person up front driving the vehicle and usually supervising fare collection. While this represents the majority of cases, considerable variation is possible.[21] One of the interesting and useful feature of buses is their ability to respond to differences in service needs, which are not limited to size and carrying capacity alone.

[21] V. Vuchic, *Urban Public Transportation,* op. cit., pp. 193–241, provides detailed descriptions of transit bus vehicles. While this information is from prior to 1980, the physical characteristics of buses in operation today have not changed much.

Vehicle Classes

Looking only at vehicle types (their physical configuration), and not at the type of service in which they may be engaged (since any model of a vehicle may perform different functions), the following groups can be identified:

- *Passenger automobiles, station wagons, and sport utility vehicles.* These are vehicles in individual, mostly private use. While they can be employed in paratransit, they do not fall under the category of bus transit.

- *Passenger vans.* They have a single large door on the side, in addition to the two front doors, and they can accommodate up to 12 riders. Seats can only be reached by crouching, and there are no interior passageways. They are widely used in developing countries as jitneys; they are in operation as point-to-point shuttles in many communities in North America and elsewhere. Applications include commuter van services, distribution of air passengers, access to hotels and institutions, and similar operations where the entire passenger load can be (mostly) assembled and discharged at one time. This is not true transit either, although commuter service comes close.

- *Minibuses.* The principal distinguishing characteristic of these vehicles is the ability of passengers to move about inside standing up. While the external dimensions are as small as they can be and still retain the appearance of a regular bus, the vehicles do have most of the same features, sometimes including two doors on one side. They can be (and they are) employed for private as well as public service (with payment of fares), in scheduled or demand-responsive operations. Gasoline engines are likely to be used, but diesel engines are becoming quite common for this class as well.

- *Midsize buses.* They are sometimes called *30-footers* or *35-footers,* in reference to their length. Because they have fewer seats than regular buses, these vehicles are suitable on routes with lower passenger loads where frequent operations are desirable. An average capacity of 40 passengers can be assumed. Midsize buses have not been particularly popular among transit agencies because they complicate the composition of the fleet, but this attitude may be changing.

These, as well as all the larger vehicles, use diesel engines almost exclusively.

- *Standard city buses.* This is the most common type by far, in wide use around the world. In North America, they are 40-footers, although 45-footers are also possible.[22] They have two or three doors of varying widths along the side. Seats have low backs, standees have to be accommodated, luggage is not considered to be a factor, and passengers keep moving inside the vehicle. There is space for advertising. The ratio between seated and standing passengers can be varied by different interior layouts of the vehicle. (A standing rider takes up less floor space.) The usual carrying capacity is 60 to 75 passengers, although 100 can be squeezed in.

- *Long-distance buses.* In the urban context, long distance means trips at the regional or metropolitan scale.[23] The vehicles have a single door and comfortable accommodations for relatively long trips, usually at premium fares, and no standees. The seats will be upholstered and have high backs. Toilets, facilities for refreshments, television, telephones, and other amenities may be included. Travel will be in the express mode, with few stops. Since these vehicles are intended primarily for daily commuter trips, no special luggage accommodations are provided, except for overhead racks inside.

- *Double-decker buses.* To place as many as possible seats in the shortest (most maneuverable) vehicle, double-deck arrangements have been employed since the earliest days of transit. Significant variations in capacity again exist, but approximately 120 passengers can usually be accommodated. There may be problems with vertical clearances on the street, low headroom on the upper deck, and passenger safety on the stairs. Contrary to popular impression, double-deckers have no problems with stability because they are deliberately designed with a low center of gravity.

- *Articulated buses* (artics). A hinged joint in the body allows horizontal and vertical bending, resulting in large vehicles

---

[22] 45-footers are rare in city use because regulations require that they have three axles. Two axles are adequate for 40-footers.

[23] Intercity motorcoaches represent a separate class outside this discussion.

with continuous interiors that can turn as well as or better than standard buses. This is accomplished by shortening the wheelbase (distance between axles) and steering the rear wheels as well. The vehicles will have three or four doors. The regular artic will be able to carry up to 120 passengers under normal full loading. It should be noted, however, that unless there are more entry channels than on a regular city bus, the overall running speed may be lower because a greater number of passengers have to be accommodated at most stops, thus delaying movement. A double-articulated vehicle with four axles and five doors is also available, but American road regulations bar its use because of its length.

- *Tractor-trailer buses.*   These are one- or two-deck passenger compartments, similar in concept to cargo containers, that are attached to tractors. Such vehicles are rare—seen mostly in India and Singapore—but the division of the bus into two components (power unit and passenger container) presumably provides some added flexibility in managing the fleet.

    Another possible option is to attach a motorless passenger trailer to a regular bus. They have been used in different countries of Europe; they appear today in a few developing countries with large passenger demands; they have never been tried in the United States, again because of regulations. They cause some control, maneuverability, and safety problems; they need a separate conductor.

Leaving aside the smaller vehicles, Fig. 8.2 and Table 8.1 summarize the basic characteristics of bus types used for urban transit. It should be noted that external dimensions of motor vehicles are governed by state and local ordinances that place limits on lengths, widths, and heights. They are quite similar from country to country, but the figures given below apply to United States specifically. All buses (except minibuses) used in American cities are 96 or 102 in (2.4 or 2.6 m) wide or less. Long-distance buses, most 40-footers, and special coaches are 102 in (2.6 m) wide.[24] A normal passenger car is 66 in (1.7 m) wide and 15 ft 4 in (4.7 m) long.

---

[24] Buses of this width cannot navigate narrow lanes with tight horizontal clearances. An example is the Holland Tunnel under the Hudson River.

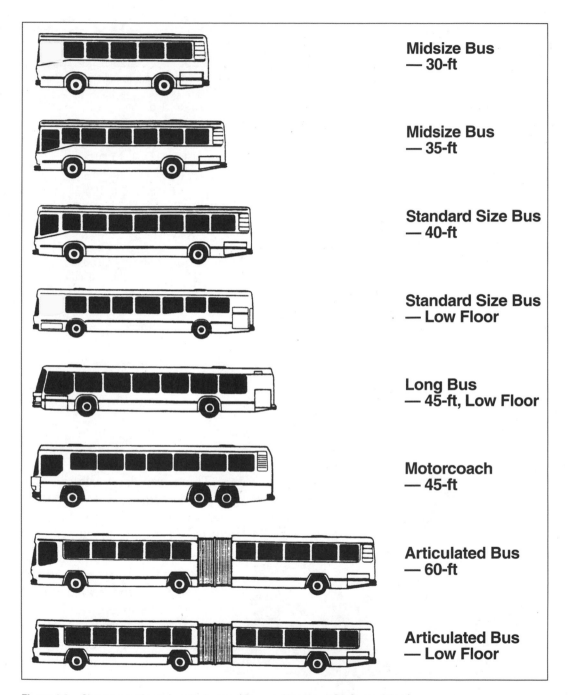

Figure 8.2 Size range of regular city buses. (*Source:* Neoplan USA Corporation.)

Table 8.1   External Dimensions of City Buses

Bus Type	Usual Passenger Capacity	Length and Height
Minibuses	12 to 20 seats Some standees	19 ft 4 in to 27 ft 6 in (5.9 to 8.4 m) long 8 ft (2.4 m) high or more 6 ft 9 in (2.1 m) wide or more
Midsize buses	25 to 35 seats 10 to 15 standees	29 to 35 ft (8.8 to 10.6 m) long (22 ft 0 in to 37 ft 5 in [6.7 to 11.4 m] possible) 9 ft 4 in to 10 ft 4 in (2.8 to 3.1 m) high
Standard city buses (see Fig. 8.3)	41 to 45 seats (35 to 53 possible) Up to 40 standees	37 ft 6 in to 42 ft 5 in (11.4 to 12.9 m) long (35 ft 11 in to 45 ft 0 in [10.7 to 13.7 m] possible) 9 ft 6 in to 10 ft 2 in (2.9 to 3.1 m) high
Suburban service buses	Up to 45 seats No standees	35 to 45 ft (10.7 to 13.7 m) long
Double-decker buses	64 to 92 seats (40 to 102 possible) Up to 40 standees	27 ft 6 in to 39 ft 5 in (8.4 to 12.0 m) long 13 ft 0 in to 14 ft 6 in (4.0 to 4.4 m) high
Articulated buses (see Fig. 8.4)	55 to 70 seats (35 to 76 possible) Up to 80 standees	54 to 60 ft (16.5 to 18.2 m) long Up to 10 ft 2 in (3.1 m) high
Tractor-trailer buses	120 total (but more are possible)	Up to 54 ft (16.5 m) long up to 14 ft 6 in (4.4 m) high

Note: In all instances, the largest number of standees *cannot* be accommodated in the same vehicle if it is provided with the highest number of seats.

Double-deckers on a street in London.

There are also various types of special buses, not in general transit use, that are *not* included in this discussion:

- *School buses* are somewhat spartan units intended only for the transport of schoolchildren and students on short trips. These are actually very simple passenger compartments usually placed on a truck chassis; they constitute a huge fleet in this country.

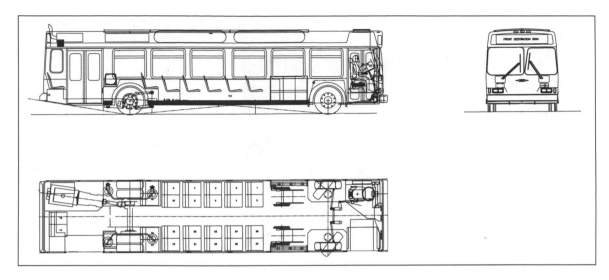

Figure 8.3 Low-floor city bus. (*Source:* Orion.)

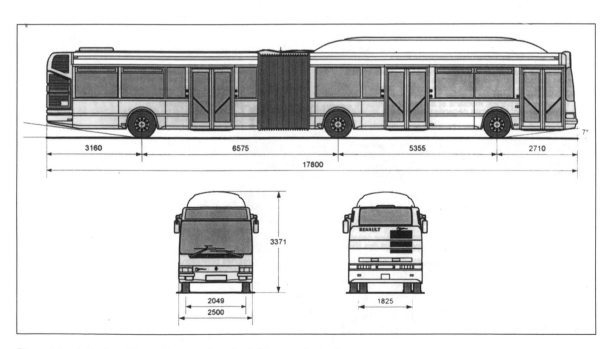

Figure 8.4 Articulated bus with natural gas tank. (*Source:* Agora.)

An early articulated bus in Cairo, Egypt.

- *Sightseeing buses* are designed to accommodate tourists and allow good visibility of the surroundings. Besides specially designed vehicles, double-deckers are particularly popular for this purpose.

- *Touring buses* are intended for extended travel, with built-in living accommodations. They generally utilize custom-modified long-distance bus shells or are purpose-built motor homes.

- *Long-distance buses* serve travelers between distant intercity points. They have to provide a reasonable degree of comfort and are almost always equipped with toilets, large luggage compartments, and other amenities. The preferred name is *motorcoach*. The new models are as wide and high as road regulations allow (102 and 148 in, respectively), and seat 37 to 59 passengers. The 45-ft units sell for $350,000 and up.

A double-deck trailer bus pulled by a tractor in New Delhi, India.

- *Private- and special-service buses* are used for employee transport, airport access, car rental customers, etc. They are frequently of the *minibus* type.

- *Customized units* are used for luxury travel by individuals or groups, as exhibition space, as dressing rooms for performers, and for other possible functions. They are called *luxury coaches,* and prefer to be classified with stretch limousines, not buses.

- *Trolley replica buses* are pure examples of kitsch, but they have established a strong presence in downtowns and historical districts of American communities, primarily for tourist use. They are encountered frequently and are popular enough to constitute an identifiable type today, although in terms of standard types they are *minibuses*. Actually, they do not replicate anything, but try to resemble early streetcars and cable cars. Several manufacturers are in the business of building these colorful boxes with traditional details, usually atop a truck chassis.

- *Special vehicles* of an infinite variety that still have some claim of belonging to the bus family may be found around the world. For example, there is an amphibious sightseeing bus in Ottawa, Canada, that can carry its customers along the local waterways as well. Mobile lounges that can adjust their height to match the floor elevations of airport terminals and aircraft are somewhat buslike in purpose and appearance.

Applicability of Vehicle Types

The one overriding factor in the planning and management of bus service is the obvious fact that every vehicle needs a driver. Since drivers' salaries are the single largest expense item in the industrialized countries, it makes sense to utilize units that accommodate the largest number of riders. The additional cost of a regular bus as compared to a smaller vehicle is not directly proportional to the number of seats, and neither is power consumption; therefore, the tendency is to avoid having midsize and smaller units in the regular fleet. There is managerial convenience for an operating agency in having a single model or only a few in the fleet, thus avoiding complications with maintenance and spare parts inventory, as well as having a simpler system to worry about.

With a fixed volume of riders on any given line, the larger

Minibus providing feeder service in Liepaja, Latvia.

Record Sizes

There is an interesting question, although not of critical importance, as to the largest bus ever made or used in regular operations.

- The first such claim could be made by the special vehicles used during World War II to move troops at Army bases in the United States. These were actually converted automobile carriers with a tractor, and probably do not count as real contenders.

- In 1977, the Wayne Corporation built a 62-ft (19 m), two-level tractor-trailer rig that was delivered to Egypt to transport 187 oil field workers at a time. Again, this was a special purpose, one-of-a-kind effort.

- Singapore operates a *superbus*—a three-axle double-decker—and has claimed the world's record for regular service. Orange County, California, has also introduced a *superbus*—a tractor with a passenger module 46 ft long that accommodates 67 seated and 60 standing passengers. The driver in the cab has three TV monitors to watch what happens on board and behind the vehicle.

- Renault experimented in the late 1980s with a *megabus*—a double-articulated single-level vehicle 79 ft (24 m) long that could carry 200 to 220 passengers.

- The gold medal at this time has to be awarded to the double-articulated bus produced by Volvo du Brasil, which can carry 270 passengers in its 80-ft (24-m)-long body. These units, used in Curitiba, constitute the largest fleet of this kind.

- Reference can also be made to the extraordinary vehicles used in several airports, mostly in Europe, to carry air passengers between terminal buildings and parked aircraft. They are the size of a tennis court, and they can deliver a full jumbo jet passenger load. Needless to say, they cannot even drive off the airport aprons and taxiways; therefore, they really do not count as buses but rather as moving parts of airport terminals.

vehicles can run at greater intervals and still accommodate the overall demand. While all this improves the efficiency of operations, it will be perceived as a reduction in the quality of service by riders. A specific example is the program of the bus operating agency in New York City—the Metropolitan Transportation Authority (MTA)—to replace regular buses with articulated units on crosstown lines, promising thereby an augmented capacity. The neighborhood groups, however, see this step as a threat of longer waiting times at bus stops.

One result of these considerations is that minibuses, being the most expensive to operate on a per-passenger basis, will be seen only in service areas where fully responsive service is necessary and where high loading can be ensured on a continuous basis (commercial districts, perhaps). Midsize buses are shunned by most operating agencies in North America (see Table 8.2) because regular-size units are able to do the job, even if excess capacity is frequently offered.

Articulated units grow in popularity because of their capacity (Fig. 8.5) as well as their maneuverability. The latter assertion sounds paradoxical, but artics have design features that allow close turns on corners. Double-deckers are seen as problematic in North America because riders have to climb tight stairs while the vehicle is in motion, and insurance liabilities are seen as significant.

Regional express buses (or suburban buses) represent a very interesting and instructive case regarding transit development today, both as to general implications and specific service capability. Such service has been successful in most areas where this option has been attempted, particularly in the New York–New Jersey region and in service to the more remote districts of the global city. Express buses have shown an ability to keep American commuters out of their cars to a reasonable degree. This is due to the fact that the buses travel fast, with long line-haul route segments and few stops at either end of the journey. Regional expresses

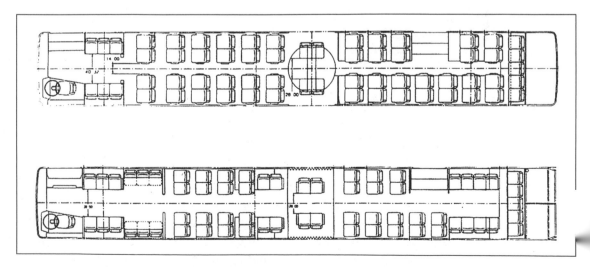

Figure 8.5 Seat layout options for articulated buses. (*Source:* Neoplan USA Corporation.)

Table 8.2　25 Largest Transit Bus Operators in North America

Rank	Transit Agency	Midsize Buses (35 ft & under)	Standard Buses (over 35 ft)	Articulated Buses
1.	MTA New York City Transit, Brooklyn, NY	—	4300	260
2.	New Jersey Transit Corporation, Newark, NJ	79	3055	105
3.	Los Angeles County MTA, Los Angeles, CA	88	2349	—
4.	Chicago Transit Authority, Chicago, IL	—	1770	120
5.	Montreal Urban Community Transit Corporation, Montreal, PQ	89	1597	—
6.	Toronto Transit Commission, Toronto, ON	149	1469	19
7.	Washington Metropolitan Area Transit Authority, Washington, DC	194	1258	64
8.	Metropolitan Transit Authority of Harris County, Houston, TX	165	1119	211
9.	Southeastern Pennsylvania Transportation Authority, Philadelphia, PA*	95	1068	155
10.	New York City Department of Transportation, private companies in New York, NY	—	1291	—
11.	King County Metro, Seattle, WA*	146	453	510
12.	Regional Transportation District, Denver, CO	313	811	119
13.	BC Transit/Coast Mountain Bus Company, Surrey, BC*	21	776	121
14.	Massachusetts Bay Transportation Authority, Boston, MA*	30	952	—
15.	Port Authority of Allegheny County, Pittsburgh, PA	145	832	49
16.	Pace Suburban Bus, Arlington Heights, IL	481	475	—
17.	Metro Transit, Minneapolis, MN	25	805	115
18.	Ottawa-Carleton Regional Transit Commission, Ottawa, ON	—	757	121
19.	Dallas Area Rapid Transit, Dallas, TX*	26	821	—
20.	Mass Transit Administration, Baltimore, MD	19	778	30
21.	Edmonton Transit, Edmonton, AB*	19	728	6
22.	San Francisco Municipal Railway, San Francisco, CA*	45	306	124
23.	Orange County Transportation Authority, Orange, CA	276	473	30
24.	AC Transit, Oakland, CA	88	620	60
25.	Greater Cleveland Regional Transit Authority, Cleveland, OH	108	649	—

* Also operates trolleybuses.
Source: Metro, September/October 2001.

work best, of course, where automobile commuters encounter high parking charges and are regularly delayed in traffic (provided that buses receive preferential treatment). An important contributing element may also be the fact that express buses offer a premium service at a premium fare, with respectable comfort features— upholstered seats, good ventilation and air conditioning, and the absence of "socially disadvantaged" riders. This last observation is a difficult statement to make, and it certainly does not represent a politically correct or equitable policy, but the realities of how people behave in the transportation environment and how they react to actual service characteristics should at least be acknowledged by transportation planners.

Since buses in urban service vary widely in their characteristics and dimensions, as previously explained, a comprehensive summary of their physical features cannot be prepared. Instead, Table 8.3 gives the actual characteristics of specific vehicles that are in wide use today in North American communities.

Network Configuration

It should not be expected that bus service planning will start with the selection of an abstract geometric pattern, and that it will then be superimposed on city districts and the street network. Never-

Table 8.3 Characteristics of Urban Transit Buses

Characteristic	Small Bus	Standard Bus	Low-Floor Bus	Articulated Bus
Manufacturer and model	Blue Bird C1FE	Nova BUS RTS	Orion VI	Neoplan AN460
Year introduced	1994	1977	1997	1978
Length	25 ft (7.6 m)	40 ft (12.2 m)	40 ft 8.5 in (12.4 m)	60 ft (18.2 m)
Wheelbase	132 in (3.4 m)	298.7 in (7.6 m)	278 in (7.1 m)	209 in (5.3 m)
Width	96 in (2.4 m)	102 in (2.6 m)	102 in (2.6 m)	102 and 96 in (2.6 and 2.4 m)
Height	116.3 in (3.0 m)	118.5 in (3.0 m)	122 in (3.1 m)	132 in (3.4 m)
Height of floor	18 in (0.5 m)	32 in (0.8 m)	15.5 in (0.4 m)	33 in (0.84 m)
Risers	8 in (0.2 m)	10 in (0.3 m)	—	—
Number of tires	4	6	6	8
Minimum (inside) turning radius	24 ft 8 in (7.5 m)	38 ft (11.6 m)	40 ft 6 in (12.3 m)	44.8 ft (13.7 m)
Number and width of doors	1–30 in (0.8 m)	2–30 in (0.8 m)	Various	2–49.2 in (1.25 m)
Number of seats	29	47	39	65
Number of standees	11	32		50
Empty weight	25,500 lb (11,600 kg)	24,500 lb (11,100 kg)	41,750 lb (19,000 kg)	41,600 lb (18,900 kg)
Maximum speed	65 mph (105 kph)	65 mph (105 kph)	63 mph (101 kph)	60 mph (97 kph)

Source: Manufacturers' data.

theless, there is some value in becoming familiar with these basic shapes and the service and operational implications of each one.[25] Network types are the following:

- *Shuttle service.* A single route between two significant points, with the frequency of service dependent on demand (Fig. 8.6*a;* see "Bus Scheduling Example" later in this chapter).

- *Radial pattern (through-running).* A common arrangement, with almost all lines running to and through the center to the other edge of the community (Fig. 8.6*b*). There may be branches that converge on the same streets near the center, thus augmenting service frequency.

- *Radial pattern (return-running).* Also a star-shaped arrangement focusing on a single center, but with the routes terminating there and vehicles returning to point of origin along the same (or parallel) path (Fig. 8.6*c*).

- *Gridiron network.* A series of approximately parallel lines in one general direction, crossed by another set generally perpendicular, providing an approximately even density of coverage over the entire service area (Fig. 8.6*d*).

- *Feeder service.* Accommodation of riders from local districts to nodes where a heavier transit mode (metro, for example) may be operating in citywide or regional service (Fig. 8.6*e*).

- *Trunk line and line-haul service.* In metropolitan areas where buses are the principal means of public transport, service can be expedited by eliminating stops along the middle portion of long routes (Fig. 8.6*f*). Expressways are frequently used for this purpose. Usually, instead of having passengers transfer from a local to an express bus, the same vehicle will continue on its run.

- *Loops and circulators.* Bus service can be run in a single direction continuously, without returning on the same path, which allows coverage of a larger territory (Fig. 8.6*g*). However, such routes have to be relatively short, because there will be an imbalance in the length of trips between any two points that are close together, depending on direction.

[25] A similar discussion of network patterns can be found in Gray and Hoel, op.cit., pp. 157–162.

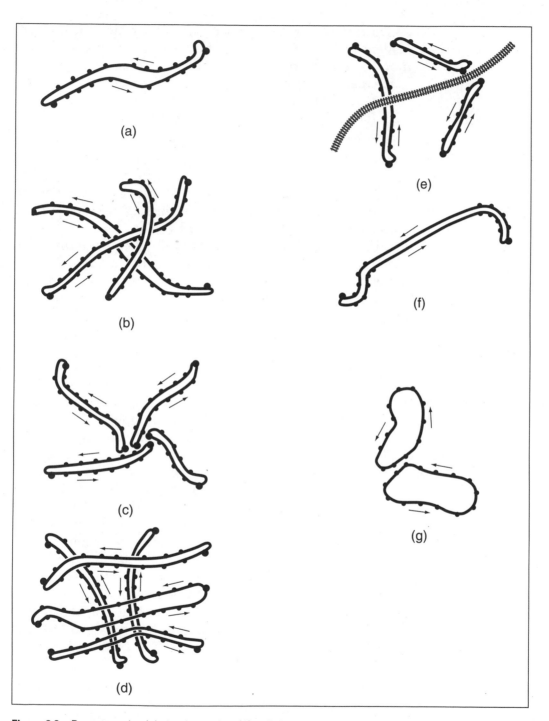

Figure 8.6 Bus networks: (*a*) shuttle service; (*b*) radial pattern, through-running; (*c*) radial pattern, return-running; (*d*) gridiron network; (*e*) feeder service; (*f*) trunk line and line-haul service; (*g*) loops and circulators.

Downtown services are frequently provided for the convenience of shoppers within the commercial district. There may be low or no fares, subsidized by local businesses or municipal promotion programs. Sometimes fares are not collected on the downtown segments of regular bus routes, also with the aim of encouraging commercial activity.

In real life, service networks have to respond to local demands, and they have to respect physical constraints, such as topography, established street configuration, narrowness of channels, etc. There is hardly any sizable city where a "pure" pattern will be found, because the total system has to grow out of the need for service as it is generated zone by zone.

The radial patterns, which are historically dominant, are predicated on the existence of a single multipurpose center, which most travelers will wish to reach as workers, shoppers, or visitors for other purposes. An inherent consequence of this arrangement is increasing congestion toward the center, if the core remains a strong point of attraction. If this singular demand orientation becomes eroded by the emergence of other important destination nodes, as appears to be the case today in American cities, the provision of adequate service becomes more difficult because of the diffusion of demand locations and the lessening of passenger concentrations on specific lines. This challenge applies to all transit modes, with buses perhaps being better able to respond to changing and decreasing demand conditions than fixed rail transit.

As a general concept, running service routes *through* central, high-density districts (rather than turning vehicles around) is a preferable policy, because it gives greater clarity to the network and does not require any space in the congested core for temporary storage of vehicles and turnarounds, which can sometimes be difficult. On the other hand, few people will wish to travel from one edge of the community to another, since bus trips tend to be rather short—3.8 mi (6 km) on the average[26]—and therefore few passengers will gain much convenience from long routes. No city route should be so long that operational control may be lost (not in excess of 2 hours for a round trip). The latter observation refers

---

[26] American Public Transportation Association 1998 data, which compares to 10 to 11 mi (16 to 18 km) for the average commuting trip in American communities, consisting primarily of automobile trips, as well as the use of buses with transfers to other modes.

not so much to the stamina of the driver as to the fact that traffic conditions on the streets can seriously impact schedules. Such dislocations would be aggravated on long runs. This is particularly the case if a bus has to cross the central business district during peak hours, when it becomes nearly impossible to predict when it will emerge at the other end, thus destroying the schedule on the downstream leg.

A natural elaboration of the radial hub-and-spoke pattern is to employ *branching*—sending vehicles from a major central node to different outlying destination points, or placing different routes with diverse terminals in the peripheral districts on the same alignment as they approach the center. Thus, more frequent service is provided on principal streets near the core where demand is higher, with a gradual lengthening of service intervals where demand is lower. Much the same can be accomplished by cutting some runs short—i.e., not running all vehicles to the terminal point outside and turning them back at locations where demand volumes start to decrease significantly (extended runs vs. short runs).

The *gridiron* pattern provides good coverage and accessibility to all parts of a community—any point can be reached with a single transfer in most instances (but, of course, most trips will require a transfer). The large systems of Paris, Chicago, and Toronto approximate this model. The basic premise of the gridiron service structure is that most destinations are not clustered in a single area, but will be spread over a large space. An operational concern is the associated waiting time for passengers at transfer points. If the connecting bus is a long time in coming, riders will become impatient and discomforted. Thus, this system calls for short headways on all routes, which may be difficult to achieve in most communities, particularly smaller ones.

In larger places that are equipped with rail service as the principal means of mobility, the employment of buses in a *feeder/distributor* mode is a responsive and effective approach. An important consideration here is the provision of convenient and effective bus loading bays at the stations to expedite the transfers. Just about all rapid transit stations constructed recently incorporate this feature, giving public-service vehicles the closest access to the rail station entry.

Given the fact that in recent decades it has been increasingly difficult to provide fully satisfactory, frequently scheduled bus service in American communities, a new operational concept has been developed and introduced with considerable success in many places—*timed transfers,* or *pulse scheduling* or *radial pulse.* This is the same concept that many airlines use with their hub systems. In order to avoid long waiting times for riders and provide interconnectivity as well as the densest possible service coverage, the vehicles serving many (or all) routes converge on a single point at about the same time. After allowing some time for riders to find and board their connecting buses, the buses continue their routes. This is a way to achieve maximum coverage with a fleet of limited size, but it requires finding space off the street or along curbs, where the transfer maneuvers can be accommodated and regular street traffic is not impeded. Since routes are of different lengths and local street conditions vary, scheduling to coordinate all arrivals and departures becomes a complex exercise.

The principal concern with this system is the fact that using a bus in a city with pulse scheduling is not so casual and routine as going to the bus stop and expecting there will be a vehicle going your way pretty soon. Instead, a prospective rider has to pay close attention to the schedules for the entire linked journey, and the windows of opportunity to travel will be specific and limited.

However, the pulse-scheduling concept is being applied quite successfully in Germany for small and medium-size cities that have started to suffer a gradual decrease in regular transit operations. These are the so-called *Stadtbus* (city bus) systems, superimposed on general networks, usually utilizing smaller vehicles on local hub-and-spoke routes, with coordinated schedules that bring all services together at central transfer nodes. The operations are deliberately kept simple and distinctive to gain ridership at the community level; they are subsidized by a variety of local funding sources.

An increasing concern in the structuring of service networks in American communities is the progressive dispersal of metropolitan development.[27] Travel to and from a dominant center is not the only requirement, but points within the suburban rings cannot be connected efficiently, because only a few nodes tend to generate significant transit demand.

[27] Since the 1960s, more jobs are to be found in the new outer ring districts than in the center.

Thus, the question for future network development is whether there are realistic possibilities of augmenting the traditional hub-and-spoke pattern with routes that are tangential and peripheral. How are reasonable ridership volumes to be generated among points in the suburban sprawl? The answer may be small units or responsive paratransit services, as discussed in Chap. 6.

A significant type of service that has been instituted in some downtown American cities is free bus service for all local riders. The effort has been directed toward recapturing business volume that frequently has moved largely to suburban shopping centers. If the suburban centers provide free parking, the central business districts can at least offer free internal mobility. These can be either separate circulator routes, usually utilizing small vehicles, or arrangements in which any trips that start and end within the central district are not charged. Seattle's Magic Carpet service (1973) was one of the first such programs among major cities.

Recognizing that regular bus service tends to be slow, and that much time is lost at stops, a simple solution from the perspective of riders who are already in the vehicle is to eliminate stops as much as possible; i.e., operate in an *express* mode. This is exactly what regional commuter buses do by collecting customers from a specific district in a limited number of local stops, and then not stopping at all until the destination district (usually the central business district) is reached, where again the stops will be few. The process is reversed in the evening, with the accompanying difficulty that service levels in midday will not be high, and many vehicles will have to be stored in the already overburdened city center. Or, the surplus buses will have to make deadhead trips to remote parking places, only to be required back in the center again a few hours later. Except for the storage problem, express service works well and tends to be growing in popularity in American communities, provided that sufficient volumes of demand can be assembled. Some rail systems have found or have developed a reverse commuting market, carrying central city residents to suburban employment centers, thereby making the round-trip peak operations economically viable and solving the midday storage problem.

A special effort is currently being made in Los Angeles with Metro Rapid service—discussed in greater detail in Chap. 9, Bus Rapid Transit—which utilizes distinctively different vehicles (low floor, fueled by natural gas, painted red) that travel on the regular routes but stop only at specifically designated stops about a mile apart. An effort to do approximately the same in New York City a few years ago was not a complete success. The reason was largely the fact that the buses so employed were identified only by a small sign, and many waiting passengers at nonexpress stops complained about apparently being bypassed. Likewise, some passengers got on by mistake and were upset when the vehicles did not stop where they expected.

A partial approach toward the same objective of expediting service by eliminating unnecessary stops is to designate a certain number of bus stops as request-only stops. A riding or waiting passenger would have to signal the driver that a stop is requested, because otherwise the bus would keep going. This is a reasonable approach, but alert drivers do that on a regular route anyway when it is obvious that nobody desires to stop or be picked up at a given place.

Reasons to Support Bus Systems

The bus possesses a number of significant advantages as a transit mode, regarding both the vehicle itself and the way in which it is operated.

The Vehicle

READY AVAILABILITY

Buses do not depend on advanced technology, and they can be (and are) produced by numerous manufacturers in many countries. At any given time, many models from different sources that offer a significant range of features are on the global market. Purchases can be made off the shelf, relying on experience records and catalog descriptions of elements. A notable constraint in this respect, however, may be the existence of laws in some countries, prominently including the United States, which require that government sponsored public transit programs rely on equipment that is largely produced in the same country. If the market is large enough, branch assembly plants can be established in the purchasing country to ensure that more than 50 percent of the value is locally generated. For example, there are currently several manufacturing firms that produce buses for large orders by cities or public transit agencies at plants in the United States, but with origins that are traceable to other countries.

The Question of Capacity

The determination of the capacity of a regular bus route along a street is an easy task, *if conditions are not constrained* by traffic congestion and high volumes of boarding and exiting passengers at principal stops.* *Throughput capacity* past a given point is a function of the bus size (number of passengers it can carry) and the spacing between buses (headway, or the number of buses that move past that point in an hour). Thus, for example, with comfortable space for 60 passengers in buses that run at a 5-minute headway, the capacity is 720 passengers per hour (12 units × 60 riders) in a lane that also carries other traffic. If the same buses are packed with 75 passengers each, the volume will be 900 passengers per hour. If, on the other hand, a headway of 2 minutes can be maintained and artics are used (120 passengers), the throughput capacity will be 3600 passengers per hour.

In reality, however, operations rarely proceed this neatly on actual streets, particularly when congestion and high passenger loads are encountered. If, for example, the dwell time at any given stop starts to approach the headway between buses on that route, not only will the first bus be delayed, but the progress of the next vehicle may also be impeded as it catches up. The schedule (and thus the throughput) will be impacted.

(This problem could be mitigated by providing more than one berth at the stop, thus allowing the second vehicle to be processed while the first is still at the curb. This is not likely to be an option in many places, because curb space is in high demand in dense urban districts. Furthermore, there will be operational problems, with some passengers becoming disoriented and most flocking to the first bus. Nevertheless, if this is a common practice, patrons will soon learn that the second bus is likely to be less crowded.)

Persistently growing dwell times are common at the start of peak periods when ridership volumes increase minute by minute. Thus, each successive bus becomes delayed more, and the uncertainty of operations increases. Capacity deteriorates. The lost time could theoretically be recovered by driving faster between stops, but usually just the reverse happens, because street congestion also tends to increase concurrently.

The true capacity of bus service on city streets under stressful conditions is determined by the ability of the busiest stop to process vehicles and passengers, not by the ability of the street to carry traffic. Such exact calculations can be made by recognizing the acceleration and deceleration rates of the vehicle, the number of passengers boarding and exiting, the number of seconds that it takes to accommodate each, the required clearance time between vehicles, and a few other operational factors.[†] Precise analyses of this type, however,

* The discussion here is limited to bus operations in mixed traffic on city streets. For situations with preferential treatment (busways, priority lanes, etc.), see Chap. 9, Bus Rapid Transit.
[†] The method was first outlined by W. F. Hoey and H. S. Levinson in their article "Bus Capacity Analysis" in *Transportation Research Board Report 546* (1975), pp. 30–43. This approach has been acknowledged in several subsequent discussions.

are seldom done, certainly not in the day-to-day routine of a transit agency. The situation is too fluid and the events too capricious to expect detailed formulas to have much immediate practical value when hundreds of vehicles have to be pushed through a crowded city every day. This is not, as they say, rocket science, dependent on mathematical precision, but rather the art of the possible.

Therefore, given a certain model vehicle, the key to capacity is the ability to maintain steady and constant headway. When considerably more advanced information and management techniques come into regular use, such proactive control may be possible and advisable. It will involve at least real-time information regarding where each service vehicle is located at any given moment and a predictive capability based on experience as to what delays at what minute are to be expected along the route. Then units would be inserted into the flow in anticipation of demand increase downstream and spacing would be adjusted along the route by centralized control or roadside dispatchers. We are not there yet.

In the meantime, it can be recorded that 60 to 120 buses per hour can operate on a street with mixed traffic in regular service, as has been shown by actual experience in many instances. This would be at most 9000 passengers per hour (120×75 riders per bus). Theoretically, higher numbers could be achieved, but only with extensive on-street traffic management and bus controls—which would bring us into the realm of bus rapid transit, the subject of the next chapter. The preceding scenario also describes a situation in which a number of routes overlap (i.e., use the same alignment) but have separate stops, so that the vehicles of different routes do not interfere with each other but operate somewhat independently, at least one movement lane next to the stopping lane is reasonably available for bus maneuvers and forward progress, and some platooning of vehicles (moving in tandem) is practiced.

Various on-street, real-time supervisory techniques have been developed and have been successfully applied in a number of instances by operating agencies. It is now generally recognized that complete schedule adherence is not consistently achievable in massively congested urban settings, and that controlling headway is the key to maintaining line capacity and the trust of the riding public. It is frustrating for patrons who voice unhappiness with the delayed arrival of vehicles or finding bunches of them backed up to be told by the driver that there is congestion—as if that were an unusual or unexpected situation.

NO RESEARCH AND DEVELOPMENT REQUIRED

While technical improvements to buses continue, and new elements are introduced from time to time, this is a relatively slow evolutionary process, with no major breakthroughs anticipated or sought in the near future. The vehicles already available will satisfy the basic requirements of almost any agency or situation; thus, demand for special performance characteristics will be a rare occurrence. There is a tendency, however, for many agencies to specify custom features (such as seating arrangements), which often precludes the achievement of full large-scale production efficiencies.

NO SPECIAL WORKFORCE OR SKILLS REQUIRED

The bus and the diesel engine have been around for a long time and represent basic technology well known to almost anybody in this field. Any truck mechanic who understands engines can take care of buses with little additional training, and there are very few places on the globe where such skills would not be available. Anyone who has a regular driver's license can learn to operate a bus with a little training and practice. (This does not necessarily mean, of course, that just anyone will also have the personal-relations skills to deal with a clientele that is frequently in an advanced state of stress.)

LOW INVESTMENT

Since buses almost always use existing city streets, there is no additional construction expense for the transit channel. It could be argued that bus operations should contribute to some part of local street maintenance, if not compensate for a proportional share of the original construction expense, but that has never happened nor is it expected. If this is a form of subsidy, so be it! There are some exceptions to this—for example, transit agencies sometimes pay for the construction of concrete pads (hardstands) at bus stops because asphalt surfaces may creep and become corrugated under heavy use during hot weather.

The cost of the vehicles themselves is also reasonable, considering the amount of work that buses are expected to perform over a long useful life, which some experts assume to be 13.5 years for regular street-service vehicles, but the general recommendation is still that they be replaced after 12 years of service.

A comparison can be made with automobiles, as well as rail vehicles. While the price depends on options, size of order, driveaway and shipping costs, and taxes, a standard city bus could be purchased for $295,000 in 2000.[28] The cost can be somewhat

---

[28] American Bus Association information. Prior to the 1970s, the purchase price of a standard bus had remained rather steady at about $40,000. There was a rapid escalation of the cost during the next decades. In the early 1980s, it approached $150,000 per unit.

less if the order is large, but the price of a 60-ft artic unit will approach $400,000. This is not a bad price, if one considers that an average passenger car able to carry five persons (but seldom achieving that loading) will cost in the range of $25,000. The regular bus can carry at least 14 times more passengers and will cost 12 times more, but it will accumulate many more service miles during its useful life.[29] The purchase price of any rail-based self-propelled vehicle is an order of magnitude larger, with only 2 or 3 times greater passenger carrying capacity.

ENERGY CONSUMPTION

The bus offers significant fuel-saving opportunities compared to other modes, due to the efficiency of the power plant and the relatively light weight of the vehicle. The official data show the following consumption rates:[30]

Vehicle	Btu/Passenger Mile
Single-occupancy automobile	8360
New heavy rail	3080
Carpool	2390
Old heavy rail (existing)	2320
Light rail transit	2590
Bus	1420

Service Operations

FLEXIBLE OPERATIONS

Since the vehicles are not tied to a track or a guideway of any kind, buses can move on any solid street surface. Routes can thus be changed and shifted without any capital expense. (There are frequently, however, institutional problems, inertia, set community preferences, and other restraints to modifications.) This characteristic is particularly significant for communities that are growing or experiencing shifts in their major activity distribution. Buses, more than any other mode, can provide an accommodating response.

---

[29] The average age of a transit bus in the United States is 8.5 years (APTA 1998 data). In large cities with intensive schedules, the average bus will travel 30,000 to 40,000 mi each year.
[30] Congressional Budget Office, October 1977, for Senate Transportation Subcommittee.

LINE-HAUL ABILITY

Buses can make frequent stops to pick up and discharge passengers, but they can also move relatively fast without stops. This line-haul ability is a critical feature for express-type operations that serve two districts far apart and bypass the intervening areas. Given an open channel, buses can approach the speeds of rail transit.

MANEUVERABILITY

While buses are large vehicles, they can negotiate almost all street configurations with narrow rights-of-way and tight turns, as long as any motor vehicle can get through and the driver is reasonably skilled (Figs. 8.7 and 8.8). In more constrained conditions, smaller bus vehicles are available.

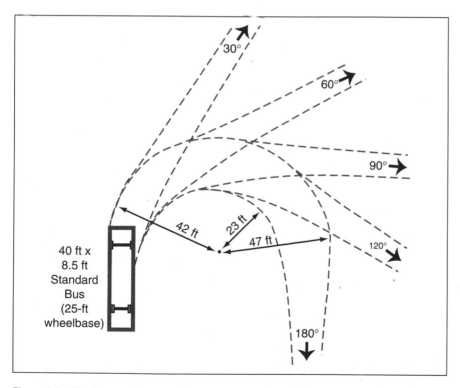

Figure 8.7   Turning requirements for standard 40-ft bus. The paths shown are for the left front overhang and the outside rear wheel. (*Source:* AASHTO design standards.)

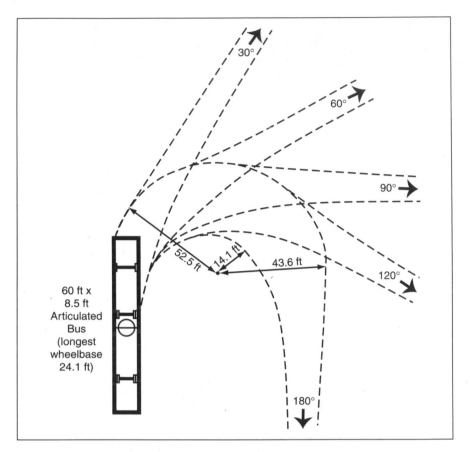

Figure 8.8 Turning requirements for articulated 60-ft bus. The paths shown are for the left front overhang and the outside rear wheel. (*Source:* AASHTO design standards.)

TEMPORARY DIVERSIONS

An obvious ability that only buses possess among the heavier transit modes is to avoid temporary obstacles that may appear on city streets and to bypass a disabled bus in front. Rail vehicles and even trolley coaches become trapped behind such a constriction, while individually operated buses are free to seek their own unencumbered path.

Reasons to Exercise Caution

Buses are splendid pieces of hardware, but, alas, they are not perfect. There are problems and constraints that have to be faced.

How critical these negative characteristics may be depends on each individual case when the suitability of various modes is examined. For example, the need for many bus drivers may be a serious problem in places where labor costs are high, but it can be considered a benefit in countries that need to develop employment and skills-building opportunities.

Labor-Intensive Operations

The ratio between operating personnel and number of passengers carried is considerably higher for buses than for those modes that depend on large units and automation. It is difficult to see any reasonable opportunities to reduce the need for drivers, given the limits of current technological development. Maintenance and administrative staff can be of normal size and composition, as compared to other forms of transit, and may command lower salaries on average because buses do not include many fragile or complicated components.

Pollution

As we all know by now, gasoline and diesel internal combustion engines can seriously damage the urban environment. While the former may have been improved to a significant degree under extensive public and legislative pressure, but not completely tamed, the latter has become a major concern lately. It has been discovered that the tailpipe gases contain many compounds that either have long been recognized as harmful to human health or have just been placed in that category. The quality of diesel engine maintenance is always a concern, since it has much to do with the volume and type of pollutants actually emitted on the street. Some enterprises and agencies are known to fall short of satisfactory performance in this area. Although buses may represent only a few percentage points of volume in city traffic streams, this situation will generate much attention in the next few years. The diesel engine can be improved in terms of its performance and emission controls,[31] but the near-term solutions are likely to be in the area of fuels used. Many communities, public interest groups, and organizations have exerted much pressure

---

[31] For example, a platinum-coated catalyst and a filter that may be able to remove 90 percent of particulate matter, carbon monoxide, and hydrocarbon emissions from diesel exhaust appears to be promising and is being tested.

lately to have operating agencies switch to nonpolluting fuels and make all other associated adjustments.

Among the alternate fuels, compressed natural gas (CNG) has moved to the forefront today, with many larger and smaller pilot operations across the country. Some potential fuels (such as methanol, ethanol, and biodiesel fuel) have faded from the scene, while other energy sources (such as hydrogen and electricity) are still in contention. Claims that low-sulfur fuel may be able to achieve comparable results are being made.[32]

The most promising engine technology currently is the hybrid system, but, again, upgraded diesel engines may prevail. Fuel cells and complete electrical power plants are being developed as well.

As of 2001, the CNG bus was about 15 percent more expensive than a regular diesel vehicle and required the support of new fueling and maintenance facilities, but these costs are coming down.

Diesel engines can be quite noisy, particularly if proper maintenance is lacking. Here, too, various improvements and better muffling devices are available.

The conclusion thus has to be that steps toward mitigation are possible, even if this progress entails higher expenditures at the start, and that public pressure will cause the adjustments to be made.

Street Congestion

Buses get caught in street congestion, and they contribute substantially to it in turn. As long as buses operate in mixed traffic, which is one of their strengths, this problem will persist. It makes bus service slow and unreliable in such instances, sometimes destroying its viability entirely. Since the total amount of street space available in almost any given area is fixed, any program to favor public transit will represent a shift of traffic space away from automobiles, trucks, and taxis. Conflicts are thus unavoidable, but there can be little

Bus powered by natural gas (tank on top).

[32] For a summary of the evolving situation, see "Searching for a Cleaner Burn," *Bus Ride,* February 2002, pp. 10–16.

question that vehicles carrying large numbers of people deserve priority. Nevertheless, under current conditions the stigma of poor reliability haunts the bus mode.

Slow Service

In addition to the congestion problem, the smooth operation of buses can be seriously retarded by fare collection practices. This happens if everybody has to enter single-file past the driver in single-person bus operations, while the driver has to collect fares or at least exercise some level of payment control. The worst scenario is one in which every rider has to pay an odd amount of cash on entry and the collector has to make change. Minutes in running time can be lost if even a few passengers create complicated transactions. The best scenario is a situation in which every rider has prepaid the fare one way or another, and everybody can move in and out through all doors without constraint. Another concern that affects the speed of operation is the adequacy of channels for boarding and exiting—the number and width of doors. If passengers can move in and out two abreast—i.e., on two channels at each door—dwell time at stops will be reduced. Door widths of about 52 in (1.3 m), instead of 30 in (0.75 m), will accomplish this task, as most European bus models do. Unfortunately, standard buses employed in American communities lack this feature.

Another method to expedite operations is not to have any delay at the doors, but to sell tickets and check passes only after the passengers are aboard. This, however, requires the presence of a conductor in each vehicle.

Comfort and Ride Quality

Transit users have an instinctive preference for rail-based modes as compared to buses. Most likely, the explanation for this attitude is largely to be found in the fact that rail provides a stable and steady ride, while the bus wobbles, shakes, and sometimes hits potholes or uneven pavement. The driver may have to apply the brakes suddenly or may have a heavy foot on the accelerator. All this means that it is difficult to walk within a bus; standees have to balance themselves continuously, and reading is difficult even if one is seated. A crossword puzzle cannot be done neatly on a bus. There is little room to move around, and the seating tends to be tight. This again is a situation in which the source of the problem is endemic due to established practice, not necessarily inherent in the vehicle or in the system itself. Some improvements can be expected, but the basic difficulties will remain.

A problem that can be solved, however, is the other discomfort feature of the bus: high steps to climb to board or exit. Since the standard bus has a level floor placed above the cross-running axles, the floor elevation tends to be at least 29 in (74 cm) above the street surface. This means two high steps, which the elderly and less-agile riders find strenuous and time-consuming to negotiate. A kneeling feature and stopping close to the curb help somewhat,[33] but here again we have a case in which a personal comfort element becomes a constraint on operational efficiency. (Low-floor buses are no longer just an exotic possibility in American communities. They have appeared in several places, and the trend will undoubtedly continue.)

Very basic human comfort features of public transit are adequate ventilation and temperature control. Beginning in the late 1960s, air conditioning has been introduced, but with mixed success during an extended early period. It took several decades until the units were engineered and built robust enough to stand up to the continuous shaking and vibration of a city bus. Another accompanying problem was caused by the insistence of manufacturers and some operating agencies that windows be sealed. When air conditioning broke down in summer the vehicles became unbearably hot, and no advantage could be taken of natural ventilation, which can be quite effective on most days.

Lower Capacity

Each bus unit is considerably smaller than any rail vehicle. Street conditions and loading demands do not allow running them as a continuous chain. Thus, the practical throughput capacity will be lower than for any rail-based system. That is a fact that places the different modes at different points on the spectrum of transportation choices. As will be shown later, it is possible to reduce the capacity difference if buses are given preferential treatment on streets, moving toward an exclusive right-of-way situation.

---

[33] The air bag that provides suspension at the front right wheel can be deflated, lowering the corner some 5 in (13 cm), and then be reinflated. Each operation takes a few seconds.

Public Image

Buses, at best, are regarded as utilitarian devices, seen every day by all members of the community. They are accepted as elements on the urban scene, but they generate no excitement. (Perhaps once in a decade, briefly, when a new model appears.) In terms of their ability to attract users in American communities, the situation is worse. Buses appear to have a negative public image; many people seem to believe that their social status would be impaired if they were to be seen using a bus. Such widely held perception is a most serious issue that cuts to the basic viability of the bus mode. This is not something that the popular or professional press particularly wishes to discuss, but it is important enough to include the following section in our review of modes.

Social Status of Buses

It is said that long-distance buses on the North American continent serve only the poor, the minorities, college students, and military personnel on leave. Everybody else flies or drives. (Yes, a few take the train.) This blatant exaggeration points to a widely held attitude toward the bus mode. These vehicles are perceived as a means of travel for those who have no other choice. You will take it, if you absolutely have to, but you would rather not, since it reflects on your social standing. Unfortunately, this attitude also extends to city service buses, with a few exceptions.

This is a subject that is difficult to discuss, because everybody knows that it should not be so and that it is a cause for some societal embarrassment. It is even more difficult to document the true situation, because people tend to respond by giving the "correct" answer rather than admit to what they believe and actually do.

The evidence is plain to see if one stands near a bus stop in almost any American city and observes the clientele. There is a clear class distinction, which most often translates into a racial and ethnic difference. The exceptions may be the large old cities on the East Coast, particularly New York, where taking the bus and using transit in general is a common and long-established practice. In Savannah, however, the people clustering at downtown bus stops on major streets during the peak afternoon period tend to be black. In Houston, office workers, mostly white, use the air-conditioned underground pedestrian tunnels to move between their desks and their parking spaces; the people getting in and out of buses on the hot surface streets are mostly service workers. In the Southwest, Mexicans are the largest cohort of bus riders. Exactly the same situation can be seen in just about any other city, and this is by no means limited to the southern states. A middle-class white male regularly commuting on a bus is a rare sight in all but a handful of cities in North America.

Indicative of the prevailing condescending attitude is the fact that a number of new or improved bus services avoid the term *bus* entirely, and coin neutral or "positive" names instead—*Metro Rapid* in Los Angeles; *Silver Line* in Boston; *Orange Streaker* in Miami; *Fearless Flyer* in Nashville; *Rabbit Transit* in York, Pennsylvania; *Magic Carpet Service* in Seattle; *Citilink* in Fort Wayne; *DART* in Denver; and many others.

There is also a practical issue. If the leaders of a community and the members of its vocal middle class do not ride the buses, there are few effective spokespersons for adequate service. An outcry when service is reduced is not well heard. With declining ridership, fares have to be increased and/or headway has to be lengthened, both of which reduce patronage further.

In reality, we are thus facing a situation in which a means of mobility is perceived to be second rate. The mode is frequently underfinanced and most often not quite adequate, but there is no question that its presence is vital to keep basic urban operations (service and production) going and to accommodate lower-income shoppers, schoolchildren, and others with no access to a car.

This socioeconomic condition has been formally recognized, for example, in the Los Angeles region, where legal action has been taken, claiming that the allocation of large funds toward the expansion of a rail system, used mostly by middle-income riders (thereby diverting resources from bus operations, used mostly by lower-income riders), is discriminatory and illegal. In the New York region, similar action has been considered, suggesting that fare increases should be proportionally lower on subways and city buses because they are primarily used by less-affluent city residents, as compared to regional rail systems serving suburban commuters.

Thus, it appears that American communities face a choice of two basic approaches. The first is to accept the currently prevailing role of bus service as a second-class service that has to be kept going at a base level so that an elementary level of mobility is available to all residents. This is probably what most places will

A satirical but widely held view of bus service in the United States, originally published in the *New York Times,* December 9, 1971.

A Fling With the Buses

Russell Baker

WASHINGTON, Dec. 8— For the first week or two, riding the bus between home and office seemed a pleasant change. The car had become hateful. For one thing it was begining to gobble bigger and bigger sums of money. It was secretly supporting a parking-lot operator in byzantine luxury, and what with the tendency of traffic paralysis to last for longer and longer periods and the tendency of gasoline barons to raise prices more and more often, the car's eight big cylinders with their habit of gargling on high-test gasoline were going through the family fortune like a plague of college-age children.

Those first few days on the bus were so cheap! Forty cents downtown. Forty cents back. A bagatelle and a pittance compared to the sawbucks on which the car gorged.

The bus also offered surprising engagement with humanity. There at the bus stop were people. Genuine people. They could have been touched and their humanity thus affirmed; except for our cultural inhibition against human contact without a license. Still, even without tactile confirmation, they were indisputably people. You could get close enough in the morning to see the rheum in their eyes and in the evening to see the smudges under their chins where fingers had paused absent-mindedly in the day's work to test the firmness of the flesh, to feel the march of fat, the creep of age there under the jaw-bone.

This was the way people used to be. Flesh, dirt, smells of life, tension or sag in the figure, warning signs up. To say it was exhilarating would overstate the fact. It was merely interesting. From within a car, people are seen as part of an outer world kept distant by barriers of glass, engine noise, metal and the psychology of privacy. A man in a car sits alone at the center of his exclusive universe.

From inside the wheeled one-man universe, someone glimpsed through a windshield may occasionally seem a curiosity or a nuisance; more often the lot of them out there are cyphers in dim masses, not much more interesting than the faceless thousands wiped out in humdrum headlines by a storm in Pakistan or a political shenanigan in Indonesia. This is probably why, when one of the dreary devils irritates our car in its progress, we feel ourselves reflexively sympathizing with the car's urge to run him down.

But the bus——. The pleasures of feeling money saved, of rediscovering that people everywhere— not just the people at the office and at home—that people everywhere were really people, just as they used to be in the mediocre old days—these pleasures were only temporary, the delights of fresh adventure. After they subsided, the reality of urban mass transportation asserted itself.

Except in rush-hour periods, service was thin. Standing at the bus stop inhaling the fragrance of humanity and noting rheumy eyes and smudged chin can be exhilarating when the bus is going to be along any minute now. Doing it for twenty minutes in a drenching downpour before the bus lumbers into view does not improve the spirits, particularly if, as usually happens, you are waiting for the bus to Eyesore Estates and the bus lumbering into view is going to Impecunious Gardens.

Seating on the buses appeared to have been designed for a race of short, small-boned people, like the Indochinese. This seemed odd because, although there are many Indochinese in Washington, most of them seem to travel in chauffeured limousines.

Space between seats was so limited that any American slightly longer than average—and what American is not in the golden age of beefeating?—was subjected to considerable discomfort. Bus seats, like men's ready-to-wear, seem to be cut only for "regular"-sized persons, or what was a "regular" size in 1900. If mass transportation is to succeed, someone will have to notify seat designers that Americans now come in lengths up to seven feet.

With great diligence, one could read on the buses. This was a decided gain over the car, where it was impossible to do anything but listen to the radio describe the weather outside the windshield, but it may have laid up future business for the eye doctor. Most of the buses shuddered and vibrated constantly with elephantine death rattles which made it painful to focus on print very long. Many had dim or flickering lights that made interiors after dark suggestive of foggy nights in the old Limehouse of Fu Manchu.

All buses required an exact fare of forty cents, a device which seems to have broken a Washington habit of murdering bus drivers for a handful of silver. Fine, but trying to assemble forty cents in exact change on a street corner nine blocks from the nearest cash register can occasionally inspire another motive for murder.

Now the first romance of the buses has faded. The people are dull and inhuman. Forty cents is a lot for such thin, ill-lit, vibrating, uncomfortable service. On rainy nights buses are hateful. Can't anybody contrive a sensible way to get around this country?

do. If the social and political implications of this approach are recognized and deemed to be locally tolerable, this is a workable policy—but not an element of civic pride. Nevertheless, public funds will have to be spent to keep the operations going, and public policy should ensure that improvements are made to the base urban mobility service.

The second approach would be bolder, involving constructive steps to bring bus service to a higher level of responsiveness and convenience—i.e., *European* quality. This will call for considerable effort and expense, with the aim of recapturing not only the effectiveness but also the attractiveness and public standing of bus operations, particularly in those places where no other transit services are present or feasible. It will be an uphill battle, but the assumption will be made here that it is worthwhile doing. It may very well be that this is the overarching quality-of-life decision regarding transportation services that American communities will have to make in the foreseeable future.

Application Scenarios

The overall contention here is that buses are employable as a form of transit in *all* urban situations, and remain the most affordable choice. The exceptions are low-density districts where sufficient demand cannot be generated for any regular public service, even under the best of circumstances (see Chap. 6, Paratransit). This limit may be 4 housing units per acre (10 per hectare) or less.[34] Nevertheless, some publications that are widely used by transportation professionals[35] still suggest that districts with less than 3 dwelling units per acre (5800 people per mi^2; 2200 people per km^2) should have a basic bus service in place, and any community with higher density should have a good network of routes. That would certainly be commendable, but may represent wishful thinking under current development patterns and municipal budgets in American communities. It can be done—and perhaps it *should* be done—but substantial financial assistance from

[34] This limit was identified by B. S. Pushkarev and J. M. Zupan in *Public Transport and Land Use Policy* (Indiana University Press, 1977) and has not been seriously challenged since that time. See Chap. 11, Streetcars and Light Rail Transit.

[35] Such as J. D. Edwards, *Transportation Planning Handbook,* (Institute of Transportation Engineers, 1992), p. 153.

some non-fare-box sources would be required to achieve such standards.

A difficult question is the total ridership that is necessary to justify the implementation of bus service in any given corridor. More precisely: since all public transit is subsidized to a significant level in the United States, what is the minimum ratio of expected revenue to operation and maintenance costs that a route has to generate for it to be viable? The answer should emerge from a review of how serious a lack of public transport would be to the residents in any given corridor. Is the local population entirely dependent on public service, or are other adequate and affordable means available (paratransit and taxis, perhaps)?

One suggestion—which sounds reasonable, but can only be regarded as an approximate starting point for a rational examination—is that total daily volume in both directions of 1800 to 2000 passengers should be expected, with no less than 150 to 200 passengers in the peak hour.[36] (The latter demand can be carried by three or four buses, which would result in only a couple of vehicles per hour in a single direction.) But even if this number is not achieved, there would be a strong reason to run at least one bus every hour on significant arteries, thus not leaving neighborhoods stranded without any means of some mobility for everybody. Even in prosperous areas, schoolchildren and service workers will not drive cars.

In this context, two basic situations can be distinguished:

- *Buses as the only (or dominant) transit mode.* If buses constitute the base service—i.e., the bottom-level means of mobility for a community—it can be argued that all households and all establishments should be within a reasonable distance of bus stops, and that all routes should be interconnected to each other at some point. This is thus a matter of geographic coverage, and the ideal situation would be a network of routes that are never further than 0.5 mi (or 1 km) apart. This would result in walking distances to and from stops not exceeding 1500 ft or 5 minutes for all residents. Unfortunately, such a scenario would necessitate considerable expense in running very lightly loaded vehicles, particularly in the low-density districts of American suburbs, and it is not likely to be a supportable and prudent public policy.

[36] G. A. Giannopoulos, op.cit., p. 107.

Then the key question becomes the definition of an acceptable walking distance. Evidence shows that anything beyond 0.5 mi or 3000 ft (1 km) would not be practical, since Americans walk that far only in special cases (to get to a seat in a football stadium or to an office desk in Midtown Manhattan). Thus, spacing routes 1 mi apart can be considered the maximum limit if the bus service is to have reasonable utility. Unfortunately, this is not universally achieved either, in the face of the political realities of transit budgets.

A further refinement to this process would be to recognize the presence of higher-density corridors within the urban fabric and to place routes on such alignments. For example, one such suggestion is to ensure that 75 percent of the population in urban districts resides within 400 m (1300 ft) of a bus stop, and that in suburban areas 50 to 60 percent of residents live no further than 800 m (2600 ft or 0.5 mi) from a bus stop.[37]

The ultimate achievement would be to follow *performance* specifications and structure a public transit system that allows almost all commuting (work) trips to take no longer than, let us say, 45 minutes. This would require intricate analyses of the community and its structure, and result in a high-quality but probably expensive service network. That is the approximate practice in Scandinavian countries, but it is encountered hardly anywhere else.

- *Buses supplemental to a rail network.* If the principal transit system of a city or metropolitan area is rail-based, the primary purpose of most bus routes would be to act as feeder and distribution services. The stations of commuter rail, metro or subway, light rail, or even express bus operations become the nodes that focus local bus lines fanning out into the surrounding neighborhoods and districts. The considerations of geographic coverage are the same as previously described, but the principal need is to provide as unconstrained connection as possible between the modes. This is a physical and operational task: bringing the loading and unloading of the buses close to the station platforms, providing mechanical assistance and weather protection for

[37] G. A. Giannopoulos, op.cit., p. 112.

transfer movements, instituting joint tickets, and maintaining attractive surroundings and service amenities. Perhaps even more important is coordinating the schedules of all services so that waiting times and transfer frictions are minimized.

Crowds of prospective riders in a Shanghai bus terminal.

In all the preceding scenarios addressing the service network configuration, there is another dimension—schedules and frequency of service, which to a large extent are governed by the volume of demand (see "Scheduling of Bus Operations" later in this chapter). It is also a question of the size of the vehicle to be used. Given a certain passenger volume, it can be accommodated by smaller units running frequently or by larger units at greater intervals. Unquestionably, the patrons prefer the former as a matter of responsive service; operating agencies would tend to favor the latter as a matter of cost-saving efficiency. It can be suggested that a 15-minute headway can be regarded as reasonably good quality urban service, no matter what the loading might be, while an interval exceeding 30 minutes certainly represents an inferior situation, no matter what the size of the vehicle might be. On the other hand, nothing much is gained in consumer satisfaction if the interval is reduced below 5 minutes. In such instances the use of increasingly larger vehicles would be appropriate to maintain adequate capacity while also maintaining efficiency and reducing street congestion.

Overloaded bus in Kathmandu, Nepal.

General Operational Characteristics of Bus Transit

Capacity of standard bus   60 or 75 passengers (depending on comfort level)

Boarding time per passenger per channel

   Free entry   2 or 3 seconds

   With tokens or magnetic cards   4 seconds

   With other transactions   10 seconds or more

Alighting time per passenger per channel   2 seconds

Dwell time at major stops   30 to 60 seconds (and more at key locations)

Dwell time at minor stops   10 seconds

Overall running speed (⅔ of prevailing speed by general traffic on street)

   In average districts   13 mph (21 kph)

   In dense districts   6 mph (10 kph)

Note: All numbers are approximations only.
Source: Field tests.

Components of the Physical System

Since the channels for movement for regular bus service are city streets and highways,[38] the types and diversity of the system elements are rather few.

[38] Buses as principal components of a *bus rapid transit* system involve public (possibly private) rights-of-way, rolling stock, minimal wayside improvements, maintenance and storage yards, and personnel with appropriate facilities.

Suggested Physical Dimensions of Bus Facilities

Width of loading/unloading lane at bus stop	10 ft (3.0 m)
Length of loading berth	
Near side	Length of bus in use plus 65 ft (20 m)
Far side	Length of bus in use plus 40 ft (12 m)
Midblock	Length of bus in use plus 100 ft (30 m)
Length of second (tandem) berth	Length of bus in use plus 5 ft (1.5 m)

Rolling Stock

The types of vehicles, primarily classified by size, have been listed previously; the task here is to discuss various elements that constitute the vehicle and influence its utility and comfort levels.[39]

The interior *seat arrangement* of a regular bus can be easily arranged in any configuration that is appropriate for a given service demand; i.e., it can be specified in the purchase order without a cost premium. The basic question here is the number of seats to be provided. There can be as many as 50 (even a few more), if the vehicle is to be used for relatively long trips with no standing passengers. Or, there can be only a few dozen seats, leaving the floor space open to accommodate as many standees as possible, if it is to be used for short trips with frequent boarding and exiting. The central part can be left entirely open with, let us say, 30 seats arranged along the side walls. Associated considerations would be the type of handholds provided and their placement. (A seat requires a floor space of about 3.8 ft^2 [0.35 m^2]; an average standee will occupy approximately 3 ft^2 [0.28 m^2], which is not yet crush loading.)

A critical question is the number and size of *doors*. Again, a number of possibilities exist, keeping in mind that larger openings will allow quicker exiting and boarding by passengers. Some seats will be lost thereby, and this may also somewhat complicate the structural design of the bus body in order to retain sufficient strength and rigidity. Nevertheless, there are successful city service buses that have three double-opening wide doors on one side, which cut down the dwell time at stops to a few seconds even with heavy loads. The catch is that this presupposes that there is no requirement for paying a fare or even showing a valid pass on entry. If there are two normal-size doors, and the driver has to collect fares, use of the front door by exiting passengers delays boarding and, in turn, stretches out the total running time on a route.

The existence of *low-floor* buses and the advantages that they offer are relatively new discoveries in the United States. This is old news in the advanced countries of Western Europe, where, it appears, just about any new bus placed in city service today is a low-floor vehicle with the floor 12 in (30 cm) above street sur-

---

[39] G. A. Giannopoulos, op. cit. pp. 149–197 (Chap. 6) provides an extensive review of vehicle elements, particularly choices in interior arrangements.

face. While many transit agencies are examining and even trying them, only about 5 percent of the U.S. transit bus fleet (not including minibuses) is of this type so far.[40] Since the floor height of the standard bus is 33 in (0.84 m) above pavement, three steps are required. The early efforts in the 1970s to design a "low-floor" bus concentrated on achieving a two-step configuration, or 17 to 24 in (0.4 to 0.6 m) vertically.[41] Various efforts, such as smaller wheels and sloping floors, created mechanical and safety problems, and considerable climbing was still required, even with a kneeling feature.

European manufacturers have succeeded, however, in developing a true low-floor vehicle, and models are now also available from several U.S. manufacturers. There is an additional cost involved because the structural design of the vehicle becomes more complex: since there is no axle running across the width of the chassis from wheel to wheel, a heavy separate box is used to hold each wheel in place from one side. This also reduces the usable floor space. This requirement can be minimized making not the entire floor low, but only the front or the middle part, while the rest of the floor steps up over the axles. The low floor allows slower customers to enter and exit with no difficulties; it also expedites the movement of everybody else.

Low-floor buses are more expensive than regular vehicles because of structural issues. They also lose some floor space because of the protrusion of the wheel housings, which are too high to be used for seating.

It is accepted today that all public-service vehicles, including buses, have to be accessible to people with *disabilities,* including those who use wheelchairs. The debate as to whether this is a good or affordable policy is over in the United States; it is now only a question of how to do it most effectively.[42] The regular wheelchair lift, added to the front or middle door, is a workable device, adding marginally to the cost of the vehicle, but its operation consumes a measurable amount of time. Regardless of what claims may be made, each action will require 2 or 3 minutes in actual practice (from the time the operator leaves the driver's

---

[40] APTA data.

[41] See the discussion of the Transbus program under "Development History."

[42] As of this writing, more than 75 percent of the transit bus fleet has been so equipped nationally (1999 APTA data).

seat, accommodates the wheelchair, and starts driving again).[43] Since every wheelchair has to be offloaded as well, there will be a 6-minute delay on the run. If the headway is only a few minutes long, the total schedule can be disrupted. It would be advisable then to institute dispatching procedures that would allow a responsive adjustment of the movement of the following vehicles and their spacing (leapfrogging the sequence, for example). The problem becomes much less severe with low-floor buses, because a simple plate can be swung out to serve as a ramp for the wheelchair and quickly retrieved.

Current expectations of riders include proper ventilation, heating, and lights. In most parts of the country, air conditioning is also just about mandatory.[44] It is fortunate that the units available today are reasonably reliable and sturdy, after a long period of suffering with devices that could not stand up very well under the constant vibration and shaking that characterize bus movement.

Besides the various physical improvements and feature upgrades currently under consideration in bus design, ranging from lightweight materials (carbon fiber) to steerable rear wheels, there is also the question of *size*. Articulated, double-deck, and trailer vehicles do provide greater capacity than regular buses, but each is accompanied by various cost and performance issues. The question is whether a single-body vehicle of 45- and even 49-ft (13.7- or 15-m) length, might not be a superior choice in high-demand situations, particularly if it is possible to build it without a third axle.

Bus Stops

Bus stops are obviously for the convenience of the riders, but their location has to also respect the efficiency objectives of the operators.

The first critical consideration is the spacing between stops. The shorter the distance, the easier will be direct access on foot by users, but the total running time of the vehicles on the street will increase. If a quarter mile (about 1320 ft; 400 m) is taken as

---

[43] It can be accomplished in 70 seconds under ideal conditions, but may consume 4 minutes if there are snags.

[44] As of this writing, about 90 percent of the transit bus fleet has been so equipped nationally (1999 APTA data).

a convenient walking distance for Americans, with a bus line servicing a corridor half a mile wide, stops might be placed rather
close to each other—i.e., 500 to 700 ft (150 to 215 m) apart—
to achieve maximum walking accessibility.[45] This means stops
every second or third block (in the short blockfront direction), if
reference is made to a standard city street grid. This can be considered only in very high density districts with good volumes of
ridership and not excessively long total travel distances, since the
overall operational speed will be slow. Stops any closer, say 250
ft (76 m) apart, could be applied to downtown distribution services where much boarding and exiting takes place along the
entire route or loop.

In more typical urban or suburban cases, where passengers
would be drawn from corridors a mile or more in width and the
total route length is many miles, stops cannot be closer than
1000 ft (300 m) apart, preferably at a spacing of 1500 ft (500
m) or even more, to maintain reasonable service efficiency.
Express service can stop every mile (1.6 km) or so at important
nodes, but transfer arrangements will be necessary.

The second principal concern associated with bus stops is their
placement with respect to cross streets. The distinction is
whether the stop is placed *before* a cross street (near side), immediately *after* it (far side), or in *midblock*.[46] The ruling consideration, no doubt, should be user linkages to origin and destination
points. The stop should be placed directly in front of buildings
that generate many trips, near other stops to which transfers are
made, and at points where a number of paths converge.

The question of far or near side, consumer desires being neutral, involves operational convenience. The ideal situation would
be to find a pattern under which the bus accommodates passengers while the traffic signal is red anyway, but finds it green every
time it approaches a cross street. Under precisely managed conditions, this is theoretically possible, but under the real, continuously changing, and fluid conditions on the street, there is not
much point to such an effort.

Near side location allows easier vehicle departure and merging
into traffic; it is almost mandatory if the bus is to make a right

[45] See also ITE, *Transportation Planning Handbook,* op. cit., p. 154.
[46] G. A. Giannopoulos, op. cit., pp. 113–132 explores this question in considerable detail.

turn. Far-side location is to be selected if the bus is to maneuver across traffic lanes to make a left turn at the next intersection, but there are also safety and visibility considerations, particularly with heavy street traffic.[47] Many communities, to minimize customer disorientation, select one of the options for use throughout the system.

In situations in which buses constitute the principal means of transit for a city or metropolitan area, major nodes in suburban areas increasingly require park-and-ride lots that accommodate commuters and shoppers.

Bus Shelters

Protection against the elements is a most desirable amenity for customers, particularly when the average wait may be rather long.[48] The standard bus shelter is a rather skimpy structure that offers little comfort when a wind is blowing or temperatures are low. Its principal asset appears to be an ability to accommodate large advertising displays. This is not all bad, because under the arrangements that are common today, the community obtains revenues through a franchise fee, and the shelter is built by private enterprise at no cost to the local government. Whether all this contributes to the beauty of the streetscape remains an open question.

It can be suggested that a fully adequate shelter should have the following features:

- A large enough roof overhang to keep customers dry
- Vertical transparent walls on more than two sides to control wind, but with enough entry/exit points so that users do not become trapped
- Seats for some passengers
- Litter baskets
- Lighting
- Full information displays, with schedules and maps

[47] American Association of State Highway and Transportation Officials, *A Policy on Geometric Design of Highways and Streets* (Washington, 1990), pp. 560–563, the definitive reference work for roadway design, strongly favors far-side location.

[48] See G. A. Giannopoulos, op. cit., pp. 133–138 for design details.

Bus terminal lanes in Bahrain.

Improvised on-street bus terminal in Beijing.

- In very cold climates, perhaps some heating
- If advertising is unavoidable, space for public and community announcements as well

It must be recognized that in some places problems with potential vandalism are most severe, and adequate sanitation will always require close attention.

Movement Lanes

Buses operate on regular streets and highways, and some attention must be paid to the fact that they are large vehicles. Transit buses are wider (8.5 ft; 2.6 m) than normal passenger cars, and therefore need at least 10.5-ft- (3.2-m)-wide lanes. They can operate on narrower lanes, but with much caution and loss of efficiency. Buses also make wider turns with a longer radius than automobiles; therefore, sharp corners will not work. Corners have to have an inside turning radius of at least 20 ft (6 m), preferably 30 ft (9 m).

While buses can operate on streets of any steepness (even in San Francisco), grades preferably should not exceed 8 percent, particularly if icing may occur. Unless they are equipped with special and powerful engines, buses will move slowly, create much noise, and generate more exhaust emissions on steep grades.

Loading Spaces and Turnouts

A desirable feature in bus system design is to place loading spaces *off line*—i.e., not stop the bus within a regular moving lane, but pull it out on the side in a *bus turnout*. This, of course, requires additional street or sidewalk width, but the gains in not impeding traffic flow are considerable. The bus will create some friction in the stream anyway, by slowing down to turn off and trying to find an opening to reenter the moving lane of traffic. This happens in effect if parking is allowed along the curb, except at the bus stop. The price to be paid for this arrangement is a slowing of bus movement due to lateral maneuvering.

If the turnouts are placed on high-speed freeways and arterials, deceleration and acceleration arrangements have to be provided for safe and efficient exit from and entry into the moving lane.[49] These would include tapering speed-change lanes, wide stopping areas, possibly lateral barriers, and protected passenger access from the "land" side.

The opposite concept is *bus bulbs* (or extended bays)—the bulging out of the sidewalk waiting area across a parking lane to meet the movement lane directly.[50] These are simple devices that expedite bus operations, but constrain general traffic flow. They have rarely been used in the past because the bus simply blocks one traffic lane; an off-lane bus bay within the parking lane has been much preferred. The construction of bulbs, therefore, is emblematic of effectuating a bus-priority policy—it allows the bus to reach the curb directly, positions all doors without large horizontal gaps from the curb, gives waiting patrons more space, and in many instances shortens the crossing width of the adjoining street for pedestrians.

Fare Collection Systems

This is not the place to discuss the level of fares or the amount of subsidy that transit systems should enjoy, since those subjects are independent of how systems are to be selected or structured. It is important, however, to consider the fare collection mechanism, because—as has been pointed out before—that aspect determines to a significant degree how expeditious the whole transit operation will be, as experienced by the consumers. Up until recently this was a difficult and complicated subject, because traditional (manual) procedures were time consuming—particularly if tickets had to be bought onboard, transfers had to be dealt with, and drivers had to track each passenger across several fare zones. Electronic fare cards offer much better options.

---

[49] AASHTO, op. cit., pp. 405–407, 563.
[50] Transit Cooperative Research Program, *Evaluation of Bus Bulbs,* TCRP Report 65 (Transportation Research Board, 2001, 31 pp., with appendix.)

The ideal situation is to have no delays whatsoever for incoming passengers as a result of any action associated with fare control. This leads to the so-called proof-of-payment or honor system, which makes all riders responsible for having a valid ticket or pass when they cross a well-marked line. (Inspectors, of course, would ensure compliance.) This method has been used by most agencies in Western Europe for some time; a few cities in the United States have now also adopted this approach with reasonable success on selected transit routes, but not so much on bus systems.

A person in free motion (assuming no steps to climb) can move past a point in a couple of seconds; if a token has to be dropped in a box or a magnetic card must be validated, the time consumed will be about 4 seconds on average. If tickets have to be purchased or transfer slips must be obtained, a minimum of 10 to 15 seconds will have to be allocated.

Thus, today, to achieve the quickest boarding and exiting times, with full control over proper transit usage, the following methods are recommended:

- No fare at all.

- The proof-of-payment or honor system, with roving inspectors.

If the driver has to give change for purchases, the bus will never get anywhere. In this area of individual human behavior, Murphy's law is in full effect: with some frequency, prospective bus riders will appear who will fumble every part of this series of actions and react with indignation if reminded that hundreds of fellow travelers may be inconvenienced. There should be the fewest possible chances for this to happen in the path of passenger entry and exit. Even with free entry, somebody will inquire and debate whether he or she is on the right bus and delay everybody. The bus cannot move until the doors are closed, no matter what technology is used.

Design handbooks tend to suggest times shorter than the 4 to 15 seconds given here. They can be, but only if the performance of riders is efficient and impeccable. A smooth retrieval of the fare card in one motion from a pocket, for example, unerring insertion into the slot, and an elegant retrieval and deposit back into the pocket while walking on. That would be 2 seconds, but such skillful performance certainly does not happen every time. This is an area in which extensive detailed behavioral studies might give better guidance than the approximations we rely on today.

- The wide use of prepaid monthly or daily passes, which can be electronically checked by swiping or proximity readers upon entry.

- Purchase of magnetic cards that record an account balance from which a fare is subtracted on entry or exit, with the ability to also handle transfers. The automatic reading and recording machines are now quite reliable, and it can be assumed that not so far in the future all residents will carry a universal electronic credit and identity card through which all purchases of goods and services will be made. It might be attached to a personal communications device.

Maintenance and Storage Yards

At the end of a shift constituting several trips along a route, the driver has to bring the vehicle back to an operations base. The base accommodates a number of functions: removal of collected money and recording of passenger activity completed, provision of rest facilities for the drivers, cleaning and refueling of the vehicles, and parking of the buses overnight or until the next run. Dispatching control and a wireless communications base may also be located at the yard, although other locations are possible. There will usually be an administrative office for the entire system or the division that deals primarily with scheduling, driver assignments, and dispatching and control of vehicle operations.

The location of such a facility is of some importance, because it is most desirable to minimize the aggregate volume of deadhead runs (trips carrying no passengers, to start or end a shift at the terminus of a route). The site selection, however, is in reality most often influenced by the availability of a buildable parcel of land of sufficient size, particularly because residents do not regard these facilities as good neighbors. The convergence of many large vehicles on local streets at all hours of the day, the presence of large volumes of fuel, and the semi-industrial internal operations are seen as problems. Bus yards are often classified by residents as locally unwanted land uses (LULUs), only a notch less objectionable than yards for sanitation trucks.

The human activities will be housed in a building that can be made visually attractive; the vehicle spaces will consume large areas and are difficult to camouflage. In places with very intensive sunshine or extreme rainfall, the parking areas should be sheltered. In cold climates, it is advisable to build enclosed garages or *barns*.[51] Diesel engines are difficult to start at cold temperatures; therefore, it was formerly normal practice to keep them running during cold nights in northern cities, which greatly exacerbated the pollution issue. Today, heated barns (maintained at 50°F; 10°C) or vehicles with built-in insulation and heating features will be used.

It is normal practice to include in each yard facilities for regular housekeeping or *light maintenance* of vehicles. This encompasses washing machines, storage of spare parts, and a shop that can deal with most routine mechanical and electrical problems, tire changing, and interior and exterior repair. Fuel tanks, a maintenance pit or hoist, simple body and paint shops, and other routine facilities and equipment will be available. *Heavy maintenance*—overhaul of engines and transmissions, major repainting, etc.—has to be done in a larger special facility, which may or may not be attached to one of the yards. There probably will be only one such installation for the whole system; this work could also be done under a contractual arrangement by a commercial heavy motor vehicle maintenance enterprise.

Experience suggests that bus yards should not be excessively large to maintain good supervision, to manage logistics, and to achieve reasonable geographic distribution (access to routes). Approximately 250 vehicles per facility appears to be a good number. Keeping in mind that each bus requires about 540 ft^2 (50 m^2) for parking, the total space needed becomes sizeable.

The final consideration is *security* since valuable equipment is stored and the large blank sides of vehicles represent a temptation for graffiti artists (or vandals). A reliable fence, effective lights, and the presence of guards (even dogs) are mandatory.

The quality of the bus maintenance facilities, the skill of the personnel, the availability of resources to do the required job, and the commitment by the management to show results are key ingredients in operating a responsive and reliable public service—particularly when the industry target is to have at least 85 percent of the fleet in operable order at any given time. All agencies do not achieve this all the time.

---

[51] This designation can be traced back to the horse-trolley barns that were converted to streetcar and then to bus use in many old cities when the modal switches were made.

Bus Terminals

Regular city service buses do not have (nor should they have) buildings with off-street loading facilities as terminals. Their operational aim is to drive through their service corridors as rapidly as possible.

The exception to this condition would be a pulse-scheduled operational system, which does require space on streets or adjoining parcels where many loading bays can be located in close proximity to each other for easy transfer by

Bus depot and yard in San Juan, Puerto Rico.

passengers in every direction. An ideal solution might be an open city block, although such opportunities near core activities may be rare (i.e., assuming that a significant percentage of the riders have trip destinations and origins in that district). A small building with a waiting area, information counters, public rest rooms, and some concessions would be appropriate.

Regional buses have different demands.[52] The vehicles indeed terminate their journeys at some central location, even though the passengers continue to their final destinations on foot or by some other means. Since commuting demand dominates this type of operation, there will be heavy one-way inbound travel in the morning peak period, with the reverse in late afternoon. Space for a good part of the fleet to lay over has to be found. Boarding operations for the two periods are also quite different. Exiting is quick and should be unconstrained since the customers are eager to reach their workplaces. On the return trip, however, the empty vehicle will stand at a loading platform for at least several minutes to assemble a load and keep to schedule. Since the departures to any given destination may not be that frequent, some

---

[52] Discussion of the interior arrangement of a bus terminal is outside the scope of an urban transit review. Sufficient information can be found in Vuchic, *Urban Public Transportation,* op. cit. pp. 275–284. Extensive review of terminal layouts and the organization of channels around them can be found in ITE, op. cit., pp. 218–225. See also Chap. 18, Intermodal Terminals.

passengers will have to wait, which means that they will make some purchases and enjoy some refreshments. All this suggests a simple terminal building, unless these activities are combined with commercial intercity operations, in which case a joint terminal will assume larger dimensions.

Since this facility becomes one of the principal gateways to the center of a metropolitan area or a city, its placement as close as possible to the focal point of the core is indicated. Proper architectural design, reflecting the civic importance of this facility, should be expected. A serious social problem associated with bus terminals (and railroad stations) in American communities is the fact that they attract homeless individuals and other persons who do not practice "normal" behavior because bus facilities have the longest daily hours of operation, and they offer inexpensive services and eating places. The extent to which bus terminal operators or social service agencies have to take responsibility for this situation remains a point of debate. Terminals are, after all, public places, and selective exclusion is rarely morally or legally defensible.

Scheduling of Bus Operations

Since not much planning and design effort is needed to establish physical facilities for a bus system, principal attention can first be devoted to the layout of routes, so that good geographic coverage is achieved and the largest number of potential riders can gain convenient access to the service. The second major task is to schedule service[53] so that demand is reasonably satisfied with the least consumption of resources, and to determine the total fleet of vehicles needed.

Such analysis should start with the determination of the *policy headway* for the entire system or for a particular route. This is the base interval between buses that is to be maintained during all operating hours no matter what the loads might be. This headway would prevail during the middle part of the working day, late evenings, and perhaps on weekends.

---

[53] The scheduling of service is a rather intricate process that utilizes various elaborate programs, including computer based procedures. The discussion here intends only to outline the basic principles and assumptions that drive these tasks and provide examples. This information is sufficient to achieve workable results for simple situations.

The Case of the Broadway Bus

Field surveys were undertaken for the specific purpose of verifying the basic characteristics of bus operations as described in this chapter.

The M104 bus route in Manhattan, operated by MTA, was selected. This is by no means an average or typical line as found in American communities, but the deliberate choice was to examine a busy route in an intense urban setting. M104 carries a heavy passenger load, and it runs along the spine of Manhattan. It ranks in the 11th spot among the approximately 200 bus routes in New York City.* Its annual ridership is 10.3 million passengers (1999), and daily volumes in recent years have ranged from 26,200 (July 1998) to 33,600 (October 1999).

The route connects Morningside Heights (a neighborhood with many major institutions) and the United Nations, utilizing Broadway for the north-south portion of the line and 42nd Street for the east-west leg. It sits on top of subway lines (Numbers 1, 2, 3, and 9 below Broadway and the Shuttle below 42nd Street). Most trips on the buses, therefore, are short and of a local character. Not too many tourists know about this route. All street intersections are equipped with traffic signals.

The surveys were done in January and February 2001, on midweek days, and not during rush hours. A significant percentage of riders were senior citizens and mothers with small children. During the survey hours, no unusual events took place in Manhattan; the weather was not excessively cold; there was no precipitation. Traffic moved smoothly by Manhattan standards; i.e., considerable jockeying for space was involved, but there were no instances of gridlock or inability to get through a single signal phase.

There are 39 seats on the regular bus and a reserved place for wheelchair tie-down. All buses have wheelchair lifts and kneeling capability at the front door. Entry onto the vehicles is controlled by a combination fare box and magnetic reader that accepts coins and magnetic cards, which almost everyone now carries.

Bus stops are generally placed on every second block in the north-south direction (a block is 200 ft [61 m] wide; cross street are 60 or 100 ft [18 or 30 m] wide) on the far side; and on every block in the east-west direction (usually 800 ft [244 m] long, with avenues 100 ft [30 m] wide).

Seats were available at almost all times during the survey. The headway on published schedules during these hours is labeled "frequent"; i.e., buses are never (well, hardly ever) more than 5 minutes apart, and sometimes they are close enough to leapfrog each other. There are 200 bus runs in each direction during the 24 hours of operation. The average interval during the 3 to 7 P.M. hours is 4.1 to 4.4 minutes.

The on and off movements were distributed fairly evenly along the route. On the northbound runs, the busiest entries were along 42nd Street between First

---

* MTA data for 2001, counting everybody who dropped a token in the box, swiped a Metrocard, or used a special pass on entry. Detailed entry/exit data are obtained through field counts during a few days each year, with surveyors riding each bus.

and Park Avenues, at 57th Street and Eighth Avenue, and Broadway and 73rd Street. The highest number of exits were at 42nd Street and Eighth Avenue (Port Authority Bus Terminal) and 96th Street and Broadway.

On the southbound runs, the entry/exit volumes were even more uniform. The highest number of entries were recorded along Broadway at the principal cross streets, such as 86th and 79th; the most exits were on 42nd Street at Second and First Avenues, and Broadway and 68th Street.

The run from West 121st Street to First Avenue is 6.53 mi (10.5 km), and the time consumed may be 2 minutes below an hour or 10 minutes above an hour under normal midday conditions. Thus, the overall running speed ranges from 5.5 to 6.7 mph (8.9 to 10.7 kph). A reasonable average assumption would be 6 mph (9.7 kph). During rush hours, the speed is significantly slower, when the whole street traffic situation becomes very fragile and largely unpredictable, particularly at the major nodes in the southern portion.

While the dwell time (number of seconds from full stop of the vehicle to when it starts moving again, but not including deceleration and acceleration intervals and time consumed in trying to get back into the traffic stream) can be as low as 5 seconds with one passenger exiting, but this is a rare accomplishment. Since NYC buses allow exits through the front door, where all entries have to be made along a single channel, there is no flexibility or redundancy in these simple but critical operations.

With several passenger movements through the front door (up to three riders getting off or on), the dwell time is 10 to 20 seconds. With a volume of about 6 movements, the dwell time becomes 45 to 60 seconds. All this is under regular conditions. The times can become considerably longer if an entering passenger starts asking for directions while standing in the channel, a card is misinserted, an exiting rider misses his or her turn and tries to get off while others are getting on, an accommodating driver waits for a running entrant, a prospective patron asks for change for a dollar from other passengers, or other human frailties emerge.

Leaving aside the traffic congestion problems, which are not under the control of the bus operators, there are some obvious actions that would substantially expedite this transit service:

- Allow exits only through the rear door (except for infirm riders)—a no-cost improvement.
- Provide effective communications systems with drivers and/or automatic vehicle location devices that would allow schedule and sequence adjustments along the route.
- Acquire buses with two entry channels at the front door.

If all this were done, the running time of the M104 bus during off-peak hours could probably be reduced by more than 5 minutes, reaching an overall running speed of 7.3 mph (11.7 kph) even in Manhattan. If low-floor buses were used, a few more minutes could certainly be shaved off that total running time.

Reported by Bill McKibben in "Busman's Holiday" (an article in *New York* magazine, September 25, 1995, pp. 64–66, 107–109) as his experience during a rush hour. The distance covered is actually 2.3 mi (3.7 km), thus giving a speed of 3.5 mph (5.6 kph) if the timing is correct. It is fortunate that he could not get on the M27 bus, because it makes a turn in the opposite direction from the Public Library.

An Alternate View of M104

Bill McKibben

8:54: I'm waiting for the M104 at 79th and Broadway. When it arrives, every seat is taken. It takes about a minute for seven of us to board.

8:59: 72nd Street. The bus is so full, in fact, that passengers can't thread their way through the aisles. A father is trying to disembark with his son, but the bus pulls back out into traffic before he reaches the door. A ruckus being raised, the driver strops in mid-street and the two, grimly clutching briefcase and Power Rangers backpack, pile out.

9:11: We push, finally, through Columbus Circle, and begin to pick up speed.

9:16: Pulling out from the 50th Street stop, we collide with a fruit truck. It knocks the side mirror off the bus and the driver immediately cruises to the curb, announcing in a bored voice that we are "out of service." We collect transfers, and wait at the curb across from a large marquee for *Sodomania*.

9:23: An M27 rolls by without stopping, and then another Not in Service bus. Most of the passengers have walked away, muttering; only the older ladies, who clearly don't relish a walk through Times Square, remain.

9:27: Two M104s arrive simultaneously, the first one jammed, the second one empty.

9:31: We turn east on 42nd Street.

9:32: A lady tells the driver she has boarded the M104 mistakenly, intending to take the M10 downtown. May she have a transfer? No.

9:35: We reach Fifth Avenue and the Public Library, which had been my goal all along. We've covered a distance of perhaps four miles in 40 minutes, which is not so far off the TA's average for midtown. I've been on better bus rides and I've been on worse. This was a lurchingly typical forty minutes spent in Bus Time, an alternative dimension that moves at half the pace of New York life—one reason, certainly, that ridership has dropped 48.5 percent in the past three decades.

It can be argued that only a headway shorter than 10 minutes can be considered to provide good and responsive service, i.e., to establish a performance level that allows any rider to walk to a bus stop and be able to board without having to plan the time of arrival beforehand. If the headway exceeds 15 minutes, passengers will have to consider the exact schedule (to the extent that it is actually reliable) in order to arrive at the stop at the proper time. Unfortunately, due to low ridership levels in most situations and the cost of running frequent but underutilized vehicles, the desirable short headway conditions are achievable only in relatively few cases in American communities. The service objective then is not to make operations very good, but to achieve the best within the bounds of affordability. Consequently, a frequent suggestion in standard reference works is that the policy headway

The Case of the Airport Bus

For contrast with the highly urban situation experienced by the M104 bus, field observations were done also for the M60 bus route, which operates in a semi-express mode and connects Morningside Heights to LaGuardia airport in New York City. It starts on Broadway near Columbia University, crosses Harlem along its busiest commercial street (125th), utilizes the Triboro Bridge complex to reach Queens, runs along Astoria Boulevard on either side of the parkway, and makes a circle of all the principal terminals within the airport.

Regular fares are collected, and the bus stops at all existing stops along the way. Thus, many people use it for local service; the patrons arriving at the airport are overwhelmingly airport workers, with only a few air travelers, who tend to be students with backpacks. There are no luggage accommodations.

The service operates from 4:15 A.M. to 1:45 A.M., with an interval of 15 minutes (or 12 minutes during peak periods).

The dwell time ranges from 10 to 60 seconds (or more), depending on entry/exit volume, but a 5 second dwell time on the Queens side is not unusual.

The overall running time can vary widely, depending on congestion along 125th Street (all during the business day and evening) and on the Triboro Bridge (extended peak periods). The lengths of the inbound and outbound runs are different—11.0 mi (17.5 km) and 9.5 mi (15.5 km), respectively—because different paths are followed to and from the main terminal, but on a normal day the overall speed will be about 13.5 mph (22 kph). There are great differences among the several segments:*

In the dense city districts (2.5 mi; 4.1 km)	7 mph (11 kph)
On the limited-access ramps and bridge (3.2 mi; 5.1 km)	Up to 40 mph (62 kph) or much less
Through low density districts (3.3 and 2.0 mi; 5.2 and 3.3 km)	15 to 17 mph (24 to 27 kph)
Within the airport (1.8 mi; 2.9 km)	13 mph (21 kph)

* Similar numbers are given in ITE, op. cit., p. 154: in central districts, 6–8 mph (10–13 kph); in urban districts, 10–12 mph (16–19 kph); and in suburban districts, 14–20 mph (22–32 kph).

during peak hours should be 20 minutes in urban areas and 30 minutes in suburbia; and 30 and 60 minutes, respectively, in the evening.[54] Not really good enough, but many places in the United States would be glad to achieve even that.

The next basic decision concerns the length of the bus working day, with most smaller places opting for a complete suspension of

[54] ITE, op. cit., p. 154.

service during night hours (midnight to 6 A.M., or 1 to 5 A.M., for example). Larger cities may be able to institute special nighttime services, operating on selected routes only, with long intervals between buses. This is often done in European cities, with well-publicized information and schedules, particularly if rail and metro services are also closed. It is also a reasonably common, but regrettable, practice in the United States, particularly in smaller communities, to provide no Sunday service at all, and to operate some routes only during peak periods and have adjusted Saturday service.

The real scheduling tasks address the peak-period conditions—say, the hours from 6 to 9 A.M. and 3 to 6 P.M. These are the times when ridership should reach demand levels far beyond the base service capacity. The analysis process is outlined here ("Bus Scheduling Example"); it fundamentally requires knowing the greatest accumulation of riders at the highest demand point during the peak hour in a single direction. In the morning, that is likely to be at the edge of the core district as buses cross into that zone. If this hourly ridership volume is divided by the holding capacity of a bus (60 or 75 passengers), the number of vehicles that should pass that point during that hour is determined. This sets the peak headway for the entire route (60 minutes divided by the number of buses), but not yet the total fleet requirement. An obvious reminder is the need to round off the calculated headway to an even number, preferably one that divides into 60, so that dispatchers can easily keep track of movement spacing, drivers can remember it, and customers do not become confused.

For existing operations, the ridership loads can be monitored section by section, and headway adjustments can be made from time to time to maintain reasonable crowding levels on board vehicles. A further elaboration of the daily schedule can be the establishment of "shoulder" hours that provide a transition between the peak and the base headways.

The next task is the calculation of the number of vehicles that are required to provide service on any specific route. To understand this process, it is useful to visualize each route as a continuous chain, with each bus being a single link, and the spacing (the length of each link) being measured not by distance but by time. A basic input in this analysis is the total time that is required to make a complete round trip—the total time length of the chain.

A critical consideration here is that the vehicle will not be in continuous service on the round trip. The operations have to incorporate *recovery time* at both ends of the route. There are two reasons for including this factor. First, the drivers need a rest break—to catch their breath, get some coffee, go to the rest room, etc. Second, a slack period will allow schedule adjustments, since any given vehicle may be delayed en route and thus not be able to get back into the next allocated service slot. Also, minutes may be added to the recovery time to maintain an orderly schedule (say, service every 10 minutes, instead of 9 or 11). The rule of thumb for the total recovery time is about 10 percent of the round-trip time. It may be 10 minutes at the outer end and 5 minutes at the inner end; if the downtown area is badly congested, there may be no recovery time at all there, with the vehicles turned back immediately.

Dividing the round-trip time (including the recovery times) by the headway gives the number of vehicles that will be in operation. For example, if it takes 2 hours to make the circuit, and the prevailing headway is 4 minutes, the operating fleet would be 30 buses (120 min ÷ by 4 = 30). These vehicles are assigned to specific trips (or runs) according to the schedule. It could be argued that in theory stand-by vehicles should be available at all times to fill any gaps in the sequence as they develop. This, however, is not likely to be affordable and would generate management and labor complications. Dislocations will be handled, at best, through efforts to maintain balanced headways along the route with the equipment that is available. If there are sections of a route with recurring overloads, short-turn trips may be built into the schedule; if a major demand impact is anticipated (a sports event at a stadium), a few extra vehicles may be inserted.

An unfortunate but inescapable concern in transit scheduling is the fact that the real operations day extends over at least 12 hours (6 A.M. to 6 P.M.) and perhaps longer, since vehicles have to be brought from the garage to the start of the route and back as well. In any case, the working period for drivers is more than the regular 8-hour working day.

This condition can be tackled in several ways, always keeping in mind the prevailing labor rules and practices in a given state or community or the agreements between an agency and the unions.

- The drivers put in a long day and receive overtime payment. Many of them will not be busy in the middle of the day and should have rest facilities at the operations center.

- Part-time drivers are employed who work, for example, only 4 hours as required.

- Drivers work *split shifts,* which means that some drivers will spend the middle hours resting, reading, or sleeping in the "swing room" at the facility or doing something else. They will be off duty during that time and not paid (or sometimes paid at a token rate).

Another consequence of the existence of separate peak periods is a partially idle fleet. If all vehicles are in use when the headway is short, by definition, a number of them will not be needed when the headway is longer. This is one of the principal reasons why transit service has become expensive and requires external subsidies. If all the vehicles and the entire staff could be productively employed around the clock, most transit services would probably be fully self-supporting and profitable.

To conclude the discussion of bus scheduling, the problem of vehicles bunching up or *platooning* has to be mentioned. This is a common phenomenon that plagues high-frequency operations and annoys patrons who have to wait a long time for the next bus on the street and then see a number of them arriving all at once. This occurs when a specific bus starts to encounter high volumes of passengers on a run. The extended boarding time delays the vehicle, which allows even more passengers to accumulate at the next stop, and the following bus to catch up, thus leaving it with fewer and fewer boarding customers. The conditions become progressively worse with each stop until a platoon of buses becomes assembled.

The way to deal with this schedule dislocation is to anticipate the demand and insert additional vehicles midstream, to instruct drivers to bypass certain stops, or to allow units to leapfrog each other. Two-way radios and automatic vehicle location devices can materially assist such remedial efforts, provided that adequate supervisory intervention occurs.

Possible Action Programs

The initiation of a bus service system—or, more likely, the initiation of a new route within an existing network—may emerge from an overall planning exercise that projects and shapes the local development patterns and deals with accompanying service needs, or from the ongoing review of conditions and demands by a transit operating agency within its service area, or from the

Bus Scheduling Example

Let us assume that a bus route runs for a total distance of 5 mi from a suburban terminus to a node in the central business district (CBD). The average running speed, including all stops, is 6 mph (10 kph) on the city streets used by the service. The recovery times are set at 20 minutes at the outside end and 6 minutes in the center. The total time in hours thus consumed for the round trip is:

$$\frac{5 \text{ mi} + 5 \text{ mi}}{6 \text{ mph}} + 0.33 \text{ h} + 0.1 \text{ h} = 2.1 \text{ h} \qquad \text{or } 2 \text{ hours } 6 \text{ minutes}$$

The exiting and boarding movements of the passengers along the line during the peak hour are shown on the schematic map for each stop:

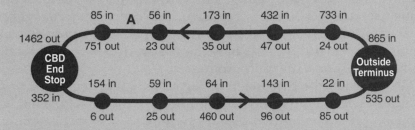

The fact that the ins and outs do not balance exactly is of little concern, because it cannot be expected that all the counts were done during a single hour under absolutely controlled conditions. And even if they were some riders would have been on the system before the study hour started and some will remain on it after it ends.

The greatest accumulation of passengers occurs at point A (as the route enters the CBD), which is passed by 2130 passengers during the hour. If passenger loading were to be kept at a comfortable 60 per unit, 35.5—say 36—buses would have to pass this point during the hour. If we assume that under peak conditions a load of 75 is acceptable, 28.4—say 29—buses will do the job. This would result in a headway of 2 minutes 4 seconds, which is then rounded off to 2 minutes.

Since the total time distance for the round-trip route is 2 hours 6 minutes, with vehicles 2 minutes apart, 61 vehicles have to be in operation.

identification of mobility and access shortcomings within a given district by community groups, organizations, or individuals. In any case, there is a follow-up need to gauge the level of anticipated demand, for which an appropriate and properly sized response should and can be made.

This task can sometimes be accomplished by examining usage levels on existing networks and assuming logical diversions to new, somewhat parallel, routes as relief or supplemental operations. There might be a generation of new traffic, too, by tapping underserviced areas. More precise and reliable methods, however, are also available to predict the likely utility of any service proposal.

These procedures, which serve for the planning of any transit system by any mode, are described elsewhere in considerable detail,[55] and only a brief summary can be given here.

The prospective line is placed on a map, and its tributary corridor is outlined (say, a 0.25-mi [0.5-km] band on either side for easy walking access and a 0.5-mi [1-km] band for reasonable walking access). Sociodemographic data are assembled for the corridor to describe the target population and estimate its trip-making propensity and characteristics. Sufficient understanding has been gained after decades of intensive empirical studies, and cause-and-effect relationships have become defined, enabling analysts to forecast with reasonable confidence what a given population will do in terms of transportation use.[56] This refers to a quantification of how many trips will be made, for what purpose, at what times. Such information can also be gained instead (or additionally) by undertaking sample surveys within the area to inquire what the residents will do (or at least what they say that they will do) under well-defined conditions.

All this pins down only one end of the trips to be made—the home base. It is also necessary to estimate the destination ends by type, particularly their locations and attractiveness to the local population. These would be workplaces, schools, shopping areas, institutions, recreation areas, and similar destination points significant to the study area residents. (These nodes are also, of course, the origins of return trips.) This step allows travel *desire lines* to be drawn, connecting origins and destinations in two-dimensional space.

The next step is more difficult and less certain in its accuracy, but it is nevertheless vital to the whole procedure: an estimate of how many of the various types of trips will be attracted to transit

[55] For example, Gray and Hoel, op. cit., pp. 162–166 outline a basic procedure.
[56] The problem with transportation models is that, while they are quite reliable in presenting an overall picture, they are based on a series of assumptions, and at the local level great accuracy cannot be expected. This is the level where bus services usually operate, and, therefore, much caution is advisable.

service, buses in this case, as compared to other options. The modal choice is influenced by the differences in cost and speed among the transportation options available, but also increasingly in American communities by comfort, convenience, safety, reliability, and possibly other factors related to quality-of-life issues. The seductive competition of the private car is a powerful consideration in all instances, although its influence is different in various types of urban situations.

The result of all these explorations will be an estimate of how many bus riders can be expected on the proposed route. This would lead to a decision as to whether it is reasonable to implement the program and whether certain modifications should be made to the original concept.

A warning is again necessary that the preceding is only a sketch, but it is also a reminder that well-tested procedures are available that can provide a solid base for transportation improvements. The steps can be completed with a series of quick approximations and assumptions; elaborate computer-based simulation models are available to do the job more precisely (and with a considerable input of study resources).

It can also be observed parenthetically—with some danger that the assertion that transportation planning should be done rationally and systematically will be impaired—that the development of bus services can benefit from experimentation and trial programs. Unlike rail modes, which involve the irreversible expenditure of large capital funds, a bus service can be placed along any promising path to see whether it will work. If it does not, the experiment can be terminated without too much of a loss, because the vehicles—the only real investment—can be reassigned to other routes. If it attracts much usage, the service can be augmented easily enough. The principal caution here, however, is to make sure that the experiment has a sufficiently long duration so that customers can make proper adjustments to their transportation patterns. A few weeks or even months probably will not be enough to reach a convincing conclusion. A year—i.e., a time span encompassing seasons—would be advisable, but real successes and failures might be obvious in a shorter period.

Assuming that the demand analyses show positive results, the action program would continue with the detailed planning of schedules and other elements as described throughout this chapter. To the extent that sizable physical improvements are involved, and certainly if a fleet of new vehicles has to be purchased, federal assistance programs are available. Obviously, any community would be well advised to look into this possibility, even if assistance toward operational expenses is reduced as public policy continues to change.

Express bus layover on Madison Avenue in Manhattan.

Another potential source of financial help is related to the fact that current federal programs show great concern with congestion mitigation and air quality improvement. To the extent that buses can contribute significantly toward these objectives, there is the possibility of gaining further assistance.

Cost Considerations and Land Development Effects

The great cost advantage that bus service enjoys is due to the fact that it almost always utilizes already available movement channels—city streets and highways. To be sure, some physical construction expenditures may be involved, but they will be marginal. Some loading bays may have to be created, some curbs moved, shelters constructed, traffic signals augmented, street striping repainted, and some pavement repaired or reconstructed, but these are not big-ticket items.

The problem is that in many places there is no unencumbered street space that buses can use without adding to the congestion overload. Thus, they impose a cost on the rest of the street users, and this is not unnoticed. Occasionally, a suggestion is made that buses should contribute to the creation and upkeep of streets, but this idea is not likely to gain any political support or achieve practical reality. The free use of street surfaces by buses is a form of subsidy, and it constitutes a responsible and equitable policy. Public-service vehicles are entitled to enjoy a higher level of priority than private cars and commercial trucks.

Significant capital expenses would be involved in the construction of a possible terminal building, the acquisition of land for concentrated loading operations (a timed-transfer place, for example), and the construction and equipping of a bus yard and barn for the storage and maintenance of vehicles. These costs are highly variable depending on local circumstances and attitudes toward appearance.

The final and largest-initial-cost item is the purchase of rolling stock. For the acquisition of a larger fleet, this would go through a bidding process and would be classified as capital expenditure. Replacements and additions of a few vehicles at a time would have to be a part of the annual regular expenditure budget. While many different models of vehicles are on the market, and various rearrangements and additions of elements can be made, a regular city bus can be bought for $280,000 to $320,000 (2001 prices).

The last observation concerns the perennial planning question: if land uses determine the need for transportation services, does the presence of service systems influence, attract, and shape land use patterns? No evidence can be found that buses have this ability, since bus service is invariably seen as a purely secondary service response, not as the initiator of any development activity. If such action is associated with substantial public investment (a busway, for example), attitudes may be different. No developer, large or small, will decide to build on a site solely because a bus route runs past it. If there is no service, operations can be readily initiated, provided that the demand is strong enough. This is one of the great advantages of the bus mode—transit service can be created quickly.

Conclusion

Some transportation planners have made the unwarranted assumption that bus-service planning can be an exact science. Regrettably, most operations are more subject to chaos theory than to precise laws of

European city bus with multiple wide doors (Vicenza, Italy).

physics. The street environment in which buses find themselves is frequently overloaded, stressful, unpredictable, and teetering on the edge of gridlock; the users are only human, often in a confused and fragile state. The ultimate service performance therefore has to be robust, ready to adapt and be altered; redundancy in everything will be a desirable feature, enhancing reliability.

Given all those caveats, however, the bus mode remains the reliable workhorse of public transportation. It has the ability to respond to a great variety of demand conditions and markets, it is readily implementable, and it is more affordable than anything else currently in the offing as an efficient communal mobility service to just about every urban community. We may be at the point of having comfortable climate-controlled vehicles, with efficient and nonpolluting power plants, low-weight construction, wide doors, and low floors, available for general use. When that happens—actually, not that much to ask for—and the vehicles are operated in a responsive and positive fashion, the bus can be the urban mobility solution in more instances than any other transportation option.

Bibliography

Bus Ride (monthly magazine), Friendship Publications, Inc., Phoenix, AZ. Articles of current interest covering all forms of bus operations and service, with regular surveys of all manufacturers by vehicle type.

Giannopoulos, G. A.: *Bus Planning and Operation in Urban Areas: A Practical Guide,* Aldershot, United Kingdom: Avebury/Gower Publishing Company, 1989, 370 pp. The most detailed bus handbook available, concentrating on European practice and examples.

Gray, George E., and Lester A Hoel (eds.): *Public Transportation* (2d ed.), Englewood Cliffs, NJ: Prentice-Hall, 1992, 750 pp. A basic reference work, including bus transit (Chap. 6), stressing planning considerations and service aspects.

Institute of Transportation Engineers (John D. Edwards Jr., ed.): *Transportation Planning Handbook,* Englewood Cliffs, NJ: Prentice-Hall, 1992, 525 pp. General reference work that includes basic material on bus systems and facilities, particularly in Chap. 5 and 7. Companion volume is *Traffic Engineering Handbook.*

Transportation Research Board/National Academy of Sciences: *Transportation Research Record,* various issues, particularly nos. 546, 746, 798, 1571, 1623, 1666, and 1731.

Transportation Research Board/National Academy of Sciences: *National Cooperative Highway Research Program Report.* Numerous articles and reports on various issues and situations associated with bus system planning and management, including Synthesis 69, "Bus Route and Schedule Planning Guidelines," 1980.

Vuchic, Vukan R.: *Urban Public Transportation: Systems and Technology,* Englewood Cliffs, NJ: Prentice-Hall, 1981, 673 pp. A full review of all transit modes, including bus transit (Chap. 4), stressing technology and hardware aspects.

Bus Rapid Transit

Background

Bus drivers, transportation planners, transit officials, and members of the riding public, if asked, would be able to suggest better ways to provide bus service than that experienced on most existing routes today. Such recommendations are quite obvious and include better vehicles (easier to get into and out of), quicker stops (less time consumed by boarding passengers), fewer delays (avoidance of chronic street congestion), faster movement (use of preferential channels), and more responsive service (balance between supply and demand). There is not yet a universally agreed-upon listing of what these improvements might be, but an acronym for them jointly has been coined already: *BRT* for *bus rapid transit*. Many specific BRT actions are not capital-intensive and they do not cost large additional funds, but, because most such efforts are beyond the regular procedures, they do require the expenditure of personal and institutional energy to achieve implementation.

BRT under its current definition encompasses all those programs and actions that allow urban bus service to operate faster, but also (it might as well) includes those that offer better reliability, safety, and human amenities, such as good ventilation, comfortable seats, and secure waiting spaces. Many of the program elements have already appeared in the previous chapter, and BRT is not to be seen as a separate transportation mode, but rather as an advanced variant of the basic bus mode, as perhaps something that all bus operations should gradually move toward and eventually become, at least in those situations where significant volumes of passengers have to be accommodated. BRT is not so much a concern with the vehicle itself as it is a matter of how it is operated and to what extent it receives full or partial priority on public rights-of-way.

If BRT is so obvious and desirable, why has there not been more of it and why has it not started earlier? Perhaps because the bus mode has not yet been able to shed its inferiority complex, at least in the perception of the general public, and few people have expected or demanded much from it. A reason may also have been the fact that most BRT actions constrain to some extent the unbridled movement of automobiles, and the required political boldness to do that comes only when street conditions become truly desperate. Buses have been there all the time, they have operated without too much fuss, and only rail-based modes have been seen in the last few decades as having the potential for significant upgrade and sufficient attraction to combat the overwhelming force of the private car.

These attitudes are changing. Federal agencies, municipal officials, transit operators, and community groups have started to recognize that bus systems in most places are underused and underdeveloped. Buses could be doing much more, with higher levels of consumer satisfaction; carrying capacities of this form of transit can be expanded with relatively little capital expenditure. Bus systems cannot be simply left alone to do the best they can on congested streets; the service needs recognition and active support to do a better job. A large part of transit's future may be found in this direction.

The current definition of *bus rapid transit* is "a systematically coordinated service, fully integrated with other modes in a community, that provides faster speeds, improved reliability, and increased convenience compared to conventional bus operations." BRT involves the following:

- Rolling stock of improved design
- Expanded physical facilities, including possibly preferential or exclusive lanes

- Upgraded operational procedures, ranging from fare collection to traffic signals

- Advanced information and control methods, relying mostly on *intelligent transportation systems* (ITS)

The BRT chapter is significantly shorter than Chapter 8 (Buses) because there is no need to repeat the general history, vehicle types, operational patterns, and many other features of the bus mode. The discussion in this chapter concentrates only on the specific elements that characterize BRT at this time and in the foreseeable future.

Development History

Transit records show that the first exclusive bus lane on a city street was initiated in Chicago in 1939. Follow-up efforts were few and rather sporadic since during the next several decades principal transit actions in American communities consisted of replacing streetcar service with simple bus operations that lacked any refinements. Gradually, however, increasingly more elaborate but highly localized and incremental improvements appeared. Almost invariably, they were efforts to address bottlenecks that severely constrained bus movements in congested areas—not components of larger plans to organize an effective overall bus system.

The history of these developments over the past three decades is not marked by major events. Year by year, more examples appeared everywhere as city managers and transportation officials tried to cope with growing traffic congestion and bus service strangulation. The only newsworthy occasions were periodic high-profile announcements in large cities of reserved- and priority-lane programs of significant size within the regular street network, accompanied almost always by protests from motorists and shopkeepers. Paris, London, and New York were the more visible instances. Today, however, preferential bus-turning arrangements and dedicated lanes can be found at many locations in just about every large and medium-size city in North America. They are becoming common in Europe and other continents as well, particularly where street traffic volumes choke circulation and dramatic actions are needed to maintain internal mobility (e.g., Bangkok and Shanghai).

The United States has experienced a period during which significant modifications to many downtowns were made—the creation of malls and pedestrian zones. The principal reasons to embark on these programs were the desire to recapture business activity and to generate attractive spaces for shoppers and visitors. In almost all cases, however, there has been a strong accompanying bus element, not limited to priority access to the principal destination points, but including transitways, too—roadways running through these precincts that accommodate only public-service vehicles. The first such major effort was the Nicollet Mall in Minneapolis (1967), followed soon by 63rd and Holsted Streets in Chicago, Chestnut Street in Philadelphia, and transit malls in Portland, Oregon. Auto-restricted zones (ARZs) are a further expansion of the same concept, but they are rather rare in North America, downtown Boston being the prime example. (See Chap. 5, Automobiles.)

There were, in parallel and at a growing rate, determined efforts to add to cities new major bus-oriented elements with significant investments and use of space, not just reassignment of existing street surfaces. The history of those programs can be easily traced, having been well recorded in the media and technical reports. They were quite unusual at the beginning.

A large step forward in the United States was the development of *busways* as components of limited-access highways within metropolitan areas. This action went hand in hand with the progressive changes in federal capital funding programs allowing highway construction monies to be also used in limited ways for public transit purposes. In the early days, such actions had to be closely associated with specific highways, and a busway in the same alignment responded well to that administrative mandate.

A *busway* is a roadway (a set of traffic lanes) to be used by public-service (and emergency) vehicles only. It may have a separate alignment or be associated with a conventional highway. An *exclusive* busway allows only bus use.

High-occupancy vehicle (HOV) lanes are similar in almost all aspects to busways, except that they are designed for and allow private vehicles carrying at least two (or three) persons. Buses, of course, fall within this category.

A *transitway* or *transit mall* is a set of designated and reserved lanes on surface streets, usually as a component of downtown improvements, accommodating not only buses but frequently also taxis.

The first true busways in the United States were placed on the Henry G. Shirley Memorial Highway (I-395) leading southwestward from the center of Washington, D.C., and on the San Bernardino Freeway (I-10) running due east from downtown Los Angeles to El Monte.[1] Planning for both of these facilities started in the late 1960s; the first opened for service in 1969, the second in 1973. Both are still in full operation today.

Entrance to the El Monte busway.

The Shirley facility consists of two reversible lanes located in the median of the highway. It has been extended in several stages, and it runs now for 12.5 mi (20 km) and is equipped with elaborate on- and off-ramp arrangements. The original buses-only restriction has been changed to also allow carpools and vanpools. The El Monte Busway is 11 mi (17.7 km) long and has two permanent bidirectional lanes in the center or along the side of the freeway in a separate right-of-way. The lanes are separated from regular traffic by barriers and wide pavement buffers. This right-of-way contained railroad tracks, which were removed. It has on-line stations, and a major bus terminal and transfer node is in place at the El Monte end.

In subsequent years, the general conclusion was that the creation of priority channels was a good idea, but also that buses alone could rarely fill the available lane capacity along these corridors in most American communities. Thus, the concept gradually emerged of allowing high-occupancy vehicles (HOVs)—those that carry more than a single

Entrance to the Shirley busway.

[1] S. Grava, "The Role of Busways in Cities," conference papers *Public Transport Systems in Urban Areas,* Göteborg, Sweden, 1978. Both the Shirley and El Monte busways, as pioneering efforts in the United States, have been analyzed and scrutinized in numerous follow-up reports.

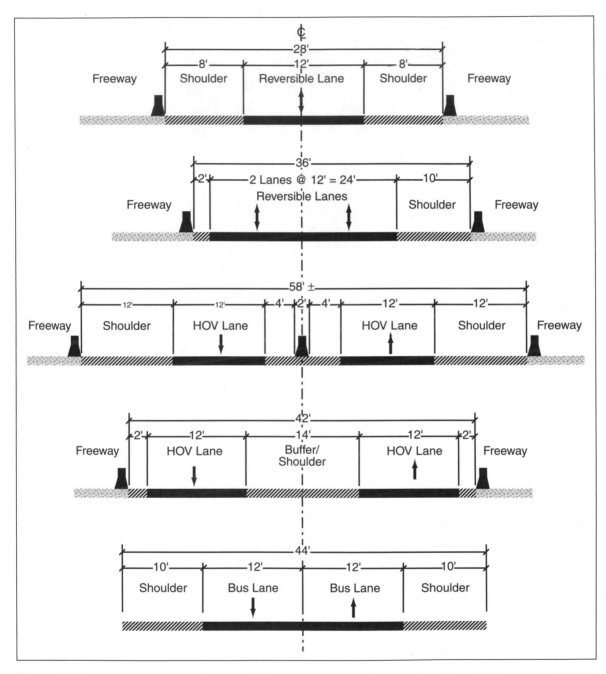

Figure 9.1   Recommended cross sections of HOV lanes and busways. (*Source:* C. A. Fuhs, *High-Occupancy Vehicles Facilities,* Parsons Brinckerhoff, 1990.)

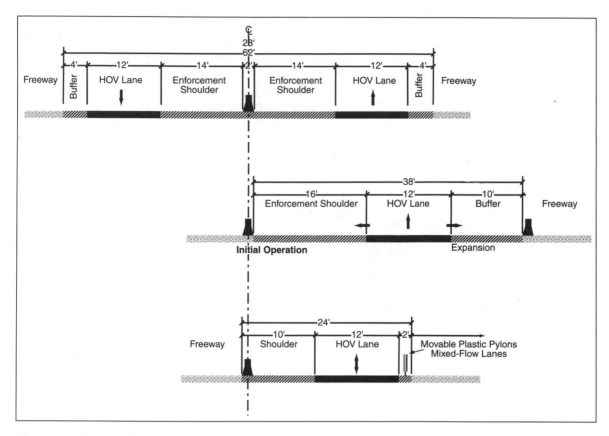

Figure 9.1 (*Continued*)

person—to travel on the dedicated lanes. As such projects became more commonplace, the term *busway* gradually faded from use in association with freeway improvements. The next group of programs, implemented in the early 1970s, encompassed the contraflow HOV lane on the Long Island Expressway (1971) and the SR 520 HOV lanes in Seattle (1973). These were followed by efforts in Atlanta, Chicago, Dallas, Kansas City, Milwaukee, and other cities as the development of HOV lanes became almost routine across the country.[2] Along the way, the now universally recognized lane-marking symbol—a diamond—became accepted to designate priority lanes. Active HOV lane programs can now be found in Australia, Great Britain, and elsewhere around the world. (See Fig. 9.1.)

[2] A comprehensive listing of all projects up to 1998 is found in NCHRP Report 414, *HOV Systems Manual,* pp. 2-11–2-14 and 4-49–4-50.

Busway with pedestrian underpass in the British new town of Runcorn.

One of the most interesting examples of applying a full-scale busway concept to an entire community came rather early—the entire urban structure of the British new town of Runcorn, authorized in 1964, is based on a central roadway that allows only buses (and emergency vehicles) to operate on it.[3] There is a figure-eight loop that has a total length of 12 mi (19 km) and connects the centers of various neighborhoods, with the town center at the point of convergence. The expectation was that most of the internal trips would be accommodated by this facility, because residential uses cluster along the alignment and the industrial district is on the loop as well. However, this has not worked out to the complete satisfaction of planners because automobile use has grown here as anywhere else, but the street system, which is oriented to the outside of the community, has been able to maintain smaller dimensions than would otherwise be the case. The busway right-of-way is completely separated, and numerous pedestrian under- and overpasses protect the network of pedestrian walkways. The design target of the town was a 5-minute walking distance (500 yd) from residences to all significant destination points, including bus stops. Some of the neighborhood centers actually bridge the busway, and the town center is a multistory structure with buses touching it directly on two sides on the second floor.

The Runcorn model has not been repeated elsewhere, not even in subsequent British new towns, presumably because it is too single-minded and tries to place all the circulation eggs in one basket. It is largely forgotten by now that the Columbia, Maryland, new town, developed by the J. Rouse organization in the early 1970s midway between Washington, D.C., and Baltimore, originally attempted an internal busway serviced by minibuses. It never caught on among the multicar families living in a basically suburban environment.

[3] S. Grava, "Busways in New Towns," *Traffic Quarterly,* October 1977, pp. 657–672.

The city of Curitiba in Brazil is the sine qua non of all bus priority schemes and is often presented today as the mother of BRT.[4] That is not exactly the case, as the previous discussion has shown,[5] but this city has unquestionably the strongest and most encompassing bus system anywhere in the world. Almost all the programs that urban planners have advocated in recent decades have been implemented in Curitiba, ranging from accessible green spaces to waste recycling. These programs certainly include transportation, and some remarkable results have been achieved here with a "surface metro" system—as shown by the simple fact that a metropolitan area with only slightly more than 2 million people and high automobile ownership generates 1.2 million bus riders each day.

The core of Curitiba's Integrated Transit Network (ITN) is a system of exclusive busways, which have expanded from when the first line opened in 1974 (after Jaime Lerner became mayor in 1971) to five radiating corridors today. But these are not just ordinary busways. Each of them is the central spine of a "trinary" roadway concept:

- Two restricted lanes for high-capacity buses are placed in the center, with stations connecting to feeder and crosstown buses.

- Auxiliary one-way lanes are located on the outside of the separated busway, providing access to tall buildings.

- A block away are one-way streets that are intended to accommodate limited-stop parallel bus service as well as regular traffic. These streets delineate the high-density linear city, which, in turn, is bordered by housing zones.

[4] Just about every journal and newspaper that has any concern about the contemporary urban situation and the built environment has had an article on Curitiba. There are reports and analyses that cover the experience from every perspective. The comprehensive book published by the Prefeitura Municipal de Curitiba (1995, 164 pp.), the article by J. Rabinowitch and J. Leitman, "Urban Planning in Curitiba," in *Scientific American* (March 1996, pp. 46–53), and Chapter 10 of R. Cervero's *The Transit Metropolis* (Island Press, 1998) can be recommended for further information.

[5] The elements and devices that we regard today as components of BRT have been known and used for some time. A report published in 1975—*Bus Rapid Transit Options for Densely Developed Areas* (prepared by Wilbur Smith and Associates for the U.S. Department of Transportation)—lists and analyzes just about all of them.

In the early planning stages, the use of rail-based modes was considered, but Curitiba opted for buses as more appropriate in scale and service capability for the size of the city and as an economical, easily implementable and adaptable approach. The integrated system relies heavily on the presence of interdistrict or crosstown service that feeds the busways and accommodates noncentral movements. Further significant refinements were a single-fare arrangement and the building of convenient and comfortable transfer facilities. The need for flexibility in transportation systems for a developing city was brought home in the mid-1980s when growing ridership started to overwhelm the service then in place. The demand approached 10,000 passengers per hour per lane on the busway. The response was to institute "direct-line" or "speedy" service on the parallel lanes with high-capacity articulated buses (110-rider capacity), making only a few stops, and the use of quick-boarding elevated platforms. A few years later, the throughput capacity on the busways was augmented as well by placing in service double-articulated buses (270-rider capacity).[6]

The brilliance of Curitiba's operational and technical bus service arrangements should not overshadow the system's principal accomplishment: a superior transit service driving and supporting an effective city structure and land use distribution. Curitiba has earned its standing as the Mecca and the Lourdes for transportation planners in "normal-size" cities (not quite megacities). The ultimate step in system integration may be the use of old buses (after their service life is over) as classrooms for various courses that are brought to the neighborhoods. (It should be noted that

Curitiba bus stop cylinder (on display in Lower Manhattan).

[6] Volvo established a subsidiary plant in Curitiba to serve this and other markets. Road regulations in North America preclude the use of vehicles longer than 60 ft (18.3 m). The double articulated buses extend 80 ft (24 m). The direct-line buses are painted gray, the express buses on the busway are red, the feeders are orange, and the interdistrict vehicles are green.

busways have been developed also in other Brazilian cities, particularly São Paulo.)

In Germany, M.A.N. and Mercedes-Benz caused considerable excitement in the transit world with the development of a guided bus that uses mechanical roller arms in contact with continuous vertical curb surfaces along the lane's edges to control steering. High speeds on narrow channels and precise positioning of the vehicles at elevated platforms are thereby made possible. Drivers maintain control of acceleration and braking, and the vehicles can be operated in the conventional mode at either end of the journey. This is the so-called O-Bahn, opened in 1980 as a pilot project in the city of Essen to bring buses into the underused but narrow tramway tunnels under its downtown. Two route segments were constructed (one jointly with a streetcar line), and the system proved workable. Experimentation with dual-mode buses to cope with ventilation problems, however, was terminated in 1995. There have been few follow-up projects with the O-Bahn elsewhere because the advantages are not quite compelling.[7] (Further development of guided buses—to the extent that the concept can be shown to have realistic benefits—may be found in electronic, not mechanical, guidance systems, as explored under the U.S. Department of Transportation Automated Highway System demonstrations started in 1998, and by programs in Europe, particularly in France.)

Given all this activity with large and small bus service improvement projects, and gradually recognizing that buses offer the cheapest and most flexible transit options, the bus rapid transit concept emerged and became formalized in the mid-1990s. Like Athena, BRT came to life already developed and fully armed. Image counts for much in our society, and thus giving a recognizable label to worthwhile programs should help to legitimize and popularize them in the public forum.

In 1999, the Federal Transit Administration of the U.S. Department of Transportation selected 10 operating agencies (in Boston, Charlotte, Cleveland, northern Virginia, Eugene, Hartford, Honolulu, Miami, San Juan, and Santa Clara County in California) to be members of a BRT demonstration program. Furthermore, a BRT consortium made up of these places as well as seven others has been established to share information and experience among them.

---

[7] See section on special vehicles later in this chapter.

Reasons to Support Bus Rapid Transit

As discussed in the previous chapter, buses have great intrinsic advantages as an all-purpose transit mode. It is a service that can be implemented at less cost and faster than any other means of public transport. Yet buses suffer from a poor public image; they have not been seen as having great potential in the development of attractive service; and they seem to represent the choice of last resort for potential users. The general perception is that, yes, every community should have them as the rock-bottom public mobility service, but the minimum effort will suffice. Much of this negative attitude stems from the fact that city buses, left to their own devices, suffer from street congestion and inefficient operational procedures, thus providing extremely slow and frequently quite frustrating means of mobility. Experience in the past shows that only when services break out of the standard format and low expectations (as regional express buses do) do bus operations receive favorable attention and a loyal clientele.

Not only is there a significant potential to upgrade bus operations, but there is also the fact that rail-based modes, as glamorous as they frequently are, involve extraordinary expenditures that many communities are hard-pressed to marshal, and the implementation time for them stretches beyond a decade, which is a time span long enough for the demand patterns to change, sometimes dramatically. Rail services can never have the same flexibility as bus operations, either in the long term or the short term. (For example, buses can change network patterns or drive around stalled vehicles.)

Bus rapid transit concentrates on addressing these issues, offering quick implementation and fast service once the systems are in place. To be truthful, BRT copes only with one dimension of a single mode, but it is a critical dimension of a seminal mode.

The programs are intended to achieve the following:

- *Reduced travel time,* by saving time at stops and while under way

- *Improved reliability,* by minimizing all factors that can interfere with vehicle flow and providing responsive management controls

- *Upgraded human amenities,* by providing attractive facilities and spaces inside and outside the vehicle and offering useful information to riders

- *Improved safety,* by providing monitoring systems, removing potentially dangerous features, and *bringing many riders on the system*

Reasons to Exercise Caution

Bus rapid transit, when applied to existing urban areas, almost always encounters a space conflict—its movement channels and facilities have to be placed on land that is already used for some other purpose. Most often this is a question of taking away lanes from regular motor traffic; sometimes it may involve the use of green space that is not exactly a park (e.g., landscaped highway medians) but nevertheless provides some visual relief. These issues may affect the implementation of other modes on guideways as well. (See Fig. 9.2.)

Thus, there will most likely be opposition. Even though BRT and public service are politically correct and defensible policy concepts, realpolitik in communities suggests that taking away established "rights" will not pass unnoticed. After all, motorists greatly outnumber bus riders. Experience has shown that success in implementation is more likely if no existing circulation space is taken away, and success in continued operation can be expected only if the new facility is in reasonably full use. If the latter cannot be achieved, the situation can lead to "civic disobedience" (i.e., motorists intruding on busways and HOV lanes if they see them in very light use).[8] Corrective measures include easing the car occupancy level (e.g., from three to two persons per car) and selling the privilege of using the fast lanes.[9]

Given the frequently prevailing attitudes among some motorists within our society who regard the strict observance of traffic regulations as a matter of individual choice, restricting scarce street space and movement time to only one type of vehicle at the

[8] The best-known case is the Santa Monica Freeway in Los Angeles, which began HOV lanes in 1996. The volumes were quite low, leaving unused space that irritated the stalled motorists on the adjoining lanes to the point of flouting road regulations and initiating political action. The lane designations were removed after 21 weeks.

[9] Called HOT (*high-occupancy toll*) lanes, these allow motorists who are willing to pay a variable charge to enter, thereby generating revenues and filling up the space. Nothing much can be said about this as a social policy concept, but it does provide additional funds that can be used for good overall transportation purposes.

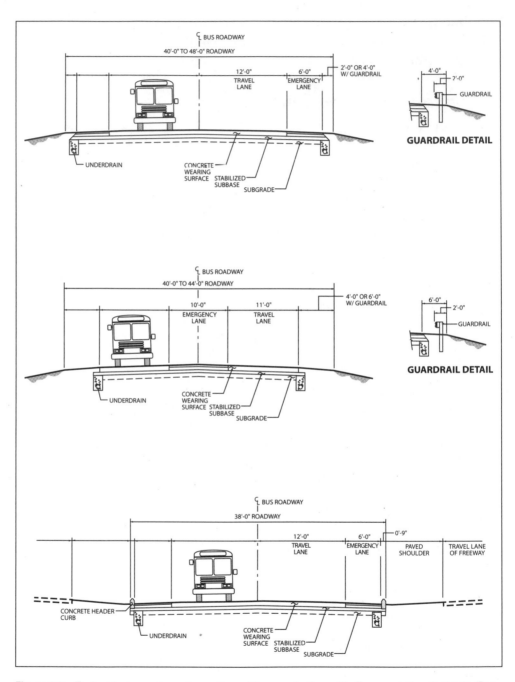

Figure 9.2 Typical busway at-grade sections. (*Source:* Jacksonville Transportation Authority/Parsons Brinckerhoff.)

expense of all others requires police supervision. In some countries, signs and pavement markings suffice, but that will not be fully effective in North America. Deliberate and willful violations of regulations and ingenious efforts to bypass constraints are not always to be condemned in a free society if the designs and programs are not clearly justifiable. Physical constraints, such as stanchions and barriers, are desirable, but police presence and vigorous enforcement actions will be called for. In some places, particularly the larger cities, the police force may have to adjust established attitudes that traffic control is not a significant priority, and the courts may have to take a tougher stance toward those who impair the effectiveness of public services.

Application Scenarios

Unlike the other modes described in this book, which rely on fairly standard equipment and operating procedures and allow the use of simple summary tables of actual cases, BRT programs are all different. They are tailor-made for each situation, and the various efforts present possibilities largely on their own. Thus, each case requires a separate description to outline its specific characteristics and probe the implications.

Quick Passenger Exiting and Boarding

Extended waiting times at bus stops (a critical issue discussed in the previous chapter) substantially slow service and impair the competitiveness of buses. After all, from 30 to 40 percent of the total time on any given bus run may be spent standing at a stop to accommodate entering and exiting passengers. To summarize, the possible upgrading actions would encompass the following:

- Have many doors along the side of the vehicle, thereby losing some seats, but giving many channels for moving people in and out.

- Do not collect fares or check passes at the door. Do that inside when the vehicle is already moving.

- Rely on prepaid fares or passes on the honor system, with roving inspectors to check compliance occasionally.

- If fares are to be paid on entry, use "smart" magnetic cards (that can be swiped across a sensor without physical contact) to expedite boarding.

- If there is a fare box or magnetic reader, have several entry channels and do not allow exits there.

- Reserve one door for access by handicapped persons (with lift or ramp) and slow boarders (mothers with children, the elderly).

- Use low-floor vehicles, which allow very rapid boarding and exiting.

- If high-floor buses are used, have elevated platforms alongside.

Examples

Most European bus systems employ the honor system and use all doors for exiting and boarding. In less prosperous countries where labor costs are low and voluntary payments cannot be relied upon, conductors in each vehicle perform their traditional tasks.

In Curitiba, platforms at the height of the bus floor are provided, and entering passengers pay their fares as they climb to the platform. The waiting space is enclosed in an attractively designed tube with a sidewall and doors on the boarding side; it is provided with a wheelchair lift. A full busload, if necessary, can be assembled on the platform. When the bus pulls in, it can be positioned precisely along the edge because the wide doors are on the driver's side. All doors open, a bridge plate comes down, and the in and out movements can be accomplished in seconds as on the subway.[10]

Priority Treatments on City Streets

A vehicle carrying many people should enjoy preference in the use of public rights-of-way over private cars with only a few occupants. There are several means to accomplish this:

- Allow only buses to make turns at critical junctions.

- Provide reservoir pocket lanes for buses making turns; establish shortcut links ("queue jumping"); reserve other localized movement or waiting space for buses only.

- Give buses the right-of-way when they are changing lanes, reentering the traffic stream, or executing any other traffic

---

[10] A demonstration project in 1992 showed the Curitiba tubes and vehicles in Lower Manhattan. It received good press, but there was no follow-up.

Mind the Gap!

This warning seen frequently in London Underground stations causes merriment among American visitors, but it highlights a serious safety issue—the possibility of getting a foot caught in the space between the platform's edge and the sill of the car floor when riders disembark or enter the vehicle. This concern has never been associated with buses because passengers climb up and down several steps anyway, and the problem was (and still is) the ability of every patron to surmount this difference in elevation.

With conventional vehicles, if the passengers have to start from the street surface (the bus has not pulled up to the curb), the climb will be 32 in (81 cm) with three steps, the first being rather high (at least 12 in). This can be reduced by some 3 or 4 in if the bus has a kneeling ability at the front door.

If the bus does pull up near the curb, which is usually 6 in (15 cm) high, a "gap" issue emerges. Some riders will step across the gulf, and others, being somewhat apprehensive, will not. But we never worried about this situation with regular bus operations.

If, however, safety is to be improved and service is to be expedited (by a few seconds at every stop), as BRT is supposed to do, it becomes important to position the vehicle only inches from the curb. If high platforms are being used, this becomes mandatory because of the threat of a fall several feet deep. If low-floor buses are employed, being near the curb significantly enhances the ease and quickness of in and out movements.

Acceptable horizontal and vertical gaps (those not requiring any assisting devices) are shown in Fig. 9.3.*

A further refinement in addressing the gap issue would be to consider the situation at the middle and rear doors. A bus normally approaches a stop at an angle, and even if the front door is near the curb, the rear will "hang out" several feet. To make the passenger movements comfortable and effective at the rear, the following responses can be considered:

- Make the approach angle small and the approach distance long so that the side of the vehicle remains largely parallel to the curb.

- Move the waiting space out (use bus bulbs or extended bays, as discussed earlier) so that the bus remains in the travel lane and does not change alignment direction.

- Use vehicles with steerable rear wheels that turn in the same direction as the front wheels, causing the bus to move essentially sideways.[†]

* *Transportation Research Record 1666* (1999), p. 86.
[†] Such a vehicle has been designed and built by Neoplan.

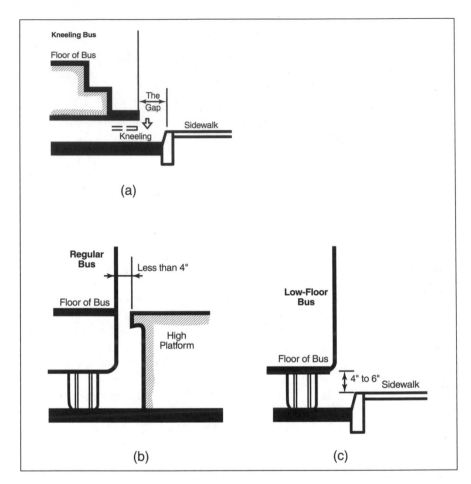

Figure 9.3   Gaps for (*a*) kneeling, (*b*) high-floor, and (*c*) low-floor buses.

maneuver. This is contrary to current traffic regulations, except that such a law has been passed in two California counties on a temporary basis as an experiment.

- Install signal priority or preemption devices—radio signals or automatic sensors that request a green phase when a bus approaches a traffic signal or weighs the signal timing in favor of buses. Admittedly, this is not fully effective when traffic and bus volumes are high in all directions within a district.

- Designate exclusive bus lanes on existing streets if the volume of buses is high enough to reasonably fill the capacity (say, a bus at least every minute[11]). These may be along the curb in the *same direction* as street flow, with cars turning right at the next intersection allowed on them as well. Extensive police supervision may be required. The other choice is *contraflow* lanes on one-way streets—buses running on the left side in the opposing direction. This solution has the advantage of being largely self-enforcing because motorists will not be tempted to risk head-on collision with a large bus, but there are some safety problems, including the necessity for pedestrians to exercise more care when crossing the street.[12] Deliveries along the curb may have to be prohibited.

There are several ways to delineate and separate the reserved lanes. In communities where rules are universally respected, lane markings (painted white double lines) will suffice. Stanchions that break without damaging cars, placed at intervals, are more authoritative. Solid barriers are, of course, fully effective, but they can be recommended only for very high traffic volumes and should have some buffer space on either side to minimize scraped doors and fenders. Another solution may be rumble strips that constitute few dangers (except to bikers) but send an unmistakable signal not to cross.

Examples

Since 1981, "red lanes" for exclusive bus use have been instituted in Manhattan on central sections of busy avenues and major cross streets. There are double lanes on northbound Madison Avenue and 11 other instances with single reserved lanes. They carry sufficiently high bus volumes to exclude all other vehicles, except emergency vehicles and regular cars making a turn within the same block. Having two lanes in parallel, of course, expedites bus movements considerably because passing vehicles do not have to reenter the congested regular traffic lanes. The exclusive

[11] The problem here is that it is generally accepted to have exclusive rights-of-way for LRT service even when there is vehicle only every 5 minutes. If this were done for buses, the lane would appear to be empty most of the time, and severe complaints from the motoring public can be expected.

[12] Pedestrians may acquire the habit of looking in only one direction when crossing what they believe to be a one-way street.

Dedicated bus lanes on Madison Avenue in Midtown Manhattan.

lanes carry the famous signs stating, "No Parking, No Standing, No Kidding, Don't Even Think of Parking Here!" The average bus speed was increased from 4.7 to 5.5 mph (7.6 to 8.8 kph). The Madison Avenue facility in operation from 2 to 6 P.M. and from 42nd to 59th Street accommodates an average of 218 buses per hour in the peak period and 136 per hour overall.

Paris has gradually established an extensive network of contraflow bus lanes that in the aggregate extend for many kilometers. They are all well marked and designated, including even rows of stanchions. It is not particularly easy for a private car to blunder into them, and violations are quite rare. The guiding rule has been that they are justified if the bus volume exceeds one vehicle every 3 minutes. This represents quite a bold public policy, giving distinct preference to what could be regarded as somewhat modest bus volumes. Such a standard stands little chance of being acceptable in the United States.

Separated exclusive bus lane in Paris.

In 1969, a morning-only contraflow exclusive bus lane (XBL) was opened on the limited-access highway (I-495) in New Jersey leading to the Lincoln Tunnel and hence to Manhattan. This was an early BRT effort, and it runs for 2.5 mi (4 km) from special entry ramps to the toll plaza. Every morning at 6:15 A.M., breakable posts are manually placed to segregate the inner lane of the westbound carriageway for the use of eastbound buses. The cones are removed every day at 10:00 A.M. This program is a striking success, accommodating on average 1713 buses carrying about 62,000 passengers each day. It has set what is almost certainly the world record for high-density bus operations—730 units in an hour on a single lane. More than 700 buses is not a rare occurrence, although the volume drops below 650 on about a third of the working days. It is a sight to see, with the buses acting for all practical purposes as a continuous moving belt.

Contraflow express bus lane at the Lincoln Tunnel leading to Manhattan.

Under these supersaturated conditions, extreme traffic control vigilance is required, and there is standby emergency equipment. The riders save 15 to 20 minutes on each trip, and, were they to switch to cars, they would generate some 45,000 waiting automobiles, which would have no hope of getting anywhere because the tunnel is operating at more than full capacity. There is no corresponding afternoon effort because volumes then are approximately equal in both directions, and no slack space can be found to borrow a lane.

In Los Angeles, Metro Rapid service has been placed on two busy corridors—along Whittier-Wilshire Boulevard to Santa Monica and along Ventura Boulevard—for a total length of 42 mi (68 km). The 100 vehicles (made by North American Bus Industries) are distinctive and quite different from regular buses that operate on the same right-of-way: painted red, with low floors in the front, propelled by natural gas engines. They stop at approximately 1-mile intervals; the units run

Rapid bus service on arterial street in Los Angeles.

at 3- to 5-minute headways. The time savings are about 25 percent compared to the conventional bus service along the same routes in mixed traffic. The expedited flow is achieved by skipping stops and applying signal-priority devices.

Exclusive Channels (Busways or HOV Lanes)

This group includes improvements that provide separate, designated facilities for buses on or along freeways and arterial streets and completely exclusive busways introduced within the urban fabric as separate roadways.

- High-occupancy vehicle (HOV) lanes as parts of limited-access highways are frequently placed in the median between the two carriageways. Their use is restricted to vehicles that carry at least three people (or two, if the volumes are not very intensive), which, of course, includes buses and other public-service vehicles. They may be separated from the regular lanes by prominent pavement markings or by raised barriers that permit entry and exit at strictly controlled locations. The lanes may operate in *two directions* as a central roadway, or the facility may be *reversible,* operating in one or the other direction during the respective peak periods.

  The HOV lanes may also be the inner (fast) lanes attached to each carriageway, typically added as new construction on both sides of the median. In some instances, contraflow lanes are borrowed by taking an existing off-peak direction lane and designating it for HOV use. No HOV lanes, except for contraflow treatments, have been created in recent years by taking existing general-purpose lanes. Any new HOV facility built today consists of *added* lanes.

  These lanes will usually be separated by buffer strips or just prominent lane markings (paint). A problem associated with the use of HOV/bus lanes on highways is the difficulty of providing bus stops along the way (i.e., pedestrian access to them across fast traffic). Thus, their primary utility in a BRT system is to offer fast express runs with few interruptions in the flow.

  There are many examples throughout the United States and in other countries, and extensive experience has been gained in planning methods, design standards, and enforcement and safety procedures.

- Queue bypass lanes or ramps can be provided at critical junction points that allow buses and other priority vehicles to move around automobiles stacked up at congested locations. They are frequently placed at highway entries leading to HOV lanes.

- True busways are roads that accommodate only public-service vehicles (and emergency vehicles). They have separate alignments, and they become special components of the street network. In some instances (Pittsburgh, Pennsylvania, and Essen, Germany) they may be combined with light rail operations.

Examples

The first HOV or busway facilities in the United States on Shirley Highway and San Bernardino Freeway (1969, discussed previously) have been followed by a number of other examples.

Reasonably typical but of considerable scale have been the efforts in Houston. This sprawling young Texas metropolis with many service centers, no zoning, and an automobile-dominated environment may very well be the vision or the threat of the future American city. Its voters have defeated rail transit proposals repeatedly, and only recently has an LRT project received support. In the meantime, however, the massive investment in highways has been accompanied by an extensive HOV-lane program. Starting with the North Freeway (I-45) that installed a long contraflow lane in 1979 and converted to a reversible transitway in the early 1980s, there are now six separate facilities totaling 98 mi (158 km). This system is programmed to grow substantially. Curiously, given the dispersed nature of the Houston region, all the HOV lanes converge on the downtown and are reversible to accommodate commuter flows. Express buses are significant users of these facilities, as components of an overall service network taking advantage of well-equipped transfer centers (four regional and nine at the neighborhood level) and 23 park-and-ride lots, many with direct access to the reversible HOV lanes. This network of facilities serves now over 40,000 express bus riders and 45,000 carpoolers daily.

The most prominent example of exclusive busways in the United States is the system built in Pittsburgh over several decades. The South Busway was the first exclusive busway in the United States with its own right-of-way and bus stops, opening in

1977. It runs for 4.5 mi (7.2 km) from the South Hills suburbs to downtown and saves at least 12 minutes for all commuters. It includes a long viaduct, joint use with LRT vehicles of the Mount Washington Transit Tunnel, and special on- and off-ramp arrangements.

The initial success led immediately to the next effort—the 6.8-mi (11-km) East Busway (Martin Luther King Jr.), which opened in 1983 using largely former railroad properties. Here, too, ridership boomed and has remained at a high level. The East Busway is currently being extended by another 2.3 mi (3.7 km).

The third component—the 5-mi (8-km) West Busway—was opened in 2000 in the airport corridor. It includes nine park-and-ride lots and has ADA-compliant stations.

Ottawa, the capital of Canada, made the bold and rare decision to structure its transit system using buses as the principal passenger carriers, expedited to a significant degree by exclusive-use busways. These channels constitute the backbone of the service network, with most other bus routes acting as feeders to it.

Starting in 1983, several completely grade-separated alignments have been built that lead to and from the city center and penetrate into the low-density peripheral communities. The facility consists of a lane in each direction, with shoulders and rather elaborate stations, mostly with side platforms, reached by stairs and elevators from the street level. A strong architectural image is carried throughout the system: glass structures with curved framing members painted red. Pedestrian overpasses are seen frequently, and passengers are usually prevented from crossing the bus lanes at grade.

Within the central business district, the busway is interrupted, and the flows are accommodated by two one-way streets. At the outside ends, the bus lanes are placed within shoulders on existing limited-access highways. Several bus routes never leave the busway; many others service various neighborhoods and communities and use the busway for

Exclusive busway in Ottawa, Canada.

parts of their journey. Regular bus vehicles are in use, although low-floor units are being introduced gradually.

The one city in Europe that is planned and built entirely to depend on a busway—the British new town of Runcorn—was discussed under "Development History" earlier in this chapter.

Among the other prominent busway examples found around the world, besides Curitiba, which was also discussed in an earlier section, are São Paulo, Belo Horizonte, and Pôrto Alegre in Brazil.

Entrance to busway station in Ottawa.

Access to Major Destination Points

Public-service vehicles can be singled out to provide access to places and to operate within areas that exclude other motor vehicles so that pedestrian amenities and safety are preserved and strangulation by cars is prevented. (See additional discussion in Chap. 5.) Such situations may be the following:

- In neighborhoods that have received *traffic calming treatment* (various devices that slow down motorcars and preclude through traffic), buses may be allowed to cross the barriers and maintain efficient routing patterns. The same approach can also be used in commercial districts where constraints may have been implemented to discourage internal cruising by regular motorcars. The barriers obviously have to be easily mountable, and police supervision is advisable.

- Pedestrian precincts or automobile-free zones may be provided with transit service and include bus routes that penetrate the areas and represent the only motorized traffic that is allowed inside (possibly also taxis carrying fares). This action in particular would emphasize the advantages offered by transit—its superior ability to reach destination points directly.

Examples

The central district of Göteborg (Gothenburg), Sweden, has been subdivided into five cells since 1970 that allow car circulation only within each of them, with access from the periphery. The barrier lines may not be crossed by automobiles, but public-service and emergency vehicles can move unimpeded. The scheme has been effective in curtailing car use locally and giving a boost to transit operations.

Transit malls exist in many cities in Europe as well as in North America. The most extensive and best-known examples in the United States are the Nicollet Mall in Minneapolis, Minnesota, Chestnut Street in Philadelphia, and the Portland Mall in Portland, Oregon, on a pair of one-way streets.

The eight-block Minneapolis facility was an early effort (opened in 1967, expanded in 1981) and remains one of the most successful examples in the United States. Nicollet Mall has become truly the Main Street for the entire region, maintaining a strong commercial and service base. The two-lane, two-way transit street extends for eight blocks (0.7 mi, or 1.1 km), and it carries 45 to 60 buses per hour in both directions. The mall is a gathering place offering an attractive and secure urban environment. It is the core of further improvements that encompass skyways, well-located garages, indoor spaces, and organized access systems from the surrounding communities.

Chestnut Street transitway in central Philadelphia.

Chestnut Street is a regular street of the old downtown grid of Philadelphia, with extensive retail and business activity along it. It was closed to all car traffic in 1976 and rebuilt as a two-lane, two-way transitway for 12 blocks with sidewalk improvements and a generous array of street furniture. Taxis can enter only to service hotels on any given block.

Portland has been doing many things right, and the transit mall on Southwest Fifth and Sixth Avenues is only one element of a restructured pedestrian-friendly downtown. The 15 blocks have completely rebuilt pavement, widened brick sidewalks, and attractive street furniture, such as fountains, kiosks, benches, landscaping, sculpture, and shelters. Only buses and taxis are allowed on the mall, which concentrates a number of city-scale bus routes and is crossed by the LRT system.

Advanced Communications Systems

Recent developments in *intelligent transportation systems* (ITS) offer a number of opportunities to manage operations with enhanced responsiveness to service demands as they occur on streets at any given time and to achieve greater efficiencies in the use of rolling stock and deployment of personnel. This includes, in particular, procedures and devices that can track the location of vehicles continuously,[13] monitor vehicle performance en route, immediately identify emergency conditions, provide up-to-the-minute information to patrons, achieve automated fare collection and checking of passes, and even position vehicles precisely at bus stops.[14]

The simplest of these technologies, radio voice links between every driver and control center, are now fairly routine. Their utility is primarily to respond to emergency or unusual conditions, which is quite important in removing bottlenecks and adjusting service incrementally, but such manual methods are not particularly suitable for managing complex systems on a continuous basis. Good communications with roadside dispatchers, however, can be effective in speeding up and delaying vehicles or inserting supplemental units in the stream (if such reserves exist).

The next step up is automatic vehicle location (AVL) systems that provide continuous information at the control center regarding where each service unit is to be found within the network. Coordinated responses are then possible. These devices have been around for some time, and they rely on roadside sensor stations

---

[13] Among the several possible components of ITS, the *automatic vehicle locator* (AVL) tied to a *global positioning system* (GPS) appears to have the greatest utility in BRT management, providing a full picture in the control center at all times showing where service units are located and how they are performing.

[14] This technology is currently under development and experimentation. Its designation is *intelligent vehicle initiative* (IVI) that is able to detect and react to obstacles and stop a vehicle at an exact location, thus expediting alighting and boarding.

that register the passing of each specific vehicle or on global positioning satellites (GPS) that track each unit and assemble information on locations. This capability would replicate control systems on the street that so far only modern rail-based services within a closed network have enjoyed. Some operating agencies have already deployed such systems, but considerable startup and operating costs are involved.

To operate systems very precisely, data on the number of passengers found in each vehicle would be of value so that supply of service could be adjusted precisely to demand. Although a number of rather ingenious devices have been developed (light beams and pressure-sensitive plates on stairs) to count in and out movements from vehicles, none of them appear ready for full-scale and reliable deployment.

It remains to be seen how reliable and cost effective these high-technology improvements will be in operating a responsive, rapid, and robust bus service. Such development efforts are certainly promising, but whether they will be practical can only be determined after they have been tested on the streets for an extended period.

Examples

This is an exciting period in the development of transportation systems, with a potentially revolutionary capability being introduced. This is happening in many places, gradually and incrementally, with much probing and experimentation to explore various possibilities. Out of this, most assuredly, some basic conclusions will soon emerge.

Essex County in the United Kingdom is one such instance where GPS-based arrangements, provided by the Essex County Council, give continuous and fresh information to waiting bus passengers (time of arrival of next vehicle). This services routes operated by private bus companies, but it also allows the county to monitor performance quality and adherence to published schedules.

System Integration

BRT should not exist in isolation. The basic mobility objective is to achieve a well-functioning service over the entire territory of an urban area. If the community is to designate and develop the bus mode as its principal means of public transportation, feeder lines to the core BRT system need to be tied into it by effective physical connections and schedule coordination.

Examples

There is no doubt that the most fully developed and internally coordinated bus system today is to be found in Curitiba. However, there are many less-comprehensive efforts around the world that move in the direction of achieving better coordination among services and exploiting the advantages of the bus mode. This is being done piece by piece, from inside out, so to speak.

The Silver Line in Boston is a BRT effort, currently under construction, that has received much attention lately. It is presented as a temporary bus facility with dedicated lanes, eventually accommodating LRT, for which there has been much local demand. If and when such conversion takes place remains a decision for the future. In the meantime, a hybrid channel design is being implemented that responds to local conditions segment by segment. A significant feature here will be the monitoring of ridership buildup and the gauging of demand levels before further major capital investments are committed.

The Silver Line will run from Dudley Square in Roxbury to the city center, then through the Ted Williams Tunnel to Logan Airport. Articulated low-floor buses will be in service employing signal preemption devices, with compressed natural gas as fuel, except that overhead electrical pickup will be used in tunnels.[15] Stations, not stops, will be provided along the route, with real-time schedule information, including countdown time until the next bus.

First phase: Improved and reconstructed Washington Street with a dedicated lane all the way to downtown

Second phase: Underground along Tremont Street tunnel to South Station

Third phase: Continued underground to South Boston Waterfront, World Trade Center, and so on, and to Logan

[15] The vehicles are produced in Europe, and Cleveland joined Boston in assembling a large enough order for a manufacturer to be able to respond. The Buy America clause was waived because no responsible bid was possible from United States sources.

A similar example is the busway in South Miami that runs as an extension of its heavy rail system for collection and distribution of customers. It is being expanded as well.

Special Vehicles

Given the long-recognized problems that impede rapid and efficient bus operations, there have been from time to time efforts to improve performance through technological devices, sometimes of a rather revolutionary nature. In most cases they have not proven themselves to be fully workable or to have real practical advantages. Several such efforts can be outlined for the record; a few are entering service or being considered:

- Buses can be equipped with devices that keep them within a lane and preclude the need for a driver to steer the vehicle. What is gained thereby? The driver has to remain in his or her seat anyway, if for no other reason than to assure the passengers that they are not at the complete mercy of a machine. The lateral guidance also assures a smoother ride, and the "fixed" feature of the operations suggests that the service is of a permanent and reliable nature. The principal advantages of this technology, thus, are to be able to operate buses on channels with very tight clearances where a human driver may not be able to avoid scraping the sides (within narrow tunnels, for example) and to provide a safety factor in keeping the vehicle on the road when speeds are high. The German O-Bahn, using a mechanical arm for guidance, has been available for some time now.

  The guidance task can also be accomplished by optical scanners that "see" the roadway in front of the vehicle and generate the proper control signals. Such devices are currently being deployed in France and might be tried in the United States.[16] The Advanced Highway Demonstration project of the U.S. Department of Transportation experimented with electronically controlled driverless buses on the HOV lanes in Houston a few years ago. It worked, but there has been no specific application of the concept in actual service yet. It will be interesting to see whether vehicles under automatic control can capture a meaningful and

---

[16] This is the Civis bus, an advanced-design vehicle with hybrid propulsion and optical road-sensing devices.

cost-effective role in the transit field (once the equipment becomes tested and completely reliable). An early application of the self-drive ability might be to guide vehicles through maintenance stations in yards.

- Another apparently clever idea has been to try to take advantage of some of the underutilized rail rights-of-way that crisscross metropolitan areas and place bus service on existing tracks (in the express mode). This presumably can be done by equipping a bus-type vehicle with additional retractable flanged rail wheels that can be deployed when the bus drives onto the tracks. At least one such vehicle has been built, but the mechanical arrangement proved too cumbersome, unreliable, and time-consuming to envision any reasonable application in actual service. This proposal also begs the question of whether light vehicles of this type, including light rail transit, can ever be considered safe enough to mix with regular rail operations.

- The concept of the *bus-train* emerges repeatedly as a "new" idea when possible improvements to urban transit are envisioned. Regular buses could circle in residential neighborhoods or business districts picking up passengers, converge on a designated node, then assemble in a train for fast line-haul to a destination node, where on-street distribution by individual units would take place. The buses could be coupled directly and then moved as a single train along HOV lanes or busways; they could be driven on railroad flatcars and moved on tracks; or they could be designed as convertible units with flanged wheels. In each instance the benefits would be the employment of only a single driver for the train, fast movement along an unobstructed channel, and a one-seat ride (no transfers) for most riders. If passages were to be provided longitudinally from one vehicle to another, the patrons could transfer en route to the unit that operates on a loop touching his or her desired destination.

Such schemes have never been attempted in the field and remain only material for Sunday supplements and popular technology magazines. It is conceivable that a large city could transform its base transit system to this form by developing and acquiring new rolling stock, creating a network of local and express route alignments, and reforming labor practices. It is difficult to see all these complications being faced and surmounted by only a small or partial effort. The benefits even under the best circumstances would not be overwhelming.

Examples

The O-Bahn has been in experimental and regular operation on several routes in Essen, Germany, since 1979. There have been steady efforts to sell the idea to other communities around the world, and the concept has been explored in countless transportation planning studies. This has led only to a few experiments and tests in Great Britain (the Millennium Dome, Leeds, Ipswich) and Germany (Mannheim), but there are no plans to extend the system further in Essen.

The issue is basically whether a compact guideway lane has advantages over a conventional bus lane. On the plus side there is the ability to deal with tight clearances, drive safely at high speeds, and effectively use high platforms. The negatives include higher costs, greater mechanical complexities, and the problem of removing disabled vehicles that may block the guideway (only if vertical curbs are used to guide the vehicles).

The only full-scale application of the O-Bahn concept so far is found in Adelaide, Australia.[17] A 7.5-mi (12-km) guideway has been built in two stages (1986 and 1989) as the high-speed segment of several trunk routes. This channel is incorporated in the metropolitan network along the northeast corridor that was originally intended for LRT service. It accommodates buses, including articulated models, that enter it from several suburban neighborhoods and leave the facility to operate in a conventional mode as the vehicles approach the city center. Since

The O-Bahn (guided bus) track and stop in Essen, Germany.

[17] The most complete and readily available account of the system is, again, R. Cervero. op. cit., Chap. 14. The author has high praise for this facility.

a number of bus routes converge on the guideway, the average interval between vehicles during the peak periods is about 1 minute. The operators believe that a 20-second separation would be workable. The roadway is built of precisely fabricated concrete units resting on structural piles.

The O-Bahn busway is justified in the opinion of local officials by the following factors: construction costs have been less than those for an LRT alignment; buses provide greater service flexibility and closer access to origin and destination points than rail-bound operations; high speeds (60 mph; 100 kph) can be maintained on the channel; the ride is very smooth on the carefully constructed guideway; and a very narrow strip of land is required (24 ft; 7.3 m) within what is basically a riverside park.

There is no doubt that the guided busway has bolstered the international image of Adelaide and that riders are most pleased with the service. The question has not been answered, however, whether the same operational benefits could not have been attained by a conventional busway—say, 33 ft (10 m) wide— with drivers remaining in full control at all times. Other possible O-Bahn alignments have been studied in Adelaide, but no specific expansion plans exist at this time; although many other cities have sent delegations to inspect the system, none have decided to follow suit.

(It should be noted that a trolleybus system that will utilize the O-Bahn guidance devices is being developed in São Paulo, Brazil—see Chap. 10.)

Components of the Physical System

The principal design determinant of BRT facilities is the need to expedite the movement of various vehicles (not only buses, but also emergency equipment and regular passenger cars) and to do this in as safe a manner as is reasonably possible. Recognizing, however, that circulation requirements and transit systems have changed repeatedly and will continue to do so from time to time, it would also be wise to keep future conversion possibilities in mind. This includes not only as-yet-unknown bus-type vehicles, but light rail options as well. All this argues for generous standards and dimensions, not penny-pinching.

There is also a difference depending on whether the persons at the wheels are professional drivers or amateurs like most of us.

The Question of Capacity

The highest bus service throughput that can be identified or even envisioned is the XBL (contraflow exclusive bus lane) at Lincoln Tunnel from New Jersey into Manhattan. With 730 buses per hour, each carrying a full load of 50 passengers, the achievable capacity is 36,500 riders per lane per hour with no stops on a limited-access highway. This situation cannot be reliably sustained, even though such a number of vehicles per lane in 1 hour has been recorded repeatedly. It represents a headway of 4.9 seconds! It cannot be expected that each bus will be filled to capacity, and the scenario represents a fragile state in which the slightest disturbance or lapse in attention by any driver can cause a possible breakdown. (A chain of rear-end collisions has not happened on a major scale in the XBL—attesting to the skill of professional drivers.) This possible number, therefore, is one end of the bus capacity range, exceeding the capabilities of most rail-based operations.

If the average loading of 36.3 passengers per bus were to prevail, and 10-second headways were to be maintained, the volume would be 13,000 riders per lane per hour. In Curitiba, with the direct-line service that employs double-articulated buses (270 riders) operated regularly at 72 second headways, the achieved throughput capacity is 13,500 riders per lane per hour. On multiple-lane highways with light traffic and no signals, a throughput of 9000 is claimed. This is still within the range of most LRT operations. On São Paolo's busway (Avenida 9 de Juhlo), 18,600 passengers per hour during the morning peak hour and 20,300 during the afternoon peak hour are achieved.

Beyond that, the volume level is not only a function of the size of the bus and its operation at close intervals, but the length of required dwell times at bus stops, traffic signal timing, and congestion on the street. Obviously, if the dwell time and the red phase of traffic signals, exceeds or even approaches the headway, buses will bunch up and the schedule will be dislocated, as was discussed in Chapter 8. Thus, if regular buses with 60 passengers each could be run every 15 seconds, a capacity of 14,400 riders per hour could be envisioned. This, however, could only be done with ample off-line stops, the availability of an adjacent free lane for the movement of vehicles, perhaps skip-stop service, and few signal interruptions to allow the flow of 240 buses per hour on the roadway itself. The present Curitiba procedures do not achieve this level, but on Madison Avenue, two bus-reserved lanes approach that capability, with 220 vehicles per hour—and perhaps a few more. A modifying factor here, however, is the presence of many express buses going to suburban destinations that make no stops on these lanes at all.

The general conclusion, therefore, has to be that a throughput volume of 10,000 riders per hour is possible on a street network with buses if a whole array of BRT concepts and devices is employed. With exclusive busways (and off-line stops), an hourly capacity approaching 15,000 passengers per hour is conceivable on a regular basis (recognizing that higher volumes are possible).*

If congestion and a multitude of interfering automobiles are precluded, a very high throughput of transit passengers on individually operated service vehicles can be achieved. That is why we started to examine BRT in the first place.

* It should be kept in mind that the discussion here is not so much directed toward a search for the maximum possible volume that can be achieved, but rather identifying reasonable and sustainable capabilities within acceptable levels of service and reliability.

For example, on true busways, adjoining lanes running in opposite directions present no problems for bus drivers, whereas for HOV operations at high speed, some lateral separation is most advisable.

The movement lanes on busways or HOV channels have to be 12 ft (3.7 m) wide, with 13-ft (4.0 m) lanes frequently recommended.[18] Ample shoulders should be provided, both for safety reasons and to store disabled vehicles out of the traffic stream. Because many delays on highways are caused by blockages due to accidents or vehicle breakdowns, arrangements to bypass such obstructions before they can be removed are of critical importance. Therefore, even if shoulders are not possible, some lateral clearance between the outside edge of a lane and any obstructions along the side is mandatory. (See Fig. 9.4.)

Stops or stations should be off-line, or there should be provisions for a moving vehicle to proceed past a unit stopped at a station. The loading/unloading lane should also be protected from the through movement and may have several berths. If it is placed on a high-speed highway, there will have to be deceleration and acceleration lanes with proper taper to safely accommodate speed changes. Such stations may be placed within grade-separated interchanges as long as their location and arrangement does not interfere with regular traffic flow and as long as safety requirements are scrupulously observed.[19]

[18] NCHRP Report 414, *HOV Systems Manual* (1998) provides design standards and explains planning requirements for all conceivable situations in careful detail on some 800 pages. Summaries of the recommended standards are also found in ITE, op. cit.

[19] See NCHRP Report 414, particularly Chaps. 6, 8, and 9.

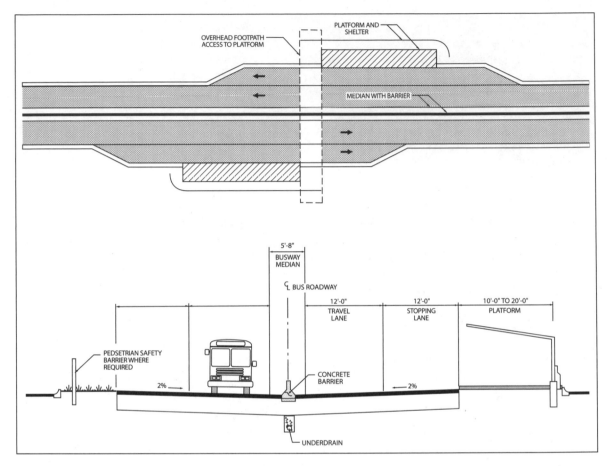

Figure 9.4   Busway at-grade station plan. (*Source:* Jacksonville Transportation Authority/Parsons Brinckerhoff.)

In all the preceding instances where buses do not reach regular streets and fast traffic is involved, special arrangements have to be made for passenger access to loading platforms, including overpasses and barriers preventing the crossing of movement lanes on foot.

Standards may be quite different if lower bus volumes are to be accommodated or if BRT is to operate in reserved lanes on arterial streets. In those instances, passing lanes at stops may not be necessary, shoulders would not be needed for disabled vehicles, and other reasonable shortcuts can be tried. There is much room for local experimentation and commonsense application of concepts. (The design standards in Table 9.1 are geared principally toward major BRT facilities associated with freeways.)

Table 9.1   Design Standards for Bus Priority Lanes or Busways

Design speed:	30 mph (48 kph) minimum 60 mph (97 kph) maximum
Moving lane width:	12 feet (3.6 m), more on short-radii turns 13 feet (4.0 m) with fast and high volumes
Shoulder width:	10 feet (3.0 m), if provided
Width of bus turnout lanes for off-line stops (loading and unloading):	
On expressways	20 feet (6.0 m)
On streets	10 feet (3.0 m)
Length of berth:	Length of largest bus in use plus 40 feet (12 m)
Length of second berth:	Length of largest bus in use plus 5 feet (1.5 m)
Lateral clearance from edge of lane to fixed obstruction:	
	2 feet (0.6 m) on left side, minimum 3 feet (1.0 m) on right side, minimum
Minimum horizontal curve radius	300 feet (90 m) with 30 mph design speed 1100 feet (330 m) with 60 mph design speed
Minimum vertical clearance (to accommodate double-deckers):	
	14.5 feet (4.4 m)
Maximum gradient:	6% (4% if to be converted to rail)

Source: American Association of State Highway and Transportation Officials; Transportation Research Board; Institute of Transportation Engineers; designs of existing facilities.

Conclusion

The acronym BRT may be new, but the concepts of rapid bus operation are old and well tested. It is encouraging, therefore, that they are receiving special attention today and their positive capabilities are recognized. As American metropolitan areas continue to disperse and as rail systems become increasingly more expensive, bus-type communal transportation services are becoming more prominent within the spectrum of choices. With fewer overall options, the array of BRT tools

Highway in Miami with signs pointing to HOV lanes.

HOV lane in center of highway (in Houston).

emerges as a logical and still promising set of programs toward an upgrading of the urban transportation situation.

Bibliography

American Association of State Highway and Transportation Officials: *Guide for the Design of High-Occupancy Vehicle Facilities,* 1992, 104 pp. Planning and design criteria for HOV operations on high ways and surface arterial streets.

Cervero, Robert: *The Transit Metropolis: A Global Inquiry,* Island Press, 1998, 464 pp. Successful transit cases, with full chapters on Ottawa, Curitiba, and Adelaide, plus references to other bus efforts.

Fuhs, Charles A.: *High-Occupancy Vehicle Facilities,* Parsons Brinckerhoff Quade & Douglas, Inc., Monograph 5, 1990, 126 pp. Comprehensive review of HOV facilities with planning considerations and design standards.

National Cooperative Highway Research Program: *HOV Systems Manual,* Report 414, Transportation Research Board, National Academy Press, 1998. (See also earlier Reports 143 and 155.) Exhaustive assembly of all considerations and design standards for every situation with a variety of applications.

Wilbur Smith and Associates: *Bus Rapid Transit Options for Densely Developed Areas,* prepared for the U.S. Department of Transportation, February 1975, 187 pp. An early compendium and analysis of possibilities, before much experience had been gained.

Trolleybuses

Background

The trolleybus (or trolley coach or trackless trolley) as an electrically powered transit mode running on streets has several interesting features, but it has never reached the top ranks among service choices. It is a cross between a bus and a streetcar, and not necessarily only the best characteristics of those two are to be found in the resulting vehicle. It looks and acts almost like a bus, except that it is tied to an overhead network of wires for power supply; it operates somewhat like a streetcar, but the reach of the power pickup poles allows it to move across several lanes.

In 1998, there were 880 active trolleybus vehicles operating in the United States, accommodating 182 million passenger miles that year.[1] The number of vehicles represent 0.7 percent of the national transit fleet, the passenger-miles only 0.4 percent of the total. The corresponding figures in 1984 (an interim peak) were 664 vehicles and 364 million passenger-miles. (The all-time peak for trolleybuses in the United States were the years 1949 and 1950, with ridership dropping steeply during the following decade.)

[1] American Public Transportation Association (APTA), *Public Transportation Fact Book,* (APTA, 2000).

The trolleybus has service capabilities that are almost identical to those of a regular diesel bus, and, therefore, only those elements that are different will be described and discussed in this chapter. (For anything else, please refer to Chap. 8.)

Development History

It did not take very long after a practical electric motor was developed to think of its placement in a vehicle. This would have required a rather long extension cord to supply power, but that problem could be solved by running a pair of live wires parallel to the path with rolling or sliding contacts linking them to the vehicle. Obviously, bare high-voltage wires could not be placed where people might touch them, but overhead was safe enough. That's all there is to the trolleybus concept (except that the moving power pick-up arrangements appear to have generated the need for more engineering attention than anything else in the early days).

Experiments with such transport devices started in the 1880s,[2] with working models being built in both Germany and France. They were not reliable or sturdy, but they proved that it could be done. The first regular services were opened in Germany—in Königstein-Bad Königsbrunn, as developed by Werner von Siemens (1901), and in Bielatal by Max Schiemann (1902). This was followed by a number of other lines not only in Germany, but also in Italy, Switzerland, Great Britain, and Denmark. In the United States prior to World War I, besides some demonstrations, a short trolleybus route was in operation in Hollywood (Laurel Canyon, 1910), and for a very brief period in Merrill, Wisconsin (1913). All these efforts remained very much in the shadow of streetcars, which at this exact time were expanding explosively in most cities, offering reasonably reliable and responsive service. Trolleybuses were attempted only in those instances where the demand was so low that the construction of costly tracks could not be justified. Since the streets at that time usually were in a sorry state and the vehicles not particularly resilient, the service was decidedly not attractive.

[2] Historical material on trolleybuses can be found in various specialized publications; summaries are provided in TRB Special Report 200, *The Trolley Bus: Where It Is and Where It's Going*, 1983, chap. 1, and V. Vuchic's *Urban Public Transportation: Systems and Technology* (Prentice-Hall, 1981, pp. 37–41).

Starting in the early 1920s, a series of efforts, while still separate and incremental, were initiated again to explore the possibilities and implement trolleybus routes in several cities of North America and Europe. Each was, in effect, a pilot project, and experience was gained and lessons learned. Toronto, for example, instituted operations utilizing vehicles built by Packard; Staten Island in New York City had a fleet of trolleybuses manufactured by the Atlas Truck Company. In Great Britain, Birmingham experimented with double-deckers, and petrol-electric vehicles—true harbingers of a distant future—ran between Middlesborough and Easton. The latter had an auxiliary internal combustion engine, allowing the vehicles to leave the power line. Several other cities in North America attempted trolleybus service, but they all faded with the exception of Philadelphia, which started this mode then and still has it today. The Staten Island effort is deemed to be the first truly successful trolleybus operation in the United States.

Toward the end of the 1920s, technology was sufficiently advanced to develop new models from the ground up that could offer fast and smooth running, good and quiet acceleration, and the use of low-cost power. Much of this was achieved by designing the trolleybus as a light over-the-road vehicle with pneumatic tires, rather than a sturdy streetcar. Better brakes and a workable power pickup (from under the wires) resulted in a suitable vehicle. Particularly successful was the design by Guy Motors of Great Britain, first introduced in Wolverhampton and then used widely in London.

This was also the period when streetcars started to be seen as obsolete transportation devices, candidates for wholesale replacement. (See Chap. 11.) This happened in London, but also in several cities in France, and on a massive scale in American communities. In most cases the replacements were motor buses, but trolleybuses were seen as a "modern" approach as well, particularly in places that wished to preserve the investment made in electrical power distribution networks but could no longer afford to lay and maintain track and wished to be relieved of the obligation to maintain street pavements that streetcar companies carried. (Many streetcar franchises placed considerable burdens of street maintenance on rail operators.)

The first large-scale effort in the United States was the implementation of an extensive trolleybus system in Salt Lake City (1928) employing the new, more efficient vehicles. A contributing factor in this and several other instances was the opportunity to use public streets at no additional costs, because street surfaces by this time had been improved considerably in response to the demands of automobile owners. Other communities monitored the Salt Lake City experience and reached favorable conclusions. Chicago followed next (1930) with a sizable network and several routes that accommodated large passenger loads previously not considered feasible (50,000 daily patrons on a route, some with 45-second headways).

The 1930s were a significant expansion period for trolleybuses in North America, boosted by transit demands during World War II. Notable among the many communities that embarked on this path is Seattle, which made a complete conversion starting in 1939 and built a system with 100 route-miles and 300 vehicles. That service is still basically in operation. The other large effort of that period was found in the old urbanized areas of northeastern New Jersey. The Public Service Coordinated Transport Company there established a complex network of routes and a diverse fleet of rolling stock that included trolleybuses with gasoline engines to reach sections of routes without power lines. They were manufactured by Yellow Coach Company and operated from 1935 to 1948. No trace remains of these operations, replaced by areawide bus service.

By 1940, some 60 communities in the United States had trolleybus service, accommodated by 2800 vehicles. In the early 1950s, which represent the peak period for this mode, there were more than 6500 units in operation. Thereafter, a period of decline commenced. After the war years, which were characterized by deferred maintenance, the infrastructure and the vehicles had worn out, but, with the onset of a precipitous drop in transit ridership, no capital-intensive efforts could be supported. Acquisition and operating costs of trolleybuses started to escalate, particularly in comparison to regular buses—presumably because of the smaller size of these operations and lack of any economies of scale.

There had been no particular incentives to upgrade the simple technology and the vehicle itself, which had not changed for decades. Above all, the service was seen as inflexible and the wires as unsightly. There were several technical improvements in the late 1960s, but they came from general upgrading of electric and electronic elements by the basic industries in Europe and

North America. Chopper control, for example, reduced power consumption considerably and assured smooth changes in speed. Regenerative braking and better power contacts were also introduced. None of this made much difference, and the decline continued.

There were no effective spokespeople for trolleybuses until concern with air pollution on city streets became a pervasive public issue. But by that time, however, it was too late to generate significant momentum back to a mode that had lost its general appeal. The petroleum fuel crises of the 1970s did not change matters either, beyond generating some discussion.

Despite all the early important development work in Great Britain, all trolleybus services were abandoned in that country. This almost happened in North America too. After all, these vehicles do constrain automobile flow on streets. The last trolleybus ran in New York City (Brooklyn, to be specific) in 1960. Even in Seattle the route miles dwindled down from 100 to 26. A watershed event was the closing of trolleybus services in 1973 in Chicago, which once had the largest system in the United States. Toronto stopped in 1961 and Calgary in 1975.

At this time (since 1973), only five American cities have trolleybuses: Boston, Dayton, Philadelphia, San Francisco, and Seattle. There are two more in Canada (Edmonton and Vancouver), and two in Mexico (Mexico City and Guadalajara). At the peak of their operations in the early 1950s, trolleybuses represented about 10 percent of the transit activity in the United States; today they accommodate less than 1 percent of the national total.

The events were not quite as dramatic in the rest of the world. Some countries in Western Europe, particularly Switzerland and Germany, have upgraded trolleybus technology and have strong operations in several places. A number of developing countries, particularly those that are reluctant to import expensive petroleum-

Trolleybuses on a street in Lausanne, Switzerland.

Trolleybus in front of the Belorussia Hotel in Moscow.

based fuels but can produce sufficient electrical energy, have turned to this mode.

The largest systems with the greatest number of applications, however, are found within the former socialist bloc. As is not uncommon, the claim has been made that a Russian engineer produced the first trolleybus—an electric autotrain with six cars at the beginning of the twentieth century. The USSR had a single-minded policy of promoting this hardware within all the countries and cities under its rule because electric energy was considered to cost only half as much as petroleum-based fuels. All of the more than 26,000 vehicles (several ZIU models) that were in operation at one time across the empire and its satellites were produced by the Uritski Works. The technology was not particularly advanced, but the vehicles were robust. Several hundred cities received trolleybuses, and, regardless of the recent political changes, they are still there by and large. Because of economic constraints, new replacement vehicles are scarce. Thus Eastern Europe and China are the places to observe full-scale trolleybus operations, if not always at the best service level.

Types of Trolleybuses and Their Operation

Trolleybuses at this time are basically buses with a different power plant—an electric motor and power pickup poles on the roof. They are usually made by diesel bus manufacturers, adding the electrical components to a regular bus. At one time, for example, a GMC New Look bus equipped with Brown-Boveri electric components was on the market. The choices of models are limited because the market is small. Basically, standard size and articulated units are available. The external differences, besides the motor and all associated internal controls, are the two side-by-side poles that tap the power lines from below with sliding, grooved, swiveling carbon shoes.

Unlike buses, however, there is a need for rather extensive infrastructure represented by the power supply system, which has to cover the entire length of all routes as well as storage and

Trolleybus Systems Around the World

Argentina	3	Georgia	9*	Norway	1
Armenia	2*	Germany	4	Poland	4*
Austria	4	Greece	1	Portugal	2
Azerbaijan	5*	Hungary	3*	Romania	15*
Belarus	7*	Iran	1	Russia	89*
Belgium	1	Italy	12	Slovakia	5*
Bosnia	1*	Kazakhstan	8*	Switzerland	15
Brazil	6	Kirgizia	3*	Tajikistan	2*
Bulgaria	16*	Korea (DPR)	7*	Turkmenistan	1*
Canada	2	Latvia	1*	Ukraine	44*
Chile	1	Lithuania	2*	United States	5
China (PR)	25*	Mexico	2	Uzbekistan	8*
Czech Republic	13*	Moldova	4*	Yugoslavia	1*
Denmark	1	Mongolia	1*	Total	348
Ecuador	1	Nepal	1		
Estonia	1*	Netherlands	1		
France	6	New Zealand	1		

Source: *Jane's Urban Transport Systems,* 1999–2000.
* Formerly included in the Soviet sphere.

maintenance yards. The double wires, usually placed 18.5 ft (5.6 m) above the pavement, have to be held in place by insulated support cables from roadside poles or be attached to adjoining buildings. The elevation may range from 12 to 20 ft (3.7 to 6.1 m). A constant elevation has to be maintained for the bottom of the power wires, which requires rather elaborate catenary[3] arrangements. Substations are needed at regular intervals to step down the voltage to 600 to 650 dc volts. Power has to be purchased from utility companies or generated separately by the operating agency. All this represents a considerably lower capital cost than for light rail transit (no track), but is higher than for regular bus systems, even if no fueling facilities are needed in storage and maintenance yards.

The vehicle does not have to be driven directly under the wires, but can deviate as much as 13 ft (4 m) from the center line

[3] A freely hanging cable from two points, supporting the power line at a fixed elevation along the entire route.

on both sides, i.e., can move in the adjoining lanes. Temporary obstacles can thus be bypassed, as long as the power poles are not blocked.

Even in the early days, an auxiliary gasoline or diesel engine or a bank of batteries were sometimes added to allow the vehicle to move "off wire," at least for short distances to bypass obstacles or navigate inside maintenance yards. These options continue to be available, and experiments with rather esoteric auxiliary power sources have been attempted (flywheels, for example). Such efforts to develop a more flexible vehicle are becoming more common. This may be a trend for the future.

Trolleybus Systems in the United States

City/Responsible Agency	Number of Routes	Route Length, mi (km)	Passenger Boardings per Year (millions)	Fleet Size	Vehicle Model(s)
San Francisco San Francisco Municipal Railway (Muni)	17	98 (158)	78.8	331	Flyer (1975) New Flyer articulated (1992)
Seattle King County Metro	14	124 (199)	Not available separately	147	AM General (1979) M.A.N. (1987) Breda dual (1990)
Philadelphia Southeast Pennsylvania Transportation Authority (SEPTA)	5	21	6.6	66	AM General (1979)
Dayton, Ohio Miami Valley Regional Transit Authority	7	16 (25)	3.4	46	Skoda (1995) ETI (1996)
Boston Massachusetts Bay Transportation Authority (MBTA)	4	16 (25)	3.4	46	Flyer E800 (1976)

Sources: Jane's Urban Transport Systems, Jane's Information Group, 1999–2000; *Metro,* September/October 2001.

Note: It is very regrettable that a number of transit agencies report in their official statistics that they operate "trolleybuses," which are actually the newly minted "heritage" (fake) trolleys—quaint passenger boxes intended to resemble a vintage streetcar mounted on a truck or bus chassis with a diesel engine. There is no accounting for taste or historical integrity, but the tourists apparently like them. But at least the terminology should not be misused—there are only five real trolleybus services in the United States.

Reasons to Support Trolleybus Systems

The positive features of trolleybuses are quite significant, and they stem almost entirely from the direct use of electrical power. It has even been suggested by dedicated advocates of this mode that trolleybus drivers are more friendly, or at least laid back, than other transit workers because they operate environmentally friendly vehicles.

- *No exhaust* is emitted by the electrical motor, and thus no air pollution is generated. A central power plant is needed, of course, but that is usually placed at a remote location and can be properly equipped and managed as a controlled large-scale operation. After passage of the Clean Air Act of 1990, commitment to clean vehicles became mandatory, and studies in several communities were undertaken to explore the feasibility and pollution control capabilities of trolleybuses. Los Angeles in particular, under the strict California state requirements, looked closely at this option, but no conversions happened. While cleaner air can certainly be attained, the amount of benefits gained by such action has not been a compelling argument in the larger environmental debate in any metropolitan area.

- *Quiet running* characterizes trolleybus operations because of the nature of pneumatic tires and electrical motors, which are not noisy even when surge power demands are placed upon them.

- *Acceleration* is quick because of the traction of rubber tires, and there are sufficient power reserves to climb steep grades, beyond the capabilities normally shown by regular buses. Advanced models incorporate regenerative braking, which feeds power back into the system instead of wasting it through brake friction or heat generation.

- Claims are being made that standard trolleybuses are *durable* and *easy to maintain* because of the simplicity of the components. That is not necessarily the case with advanced models, but the propulsion and control systems are less complex than those of comparable regular buses. However, any operating agency that already has diesel buses will want to keep the composition of its fleet as simple as possible, with not too much variety requiring special equipment, spare parts, and different skills. While it is true that the average age of a trolleybus in the United States is considerably older than that of a regular bus (16.2 versus 8.5 years),[4] it is not entirely clear that this is due to the greater durability of trolleybuses rather than to delays in replacing the fleet.

- *Petroleum-derived fuels* are not used, and thus the scarcer energy resources are conserved. Depending on the energy supply market at any given time and any given place, this may represent a significant savings in fuel costs. Switzerland, for example, has maintained a strong national policy of minimizing dependency on fuel imports. Nepal and Canada are also rich in hydroelectric resources and try to hold on to their trolleybuses.

Reasons to Exercise Caution

Most operating agencies in North America do not particularly favor trolleybuses, which explains to a large extent these vehicles' lack of prominence in the transit sector in this part of the world. The crux of the matter appears to be that most of the positive features resonate well with users and communities, who do not see the expense sheets, while the shortcomings directly affect the efficiency of agency operations, which is always under public scrutiny. The need for overhead wires is the principal drawback of trolleybus systems that generates most of the specific negative features. They represent a significant capital investment (particularly the copper wire itself, which wears out), and there are considerable engineering and construction efforts involved in keeping them on top of busy streets at an even and constant elevation.

- Unsightliness is the most often cited problem in public evaluations of this mode, as expressed by the overhead wires. At a large intersection where several routes converge and make turns, the spiderweb above can be a structurally heavy and visually oppressive presence. Even on simple straight runs

[4] APTA, *Public Transportation Fact Book,* 2000, p. 84.

there will be span and support cables, electric insulators, junction elements, poles and anchors, and feeder cables. On the other hand, perceptual surveys of city residents frequently indicate that people do not "see" the wires, i.e., they fade to the background in the total urban scene. Nevertheless, once alerted, most everybody will notice them and complain about violations of their aesthetic sensibilities. Some screening can be provided by trees, provided that they are properly trimmed to avoid contact. Feeder cables can be placed underground.

- The vehicles are tied to the lines without much flexibility in selecting a path. Trolleybuses can usually drive around small obstacles, but this mobility is limited to the next lane on either side. Temporary diversion of a route to a different street (to repave or do major utility work) involves considerable effort and expense in moving and replacing the overhead wires.

- The wires may be obstacles to other activities, such as vehicles with high loads, fire ladders, parades, etc. Running a route below structures with low clearances may also be a problem.

- The power pickup shoes frequently lose contact since there is little to keep them in place except a groove and the pressure of a spring on the pole. The replacement can be done quickly enough, but it does require the driver to leave the seat and walk to the back to fit them back manually, thereby losing at least a few minutes on the schedule. Mechanical devices have been invented to do this job, but they do not appear to be worth the trouble in normal situations. Snow and ice under extreme weather condi-

Operations on a snow-covered street (Riga, Latvia).

tions can interfere with power pickup arrangements. If the shoes are maintained properly, and switches and sharp turns are negotiated at reduced speed, problems should be minimal.

- The purchase price of a trolleybus is high as compared to a regular bus. A few years ago, a 50 percent premium was not uncommon for a vehicle of the same capacity. Currently, the price of an electric trolleybus is $642,000 (a 40-foot regular transit bus sells for $295,000).[5] This is certainly due to the limited market, since for all practical purposes every unit has to be individually made. With comparable production volumes, a trolleybus should cost the same as, if not less than, a regular bus.

- Costly infrastructure has to be in place, which was not a large problem in the early days when streetcar power distribution systems (including overhead wires, poles, feeders, substations, etc.) could be readily adapted. It is, however, a major consideration if a new network has to be created. There is no reliable cost experience to go by because little has been built in the trolleybus sector for several decades in North America.

Application Scenarios

In the 1980s, the general consensus among transportation specialists was that trolleybuses are and should be viable contenders in the modal spectrum.[6] They were seen as fitting in between light rail transit and regular buses, particularly for midsize cities (population 250,000 to 500,000). It was acknowledged that higher capital costs were involved than for buses, but it was estimated that with high-intensity use this expense could be readily absorbed. The construction of any rail line, or course, is more expensive still. The benign environmental characteristics were given much weight. Every time one of the existing trolleybus systems acquired new vehicles, a rebirth of the mode was expected.

[5] As reported in *Metro,* 2000 Fact Book Issue, p. 33.
[6] J. D. Wilkins, "Trolley Buses: Back on the Road in a Revival." *Mass Transit,* 1980, pp. 28–30; G. M. Smerk, "The Trolleybus Returns," in *Bus Ride,* September 1992, p. 53; J. Dougherty, "Electric Trolley Buses Are Making a Comeback," *Passenger Transportation,* September 16, 1991, pp. 6–7.

It did not quite work out that way. Basically, trolleybuses provide a service not much different than regular buses, but the systems are more expensive to implement and are constrained by the infrastructure.

In the process of preparing dozens of transportation studies for whole systems or specific corridors in American communities, trolleybuses have been included frequently as one of the possible modal choices. The final decisions, with only two exceptions, have been that this mode is not suitable for regular transit service under normal conditions today. Since the benefits of air quality improvement at the scale of regions is not a dominant variable in the evaluation equation, the determining factor has usually been the local agency's economic calculations related to the purchase, operation/maintenance, and fueling of vehicles. In a number of instances where electrical power has been especially accessible, the analyses have shown a reasonably competitive situation—except that the ever widening gap in rolling stock price has knocked trolleybuses out of contention.

Thus, trolleybuses remain a mode for special conditions: steep hills, unventilated spaces, or communities with a singular commitment to air quality or historical image. Two recent major efforts in the United States and one in Brazil illustrate this contemporary situation.

Seattle

Seattle, which has a long history of trolleybus use, reconfirmed its commitment to this mode in 1977 to 1979, when it closed dowthe entire system for refurbishment. The hilly terrain and the effection of local residents and officials for trolleybuses are significant factors in this community. The city extended the physical network and purchased a new fleet of 109 vehicles. In 1987, additional 46 articulated M.A.N. units were placed into service.

A downtown transit plan was also started in 1978 to streamline the city's public service operations. The decision was soon reached to build a transit tunnel that would provide direct access to major destination points and remove many vehicles from surface streets. City and suburban service routes would be channeled through this facility; later conversion to light rail transit could be provided for. Diesel buses in the tunnel would require tall ventilation towers, and "pure" trolleybuses would not be able to operate along several of the limited-access highways that are parts of

Articulated trolleybus (Esslingen, Germany).

the service network. After much discussion, the logical choice in 1985 was dual-mode rolling stock, admittedly rather complex vehicles, but already in use for years in Esslingen, Germany, Nancy, France, and elsewhere. Bids were received from European manufacturers, and Breda Construzioni Ferroviarie was selected—60-foot articulated vehicles with 66 seats and three doors, propelled by a diesel engine and an electric motor fed by retractable rooftop poles. Each unit cost $430,000 in 1986, and deliveries of the 236 vehicles started in 1989. Various parts and components came from different countries.

The L-shaped tunnel was opened in 1990; it is 1.3 mi (2.1 km) long and runs under Third Avenue and Pine Street. There are five underground stations with multiple berths, bypass lanes, mezzanines, convenient pedestrian access from the street, and artwork. The platforms are 380 ft (116 m) long and 13 to 15 ft (4.0 to 4.6 m) wide. Travel for passengers on buses within the downtown area is free, but the tunnel is closed on weekends. Before the opening of the facility, it took up to 30 minutes to cross downtown on the surface; the tunnel path now consumes 8 minutes. Best results are achieved when buses are moved in platoons through the tunnel.

Boston

Boston has not given up on its trolleybuses either and has found a new application for them. This is the so-called Silver Line that will eventually connect Roxbury to Logan Airport, planned as a replacement for the pending removal of the Orange Line of the metro. The original intention was to place trolleybuses on Washington Avenue, but this met with opposition from members of the community, who demanded light rail service. Buses propelled by natural gas on reserved lanes were opposed as well, and therefore the compromise reached in 1996 was a dual-mode system with the potential for the conversion of lanes later to light-rail transit. This became just about mandatory because the route was to be extended via several tunnels past South Station, through South Boston to Logan Airport.

São Paulo

It is appropriate to conclude this review of contemporary trolleybus projects by referring to the effort now under way in São Paulo, Brazil.[7] In addition to the rather elaborate transit networks that service this very large urban agglomeration,[8] a new tracked trolleybus route—the

Overhead power lines at a trolley depot in São Paulo.

Fura Fila—is being developed. The vehicle design is based on the double-articulated Volvo model used in Curitiba (four axles, 25 m [82 ft] long, 270 passengers), but it is equipped with O-Bahn-type horizontal guide wheels. High platforms will be used on a grade-separated busway, much of it elevated so that it can be added to the already built-up districts. The first line will run from the center of the city to residential areas to the southeast; there are plans for a very extensive network and a large fleet of these special vehicles.

The governing factors that led to this choice are that Brazil already has extensive experience with trolleybuses, that petroleum fuel conservation and air quality upgrading are concerns of national policy, that all the necessary elements and vehicles can be produced within the country, and that the comparative costs are most favorable. Their estimates show that the busway will cost U.S. $15 million per kilometer, while a light rail transit route would cost U.S. $40 million and a full subway U.S. $100 million per kilometer. This is a project that may very well be a crucial test case for trolleybuses anywhere in the foreseeable future.

[7] Bill Luke, "São Paulo Gets Trolleybus System," *Metro* magazine, January/February 1999, pp. 39–42.
[8] See the description of its busway in Chap. 9.

Conclusion

Trolleybuses continue to operate, but their future as a general transit mode is not particularly bright. They do have a role in special situations, but the global trends are still negative. Nobody likes the overhead wires (except copper manufacturers), and the problems of urban air quality are being attacked through means other than hoped-for massive switch of motorists to nonpolluting transit. If and when hybrid buses reach a competitive state in the market, which appears to be quite likely in the near future, the trolleybus may reach the status of cable cars—remaining in use in some places with special characteristics, but otherwise just being remembered with affection.

Bibliography

Gray, George E., and Lester A. Hoel: *Public Transportation* (2d ed.), Prentice-Hall, 1992 and Vuchic, Vukan R., *Urban Public Transportation: Systems and Technology,* Prentice-Hall, 1981. Includes trolleybuses in the comparative analyses of various transit modes.

Transportation Research Board: *Trolley Bus: Where It Is and Where It's Going,* Special Report 200, 1983, 64 pp. A reasonably complete effort to support reactivation of trolleybus service in the United States. Relevant articles can also be found in *TRB Reports* No. 1433 (1994), No. 1451 (1994), and No. 1503 (1995).

CHAPTER 11

Streetcars and Light Rail Transit

Background

The defining images of the American city in the early twentieth century were traffic-choked streets where the streetcar offered the only real promise of mobility and blossoming suburban enclaves that were accessible only because a trolley line was in operation. There never was any question about the technical quality of this mode (as was the case with cable cars) or any doubts about its environmental characteristics (as was the case with coal-burning steam locomotives) or its carrying capacity (as is the case with automobiles today). Streetcars were basic to the operations of cities for a considerable period, and they are coming back strongly in a new incarnation today. Few people are left who actually have heard the clang of the old trolley, but we all think that we did.

Yet for some five decades the streetcar became almost invisible in the United States, as one system after another closed down. Therefore, there is much joy in noting that this mode has emerged again in North America and at this time may represent the most promising public transit option in those communities that have maintained at least a moderate density and some identifiable corridors and clusters of concentrated development. We do not know at the moment what role this rail mode will play in American cities of the next period and how far it will extend, but it is back on the scene.

To achieve this renaissance, the streetcar had to shed its undeserved but damaging reputation of obsolescence by becoming *light rail transit* (LRT). The change in name has been an astute public relations move, because the similarities between a streetcar service network and a light rail system are much greater than the differences. The major changes have been technological improvements in the rolling stock, leading to much higher efficiency and operating responsiveness, as well as the use of separate rights-of-way whenever this can be done without major costs, yielding faster and more reliable service. Also, placement of routes within communities is done perhaps with a more realistic planning sense and reasoned acceptance of compromises. It is not to be discounted that LRT carries a very prominent and shining public image in the media.

The great advantage of light rail transit today appears to be the fact that it is the most economical means available to create high-volume passenger-carrying capacity with good service characteristics. The inherent mechanical efficiency of rail operations is not lost, even at this less-than-full-railroad scale. The current conclusion is that light rail can serve as high a demand as can be expected in any new corridor (i.e., one not having rail service already) of any American community today.

This mode can respond well to most service demands and still be reasonably affordable because we have learned to react constructively to various design challenges and do not always insist on absolute purity of concept (such as, for example, exclusive rights-of-way). This is not a proud observation but a pragmatic one, given today's political climate and some maturity in decision making, as well as the obsession with cost-effectiveness.

The streetcar lives! And our prospects for better communities are enhanced by this opportunity.

The low-floor Hudson-Bergen light rail vehicle and station in Jersey City, New Jersey.

Public Transit in the United States

Metro magazine, using all available sources,* projected in 2000 that ridership activity on the various public transportation modes operating in American communities for the year 2001 would be as shown in Table 11.1.

The same basic pattern is shown in Fig. 11.1, which compares the number of trips that are accommodated by the various transit modes.

Today's light rail ridership—while still small compared to buses and traditional rail—has grown almost fourfold from the 346 million passenger miles on streetcars in 1981, the lowest usage period before the emergence of LRT operations. There has been a steady increase in patronage in recent years as more systems have come on line and more commuters have found the new services attractive.

It should be noted that these statistics do not include all the transport modes that operate in this country, particularly those outside cities: long-distance buses, aircraft, intercity rail, and others. Exceeding all of them loom passenger cars, vans, and taxis, which account for about *2.9 trillion* passenger miles annually. If person-trips by cars, vans, and trucks had been included in Fig. 11.1 at the same scale, that line would extend across 30 pages. LRT ridership is about 0.1 percent of motor vehicle volume.

Table 11.1   U.S. Public Transit Ridership, million passenger-miles

Transit bus	23,849
Commuter rail	10,091
Demand responsive	1,172
Heavy rail	14,220
Light rail	1,295
Trolleybus	211
Other transit	859

* *Metro,* 2001 Fact Book Issue, vol. 96, no. 8, p. 23. See also annual issues of *Transit Fact Book* by APTA.

Development History

The direct ancestor of the streetcar is the horse omnibus (urbanized stagecoach, in effect) that started operating in some of the major urban centers in the early part of the nineteenth century when these cities became too large for people to reach their destinations on foot.[1] Placing the wagons on iron rails was a major

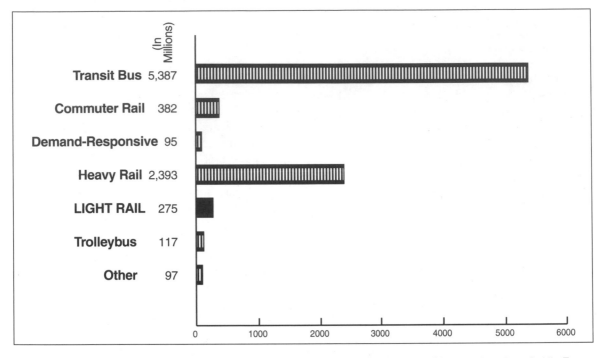

Figure 11.1 Unlinked passenger trips (boardings) in the United States in year 1998. (*Source:* American Public Transportation Association, *2000 Public Transit Fact Book,* p. 66.)

Reference to the Most Famous American Streetcar

(Blanche comes around the corner, carrying a valise. She looks at a slip of paper, then at the building, then again at the slip and again at the building. Her expression is one of shocked disbelief. Her appearance is incongruous to this setting. She is daintily dressed in a white suit with a fluffy bodice, necklace and earrings of pearl, white gloves and hat, looking as if she were arriving at a summer tea or cocktail party in the garden district. She is about five years older than Stella. Her delicate beauty must avoid a strong light. There is something about her uncertain manner, as well as her white clothes, that suggests a moth.)

EUNICE (finally): What's the matter, honey? Are you lost?

BLANCHE (with faintly hysterical humor): They told me to take a street-car named Desire, and then transfer to one called Cemeteries and ride six blocks and get off at—Elysian Fields!

EUNICE: That's where you are now.

Tennessee Williams, from *A Streetcar Named Desire,* Scene 1 (1947) breakthrough—one horse could do the job of two—and early tram systems emerged in New York, New Orleans, Paris, London, and Copenhagen. By the second half of the century, however, the situation became increasingly desperate as the industrialized cities continued to grow explosively. There was massive congestion, workers could not get to their jobs, any trip was a chore, and everybody was sloshing through horse manure.

Technology was expected to find the answer, and there was an intensive search for an appropriate urban means of mobility within the city of the Industrial Era. The steam engine was too dirty and dangerous when operating in the city, cable cars were unreliable and expensive, and pneumatic tubes were too fanciful to transport passengers. The eventual solution, at least for a long period, was the electric motor that became workable and could be placed in a moving vehicle by the end of the century. There was hope for electric batteries, but they were not powerful enough (nor are they today) to move large public-service vehicles over extended distances.

It took the better part of the nineteenth century for progressive development efforts by many inventors to transform the first crude electric motors (1830) into a practical source of propulsion for urban transport. This was done primarily in Germany and the United States, and along the way various stationary and mobile experimental devices were constructed. Thus, there is no agreement about which specific electrical machine was the progenitor of the streetcar.[2] In the 1860s and 1870s in particular, a number of elements were developed (notably the dynamo) that contributed to

---

[1] The early decades of public transit development are described in sufficient detail in G. E. Gray and L. A. Hoel, *Public Transit* (Prentice Hall, 1992, 750 pp.), pp. 4–11, and V. R. Vuchic, *Urban Public Transportation* (Prentice Hall, 1981, 673 pp.), pp. 14–32. Also, several chapters in K. H. Schaeffer and E. Sclar, *Access for All* (Penguin Books, 1975) cover the streetcar era, as does M. S. Foster, *From Streetcar to Superhighway* (Temple University Press, 1981).

[2] There are hundreds of books and publications that record the history of the entire streetcar industry or deal with specific trolley operations. This material includes, besides the aforementioned development summaries, J. A. Miller, *Fares Please!* (Dover Publications, 1960, 204 pp.); F. Rowsome, *Trolley Car Treasury* (Bonanza Books, 1956); W. D. Middleton, *The Time of the Trolley* (Kalmbach Publishing, 1967); *Ride Down Memory Lane* (Shoreline Trolley Museum, CT, 1991); J. P. McKay, *Tramways and Trolleys* (Princeton University Press, 1976); and many others.

a successful vehicle and power supply system. An American (G. F. Train!) built several lines in England as demonstrations, and E. W. Siemens showed the first electric locomotive at the Berlin exhibition of 1879. The German fairgrounds ride generated considerable attention and was believed by many to show real potential for urban passenger carriage. Various further experiments took place, including the use of different voltages, power pickup arrangements, motor types, and vehicle configurations.

The testing and tinkering culminated in 1888 in the work of Frank Sprague. This is a case where one man succeeded in combining many separate inventions of the recent past, including work by Thomas A. Edison, and single-handedly created a practical device that worked.[3] Sprague, a naval officer and electrical engineer, founded a firm that was entrusted by the city of Richmond, Virginia, to develop and build for it a complete transportation system that did not exist even in prototype form. This tremendously bold move succeeded, although many other places before and since have deeply regretted being pioneers for technical systems not yet fully tested.

The Richmond system used high voltage (750 V) for its time and consisted of 12 miles (19 km) of track that carried a fleet of some 20 cars. The rolling stock was well built and proved itself in daily service; the track was laid across rather rolling terrain, but the cars could cope with the grades; there was enough power in the cables to keep the vehicles moving with full loads. The Richmond system was an unqualified success that caused an immediate revolution in urban transportation, worldwide, and was never repeated again this suddenly and thoroughly.

Within the next few years at least 200 American communities converted their horsecars and electrified their networks, being completely fed up with unreliable and expensive animal power[4] and seeking a more efficient and faster service. In some early instances an electric motor was simply placed on the platform of

---

[3] See B. J. Cudahy, Chapter 3, "The Marvelous Mr. Sprague," in *A Century of Service,* supplement to *Passenger Transportation* of APTA, 1982.

[4] The problem was not only the accumulation of manure on streets. A whole industry had to be created to deal with the removal and disposal of carcasses, since the abused animals did not last long in the brutal environment. Another industry was the growing and distribution of hay (fuel) for the horses, on extensive fields that surrounded all cities. The last horsecar ran in July 26, 1917, in New York City.

a horse tram and connected with a chain drive to the axle. Half of the new systems were built by Sprague's firm, and most of the others were based on his designs. By 1902, there were about 22,000 miles (35,000 km) of trolley track in the country, with over 60,000 streetcars in operation. European cities followed suit very quickly, and early tram systems appeared on other continents as well, in cities such as Kyoto, Bangkok, and Melbourne.

The electric streetcar not only met the existing demand, it had an excess capability to attract new riders, thereby generating additional use. This product of invention and engineering brought city residents into a higher state of mobility compared to previous transportation means, allowing new activities to be created and making reachable a greater range of economic, residential, and social facilities and opportunities. Trolleys could operate at considerably higher speeds (10 mph, or 16 kph) and more reliably than the horsecars and cable cars preceding them. Thus, the reach of service was extended, and much land located farther from city centers became available for development. It was a means through which middle- and working-class families could leave the congested and unsanitary districts of the early industrial cities. Due to the efficiency of the mode, the longer rides remained affordable.

The fast-growing demand for service required the development of vehicles larger than the simple two-axle carriages of the first efforts. These could be double-deckers (mostly in Great Britain), or motorless trailers could be coupled to powered units (mostly in the rest of Europe). In North America, the vehicles were made progressively larger, which, in order to remain maneuverable on tight turns in city streets, had to be mounted on swiveling bogies (each a two-axle truck with motors) at each end. The basic vehicle configuration is still the same today, although some refinements—such as articulation—have been introduced.

To illustrate the diverse and intensive engineering work that was going on in the first decades to adapt the streetcar to urban needs, a brief digression can be made regarding power pickup arrangements. Before the overhead wire concept with a spring-loaded wheel (or slide) running on the underside became standard, there was some experimentation with a low-level third rail. Clearly, this was not workable, except with completely exclusive rights-of-way where people cannot intrude. Even less desirable was to use the two rails as power conduits, giving opportunities for dangerous short circuits. A number of lines were built where the car had a plowlike device on the bottom that ran along a slot in the pavement, below which the power lines were located. It is difficult to understand why something like this was ever seriously attempted, because not only did water and street debris enter the slot, but anybody with a metal rod poking around could easily electrocute him- or herself—except that some municipal ordinances (New York City and Washington) prohibited poles and overhead wires in the early days. A few such systems survived, nevertheless, to the middle of the twentieth century. Although the overhead wire remained the logical choice in power supply, the pickup pole with a wheel on top or a sliding shoe could be replaced with a bow that offered greater reliability. The latter evolved eventually into the hinged pantograph that is used most often in modern systems. Having a single power supply wire with the return to close the electric circuit provided by the rails does create some problems with stray currents and leakage.

In the freewheeling business environment of the early pioneering period, streetcars—offering a service in great demand—opened new opportunities for investment, speculation, and the creation of monopolies. In many respects streetcar company business actions paralleled those going on in the railroad industry at the same time. All services were created and managed at first by private individuals or firms, who saw attractive business opportunities. Entrepreneurs, awarded trolley franchises by municipalities, built and operated the lines, collected the fixed nickel fare, and tried to reap as much profit as possible—sometimes, it is reputed, employing questionable practices. Nevertheless, a framework for city operations was created, and a land development force established its presence. The streetcars became not only the dominant means of mobility in the early decades of the twentieth century, they were also major shapers of urban growth.[5]

[5] Streetcars have captured the attention of numerous professional and amateur historians, most of them dealing with engineering and technology, but city building implications have not been overlooked by investigators either. See, for example, D. R. Goldfield and B. A. Brownell, *Urban America: A History* (Houghton Mifflin Co., 1990), p. 262 ff. There are several large trolley museums in North America where old equipment is carefully restored and demonstrated. Documentaries can sometimes be seen on public television that touch upon the community-shaping features of streetcars. It is fair to say that just about every rail operation that has ever existed in any American community has been described in at least one monograph, usually prepared by a local rail enthusiast.

Trams could be, and were, easily and rapidly placed on any street that generated demand. Because they could bring many more people to a single point than before, streetcars allowed and encouraged the development of dense business centers where office workers could be brought together in large buildings for high productivity. Department stores could serve shoppers drawn from extensive tributary areas. The downtown became the locus of almost all citywide services and activities because the trolley lines con-

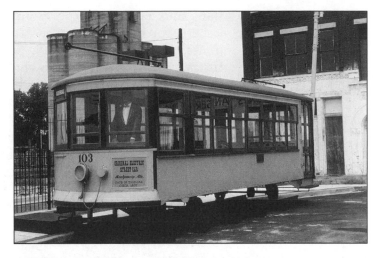

Antique streetcar on display in Montgomery, Alabama.

verged on it. Since almost all streetcar networks in American cities assumed a radial, star-shaped pattern, the mono-nucleated city with extreme concentration of activity and a drawing together of nonresidential service uses—as well as downtown apartments—emerged as the standard urban form. It is not possible to envision a successful city of the early twentieth century without streetcars.

Residences tended to cluster in significant density within walking distance of the new transportation means, forming distinct corridors a few blocks wide on each side. Local shopping extended along the service streets, sometimes in an unbroken line. All principal employment places and institutions had to be located directly on a tramline to allow worker and visitor access.

The streetcar also became a driving force toward new urban development outside the old districts under the private free enterprise system.[6] The extended reach of commuting travel brought peripheral open land onto the real estate market, an opportunity that was not lost to many energetic builders. Indeed, a rather common practice was to combine land subdivision and sale of parcels on the periphery of existing urban districts with the provision of transportation service by the same firm. This was a bril-

---

[6] See the classic study by S. B. Warner, *Streetcar Suburbs* (Harvard University Press, 1962), which analyzes the events in Boston.

liant and eminently successful idea: cheap land became marketable for housing because of the new access line, and the residents became captive riders, paying the fare twice each day. Of the many examples found in almost all American cities of that era, the best-known instance is perhaps Shaker Heights outside Cleveland, still in full operation.

A somewhat similar effort was made to create a strong attraction at the end of a line—a destination point and a reason for using the service. In a few instances this was an institution or even a cemetery, but mostly these installations were amusement parks (or *trolley parks,* as they were frequently called) or some other outdoor activity. These places were advertised as opportunities to escape the depressing city and engage in wholesome recreation, with the streetcar ride being part of it, particularly if a "breezer" (a car with open sides) was used. Bowery Bay Beach in Queens was such an attraction, which later became La Guardia Airport. Venice Beach outside Los Angeles is another example.

A major offshoot of the trolley industry were the *interurbans,* connecting communities to each other and to rural hinterlands. During the first decade of the twentieth century, an extensive network of separate lines was built covering the territory from the East Coast to the Midwest and elsewhere in the country. The cars were heavier and faster than the regular streetcar, and a passenger could travel long distances by transferring from one line to another. The interurbans filled a niche for a brief period, then disappeared with almost no trace when they became displaced by cars and buses.

The streetcar activity reached its zenith and covered its largest geographic area a few years before the start of World War I. Signs of decline became visible soon thereafter. There were several reasons for this, and none of them had anything to do with the automobile, since it was not yet a significant factor in urban transportation.

One of the fatal problems was the nickel fare, which had become such an untouchable concept in American cities that no politician or civic leader dared question it. Although 5 cents per ride may have generated sufficient revenues in the nineteenth century, even mild inflation created a point where the books of a private operation could no longer be balanced. The cost of labor and materials continued to escalate. Competition among several

Kennywood Park near Pittsburgh had been developed by the Monongahela Street Railway Company as a popular "electric park" with many recreational attractions. Neighborhood groups were able to charter vehicles to organize outings for the whole day. Here one such group in the 1940s gets ready for Irish Day.

Miners Hill

Michael O'Malley

The morning was bright, white-lit, shot with melon-colored gold light, like all good picnic mornings. Something in the air, all sorts of things in the air: the gaiety of expectation, a happy tense jiggling nervousness, a sky that promised glory and a green-cool afternoon. There was the wild anxiety to be off, to clank off rocking in a big rattling orange trolleycar, grinning importantly from the window because of the sign that said *Chartered*. There was the opulence of that *Chartered* sign, and the shouting crush of passengers— the portly Irish housewives and their picnic baskets bulging, and the huge men, broadbacked, the Irishmen, laughing and greeting each other, roughly touching their big tender callous hands, surprised and embarrassed by their joy, by their desire to embrace one another. . . .

The Rileys arrived early at the schoolyard on Main Street where they were to board the chartered trolleys. Mick and Miles went off to get the identification tags. . . . The rest of the family sat down with Bridie on the stone steps of the school, near the heavy black wrought-iron gates, to await the coming of the cars. . . .

The long line of chartered trolleycars, bells clanging exuberantly, hove into view on Main Street, and the children around them began to clap and shout. "Yay! Yaaay!" As the cars approached, Bridie's anxiety increased. "Tony! Where's that child? Oh, Holy Mother o'God, every time I go on one o' these picnics I swear I'll never set foot on another. Ye know," she said to Pat, "he gets it from yer father—yer father don't worry himsel' about nothin' atall, so Tony don't either. I could throttle the both o' them!". . . .

The first of the trolleys had drawn up opposite the gate, almost throbbing on the tracks in its eagerness to be off, the other cars lined impatient behind it, and the Irish poured down the steps and into the street and up the high steel steps, past the smiling motorman, and in the car they rushed to be seated, threw open the windows, shouted to each other and the crowd outside, and heard the hiss as the doors clumped shut, the gong clanged twice, and they were off in a great surge of cheering, the huge steel wheels of the car rumbling smoothly louder on the rails, the trolley above flashing sparks from the wire, and inside the smell of sweat and of the old straw seats and the acrid sharp oily smell of the motors below.

(Excerpt from *Miners Hill* by Michael O'Malley, published by Harper & Row Publishers, Inc., 1962. It was also included in *Streetcars in Literature* by Harold M. Englund [ed.], published by Pittsburgh History & Landmarks Foundation, 1980.)

enterprises within the same community did not help financial matters, either. The streetcar business had been initially quite profitable, but it never had very stable financial foundations.

By 1918, about one-third of the companies were in bankruptcy, with no real prospects for relief. Reluctantly, some municipal governments had to start taking responsibility for operations, because service had to be maintained. Generally speaking, they did not do a particularly good job.

The other problem, caused to some extent but not entirely by the first trend, was a general deterioration in service quality. Customers started to complain, but found no meaningful response. There were no funds or managerial energy to make improvements. This led to a general disenchantment with the service and loss of support among patrons. It was not apparent at the time, but a grave was being dug for the industry.

An effort by the operating companies to stem the tide, too late and misdirected, was to design and introduce a new vehicle—the Birney Safety Car (between 1916 and 1921). Significant cost savings were supposed to be achieved with this small and simple vehicle that could be operated without a conductor. Although passengers liked the shorter headways that were made possible, they became most displeased by its unreliability and rough ride.

Another business development in the United States that was not particularly visible in the overall turmoil was the gradual acquisition of streetcar companies by large oil trusts and power companies. The streetcar industry, because of the overall size that it had assumed, was a major business and an arena for various mergers, deals, consolidations, and other financial practices that were not always savory in those days. All this further reduced managerial concerns with user satisfaction.

Soon enough, by the late 1920s, the automobile became an increasingly stronger influence, at first affecting not so much the total volume of transit use as its composition and usage patterns. Improvised jitney operations in particular were seen as most damaging to the cumbersome tram service. By and large, work trips remained with trolleys and public transit, but social, shopping, and recreational journeys lost ground. Smaller cities in particular found it more and more difficult to maintain a critical mass of demand. The Depression and World War II delayed final decisions, but eventually the consequences of patronage erosion broke through.

In the meantime, a few other positive and negative events took place. For example, a significant blow to the established streetcar industry was dealt by the Public Utility Holding Company Act of 1935, which barred public power or fuel monopolies from owning more than one transit company serving the public. The parent companies did the logical thing and divested themselves of the financially weaker units, which meant that transit was cut adrift, having had little attraction to business by this time.

The mid-1930s saw wholesale replacement of streetcars by buses—a mode that had become practical by this time and seemed to be much more maneuverable in city traffic than fixed rail transit. Bus service could be implemented rapidly, with no physical disruption, and the vehicles carried an aura of modernity. There has been much scholarly work recently, mixed with emotion, that has documented a surreptitious and effective program by General Motors to sell its product—motor buses—and to eliminate streetcars. It is quite certain that this giant company had its own interests uppermost in its mind and corners were cut, but the after-the-fact outcry and indignation by investigators today may be a little overdone. The corporation did what most business entities practiced at that time: it engaged in a take-no-prisoners competition. It is quite telling that the public applauded the disappearance of the trolleys, viewing them as outdated and clumsy relics of a darker period that should have no place in the modern city. There were few mourners at their demise during those decades.

One bright spot appeared in the generally glum streetcar picture of the 1930s. That was the remarkable decision by a number of operating companies to design and build a better vehicle. A committee of chief executives pooled their knowledge and ideas to create the Presidents' Conference Car (PCC). This unmistakable vehicle with its streamlined ends and sleek "moderne" design incorporated a number of refinements; it was sturdy, and it was easy to

PCC vehicle in service on a street in Mexico City.

operate and to maintain. It is one of the rare instances where a committee has achieved a superior product instead of a camel. The vehicles were four-axle single units that could be coupled in trains. Some variations in dimensions were possible, and the passenger capacity varied between 90 and 140. (See Fig. 11.2.)

Between 1935 and 1952, some 5000 PCC units were built in the United States and another 6000 under license in Europe. The

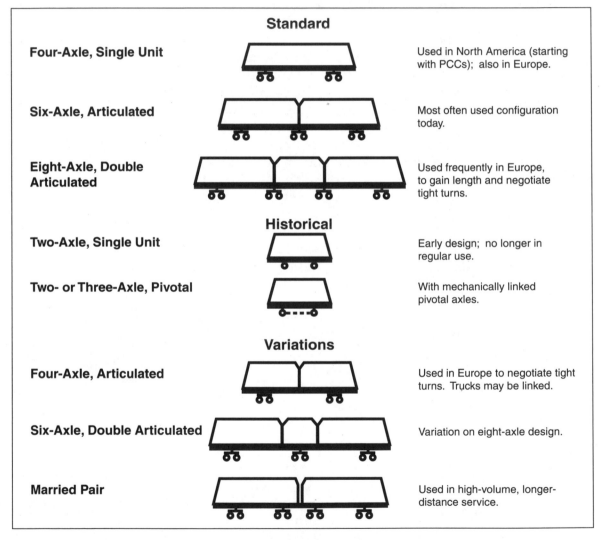

Standard

Four-Axle, Single Unit — Used in North America (starting with PCCs); also in Europe.

Six-Axle, Articulated — Most often used configuration today.

Eight-Axle, Double Articulated — Used frequently in Europe, to gain length and negotiate tight turns.

Historical

Two-Axle, Single Unit — Early design; no longer in regular use.

Two- or Three-Axle, Pivotal — With mechanically linked pivotal axles.

Variations

Four-Axle, Articulated — Used in Europe to negotiate tight turns. Trucks may be linked.

Six-Axle, Double Articulated — Variation on eight-axle design.

Married Pair — Used in high-volume, longer-distance service.

Figure 11.2 Possible light rail vehicle configurations. (*Source:* Based on *Light Rail Transit: State-of-the-Art Review,* UMTA, 1976.)

car generated much praise, and it delayed the industry's decline, but it could not reverse the tide. As lines were abandoned in the 1950s and 1960s, many PCCs enjoyed a second life as they were sold to other countries, particularly in South America. It is ironic to note that much later, when historic services were restored in a number of American cities, some of these vehicles had to be bought and brought back for restoration to undergo a third reincarnation. They are now the "antiques" that we all admire. There is at least one in each streetcar/transit museum in the country.

The last line to be abandoned in North America at the conclusion of the streetcar era was the rather interesting El Paso, Texas, to Juarez, Mexico, loop—the only international streetcar line in operation anywhere. It started in 1892 with cars pulled by mules and ended in 1973 with PCCs. There have been subsequent attempts to reactivate it. PCCs continued to operate on the few remaining regular streetcar lines in America until quite recently. For example, the Newark subway replaced them only in the spring of 2001 with Kinkisharyo vehicles ($3.2 million each).

It is of considerable interest to take a snapshot view of the state of the streetcar industry a quarter of a century ago (mid-1970s). This was a time when the old systems had been reduced to their smallest inventory,[7] when the light rail concept had been invented and had received much attention, and when a number of new projects were under way but not yet in operation. The dominant worldwide attitude at that time was most favorable toward the private car, even though serious opposition to highway building through American cities had emerged in the late 1960s. (See Chap. 5.) Public transit was not a major concern in the public forum.

At that time there were only nine streetcar systems left in North America, one of which was a small private service. They were found in Boston, Cleveland, Fort Worth, New Orleans, Newark, Philadelphia, Pittsburgh, San Francisco, and Toronto. (See Table 11.2.)

In Europe, however, except for Great Britain[8] and France, the prewar streetcar systems were in full operation, particularly in

---

[7] Streetcar passenger trips peaked in 1922 and during World War II, when the number exceeded 13 billion per year. The lowest ridership was in 1977, when it had dropped to 103 million trips.

[8] The last system to close in Great Britain was Glasgow (1962), and that country remained without streetcar service for several decades.

Table 11.2 North American Streetcar Systems in 1976

City	Number of Lines	System Length	Type of Vehicle	Size of Fleet	Remarks
Boston	5	29 mi (46 km)	PCC	294	Integrated with subway system
Cleveland	2	13 mi (21 km)	PCC	57	Between Shaker Heights and downtown
Fort Worth	1	1.2 mi (2 km)	PCC	6	Between department store and parking lot
New Orleans	1	8.5 mi (14 km)	American Standard	37	An official National Landmark
Newark	1	4 mi (7 km)	PCC	30	On exclusive ROW—old canal bed
Philadelphia	15	109 mi (175 km)	PCC and others	419	Largest remaining system in the country
Pittsburgh	5	25 mi (40 km)	PCC	95	About to undergo restructuring
San Francisco	5	36 mi (58 km)	PCC	110	All lines converging under Market Street
Toronto	11	69 mi (110 km)	PCC	389	Cornerstone of a system being expanded

Source: U.S. Department of Transportation, *Light Rail Transit,* Technology Sharing, State-of-the-Art Overview, May 1977, p. 6.

the Low Countries, Germany, and East Europe, although some of them were destined for abandonment in the next decade. Others, on the other hand, were being transformed to an LRT status. In 1975, 95 cities west of the Iron Curtain had streetcars, with West Germany accounting for 46 of them; 70 cities in socialist bloc countries of East Europe had them.[9] The Soviet Union had 76 cities[10] with streetcar service within its European portion and 32 within the Asian section. There were 45 systems in the rest of the world, for a global total of 327. The United States represented little more than 2 percent of this count.

Typical Czech-made East European streetcar (Riga, Latvia).

[9] LEA Transit Compendium, *Light Rail Transit,* vol. 1, no. 5 (1975).

[10] The author was born in one of these cities—Liepaja, Latvia—which implemented in 1899 the first streetcar line in what was then the Russian Empire.

By this time, however, after the social turmoil and oil crises of the late 1960s and early 1970s, the concept of light rail transit (LRT) was established, at least in principle and as a design approach. Basically, the LRT concept grew out of the incremental but steady efforts in Europe (West Germany, mostly) to upgrade the existing streetcar systems.[11] The professionals, the politicians, and the public in the United States were ready to embrace the "new" mode.

One of the first such actions was the desire by both the Boston and San Francisco agencies to replace their aging PCCs with a modern vehicle. They tested the German *Stadtbahnwagen* from Hannover, but found it not quite suitable. Under the auspices of the Urban Mass Transportation Administration (UMTA) of the federal government, a joint program was initiated to design an American light rail vehicle (LRV). Because parts of the old lines in both cities are underground or on exclusive rights-of-way, it was not too difficult to reclassify them as LRT operations. The large system of Philadelphia was programmed for renewal; Pittsburgh embarked on a major study to coordinate its trolley and bus services. Toronto decided not to scrap its streetcar lines after all, but to help design and acquire a new fleet of Canadian vehicles.

Thus, a sudden boom occurred in North America in the mid-1970s in rail-based urban transit, albeit at a moderate scale. Many cities had programs in the planning stage or under way, as shown in Table 11.3. While most of them did not reach construction, all of them generated attention in the national media.[12]

---

[11] The series of events during the period 1960 to 2000 regarding streetcar/LRT systems in European cities have been outlined by Michael Taplin, chairman of the British Light Rail Transit Association, in "History of Tramways and Evolution of Light Rail," 1998, found on the home page of LRTA on the Web. It describes the situation in the former Soviet Union and other socialist bloc countries, particularly Czechoslovakia; it outlines the efforts in Sweden, Belgium, Holland, Switzerland, Austria, and Germany to maintain and expand existing systems, with particularly successful programs in the latter. In addition to a review of recent events in Great Britain, the article mentions new programs in France, Spain, and Italy. Particular note is taken of the introduction of the low-floor vehicle.

Mexico, Argentina, Brazil, Australia, Japan, India, China, Hong Kong, the Philippines, North Korea, Malaysia, Tunisia, and Egypt are listed as countries outside Europe and North America having cities with streetcar/LRT service.

[12] Survey under a Columbia University master's thesis by Bernd Hoffmann, "An Old Mode with a New Potential," 1976. Other sources are periodicals and U.S. DOT, *State-of-the-Art Overview*, 1977, op. cit.

Table 11.3   LRT Projects in North America in 1975

City	State in 1975	Results in 2001
Calgary	Preliminary planning	In operation
Edmonton	Under construction	In operation
Vancouver	Advanced planning	Implemented an AGT system
Winnipeg	Proposal	No action
Aspen	Proposal	No action
Austin	Preliminary planning	No action
Baltimore	Preliminary planning	In operation
Buffalo	Continued planning	LR Rapid Transit implemented
Dayton	Preliminary planning	Rejected
Denver	Preliminary planning	First phase in operation
Erie	Proposal	No action
Harrisburg	Proposal	No action
Honolulu	Preliminary planning	Pending
Kansas City	Proposal	Rejected repeatedly
Los Angeles	Preliminary planning	Line to Long Beach built
Miami	Preliminary planning	Implemented an AGT system
Memphis	Preliminary planning	CBD service in operation
Portland OR	Preliminary planning	In operation, with extension
Rochester	Preliminary planning	Rejected
San Diego	Preliminary planning	Built and under expansion
Washington	Proposal	No action

In subsequent years, many other American cities followed the trend and initiated various studies. A number of systems were actually built, or additions were made to existing lines and networks. Much of this work is taking place in California, which has established a special fund based on sales tax revenues to assist communities in their public transportation efforts. As a matter of fact, there probably is no large or medium-size city in the United States that has not had an agency, citizens group, or organization look at an LRT possibility with various degrees of intensity during the past several decades. These efforts continue vigorously, as shown in Table 11.4.

To illustrate, note the situation in New York City. A proposal to build an LRT line across Manhattan along 42nd Street has been debated in the public arena since 1978. The chief proponent is a self-started but officially recognized group that expects to secure private financing. The ongoing discussion has included serious questions (How many tourists will wish to travel from the United Nations to the Circle Line piers? Who will rebuild the

Table 11.4 Light Rail Transit Projects in North America Under Consideration, Being Planned, or Under Construction, 2001

City	Projected Length	Target Year	Remarks
Austin	14.1 mi (22.5 km)		From north to south, passing the CBD and the University of Texas; to East Austin. Defeated in 2000 referendum; will be on ballot again in 2002.
Birmingham			Proposal from CBD to medical center and university.
Buffalo (extension)			Long-range plan for several extensions.
Calgary (extension)			Extensions of south and northwest lines.
Camden			See Trenton
Cincinnati			Northeast to Evendale; along existing rail ROW.
Cleveland (extension)			Extension of Blue Line to Highland Hills. Extension of waterfront line to form a loop.
Columbus	13 mi (21 km)		Line north from CBD.
Dallas (extensions)	12.1 mi (19.3 km) 11.1 mi (17.7 km)	2003	North central to Plano—under construction. Northeast to Garland—under construction. Southeast and northwest lines in preliminary engineering. Overall network by 2010 of 85 km.
Denver (extensions)	1.9 mi (3 km) 19.7 mi (31.5 km)	2002	Central Platte Valley connection under construction. South along I-25 in preliminary engineering. Several other corridors under consideration.
Detroit	15 mi (24 km)		On Woodward Avenue; with (5.3-km) starter line to new center.
Edmonton (extension)			Program on hold.
Grand Canyon National Park			A federal effort to accommodate visitors; with a ban on automobiles. On hold.
Hartford	9.4 mi (15 km)		Union Station to Griffin as first phase. Extensions to Bradley Airport and Windsor.
Houston	7.4 mi (12 km)	2004	LRT from CBD south, approved and under construction.
Hudson-Bergen LRT (extension)	20.5 mi (33 km)	2002 2005, 2010	Continuation northward from Jersey City to NJ Turnpike park-and-ride lot.
Indianapolis			Proposals for two routes.
Kansas City	5.1 mi (8.3 km)	Delayed	From CBD to Country Club Plaza. Negative referendum. Network under study.
Los Angeles (extension)			See Pasadena.

(Continues)

Table 11.4   Light Rail Transit Projects in North America Under Consideration, Being Planned, or Under Construction, 2001 (*Continued*)

City	Projected Length	Target Year	Remarks
Loulsville			Studies made for two corridors.
Memphis (extension)	2.2 mi (3.5 km)		Eastward extension to medical center.
Milwaukee		2003	East-west corridor studies.
Minneapolis	11.9 mi (19 km)	2003 (2004)	From CBD to airport, following years of various explorations. Feasibility studies for other lines.
Nashville			Six possible routes identified.
Newark–Elizabeth	1 mi (1.6 km) 8.7 mi (14 km)		Penn Station to Broad Street Station (Newark), with future extensions southward.
New Orleans (extensions)	4 mi (6.4 km) 13 mi (21 km)		From waterfront to City Park Avenue. To airport.
New York City (42nd Street)	2.2 mi (3.5 km)	Uncertain	River to river. No progress has been made after decades of studies.
Norfolk	14.9 mi (24 km)	2003	From Norfolk to Virginia Beach, along rail ROW. Other lines under consideration.
Oklahoma City			Early studies.
Orlando	14.9 mi (24 km)	2003	CBD to various entertainment attractions. In planning stage.
Pasadena	13.6 mi (22 km)	2002 or 2004	Construction resumed to LA Union Station. By a separate authority.
Philadelphia			Proposals for rail ROW use as LRT.
Phoenix	19.8 mi (32 km)	2006	CBD to airport and university.
Portland (extension)	5.5 mi (8.9 km) 5.6 mi (9 km)	2001 2004	From Eastside Line to airport opened. To northern suburbs, and Expo Center.
Raleigh-Durham			Use of diesel power and rail rights-of-way.
Sacramento (extension)	6.2 mi (10 km)	2003	To South Sacramento along existing rail line. North and east extensions approved.
Salt Lake City (extensions)	2.5 mi (4 km) 10.8 mi (17.5 km)	2002 (1st)	East-west line from university to CBD. To airport. Explorations of other lines.
San Diego (extensions)	6.4 mi (10.4 km) 8 mi (13 km)		Under design from Mission Valley to east line. From Old Town northward.
San Francisco (extension)			Various studies for new muni corridors.
San Jose	13 mi (21 km)		Line to Costa Mesa and Irvine. Debate continues.
San Jose	6.8 mi (11 km) 3.7 mi (6 km)		Vesona project under construction. Capitol project under construction.
Santa Ana	5 mi (8 km)	2004	Tasman East project extended.

Table 11.4   Light Rail Transit Projects in North America Under Consideration, Being Planned, or Under Construction, 2001 (*Continued*)

City	Projected Length	Target Year	Remarks
Seattle	23 mi (37 km)		From university through Seattle CBD to Tacoma. Debate continues.
Trenton–Camden	53 mi (85 km)		Economic development objectives. No federal funds. Use of diesel power. Extension southward to Woodbury and Glassboro.
Washington	4 mi (6.4 km)		Between Bethesda and Silver Springs stations.

Source: Periodicals and conference proceedings; data from agencies' Web sites.

Note: Information as of 2001, which will change continuously. The table does not give much hard data because most elements of proposed systems are under discussion and local review. The table clearly shows that a great many cities in the United States are currently looking at LRT possibilities.

underground utilities?), as well as some frivolous ones (How will the balloons of the Macy's Thanksgiving Day parade be able to get past the overhead wires?[13])

Another New York City example is the tenacious efforts by a group of trolley enthusiasts who promote the building of an LRT line on Eighth Street in Greenwich Village as the cornerstone for a citywide network.

A project that is actually being built in New York City is a system that will connect the various terminals of John F. Kennedy International Airport to the Jamaica Station of the Long Island Rail Road, where easy transfers will be possible from the regional commuter rail network by air travelers and airport workers. A branch of the AirTrain will run to large remote parking lots on airport property. Actually, this system, though called a *light rail* service, is not exactly that according to the definitions used here. It has a completely exclusive alignment, grade-separated on an elevated viaduct, and therefore qualifies as *rapid transit*. However, its name and official designation is LRT, so we will have to accept that. At least the cars will be of that type (Bombardier ART MK II). Also, since it will be completely automated, it could qualify for inclusion in the automated guideway transit (AGT) family.

[13] Clearly, the service will not operate during the parade, and a section of the wire will have to be removable.

Types of LRT Operations

light rail transit   A metropolitan electric railway system characterized by its ability to operate single cars or short trains along exclusive rights-of-way at ground level, on aerial structures, in subways, or, occasionally, in streets, and to board and discharge passengers at track or car floor level.[14]

streetcar   An electrically powered rail car that is operated singly or in short trains in mixed traffic on track in city streets.[15]

tram   A boxlike wagon running on a railway; a streetcar (chiefly British).[16]

trolley car   A public conveyance that runs on tracks with motive power derived through a trolley[17] (a wheeled device or *troller* running on top of wires). Although the *trolley* was soon discarded as the means of power supply, the name has remained in popular usage. In the early days, *streetcar* referred to vehicles in local urban service, and *trolleys* reached destination points outside.

All the preceding definitions refer to the same basic mode of transportation, and there is not much harm if they are used interchangeably—we all do that, even the technical specialists. However, to keep the discussion reasonably precise, it is useful to differentiate between a *streetcar* and a *light rail vehicle*. The fundamental difference, as the definitions suggest, is that the former runs in mixed street traffic and the latter tries to avoid that. That is entirely true, but not quite sufficient, since, practically speaking, the distinguishing characteristics are associated with the considerably advanced technology of LRT. Thus, all trolley lines built before 1960 are streetcars, and the new systems created after 1975 are light rail operations. That does not preclude the fact that there are still a number of streetcar networks in operation today that have not been upgraded, whereas many of the old systems, particularly in Western Europe, have been brought to the LRT state. Also, a few small lines have been built recently in the streetcar image deliberately to serve historic and entertainment

---

[14] TRB, op. cit. A full set of rail-based transit definitions, in a slightly different but not conflicting form, is provided in App. A.

[15] Transportation Research Board (National Research Council), *Urban Public Transportation Glossary,* 1989.

[16] *Webster's Seventh New Collegiate Dictionary.*

[17] Ibid.

districts. (There are at least three new such public lines in the United States.) In the rest of this chapter, the discussion of tram systems will refer only to the LRT configuration.

Another area of overlap in definitions is the distinction between *light* and *heavy* rail transit, the latter largely characterized by an absolutely exclusive right-of-way. But the boundary line is not always sharp. A case in point, besides the JFK Air-Train, is the Green Line LRT in Los Angeles. It has (advanced) light rail vehicles with pantographs and overhead power lines, but the track is completely grade-separated in the median of an interstate highway or on a viaduct, the platforms are high at the car floor level, and the stations are elaborate multilevel structures.

Urban Applications

Given 100 years of experience with the tram mode, the following types of applications in urban situations can be distinguished.

EXCLUSIVE GRADE-SEPARATED RIGHTS-OF-WAY

To achieve fast and safe operations, LRT can be placed on elevated viaducts or in tunnels. If this were to be done for the entire line, the system, by definition, would become *rapid transit* and be quite costly. It would not be in the spirit of LRT, but we do have the Los Angeles Green Line and the New York AirTrain. It should be noted that the vertical clearance for LRT in tunnels may have to be higher than for regular metro to accommodate the overhead power arrangement. Nevertheless, such grade separation can be and has been used effectively for limited segments, particularly in downtowns, where traffic density warrants it. Examples are Stuttgart, Vienna, Boston, Pittsburgh, St. Louis, and Edmonton. Brussels does the same, but labels its network "premetro," with the promise of eventual conversion to full metro operations. (For reasons that are difficult to explain, the line in Buffalo has been built in reverse, with the outside segments in a tunnel, but on the street in the central business district.) There are elevated sections on the new Cleveland and Hudson-Bergen lines. A covered tunnel is not always required for grade separation because in lower-density areas an open depressed cut can be used. Such examples can be found in particular on systems that have taken over old railroad alignments.

EXCLUSIVE LATERALLY SEPARATED RIGHTS-OF-WAY

To save the costs of right-of-way acquisition, it makes eminent sense to take advantage of any strips of land that are not currently in gainful use. American cities tend to have extensive railroad properties that crisscross urban territories and are frequently underused due to cutbacks in rail operations. Therefore, many of the recent projects have exploited this opportunity and placed LRT lines within these corridors. Examples include Hudson-Bergen, Edmonton, Baltimore, Dallas, St. Louis, and San Diego. There is no question that this is a significant cost-saving feature, but there are also associated drawbacks. One of them is that old railroad lines are not usually bordered by transit-trip-generating land uses—neither a concentration of residences for families with commuters nor labor-intensive offices, workshops, and institutions. Therefore, transit accessibility may be provided to bleak and semiabandoned industrial zones awaiting redevelopment, and districts on the other side of the wide alignments will not be in close proximity to the tram service. New induced development may not happen for some time.

Another serious issue is the fact that heavy rail operations (particularly freight) do not mix well at all with passenger LRT. The unfortunate term is *crashworthiness,* which is a measure of what might happen if a massive, heavy railcar or train were to come into violent physical contact with a light rail vehicle carrying people. Thus, physical segregation of channels is of utmost importance, or foolproof assurance is necessary that there will be a separation in time (freight to operate, for example, only in the middle of the night). The city of Karlsruhe in Germany was the pioneer in making arrangements between LRT and rail freight operations; the new system in Salt Lake City has extensive joint use with carefully arranged schedules.

Separation from regular surface traffic can also be achieved if the tramline can be placed within large open areas, assuming that this does not create significant public opposition due to loss of potential park territory.

PEDESTRIAN MALL TRANSIT

In downtown areas of cities, where pedestrian precincts and malls have been created, it has been found through many applications that a successful tram operation can be introduced into such zones to gain maximum accessibility for patrons to shops, offices, and service establishments. The electrically powered cars can mix with pedestrian traffic since the tracks are flush with pavement and are not an obstacle to walkers, and pedestrians are not intimidated by the large vehicles because they move slowly. It is interesting that people feel safe because they know instinctively and in fact that the tram cars cannot make sudden lateral movements, as a bus might, that may endanger pedestrians. The penalty paid for this arrangement is the slow speed of operation within the district. Examples are Zurich, San Diego, Buffalo, and many cities in Germany where operation within largely pedestrian zones is the norm for old city cores.

ALONG A CENTER MEDIAN

In some instances where a wide median strip (at least 24 ft) in a large street is available, an advantageous situation for implementing double-track LRT is present. There are some issues of concern, including the question of whether the strip of land is wide enough to accommodate the tracks and appropriate landscaping and safety barriers. Another issue is the need for passengers to cross vehicular lanes to reach boarding platforms. Pedestrians, of course, have to cross streets anyway, but at LRT stops, because of concentrated volumes, special precautions are called for that include various pedestrian control devices to minimize unthinking behavior. (See Fig. 11.3.)

Operational efficiency will be enhanced by keeping cross streets as far apart as possible or by equipping those crossings with preemptive traffic signals that give tram movements priority. There is also the basic question of whether it is acceptable to the community to lose a landscaped street feature to gain upgraded transportation service.

PREFERENTIAL ON-STREET ALIGNMENT

A very common and rational approach is to place LRT operations within a wide street, but separate the tracks as much as possible to keep them clear of motor vehicles. Pavement markings alone are not very convincing, and a better arrangement would be to distinguish the rail lanes from regular traffic lanes by different surface treatment (cobblestones, grass, or gravel, for example) or slight differences in elevation (mountable curbs along the side, for

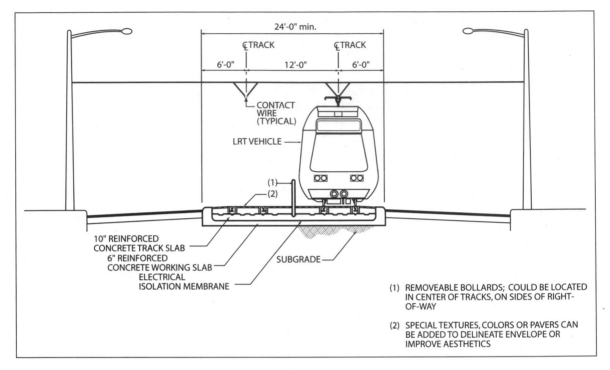

Figure 11.3 Light rail—at-grade, with embedded trackway. (*Source:* Jacksonville Transportation Authority and Parsons Brinckerhoff.)

example). The principal point is not to prevent automobiles absolutely from entering the tracks, because that may be necessary under emergency conditions, but to make it uncomfortable and obvious that motorcars are not welcome there.

A variation on this theme is to provide access and movement space atop the tracks for buses and other public-service vehicles. Crossing pedestrian and vehicular traffic, again, deserves special attention. Linear landscaping features can be introduced, which may have the additional benefit of precluding people from entering the protected track space along the way.

MIXED TRAFFIC

Historically, streetcar tracks were simply placed in the middle of existing streets, and trolleys operated together with all other types of vehicles and pedestrians, each struggling for space. This is an inefficient and dangerous situation, and it is one of the principal reasons that trams lost popular support in the previous era.

Under an LRT scenario, such conditions are not desirable, either, but if they are unavoidable or if significant cost savings can thereby be achieved, mixed traffic can be tolerated in small doses even in a completely up-to-date system. Significant precautions should be taken with boarding and alighting passengers, and there will be a penalty in the speed and, most important, in the reliability of service, as the LRT will often be caught up in the deplorable overall congestion that prevails in many city centers.

Network Configurations

Another useful way to distinguish between LRT systems is to compare their network configurations. (More will be said about this topic under "Application Scenarios.")

LINES SERVING THE CENTRAL BUSINESS DISTRICT (CBD)

Since all the recent LRT projects in the United States have built only one or a few lines, not extensive networks, they follow the traditional core-oriented pattern. They are radial, they run along preferably the highest-density residential corridors, and they wind up in or they cross the historical CBD. As a matter of fact, many of these new transportation efforts have been propelled by a civic urge and a public policy designed to help the central areas, expecting that upgraded accessibility will recapture their former strength and vitality. Whether they will achieve the full intended effect still remains to be seen. Indeed, there is no assurance that the traditional CBDs can maintain their primary role in the evolving Postindustrial Era;[18] nevertheless, the current intent is clear.

Efforts have been made to have a major destination node at the outer end, whereby more balanced usage patterns can be expected. This is the case, for example, with the airport in St. Louis, the university in Buffalo, or reaching another country from San Diego. At least the outside terminal should be an intermodal center with many regional and local bus lines converging on it.

Another variation worthy of note is the possible branching out of lines as they move into the peripheral or suburban districts to cover an enlarged tributary area. Operations under such configuration will call for precise scheduling. If, for example, three

---

[18] See S. Grava's analyses of the urban field in "The Old Downtown, It Ain't What It Used to Be," *Metropolitics,* Spring 1998, pp. 17–20, and "Mobility Demands of Urban Fields," *Transportation Quarterly,* Fall 1999, pp. 109–120.

branches, each running on a 12-minute headway, were to converge on a single track as they approach the central business district, the effective headway along that portion would be 4 minutes, with as even a spacing as possible between trains.

LINES FORMING A FULL NETWORK

Cities where LRT services represent the primary means of urban transit do not exist in North America. In Europe, there are a number of cities that are large enough, but not in need of underground metro systems. Examples include Hannover, Cologne, Zurich, Riga, Brussels, and several others in the central and eastern parts of the continent. A major example is the coordinated city networks of the Rhein-Ruhrgebiet in Germany.

A full expression of this concept will include feeder services by buses, paratransit, and private automobiles (park-and-ride). An interesting and unusual example is Liepaja, Latvia, where a relatively small but linear city with a population of just under 100,000 is serviced by a single tram line. Although all parts of the city are not accessible on foot from the rail stops, regular feeder buses are gradually being replaced by agile, privately operated minibuses and jitneys.

LINES AS FEEDERS TO HEAVY RAIL

This might be an ideal scenario in very large metropolitan areas; however, there are not any good examples in actual operation. The tram system in Toronto has some such features, and the new peripheral LRT line in the suburbs of Paris crosses radial rail corridors. The Hudson-Bergen LRT service, when completed, will have such a function among its several roles, touching rail and ferry routes.

LINES AS SERVICE TO NEW CENTERS

If metropolitan areas continue their decentralization and dispersal of trip-generating activities to new special centers, then justification for high-volume services between them and commuter access from large tributary areas will emerge. This does not suggest the disappearance of the historical center, but places this once-dominant cluster in any metropolitan area as one among a number of other destination cores. Many of the new LRT services in the United States, although they are still principally tied to the

Fact Sheet of Light Rail Transit (LRT)

The exact dimensions of the channel depend on the specific vehicle being used. Therefore, only the most common parameters or ranges are given here.

Gauge (inside spacing between rails)
　3 ft 4 in (1 m)
　4 ft 8.5 in (1.435 m) standard railroad gauge
　5 ft 4.5 in (1.63 m)

Vertical clearance (top of rail to bottom of wire)
　11 ft 6 in to 13 ft to 18 ft 5 in (3.5 m to 4 m to 5.6 m)

Lane width (dependent on the model of the car; no more than 12 ft (3.7 m)
　Width of the car + 1 ft (0.3 m) clearance on each side (the clearance may be reduced to 0.5 ft (0.15 m) within controlled spaces)

Width of reserve for two tracks
　19 ft to 33 ft (with center pole) (5.8 m to 10 m)

Distance between centerlines of track
　About 12 ft (3.7 m)

Maximum gradient
　6% (up to 10% for short segments)

Minimum horizontal radius
　36 ft (11 m) for streetcars and PCCs
　40 ft (12 m) for very slow speeds
　43 ft (15 m) preferred minimum

Minimum width of platform
　5 ft (1.5 m)

Height of platforms (above street grade)
　Low: 10 to 14 inches (25 to 35 cm)
　High: 36 to 39 inches (90 to 99 cm)

Power supply
　600 V, 700 V, 750 V DC
　759 V AC
　Diesel engine

old CBDs, at the other end or along the way touch these new concentrations of specialized development. A most interesting example, and perhaps a harbinger of future efforts is the afore-mentioned Hudson-Bergen line along the western shore of the Hudson River in New Jersey, tangential to communities and districts. This service connects a number of secondary activity centers along its path and will intersect with a series of rail, highway, and water routes that constitute the regional network.

Reasons to Support LRT

It is obvious that light rail transit occupies a position in the spectrum of transportation modes somewhere between buses and subway/metro systems. Most comparisons, therefore, identify LRT as having heavier or more intensive characteristics than buses, but less severe and extensive than metro. These are not negative or positive evaluations but simply factual descriptions.[19]

1. *Flexibility in design and implementation.* Of all the transportation modes that require capital investments along the route, LRT offers a greater capability to adapt to various constraints than any other rail-based service. There are reasonable shortcuts that can be taken in structuring a system, and there are opportunities for cost saving since standards are flexible. The right-of-way does not have to be exclusive, the vehicles can be of various sizes, the line can be built incrementally, and schedules and train sizes can be made to fit demand. If worse comes to worst and the project does not fulfill expectations, the entire effort can be scrapped without breaking the bank. On the other hand, LRT systems can also be upgraded to a rapid transit status by stepwise or comprehensive efforts. The major asset of LRT, besides being able to do a good job at transporting people, is its adaptability to local conditions and demands.

[19] For a summarizing advocacy statement in a trade publication of the industry (American Public Transportation Association), see article by E. L. Tennyson, "How Should Light Rail Success Be Measured?" *Passenger Transport,* November 13, 1995, p. 12.

2. *Mechanical efficiency and power conservation.* As a rail mode, LRT preserves the capability of moving considerable weight with relatively little power consumption (assuming that many people are to be moved). Petroleum-based fuel is not consumed, and the necessary electrical power can be produced at remote locations relying on a variety of energy sources. The thermal efficiency of electrical motors is quite good. (The same, of course, applies to heavy rail operations as well, only more so.)

3. *Reliability and safety of operations.* The track gives stability and control of movement, and, consequently, the chances for collision and running "off the road" are minimized. LRT is able to cope well in bad weather, provided that drainage systems are adequate and snow removal equipment is available. Tram operations are more controlled (i.e., contained and rationally managed) than modes relying entirely on human skill and care.

4. *Labor productivity.* Each light rail vehicle, no matter how large, requires only one person to operate it. Maintenance tasks are not difficult or too complicated; much experience has been gained; and vehicles are deliberately designed to make all components easily accessible. The maintenance of the wires and track involves no extraordinary effort.

5. *Quality and attractiveness of ride.* On a well-maintained track, with good vehicles having resilient wheels and advanced suspension systems, the movement is smooth and without vibrations. Acceleration and deceleration, unlike with streetcars, is gradual and nonjarring, drawing on a very large power reservoir. Patrons appreciate these qualities and usually find the interior arrangements nonconfining and comfortable. There is much greater acceptance of LRT by all social and economic groups, including the middle class, at this time than of any other communal transportation mode, save perhaps commuter rail. This feature becomes increasingly important in a prosperous society where riders expect significant comfort and are cognizant of their status and how their conduct is perceived by others. How we dress, what games we play, who our friends are, and what type of transport we use is thought to

matter. LRT is cool, and streetcars are cute—the public believes so at this time.

6. *Environmental characteristics.* No local air pollution, of course, is generated, although there may be issues at the electrical power plant. The voltage employed is not high enough to cause any concerns (yet) about electromagnetic radiation. A well-maintained system will be practically noiseless, but there may be local disturbances.[20] Even though the cars are large, there have been practically no complaints about their visual appearance or aesthetic intrusion on the streetscape. Note is taken of the overhead wires, but apparently people are able to ignore their presence.

7. *Image and community acceptance.* Unlike the situation in the 1940s and 1950s, tram systems at this time carry a very favorable civic image. Indeed, they are definite status symbols for any municipality that has undertaken the effort to build them. They are regarded as environmentally responsible, politically correct, and socially relevant. The opening of a light rail line is a tremendous photo opportunity for all government and civic dignitaries. There is no intention here to be cynical, but given the political and social climate that we live in, a superior public image is a major practical asset. There is a danger, however, that this perception may become too dominant a factor, leading to projects that are difficult to sustain over an extended period if the ridership base is not sufficiently large and reliable.

8. *Capacity and cost.* In most instances, the ability of a transportation system to do work and what we have to pay for it are two sides of the same coin. The first cannot be increased without the second following suit. In this con-

---

[20] Immediately after the Hudson-Bergen Light Rail line was opened, a group of residents sued New Jersey Transit for destroying their quality of life and breaking promises. According to them, the wheels screech and thump, the electrical equipment howls, bells and horns sound at every intersection, and high-powered station lights fill their apartments with glare. (Reported in the *New York Times,* June 28, 2000.)

text, LRT at this point in the history of American city development appears to fall in the range where the capacity that it offers is responsive to many demand situations, and the costs remain reasonably affordable. That may not be the case forever, but currently the stars are in a propitious alignment for trams.

Reasons to Exercise Caution

Although LRT fills a nice and sizable niche, it is not a universal solution to urban transportation needs. The most obvious limitation is the opposite of item 8 in the previous section: it cannot offer the same high *carrying capacity* that heavy rail systems can (where that is needed), nor can it be built without a sizable *capital investment*. The respective general indicators may be 10,000 versus 40,000 passengers per track per hour and $30 million versus $250 million construction expense per mile.

In addition, there are the following issues to be concerned about:

1. *Fixed character.* Any rail system is fixed in place. Although it can be moved by rebuilding, habits and patterns become established. Any major modifications are unlikely unless drastic changes in land use and activity distribution occur. (But hundreds of streetcar lines were once closed—the cars scrapped or sold, the wires removed, and the track paved over.)

2. *Interference with street traffic.* The more tram operations occur in mixed street traffic to save right-of-way costs, the more there will be a deterioration in the rapidity and reliability of the service. Experience shows that motorists are fully aware that reserving lanes for exclusive or partial LRT use will reduce vehicular capacity and constrain automobile use. They will mount opposition, which in some instances has been fatal to LRT proposals. Finding new channels on which nobody has any claims is never easy.

3. *Overhead wires.* This is always mentioned as an issue, and unless some power supply system akin to microwaves is developed, it will remain a topic for discussion. Fortunately, this visual concern appears to be a matter of first

perception rather than lasting dislike. People become used to the wires and no longer "see" them (unless they are reminded to look for them).

4. *Maintenance attention.*   Just as streetcar companies had to take care of their own track and remove snow and debris on streets, LRT operators have to exercise continuous vigilance to ensure uninterrupted service in an environment that is open to the elements, possible vandalism, and unintended abuse by patrons and nonpatrons. The technical systems are not foolproof or completely robust. Graffiti-repellent paint and unscratchable windows would be a great boon.

5. *Patronage levels.*   Recent experience with new LRT services in North America has been most encouraging with respect to usage. (The same cannot be said about all rapid rail transit.) Yet concerns remain about the propensity of Americans to embrace any public transit. The support expressed in attitude surveys does not always result in actual use, given actual opportunity. Optimism is warranted, but extreme care in patronage analysis, backed by full promotion programs, is certainly indicated. There is a suspicion from time to time that some agencies may have publicized deliberately low patronage expectations just before a line opened so that the actual numbers exceed the estimates—thus documenting a success.

When a number of LRT operations in the United States had become established and had accumulated sufficient experience, *Urban Transportation Monitor* undertook a self-evaluation survey,[21] and 17 responses were received. The features that the agencies were particularly pleased about as their accomplishments were the following:

- Ability to accommodate handicapped patrons (compliance with the Americans with Disabilities Act)
- Integration with other modes of transportation
- Integration into community
- Provision of quiet, smooth, clean, and comfortable service

[21] As reported in the *Urban Transportation Monitor,* May 12, 1995.

- The reliability of the equipment
- Successful at-grade operations in downtowns
- Low cost of construction and operation

They listed the following concerns as negative features of their operations:

- Presence of at-grade crossings that result in accidents and slowdown of operations
- Having single track sections, which constrain schedules and lower capacity
- The high costs of maintaining underground stations where they exist
- Shortages of sufficient rolling stock
- Lack of park-and-ride facilities at many demand locations

They were also ready to offer some advice to others contemplating new LRT systems:

- Keep it simple! Stick to a basic design without unnecessary complications.
- Gain as much exclusive right-of-way as possible; maintain control over all right-of-way, if possible.
- Build surface lines.
- Give high priority to the training of operators, maintenance personnel, and supervisors.
- Plan for double tracks even if they are not provided initially.
- Examine carefully handicapped access requirements.
- Plan and design a system with the pedestrian in mind, particularly for ease of access.
- Use proof-of-payment fare system (external to vehicle).
- Use low-floor vehicles.

Application Scenarios

It has been suggested by urban analysts and transit system planners at various times and in different places that LRT is appropriate for communities starting with a population size of 250,000 (preferably 500,000) and an overall density of 12 people per acre (3000 people per km²).[22] It is also said that large cities above a million in population should consider heavy rapid rail systems as their principal public transportation mode. These guidelines have some validity, but the LRT option offers too much flexibility and adaptability to be pigeonholed so strictly. Some cities in the 100,000 range have successful tram operations (largely because they have concentrated development along a corridor), and there are no reasons why very large urban agglomerations cannot have a series of LRT lines providing service along selected alignments that do not warrant metro operations (as is the case in Paris, Boston, Baltimore, Cleveland, and Los Angeles). Trolley lines have been built just for the fun of it to rejuvenate interesting districts, and civic policy has always been more important than cold and obscure numbers. It is a matter of how many people will be willing and able to get to the LRT and use it consistently and with some pleasure, leaving their cars at home.

There is considerable danger in making general and overarching observations about modal applicability, such as the preceding ones. Such statements can become simplistic, misleading, and not responsive to the adaptability of various modes, notably LRT. The basic question would seem to be whether the examination relates to an entire community (city or metropolitan area) where the planned mode would constitute the principal network and extend over the entire territory (as in Hannover and Zurich) or whether the question relates only to a single line in a special corridor (as in Paris and Istanbul). In the case of the latter, the total size of the community is of limited importance; everything depends on the potential demand as it can be concentrated along the line itself.

The record in the United States at this time shows that only Portland, Oregon, San Diego, Dallas, and perhaps one or two

---

[22] These numbers have never had any solid foundation, but stem simply from a desire to have some order and rationality in the structuring of service systems. They are based on common sense and a reasonable understanding of the urban situation. All this can be traced to the influential analysis by B. S. Pushkarev and J. M. Zupan of the Regional Plan Association in New York (*Public Transport and Land Use Policy,* Indiana University Press, 1977, 242 pp.) that arrived at nine dwelling units per acre in a corridor as the minimum density to support LRT. This number has been cited endlessly, but never actually used as the determining factor in deciding about a tramway project (to the best knowledge of this author).

other places are moving toward a pure LRT system and have decent prospects of achieving it. In all the other situations, completed or planned LRT services are either freestanding projects or the means of support for other operations. None of this implies any judgment regarding which approach is preferable; but such distinction in the scope of service analysis should give purpose and direction to planning efforts.

The ideal situation for LRT development would be a corridor, at least some 10 km long, that has not only strong destination points and trip attractions at both ends (CBD, shopping center, large medical complex, university campus, research or office park, airport, sports and recreation facilities, or similar cluster), but also comparable, if less intensive, activities along the way. This corridor should encompass within walking distance to the service line (⅓ mi, or 500 m) residential areas with at least moderately high density (preferably 40 dwelling units per acre or 100 per hectare). Beyond the central spine, residential districts of medium density (10 dwelling units per hectare or 4 per acre) could connect via convenient feeder services.

Such a multipolar pattern would assure not only sufficiently high ridership in the aggregate, but also a desirable balance of demand in both directions. It would be particularly advantageous if the public service were to be used for many purposes, not just for commuting to work. The latter condition could be achieved by devoting much attention and sufficient resources toward human amenities and comfort (and safety). In most American communities, there would have to be a campaign impressing upon the potential riders and the public at large that a new and modern, advanced-technology system with a full range of amenities is being developed or is available and to emphasize that it is quite different from the usual public transportation means, which may have become physically unattractive and socially unacceptable in public perception.

If the first line in operation is successful, and confidence is established that people will continue to use the service, further additions become feasible. Such a path has been followed in several American LRT communities recently. A consideration here would have to be whether the next segment contributes to an eventual comprehensive service network or whether the selection should simply concentrate on the next most promising corridor. If both objectives cannot be satisfied at the same time, a pragmatic approach would suggest seeking the option with the best probability of success. The network may have to take care of itself gradually, and connectivity can be provided by other modes in any case.

The danger is that a substantial LRT line may be constructed, which logically requires a restructuring of the local bus service toward a feeder configuration, but it is also accompanied by the elimination of some bus service—the large capacity of the LRT is balanced against the aggregated volumes of existing parallel bus operations. The result of this approach may be the same overall service capability in patronage numbers, but a "coarser" grain in accessibility (i.e., less responsive and convenient entry and exit options over the service area for most riders).

Public transportation advocates have to remind themselves constantly that public transit is not the preferred travel choice in North America today. Small and incremental victories, thus, are not to be scoffed at, such as those represented by a single transit operation, which could lead to a larger system if it can prove itself.

The factors entering into the examination of any one community regarding its suitability for LRT service include the following:

- *Density.*   Sufficiently high to generate enough trips to make the service practical.

- *Population size.*   Having enough critical mass to generate concentrated demands for travel.

- *Urban structure.*   A clustering of activity so that destination points are also concentrated, and a structure of corridors with intensified development along them.

- *Topography.*   Not excessively difficult for rail vehicles (slippage of wheels).

- *Available right-of-way space.*   Wide streets, underutilized railroad rights-of-way, or similar opportunities for placing track. (The nature of LRT usually precludes extraordinary expenditures for right-of-way acquisition.)

- *Civic image and local government policy.*   Proactive search for a new and advanced infrastructure system that moves beyond the ordinary. Neutral acceptance will not be enough; an aggressive stance toward enhanced mobility by suitable transit for all members of the community will be necessary.

The LRT systems that have been implemented within the past several decades in North America are described in Table 11.5, with a documentation of their principal characteristics. (Many of them, however, continue to be developed further, and, therefore, the information will certainly become dated soon. This was the situation at the end of year 2001.) During the same period, a number of similar systems have been built in other countries, including the developing world. These cities with new systems (not including the many places that have upgraded their long-established services to an LRT status) are the following: Utrecht (1983), Nantes (1984), Manila (1984), Tunis (1985), Grenoble (1987), Buenos Aires (1987), Hong Kong (1988), Istanbul (1989), Guadalajara (1989), Genoa (1990), Rio de Janeiro (1990), Lausanne (1991), Monterey (1991), Paris-St. Denis (1992), Manchester (1992), Karlsruhe (1992), Rouen (1994), Strasbourg (1994), Sheffield (1994), Valencia (1994), Saarbrücken (1995), Sydney (1997), and Birmingham (1998).[23]

The global total of streetcar and light rail systems today is 356[24]—a measurable increase over the inventory in the 1970s of 327—in other words, the construction of modern LRT systems has outpaced the abandonment of a number of old streetcar operations. Direct comparisons with the earlier set of numbers are complicated by the dramatic changes in political geography associated with the collapse of the USSR, but Russia still leads the list with 72 streetcar (not LRT) cities, or, if the former republics except Latvia and Estonia are included, the number has grown significantly to 117. Germany is in a strong second place with 57 systems, particularly because of the uni-

Regular on-street operations in Western Europe (Innsbruck, Austria).

[23] Summaries are found in *Jane's Urban Transport Systems* published by Jane's Information Group Limited UK (for example, the 18th edition 1999–2000). See also pp. 3–14 and 36–45 in vol. 2 of the *Seventh National Conference on Light Rail Transit* (Conference Proceedings, TRB, 1995).

[24] See the most recent issue of *Jane's Urban Transport Systems.*

Table 11.5   Light Rail Transit Systems Opened in North America since 1978

City	Operating Agency	Name of Line	Principal Nodes Serviced	Length of Track	Type of Alignment
Baltimore	Maryland Mass Transit Administration	Central Line	Camden Yards, CBD, airport, Hunt Valley	50.9 mi (81.9 km)	Along rail ROWs outside CBD
Buffalo	Niagara Frontier Transportation Authority	MetroRail	CBD to north	14.1 mi (22.7 km)	Transitway + 7.7 km in tunnel
Calgary	Calgary Transit	C-Train (3 legs)	CBD to south; branches northeast & northwest	18.2 mi (29.3 km)	Transitway; on street and rail ROWs
Cleveland	Greater Cleveland Regional Transit Authority	Waterfront extension	Tourist destinations	1.4 mi (2.2 km)	Interconnect with Blue and Green Lines
Dallas	Dallas Area Rapid Transit Authority	Red and Blue Lines	CBD to Fair Oaks Park; 2 extensions to south	46.7 mi (75.2 km)	Transitway; on street and rail ROWs
Denver	Regional Transportation District	Metro Area Connection	Cross access to Mall	10.3 mi (16.6 km)	North on surface streets; south parallel to railroad
Edmonton	Edmonton Transit	Edmonton LRT	CBD to northeast (Clareview)	8.5 mi (13.7 km)	Mostly on rail ROW; 2.5 km in tunnel
Jersey City	New Jersey Transit	Hudson-Bergen LRT	Bayonne, state park, JC CBD	9.3 mi (15 km)	On street or rail ROW
Los Angeles	LA County Metropolitan Transportation Authority	Blue Line	LA CBD to Long Beach	14 mi (22.6 km)	On street primarily
		Green Line	Near LAX across Blue Line to Norwalk	20 mi (32 km)	In highway median
Memphis	Memphis Area Transit Authority	2 tramway lines	Main Street, riverfront	6.6 mi (10.6 km)	Surface loop
New Orleans		Riverfront			
Ottawa				5 miles (8 km)	On railroad ROW
Pittsburgh	Port Authority of Allegheny County	T, with 3 branches	CBD to South Hills	12 mi (19 km)	Tunnel in CBD, mostly street running
Portland	Tri-County Metropolitan District of Oregon	MAX E Line + W Line	CBD with line east & west	71.9 mi (116 km)	Transitway and on streets
Portland	City of Portland	Central City Streetcar		4.8 mi (7.7 km)	North from CBD
Sacramento	Sacramento Regional Transit District	Light Rail	CBD	39.4 mi (63.4 km)	On streets
Salt Lake City	Utah Transit Authority	TRAX	CBD to south and Olympic village	29.6 mi (47.6 km)	Share rail ROW
San Diego	Metropolitan Transit Development Board	Tiajuana Trolley East Line	CBD to Mexican border	16 mi (25.5 km) 9.7 mi (15.6 km)	Transitway, mostly along rail ROW
San Jose	Santa Clara Valley Transport. Authority	Guadalupe Corridor		56.3 mi (90.6 km)	

Number of Stations	Annual Boardings, millions	Weekday Ridership	Year Open	Type of Vehicle	Size of Fleet	Total Capital Cost (by year), $ millions	Avg. Cost per km, $ millions
32	19.6 (1997)	20,000 (1994)	1992	Adtranz 6-axle	35	$364.4 (1991) $106 (1995)	$10 $9
14	6.9 (1997)	29,000 (1995)	1985	Tokyu 4-axle	39	$535 (1988)	$54
31	26.4 (1997)	100,000 (1995)	1981, 1985, 1990	DUWAG 6-axle	85	C$543 (1995)	C$19
4			1996	Breda	From Blue & Green	$48	$22
20	11.3 (1999)	35,000 (1998)	1996	Kinki Sharyo	40	$860 (1995)	$27
15	4.4 (1997)	16,000 (1994)	1994 2000	DUWAG	17	$116.5 (1994)	$14
10		36,000 (1995)	1978	DUWAG	37	$338.4 (1994)	$25
			2000	Kinki Sharyo		$1,200 (2000)	$36
22	22.7 (1996)	38,000 (1995)	1990	Sumitomo	121	$895 (1990)	$40
14		11,000 (1995)	1995			$718 (1995)	$22
28	0.6 (1997)		1993 1997	Rebuilt streetcars	14		
13	7.4 (1996)	15,000 (1982)	1987	DUWAG		$450 (1985)	$26
47	10.4 (1998)	25,000 (2000)	1986 1998	Bombard. low floor	76		
29	7.9 (1997)	23,000 (1995)	1987	DUWAG	36	$256.5 (1991)	$8
6		19,000 (1999)	2000	DUWAG	23	$312 (2000)	$13
8 2 (Miss. Va) 4	23 (1998)	12,000 (1982)	1981 1996	DUWAG 6-axle	123	$325 (1980)	
6	6.2 (1996)	20,000 (1995)	1987	UTDC	56	$540 (1987)	$14

(*Continues*)

Table 11.5 Light Rail Transit Systems Opened in North America since 1978 (*Continued*)

City	Operating Agency	Name of Line	Principal Nodes Serviced	Length of Track	Type of Alignment
St. Louis	Bi-State Development Agency	MetroLink	East St. Louis, CBD, university, airport	17 mi (27 km)	Tunnel in CBD; mostly rail ROW
St. Louis				17.4 mi (28 km)	Into St. Clair County, IL
Dallas	McKinney Avenue Transit Authority	McKinney trolley	CBD to Vineyard district	2.8 mi (4.5 km)	On streets
Detroit		Downtown trolley		1.2 mi (1.9 km)	
Galveston		Island transit		4.9 mi (7.9 km)	
Kenosha		Kenosha Transit		1.8 mi (2.9 km)	
Seattle	King County Dept. of Transportation	Waterfront	Pioneer Square, Pike Place Market	2.1 mi (3.4 km)	On surface

Sources: Urban Transportation Monitor, May 12, 1995; *Jane's Urban Transport Systems,* 1999–2000; R. Cervero, *APA Journal,* Spring 1984; periodicals and conference proceedings; data from agencies and respective Web sites; Federal Transit Administration.

fication of the two parts of the country. Western and Central Europe account for a total of 113 separate operations, and the countries that were once within the socialist bloc have 47 systems (almost entirely streetcars). The rest of the world has 56 cities with a wide range of operational tram types. The United States (23 systems) at this time accounts for 6.5 percent of the global number, all of them with rather advanced characteristics.[25]

[25] For an annual update of the LRT situation in North America see the conference issue of *Passenger Transportation* (the weekly newspaper of the transit industry published by APTA) each year with articles prepared by every agency operating rail services. The national inventory as of 1995 is documented by J. W. Schumann and S. R. Tidrick in "Status of North American Light-Rail Transit Systems: 1995 Update," and R. T. Dunphy in "Review of Recent American Light Rail Experiences," both in *Seventh National Conference on Light Rail Transit,* TRB, 1995, vol. 1, pp. 3–14 and 104–113, respectively. See also a series of articles by C. Henke in *Metro* magazine: "U.S. Begins 2nd Light Rail Revolution," November/December 1999, pp. 40–46; "Light Rail Transit Enters Its Next Phase in U.S." January/February 1998, pp. 42–52; "How Cities Can Do LRT on the Cheap," January/February 1997, pp. 29–34; "Why LRT Outlook in U.S. Remains Good," March/April 1996, pp. 28–34. A recent summary is provided by J. H. Kay in "All Aboard: Could This Be the Post-Automobile Century?" *Planning* magazine, October 2000, pp. 14–19.

Number of Stations	Annual Boardings, millions	Weekday Ridership	Year Open	Type of Vehicle	Size of Fleet	Total Capital Cost (by year), $ millions	Avg. Cost per km, $ millions
19 (3 in subway)	14.5 (1998)	46,000 (1998)	1993 1998	DUWAG	31	$464	$17
			1989	Rebuilt streetcars	5	$5.5	
3							
9	0.5 (1997)	2,000 (1982)	1982	Rebuilt streetcars	3	$3.3 (1982)	$1

Components of the Physical System

Each light rail transit system, whether it is a single line or an entire network, consists of specifically identifiable components, each requiring specialized attention, planning care, and design decisions.[26]

The Track

Physically the largest and probably the costliest element of the entire system will be the lines of track and all the civil engineering improvements associated with them. It is possible to consider for small-scale operations a single track with passing provisions at

[26] These elements will be described only at a general level in the following paragraphs. By this time, there is quite a respectable inventory of technical material on all engineering, design, and operational aspects, as shown in the attached bibliography. Particularly useful are the detailed descriptions of the recently implemented projects, each of them having different characteristics and responding to somewhat different needs. For a review of technical elements to achieve a capacity-balanced system, see D. M. Mansel et al., "High-Capacity Light Rail Transit," *Transportation Research Record No. 1623,* 1998, pp. 170–178.

Crowding into streetcars in Calcutta, India.

the end or along the line (short sections of double track) that allow running of trains in both directions. Clearly, careful scheduling and strict maintenance of safety features will be called for. Lateral reserve space is usually allocated so that a second track can be constructed without major disruptions when the demand warrants it. In most instances, with reasonable patronage expectations, a double-track arrangement would be the norm.

Since overhead wire and power feed systems, poles, or other support devices will be required, the alignment has to be evaluated with these needs in mind as well. Another major consideration is existing underground utilities, which may be vulnerable to the dynamic and static loading of trains and may need relocation. Access for right-of-way maintenance is always a factor.

Double-deck tram in operation in Alexandria, Egypt.

Terminal Points

Basically, two choices are possible. A turnaround with a tight radius can be provided at each end for the empty vehicles, which allows the use of cars with doors on only one side and a driver's[27] cab at the front end only. This arrangement offers some savings in the price of rolling stock, but consumes surface space to accommodate the loop. (See Fig. 11.4.)

[27] This person used to be called a *motorman*.

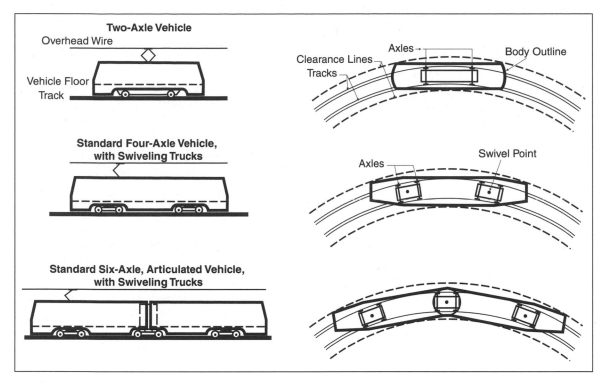

Figure 11.4 Turning clearance requirements for light rail vehicles. (*Source:* Based on V, Vuchic, *Urban Public Transportation,* Prentice Hall, 1982, p. 321)

Or stub-end tail tracks can be built, which allow the car to be brought back directly within a narrow right-of-way for the return journey. This will require the use of double-ended vehicles, probably with doors on both sides, but saves the turnaround space.

Stops or Stations

In the old days, streetcars running in the middle of streets stopped every few blocks or so, and passengers entered and exited directly from the vehicle, which meant crossing the traffic lanes. Systems running in the streetcar mode still do that, except that there are strict regulations that motor vehicles must stop and wait until the tram doors close before they can proceed. Passengers wait on the sidewalk where signs, announcements, and schedules are displayed, and sometimes a bench and a shelter are provided. This is not a desirable or safe situation, and it was one of the principal reasons why streetcar operations were abandoned in American cities.

The next step is to provide longitudinal safety islands or low platforms, still in the middle of the street, that serve as waiting space and entry/exit reservoir. This arrangement is much safer, because crossing of traffic lanes can be controlled by signals and the back of the island (the street traffic side) can be equipped with a fence. This would be the minimum expectation for an LRT system.

The length of the platform, of course, is a direct function of the length of the vehicles or trains that will be in regular service. Making the platforms longer in anticipation of an increase in demand may not be affordable, but reserving space for such a possibility is certainly advisable, as long as cross-street spacing allows this.

If the tram line is run along the curb, the entry/exit arrangements can be accommodated more directly, but parking space would be lost, the large LTVs would operate right next to pedestrians, and sidewalk space would be consumed by shelters and other passenger amenities.[28] (See Figs. 11.8 to 11.11 later in this section.)

If the LRT line is elevated or placed underground, the stations would closely resemble those designed and built for rapid rail service. (See Chap. 13 and Fig. 11.5.)

A significant challenge at LRT stops and stations is presented by the need to provide wheelchair access, which is now mandatory for all new public transit in the United States. Wheelchair lifts in cars with high floors is one option, and they can be placed either aboard each vehicle or at each stop. In either case, the operation of lifts consumes time and may interfere with the running schedule when headways are short. This is a critical concern on any rail system where flexibility in timing is limited. Another option is to build short, high platforms at the stop to reach only the front door of the vehicle via a wheelchair ramp. This solution involves no motorized devices, only a plate to bridge the gap into the car is needed, but it leaves these rather strange looking blocks along the sidewalk. Buffalo and Manchester have made this choice.

A more satisfactory solution from the operational point of view is to build high platforms (even with the floor elevation of

---

[28] All these possible configurations of stops are explored in M. Walker, *The Planning and Design of On-Street LRT Stations* (Parsons Brinckerhoff, 1993).

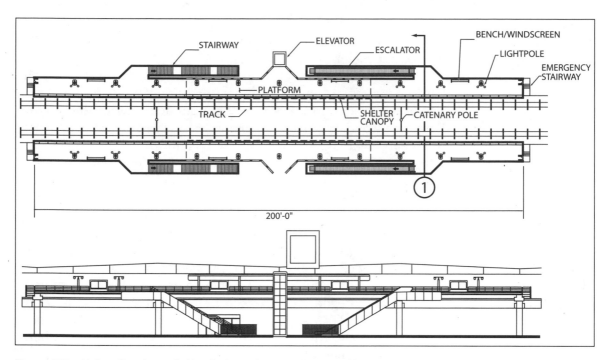

Figure 11.5 Light rail—elevated side platform plan and elevation. (*Source:* Jacksonville Transportation Authority and Parsons Brinckerhoff.)

the car) for the entire length of the vehicle or train. They achieve the safest entry/exit conditions for everybody and allow fast boarding—passengers are not slowed down by the several steps between the street surface and the floor of the car. The difference is 1 to 2 seconds per passenger to enter or exit on the same level versus 4 to 5 seconds with steps. Such large constructions require finding and allocating space, which is frequently a major problem on constrained rights-of-way. The Los Angeles–Long Beach and the Istanbul systems have followed this path.

It is quite obvious that many combinations and configurations of stops and vehicle types are possible. Different arrangements can be provided even on the same line, because it is not difficult to equip vehicles with staircases at all doors that fold down for low platforms and remain flat when high platforms are served.

There is a way out of these cumbersome complications: *low-floor vehicles*. Although there are some mechanical complications (the axles cannot run across the vehicle as they have always done) and the cars are somewhat more expensive, several models are available with floors only about 34 cm (13 in) above the street pavement. The Siemens company has even developed a car with the floor at the principal entry only 15 cm (6 in) above the top of the rail. The low section may extend over the entire floor of the vehicle or may be limited to a dropped portion (such as 60 percent or less of the total floor) at the main doors, with the remainder of the walking space over the axles reached by interior steps. The great advantage of low-floor cars is not only the relative ease with which wheelchairs can be accommodated, but the much more practically important feature of quick, almost horizontal, entry/exit by all. Very few riders are able to run up the steps or jump off a vehicle with agility; the rest of us take time to move up or down steep steps and thereby delay the entire operation at every stop. Low-floor cars are now routinely placed on LRT systems in Europe.[29] They are still somewhat of a curiosity in the United States, and so far they have been used systemwide only on the Portland, Oregon, and Hudson-Bergen systems. That situation should stop if LRT is to capture its fair share of ridership by moving along briskly. Anyone who has experienced low-floor operations will be reluctant to face an obsolescent trolley with a high floor.

Another major decision regarding stations or stops relates to their spacing. It can be as low as 650 ft (200 m), or every second short city block, which would mean that almost anybody living or working in the corridor can walk to and from a tram stop. The overall running speed, however, will be rather slow due to the many stops and starts along the line. This is the principal reason why the old streetcars in American communities became limited in the total geographic reach of their service. Instead of such a "local" service, the stations can be placed, say, a kilometer apart (3000 ft), which would require

Elevated short platform at the head of the train for wheelchair access (Buffalo, New York).

[29] The first city to do so was Geneva, in 1984.

feeder services most probably to bring patrons from large tributary areas to the tram line. The usual range for station spacing appears to be 1100 to 2000 ft (350 to 600 m), except for outlying low-density areas.

Yards

Each LRT line or system requires a reasonably flat and large space covered by track to store vehicles overnight, to keep reserve cars, to perform cleaning and maintenance tasks, and to undertake light repairs. If the prevailing climate in any one place is adverse, some, if not most, of the area may be enclosed. Heavy repair may also be accommodated on this site, but that involves the provision of rather elaborate and massive specialized equipment, as well as proper buildings. If the operations are quite small, the cars could be moved by interconnecting track to regular railroad maintenance shops within the region. Another option would be to carry the tramcars by flatbed roadway vehicles to some other major rail maintenance facility. (This, for example, is the approach to be taken with the proposed 42nd Street trolley line in Manhattan.)

Substations

These are installations occupying a small building or a large enclosed box that step down very high voltage electricity from the distribution grid and feed it into the trolley wires for power pickup by vehicles. They have to be spaced about a kilometer apart under normal conditions and can be placed at major stop locations.

Overhead Wires

The normal power supply system with overhead wires and a pantograph atop the cars is admittedly a rather primitive arrangement, but it works and is quite reliable. Besides the debate about urban aesthetics, there is also the issue of whether separate poles are to be used to carry the entire web of wires or whether, wherever possible, the support can be provided by adjacent buildings.

Feeder Systems and Park-and-Ride Lots

Any LRT line that has any length to it and intends to offer serious urban service cannot, most likely, exist relying only on walk-to patronage from the immediate corridor. The normal process would be to restructure local bus lines, paratransit operations, and dial-a-ride and taxi nodes to focus on LRT stops. This would require convenient, perhaps dedicated, traffic lanes that come in close contact with the tram platforms.

In outlying districts of a suburban character in particular, park-and-ride lots have proven themselves to be essential for any concentrated public transit service. They do consume considerable amounts of space, and care would have to be taken to assure unimpeded access by pedestrians from the adjoining neighborhoods. The siting of retail and service establishments at these nodes for the convenience of patrons is an associated opportunity (see Fig. 11.6).

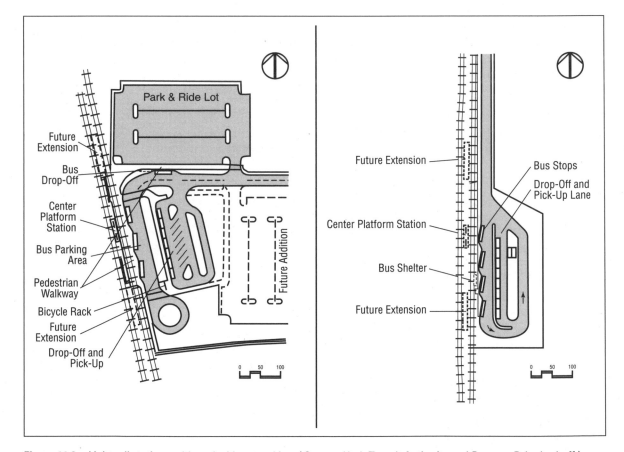

Figure 11.6   Light rail stations with and without parking. (*Source:* Utah Transit Authority and Parsons Brinckerhoff.)

Rolling Stock

Light rail vehicles (LRV) have existed and are still available in a great variety of sizes and configurations. At any given time, equipment manufacturers have models in design or production that can be acquired by operating agencies. However, it has been quite a common practice, particularly when new, large systems are initiated, to specify vehicles that respond exactly to the needs of any community. The production of LRVs, compared to other vehicles, does not serve a mass market, and therefore quite a large amplitude in design detail is demanded and can be satisfied. As can be expected, these conditions tend to increase the costs of the units.

The basic characteristics of the vehicles are shown in the box ("Fact Sheet of Light Rail Vehicles") and in Fig. 11.7. The major parameters that can be varied are the following:

- Length and width (with some very narrow vehicles able to maneuver within tight street patterns); and track gauge

- Regular high floors or low floors (extending across the entire vehicle or only partially)

- Seating arrangements (with different proportions between the number of seats and standing places)

- Single-unit vehicles, married pairs, articulated vehicles, or units suitable for train consists

- Any number of technical elements related to power supply, propulsion, and control systems

Fare Collection

The method to be used for fare collection is, of course, independent of the transit mode itself. The same considerations apply for buses, rapid transit, and LRT. If a multimodal system is in operation in any city or region, it should be expected that the same process is employed for all modes—whether it is payment upon entry into the car, the use of tokens, the validation of purchased tickets, or an honor system with prepaid fares.

From an operational point of view, there is no doubt that the preferred method is one that allows passengers to move in and out of the vehicles with no delays. The worst scenario would be the payment of a cash fare in an odd amount upon entry for which the driver may have to give change. The best scenario would be no fare at all (or prepaid validated passes) that allow quick entry/exit through all doors. (See Chap. 8.)

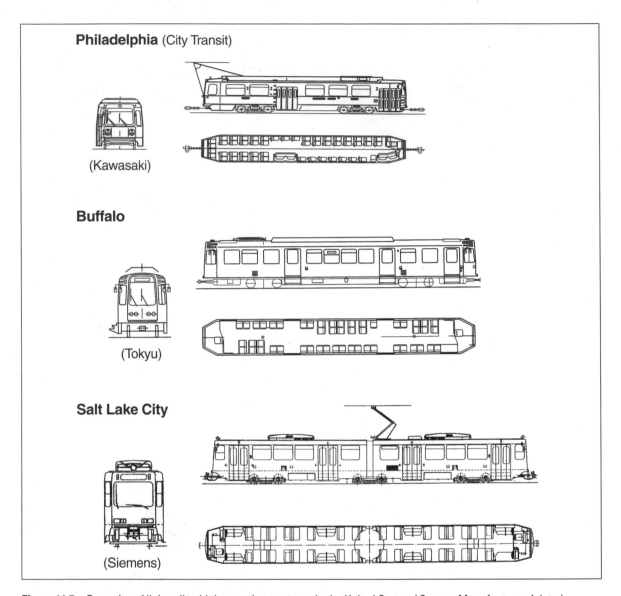

Figure 11.7   Examples of light rail vehicles used on systems in the United States. (*Source:* Manufacturers' data.)

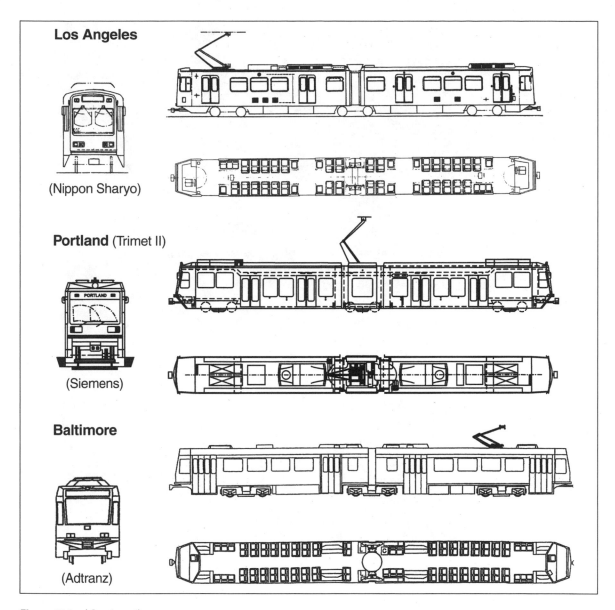

Figure 11.7 (*Continued*)

Fact Sheet of Light Rail Vehicles (LRVs)

LRVs and streetcars can be built (and have been built) in almost any size and configuration.* Following are reasonable ranges of parameters. At any given time, a certain number of manufacturers are in business and have standard models available. Today, no American firms build trams, but several international corporations have plants in the United States.

The Common Types	Approx. Capacity
2-axle, single unit (historical only)	100 passengers
4-axle, single unit	120
4-axle, articulated	140
6-axle, articulated	180 to 250
8-axle, articulated	Up to 280

Interior arrangements. Variable, depending on whether seats are favored and standing space is limited on lines that serve predominantly long trips, or the reverse is the case for trams used for in-city short trips. Capacity can thus vary considerably for the same model car.

Lengths

Single units	43 to 50 ft (13 to 15 m)
Articulated	56 to 115 ft (17 to 35 m)

Widths. 7 ft 6 in (2.3 m) to 9 ft 0 in (2.7 m)

Heights (not including pantograph). 10 ft 2 in to 11 ft 6 in (3.1 m to 3.5 m)

Speed

Maximum (on open track)	43 to 50 mph (70 to 80 kph)
Operating (including stops)	11 to 25 mph (18 to 40 kph)
In mixed traffic	5 to 12 mph (8 to 18 kph)

* A compendium *North American Light Rail Vehicles* has been assembled by Booz-Allen & Hamilton (2001), including 37 examples with diagrams and fact sheets.

Scheduling and Capacity Considerations

The literature on light rail transit mentions capacity numbers from 2000 to 24,000 passengers per hour per lane (or track) and even more. This great a range is not helpful in gaining useful answers when decisions have to be made, and, consequently, the issue has to be explored in some detail. It is not an exact science.[30]

[30] For a technically precise capacity calculation method see Gray and Hoel, op. cit, p. 403. Complete details on capacity calculations for LRT under all possible scenarios are found in Chapters 14 and 27 of the *Highway Capacity Manual 2000* (Transportation Research Board).

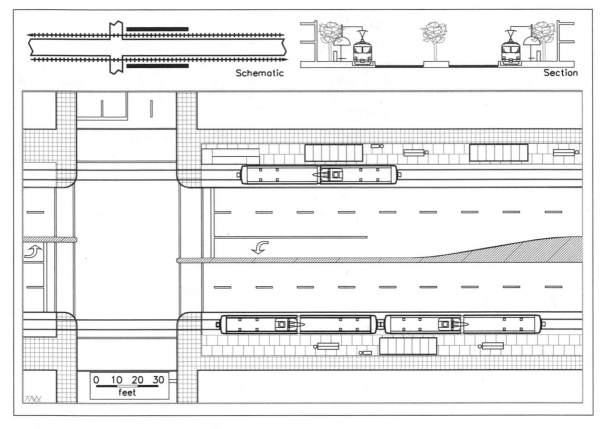

Figure 11.8 LRT station—sidewalk platforms, LRT along street. (*Source:* M. Walker, *The Planning and Design of On-Street LRT Stations,* Parsons Brinckerhoff Quade & Douglas, 1993.)

One scenario at the *lowest* level that can be visualized might be the following:

- A line operating in mixed traffic, where the street with the tram service receives 60 seconds of green in the signal cycle of 90 seconds.

- Use of small four-axle cars with a capacity of 140 passengers.

- Headways of 3 minutes that would accommodate *dwell times* (the time between the vehicle stopping at a station and starting again after passenger exit and entry is accomplished) even at the busiest stops.[31]

---

[31] If the dwell time at any one location exceeds the headway, the following train will be delayed and the schedule will collapse. The deceleration and acceleration time losses would also have to be taken into account with tight schedules.

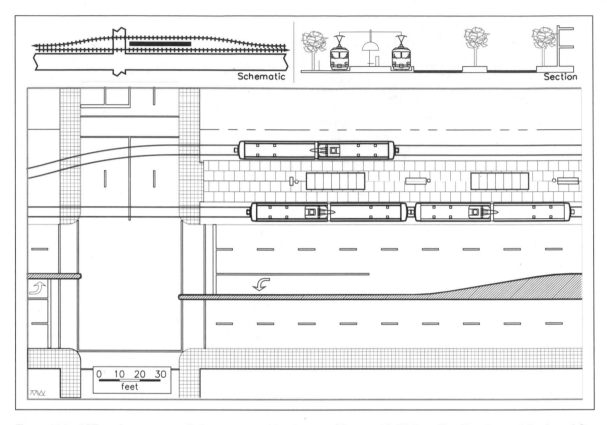

Figure 11.9   LRT station—center platform on one side of street. (*Source:* M. Walker, *The Planning and Design of On-Street LRT Stations,* Parsons Brinckerhoff Quade & Douglas, 1993.)

- Some analysts suggest from experience that on-street opera-tions result in throughput losses not accounted for in the base calculations due to the frequently prevailing traffic flow turbulence and friction. A reduction factor of 0.833 for on-street operations and 0.90 for off-street situations is rec-ommended.

This could result in 20 vehicles moving past any point on the line during an hour; however, because of traffic signals, one-third of the time will not be available. Thus, 13 vehicles per hour would move on the track, each carrying 140 passenger spaces. A reduc-tion factor of 0.83 could be applied as well. The capacity under this scenario would be *1500 passengers per hour per lane.* It would hardly seem worthwhile to build this system with so little

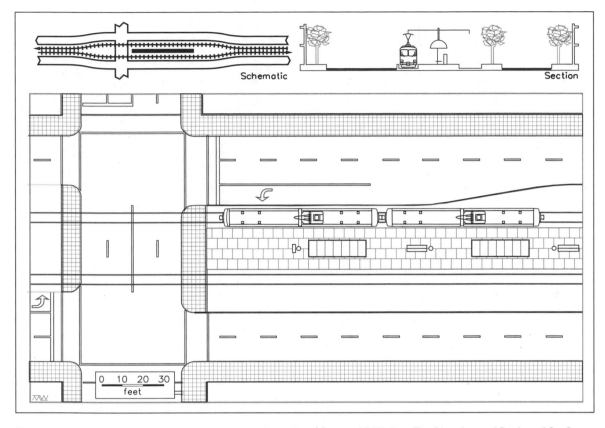

Figure 11.10 LRT station—center platform station in median. (*Source:* M. Walker, *The Planning and Design of On-Street LRT Stations,* Parsons Brinckerhoff Quade & Douglas, 1993.)

service provided. The capacity could be doubled by either cutting the headway in half or running two-car trains—assuming that dwell times, platform lengths, and other operational factors do not become constraints.

Another scenario at the *high* end of the spectrum might be the following:

- A line that has an exclusive right-of-way in the central districts and signal preemption elsewhere, thus allowing short headways to be maintained

- Use of eight-axle articulated cars with 265 passenger spaces and low floors

- Headways of 1 minute, which could not accommodate normal dwell times, unless parallel platforms were to be pro-

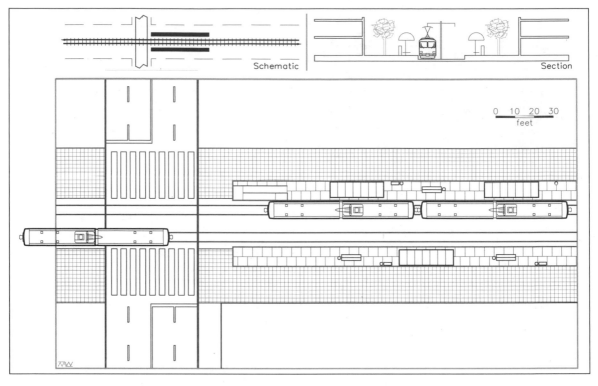

Figure 11.11   LRT station—transit mall station. (*Source:* M. Walker, *The Planning and Design of On-Street LRT Stations,* Parsons Brinckerhoff Quade & Douglas, 1993.)

vided at the busiest locations to preclude a stacking up of vehicles waiting for platform space and destroying the schedule

This would result in 60 vehicles moving 265 passenger spaces during an hour past any point on the line with a protected alignment. The capacity under this scenario would be *15,900 passengers per hour per lane.* Theoretically, multiple cars could be coupled into trains and the headway reduced to 40 seconds (most operators would be very wary seeing headways that short), thereby reaching even higher levels of throughput. This, however, would call for considerable additions to the physical facilities and would result in very fragile operational conditions. In any case, we are entering rapid transit territory as far as ability to do work is concerned and "heavy" physical elements have to be built. It can be noted that Hannover and Zurich operators claim to have actually achieved 13,000 and even 18,000 on existing lines under extreme conditions.

Another, somewhat *modest but more likely scenario* would be the following:

- A line that is semisegregated, but includes some mixed-traffic segments
- Use of six-axle articulated cars that can carry 220 passengers
- Operations at 2-minute headways, applying 0.85 as a reduction factor to account for likely delays and disturbances

The result will be 30 units per hour (corrected to the 85 percent level) times a load of 220 possible riders, or *5610 passengers per hour per lane.*

The conclusion, therefore, is that working capacities for preliminary investigations (before the actual vehicles are selected and operational procedures are defined) can be placed in the range of *5000 to 12,000 passengers per hour per lane.* This is still a wide range that illustrates the adaptability of the mode. Keeping in mind that it can be pushed higher, the findings suggest extreme care in making statements about capacities of LRT (and streetcars, for that matter). Much depends on how large an effort and what amount of resources are to be applied to expedite the smooth running of the trams.

Cost Considerations and Land Development Effects

The inescapable key question associated with any capital project, such as an LRT system, is—how much will it cost? This question has been avoided in the foregoing discussion, except for some general observations, but the time has come to tackle this issue.

No general answer can be given that would allow a quick and reliable approximation of the capital expenditures involved in any given situation. No responsible planner or engineer will venture a guess without having rather detailed information about the specifics of any proposed system. The principal reason for such caution is the great variety of features that LRT may offer and the different shapes that the service can take, as has been discussed previously. It may be a matter of placing some track on an existing street; it may require the boring of major tunnels; it may involve several construction types in various proportions. The differences in costs depending how the guideway is constructed are considerable. A comparative analysis of the actual capital expenditures for the Portland, Sacramento, San Jose, Pittsburgh, and Los Angeles LRT systems (all opened before 1990)[32] showed that a trackbed on a fill with retaining walls was 1.5 times more expensive than an at-grade construction, an elevated structure was almost 3 times more expensive, a cut with retaining walls was 5 times more expensive, and a subway was 10 times more expensive.

Another significant element is the acquisition of right-of-way. This may range from zero expenditure (if a public street is to be used) to very large capital budget amounts (if space for the track has to be purchased through developed properties). Even though LRT plans would try to avoid the latter situation, some land acquisitions are usually necessary. If underused railroad rights-of-way are available, there may or may not be significant monetary transactions involved.

Therefore, the only approach to the issue of costs is to look at actual experience and see whether certain patterns can be identified. This is not a particularly reliable method since each place is different, and closely comparable examples to a proposed project in a new place are not likely to be available. At any given time, construction costs will be different in different locations in the country, and there is always inflation to consider. Yet a general idea about costs can be obtained from past experience.

In the early days of LRT development (in the 1970s), optimistic estimates placed capital costs rather low, within a range of $3.5 million to $7 million per mile ($2 million to $4.5 million per kilometer). A single-unit vehicle was expected to be purchased for less than $100,000, and the advanced six-axle articulated LRVs were expected to cost no more than $500,000. When experience with completed systems became available in the mid-1980s, the actual numbers proved to be considerably higher.[33] The lowest costs per unit of route length were achieved by the

[32] Booz-Allen & Hamilton, Inc., *Light Rail Transit Capital Cost Study,* U.S. DOT, UMTA Technical Assistance and Safety Program, April 1991, UMTA-MD-08-7001.

[33] See *This Is LRT,* a brochure prepared for the Third National Conference on Light Rail Transit, 1982, sponsored by TRB.

Sacramento and San Diego projects, which were the simplest in their configuration, using existing street and railroad rights-of-way almost entirely. But even then the cost was no less than about $7 million per mile ($4.5 million per kilometer). Most of the other systems fell in the $12 million to $20 million per mile ($7.5 million to $12 million per kilometer) range. A significant exception was the effort in Buffalo, with 80 percent of the line in a tunnel, which cost about $70 million per mile ($44 million per kilometer).

Costs have escalated for LRT, as for anything else in the subsequent years, and the approximate capital expenditures for the various new systems are shown in Table 11.5. The general conclusion is that today it would be difficult to envision any tram system costing less than $10 million per mile ($6 million per kilometer), but usually costs are considerably higher. LRT construction costs can approach the expenses of a rapid transit line if tunnels and elevated sections are involved, which would be at least $100 million per mile ($62 million per kilometer). This is still a large range, suggesting the need for much care in deciding on desirable characteristics for a new line and identifying locally acceptable cost-saving features.

The cost situation is more predictable regarding light rail vehicles. Any project may specify tailor-made equipment with a unique configuration and unusual dimensions, but it is not likely that significant and costly departures from standard models on the market at any given time will be acceptable.[34] In the 1980s, the simplest car cost about $500,000, and advanced models remained in the $800,000 range. Today, the

Elevated station of the Los Angeles–Long Beach Blue Line.

[34] There is no PCC-like vehicle yet, but the Europeans are starting some efforts toward such an LRV. See "TRAM: The New PCC?" *Metro,* January/February 1997, p. 34. The standard models of the principal manufacturers are the following: Siemens *Combino,* Bombardier *City Tram,* Breda *VLC,* GEC Alstom *Citadis,* Adtranz *Variotram,* and Vevey *Urbos.*

respective prices are at least $1 million for a four-axle single unit to well over $2 million for a double-articulated six-axle vehicle.[35] *Metro* magazine in 2000 averaged out the price of an LRV at $2.3 million each. This is about twice the cost of a heavy rail vehicle and approaches the level of a commuter rail electric locomotive. That is an amazing development—and one not explained by inflation alone.

Possible Action Programs

The implementation of an LRT system means a commitment of capital resources (even if they are not massive), extensive legal arrangements, right-of-way reservations, training of specialized staff, and the establishment of support facilities for a special technology. A trial-and-error approach is not workable. It is also to be assumed that in almost all instances in the United States, where a regular urban transit service is considered, participation by the federal government would be anticipated in financing the project. National legislation—the Transportation Equity Act for the 21st Century (TEA 21)[36]—incorporates the objective of fostering transit development in American communities. The responsible federal agency—the Federal Transit Administration (FTA) of the U.S. Department of Transportation—does not favor any specific mode, but LRT has definitely been among the popular choices that have been made for various efforts around the country in the last few decades. Theoretically, under current legislation, up to 80 percent of the capital costs could be assumed by the national government; however, since there are never enough funds to satisfy all requests, the actual awards will quite likely be considerably lower than the permitted maximum.

To aspire toward federal financial assistance also means that a series of procedures have to be followed leading to project implementation, and various criteria have to be fulfilled. At any given time, the exact study methods are defined through applicable

[35] The Hudson-Bergen LRT system paid $2.1 million for such a unit by Kinkisharyo in 1998. The Los Angeles Green Line cars from Sumitomo, duplicates of those used on the Blue Line, cost almost $3 million each in 1993 under a constrained manufacturing schedule.

[36] See various publications, guidelines, and explanations related to TEA 21, either published by U.S. DOT, other government agencies, or transit support organizations.

guidelines, and in the recent past they have been organized as Alternatives Analyses and Major Investment Studies, with the specific requirements shifting from time to time. The basic thrust, however, has been a comparative evaluation and a search for the most effective and most responsive mode that can satisfy the transportation needs within an entire community or in any specific corridor, not the justification of a preselected means of mobility. Thus, it would not be acceptable within the federal study sequence to start with a predetermined decision that an LRT service is to be designed and implemented. It is a fact, however, that in many cases within the past decade the preferred mode after wide-ranging investigation has turned out to be LRT, endorsed by government agencies.

Without going into the details of the planning sequence, which changes over time, there are a few specific requirements that remain consistently in effect.

One of these is a careful patronage estimate. Since sizable capital investments will be involved, even if there were to be no government participation, there is a need for a reliable estimate of the use that will be made of the system, thus of the revenues that can be expected and the tangible support that will be enjoyed by the project. These studies involve elaborate procedures that use computerized simulation models to calculate the number of trips that will be generated from any given area, to estimate how they will be distributed over regional space in terms of their destinations, and to determine what proportion of them will be accommodated by or attracted to any specific mode that may be offered, such as LRT among others. These procedures are quite reliable, provided that the input data are solid and the guiding assumptions dependable.

If federal assistance is expected, it will be important to show that the investment in public transit will help to reduce dependence on private automobiles. Thus, not only should reasonable patronage be generated, but also a clear indication that there will be a shift toward resource-conserving modes, which undoubtedly include LRT, with sufficient patronage. This is not always easy to do because of the ingrained habits of automobile use for most transportation tasks in American communities. (See Chapter 5.) Credit can also be gained for other positive effects that transit would generate, particularly the shaping of communities into more effective and livable development patterns.

Another strict precondition of federal sponsorship is an environmental analysis (i.e., an investigation that attempts to identify the modes that will have the best effect not only on the quality of air and water, but also on sociocultural and historical resources and economic performance at the local and regional levels). These evaluations are assembled in reports called *environmental impact assessments* or *statements* (EIAs or EISs).

Because of the competition among communities for federal dollars, it has become increasingly important to show that the proposed project is well supported locally. This is best accomplished by increasing the local share in the total capital budget (i.e., to depend less on federal contributions for construction). A higher local financial participation will result in higher priority as the project moves forward in Washington.

Along the way, there is a practically continuous review process that involves the responsible government agencies, particularly in the transportation and environmental sectors, and, most important, the residents of the area affected, the potential users of the service, and public-interest groups that have specific concerns regarding the possible consequences of a new system. This is a matter not only of formal public meetings and hearings at predetermined, incremental, and progressive decision points, but also of close public involvement in shaping and evaluating options throughout the planning and design process.

To do all of these tasks seriously and well, considerable time and study resources are required—measured in years and in millions of dollars. Federal funds are available for this preparatory work in a phased sequence, which is intended to ensure that only those projects that survive each progressively more detailed feasibility and effectiveness analysis are allowed to proceed to the next stage. Federal capital construction monies are allocated and the exact amounts specified only when all requirements are satisfied (i.e., the proposal becomes an item of the "New Starts" program). In a few instances, mostly to expedite matters, communities have chosen to embark on a study and implementation process on their own, particularly if the state has favorable support programs. (The proposed Camden–Trenton, New Jersey, line is one such project.)

Besides the initial capital construction costs, operations and maintenance (O&M) expenses are incurred every day and have to be accounted for in every annual budget. As efficient and as simple as LRT is, a special workforce has to be employed, power has to be purchased, vehicles have to be repaired, and special facilities have to be provided. Generally speaking, with good use rates and good management, tram O&M expenditures are comparable to those of buses, and under intensive patronage they can be lower per passenger carried. The standard average in 1993[37] for LRT operations was 44 cents per passenger mile. This compares to 49 cents per passenger mile for buses, recognizing, however, that bus lines are not always placed in high-demand corridors, as has been the case with LRT thus far. The corresponding cost for single-occupancy passenger automobiles was 58 cents.

There are no instances in North America where the revenues from the public transit fare box are sufficient to cover the O&M costs (not to mention amortization of the capital expenditures). It is a matter of social policy—as it should be—to make mobility services available to all residents of a community at an affordable level. Nobody should expect a profit from public services, but the resulting deficits will be accepted only if high ridership is maintained and the service is deemed to be important to the community. A major practical consideration regarding operational costs is that, although the federal government will assist implementation and construction, no such assistance can be expected toward annual expenses. Communities have to face the O&M costs on their own, sometimes with the help of the state. This has made some cities shy away from a long-range commitment toward any rail-based system.

The issue of total costs can be looked at from various perspectives, particularly because precise cost-benefit analyses have been difficult to perform. It is most complicated to quantify and allocate secondary and indirect effects, and decision making in the public forum has paid scant attention to such analyses. It is basically a question of how a community wishes to spend its resources and what the society and the locality consider important at any given time.

Let us contemplate a hypothetical, but not unreasonable, example of an LRT line costing $500 million to build with an expectation of serving 40,000 passengers each regular weekday. That would translate into 20,000 actual patrons on average who

[37] U.S. DOT, Federal Transit Administration, *Section 15 Mass Transit Statistics,* 1993.

would commute to work (and return home), go shopping, attend school. Since some persons would engage in several of these activities during the day, there would be, say, 15,000 individuals who are steady customers of the LRT (assuming no major tourist presence). This means that each of them would be supported, in effect, by a capital investment of about $33,300. These resources would come from the general wealth of the country at large and of the entire municipality, not just the corridor to be served and the patrons to be accommodated.

Is this fair, and is it a wise expenditure?

The answer is *yes,* as a number of communities have decided recently,[38] if it is recognized that many of the nondriving patrons did not have good means of mobility before (perhaps only inconvenient bus service); that many workers now have a quick and convenient means of commuting, which could make the community more attractive for job locations; and that almost every resident and visitor can now take advantage of an alternative mode of transportation when it is appropriate. It is not solely the direct benefit to regular customers that weighs in the balance, it is the upgrading of overall mobility for the community. Support for land development is expected, as well as an organizing force toward a better urban pattern. Last, but not least, the reputation and status of the city is enhanced; whether we wish to admit it or not, this is a major contemporary factor toward implementation.

It cannot, of course, be asserted that the new rail service will substantially alleviate surface traffic loads on a permanent basis. Many of the patrons will be nondrivers to begin with, and experience shows that most new transit passengers have switched from other public modes (buses, primarily); private car users almost never give up their automobiles. Latent demand on the streets fills up any surface circulation space that may become available with cars that otherwise were not in operation. Indeed, it almost has to be hoped that operational pressures in a corridor will intensify with an LRT in place, because then the rail service will gain ridership, and additional business, residential, and entertainment activity might be attracted along the service spine.

---

[38] On the other hand, voters in Los Angeles, St. Louis, San Jose, Seattle, Chicago, Kansas City, and elsewhere have recently turned down LRT expansion or other rail-based projects through referenda, and other proposals have not been able to identify adequate demand and find support. All explorations cannot be successful.

There is another line of argument that may appear (and has appeared) in the debate about mass transit. That is the suggestion that if the $33,300 cost-per-person capital investment were to be given directly to each of the potential patrons, each could acquire a personal car and establish a permanent fund that provides for regular replacement of the vehicle, thus achieving true mobility. There is certainly no practical way of accomplishing such a program, even disregarding the fact that many of the prospective recipients may be unable to drive. Although this would be an insane public policy, it is a dramatic statement that plays well as a slogan and strikes a responsive chord among voters who take a fiscally conservative posture. It is increasingly apparent that any transit proposal in the United States will have to be able to make a case justifying fully the commitment of sizable public resources toward communal services that are accessible to everybody and result in an overall benefit.

The discussion so far has dealt almost entirely with an LRT's ability to provide good service that can cope with existing demand and accommodate the transport needs of existing development. That leaves the other dimension—the development-inducing capability of LRT—still to be reviewed. If, as history has shown, streetcars were a major force in opening up new territories around cities for housing, workplaces, institutions, and entertainment facilities, can LRT do the same today?

There are differences between then and now. The recent projects have been mostly contained within built-up areas; therefore, the potential for generating new development has not really been tested. There are, however, open lands along many of the recent tram alignments that could be built upon, and significant rebuilding of existing districts can be envisioned at higher intensity if LRT still has that power of attraction. The record has not been fully encouraging so far, but there are elements of promise.[39]

Another difference between the two time periods, and perhaps the more important one, is that in the old days the streetcar was the dominant mode and all activities depended on it. Today, LRT, even under the most favorable circumstances, will only be auxil-

[39] Many of the recent investigations regarding the ability of rail service to generate and concentrate land development are summarized by D. R. Porter in "Transit-Focused Development and Light Rail Systems," *Transportation Research Record No. 1623,* 1998, pp. 165–169.

Calgary LRT service along the center of downtown streets, with elevated stations.

iary to the principal means of mobility—the automobile—and the rail mode will carry only a few percentage points of the total transportation demand. Thus, its positive impact can easily be diluted unless vigorous steps are taken to enhance and support the capability of high-capacity modes to generate land development.

Even in Europe, where transit systems are more prevalent and more people depend on them than in North America, the conclusions have been neutral at best. A special study of transit impacts in Germany and Great Britain[40] reached the following conclusion in 1985:

> Transit investment is in large measure irrelevant to (urban growth and decline) processes, although it may affect some of them at the margin. Rail cannot save the city, if the city is going down, because the forces that are taking it down are far wider and far deeper than mere questions of accessibility. That is not to deny the potential importance of transport investments to the regeneration of a city's economy. It is to say that they would need to be planning in the context of a far better understanding of that city's malaise.

The conclusions in the United States have to be somewhat similar. There are successful LRT projects with decent ridership, but it is difficult to identify many real situations where the tram service has generated much development or redevelopment. Such searches have been made, but the conclusions are uncertain. In the early days of LRT development in this country (1984), one of the most reliable observers of the American transit scene said:[41]

[40] P. Hall and C. Hass-Klau, *Can Rail Save the City?* (Idershot UK: Gower Publishing Co., 1985), p. 169.
[41] R. Cervero, op. cit., p. 146.

. . . LRT appears to have considerable urban development potential in a number of North American cities, although other pro development forces need to exist. LRT can be an important, though unlikely a sufficient, factor in changing land uses.

A case in point at the discouraging end of the scale is the Blue Line of Los Angeles–Long Beach,[42] whose central portion runs through large districts of underused industrial properties and low-income, scattered, substandard housing. This area has seen no development activity, although the LRT stops have been in full operation for a number of years.

For the establishment of even the simplest business activity— a retail store, let us say—a series of preconditions would have to be met. The parcel where the shop would be located would have to be on the path taken every day by a sufficient number of riders and potential shoppers. Patrons would have to have sufficient disposable income that they would be willing to spend in the shop. That aggregate amount would have to be large enough to justify the acquisition of property, building a store, stocking it, hiring personnel, and gaining some profit at the end. This has not happened along the Blue Line, even with government support programs in place, and is not likely to happen at most low-volume stops, particularly where the neighborhoods are not stable.

Another scenario is the potential for new, intensive housing development. Sites on the line would have to offer prospective residents a high degree of accessibility to their regular destination points (in terms of travel time savings and convenience) that would justify the payment of full-market-rate rents, which, in turn, would be an inducement for entrepreneurs to build dwelling units on prime sites along the service route, provided that the location is not stigmatized by a negative reputation. This is a completely plausible scenario, and actual evidence can be found—for example, the intensive construction of large apartment buildings on several blocks in Jersey City near its downtown along the very new Hudson-Bergen LRT.

This idea has spawned a new concept in neighborhood design—the transit village (or neotraditional community, new urbanism, transit-dependent development, traditional neighbor-

[42] A. Loukaitou-Sideris and T. Banerjee, "There's No There, There," *Access* (University of California Transportation Center, Fall 1996), pp. 2–6.

hood development, urban village).[43] These residential enclaves would rely on transit (preferably LRT) as the principal means of accessibility and be built in a compact configuration to foster non-motorized internal mobility. A few of them have been created across the country, but none yet have any rail service. It is an attractive concept, and it should be encouraged, but it remains to be seen how many such neighborhoods will actually be implemented and marketed in the coming years and to what extent their residents are willing to curtail their automobile use.

Finally, there is the potential for positive impact of LRT projects in the development of central business districts (CBDs). Here the situation is not clear at all. Calgary is cited as an example where significant growth and upgrading has taken place since the LRT system was opened. It is not possible, however, to give all the credit to the transit service because various other development support programs were in effect concurrently. Much of this had to do with appropriate parking policies (plenty of park-and-ride accommodations) as one of the basic influences on any kind of development in North America today.

Similarly, Portland, Oregon, claims to have attracted $2.4 billion worth of development[44] along its LRT line so far. It has to be kept in mind that Portland, more than any other city in the United States, has land development controls in force that guide new construction toward efficient sites. The question remains moot, as in all similar situations, whether the new development came to the metropolitan area because of its superior mobility services or simply shifted from other possible sites to the rail corridor. Nevertheless, it is safe to say that a synergistic and mutually supportive set of efforts appears to be workable in cities

Low-platform station in the Denver LRT system.

[43] A. Duany and E. Plater-Zyberk, *Towns and Town-Making Principles* (Rizzoli, 1992, 116 pp.); R. Cervero, *Transit Villages;* and other publications.

[44] PBQ&D, *Newsletter of the Urban & Land Use Planning PAN,* October 2000.

that have reasonable vitality and good economic potential.

In most of the other CBDs served by tramways, the cause-and-effect relationships are not definitive. The services are being used, sometimes at very respectable levels, but it is difficult to point to any specific new developments as direct results of the new transit. As observed earlier, too many forces are at play to pinpoint the specific effect of only one access mode that carries only a rather small percentage of the total travel volume. In some cities, such as Los Angeles and Pittsburgh, the CBDs are so large that the LRT can only play a marginal, albeit constructive, role.

Terminal station on the San Diego LRT line at the Mexican border.

Conclusion

The development of light rail systems toward the end of the twentieth century in American communities is being judged a major success, much constructive activity has taken place, and the mass media has covered the programs thoroughly and largely favorably. The actual number of riders as a national total may not be that many, but it has been shown that workable systems can be developed. Will these trends continue, and how far will they go?

There is a general perception that recent LRT implementation efforts in the United States, although mostly commendable, have started to exceed the paying ability of local governments and voters and their willingness to incur debt. The vehicles, for example, have become extraordinarily complex and expensive, beyond the needs of many communities; the track and stations are elaborate and expensive, beyond what riders would expect and necessarily demand. Quite frankly, the current LRVs are massive and huge. They do not look at all like streetcars or even PCCs, but rather like the 20th Century Limited strayed onto city streets. Many of the recent stations are proud examples of civic architecture, outshining by far the simple shelters of the old days. Something per-

Future Shape of the Surface Transit Vehicle?

Let us go out on thin ice, and speculate about what the next transit vehicle on our streets is likely to be in the decades to come. This will not be a visionary forecast, but rather an extrapolation from what we know today, what the riding public tells us, and what the engineers are working on already. Leaving aside the heavy rail options (metro and commuter services) with their own niches, the premise is that the traditional urban surface modes that we have used for about a century—streetcars and LRT, buses, and trolleybuses—may be due for a replacement (but maybe not, since no guarantees accompany this discussion).

We have relied on them, with reasonably good results, but they are not perfect, and we could possibly do better. All three of them suffer from usually being caught in, and contributing to, general street congestion, but that is not the issue here. That problem could be alleviated substantially by giving them priority in the use of urban space and channels. There are strong and weak points intrinsic to each mode; without repeating the previous detailed review, the major characteristics are as follows:

Mode	Negative Features	Positive Features
Buses	Limited capacity of vehicle	Simple technology
	Polluting engine	Flexibility in operations
	Manual driving	Relatively inexpensive
	Safety concerns	
	Placement problems*	
	Inferior image	
Trolleybuses	Tied to power lines	Nonpolluting motor
	Limited capacity of vehicle	
	Manual driving	
Light rail transit	Heavy investment in track	Large capacity
	Tied to fixed alignment	Nonpolluting motor
		Good image

If the promises of clean fuel and/or a clean engine truly become fulfilled, the traditional bus might emerge as the victor—applying most of the bus rapid transit concepts. The bus can match the capacity and labor input characteristics of LRT with advanced vehicles, priority treatments, articulated units, and possibly bus-trains. Yet, this approach will probably not gain public acceptance in North America because of the perceived image of buses, seriously damaged in the course of recent urban history. (During a current study to develop a responsive and comprehensive transportation system for Long Island, New York, the public and the media flatly refused to consider a vehicle called a *bus* as suitable for their needs. The search is on for an "advanced rapid commuter vehicle.") Thus, what is needed is a redesigned vehicle with visual and human

---

* The vehicle cannot be operated reliably in tight spaces nor always placed accurately at the curb or at high platforms.

comfort features that reflect the space age, the limousine, the personal computer, and the cell phone.

Going back to the table of characteristics, a set of specifications can be extracted:

- Ability to operate on any street surface (rubber-tired vehicle; no tracks)
- No steps to climb (low floors or high platforms); no fare payment at the door
- Clean propulsion power (energy or fuel to be carried in the vehicle or supplied along the way unobtrusively and safely)
- Automatic operations and ability to follow a precise path (mechanically, magnetically, or optically), with option of manual operations; inclusion of crash-avoidance features
- Modularity, to be able to be assembled in large-capacity consists or to operate as small units
- Maximum amount of human and visual amenities
- Real-time information on all aspects of operations, available to service providers, passengers, and prospective patrons

Surprise! We are almost there. The Civis (by Irisbus, a joint venture of Renault and Fiat/Iveco) is a low-floor vehicle that optically follows painted lines on the pavement. It can be built in various lengths and equipped with different power sources, ranging from an overhead catenary to hybrid engines. It has been tested extensively, is in actual operation in two cities in France, and will start service in Las Vegas in 2003.

It is not yet exactly what the preceding idealized specifications called for, but it comes close. Further development work also goes on elsewhere. For example, the Ansaldo experiments in Trieste, with a power rail submerged in the pavement (and activated only when the vehicle is directly above it), have generated much attention as well.

If these efforts succeed, they might, indeed, take over the surface transit market, because the service characteristics appear to be exactly right. Perhaps we have reached a technological stage at which we can assemble the type of vehicle that we need, instead of trying to find a reasonable use for something that is being sold. Undoubtedly, these new devices can be made to work; the crucial test will be whether the very high technology input (associated with costs and fragility) can be justified in light of basic service needs. We will see, probably within the next decade.

The remaining question is—what do we call these systems? The French opt for TVR (*transport routières sur voie réservée*)—a mouthful in any language; Bombardier has floated *GLR* (guided light rail), which does not quite scan. The suggestion here is *surface urban transit* (*SUT*), an acronym that encompasses the principal characteristics, is easy to pronounce, and hopefully does not offend anybody.

haps has gone wrong with the scale, and a fragile situation in public acceptance, expectations, and attitude has been created. It has to be feared that, unless ridership becomes very strong and American communities decide to place great value on the image that their new services confer, a negative public reaction may emerge. The Europeans are building streetcar/LRT systems for much less, carrying significantly larger percentages of total metropolitan travel.

The constraints and bureaucratic procedures associated with any project in the United States that involves government assistance have become most burdensome and expensive.[45] Something needs to be done, and one response would be to use standardized elements and components as much as possible. The other major thrust currently under way appears to be reliance on turnkey procedures—engaging private enterprise in design-build-operate-maintain (D-BOM) efforts. This is being done on the Hudson-Bergen LRT and the San Juan heavy rail projects and will be employed on the Trenton–Camden LRT project in southern New Jersey. The D-BOM approach is practiced widely in Great Britain and France. It basically does not require local government to assemble large capital funds at the beginning, but to pay gradually to amortize the investment and cover the annual O&M deficits (assuming that no break-even financial performance is to be expected).

The conclusion regarding the potential influence of LRT in attracting development and shaping urban districts is one of scale. A small line with limited ridership will not turn around a severely depressed economic situation, nor will it be possible to identify specifically its contribution to development that is already flourishing. There will be an effect, however, if the new service can respond to real needs, such as bringing commuters to concentrated job locations and entertainment seekers to recreational and cultural clusters. The "accessibility index" of sites along the routes will be enhanced, and, if the market is not constrained by other forces, induced development can be expected. It is more likely to occur if appropriate regulatory means and devices of encouragement are also in place.

Above all, an LRT system—large or small—will give an added means of mobility to the local population that is available to all

---

[45] See C. Hanke, op. cit., November/December 1999, p. 44.

Situation on the Hudson-Bergen Light Rail line three weeks after the attack on the World Trade Center.

Shaved Minutes and Frayed Nerves in the Trolley's New Morning Rush

Michael Winerip

JERSEY CITY—Ray Opthof scanned the faces. It was 7:40 a.m. Friday. The commuters, many of them New Yorkers suddenly working in New Jersey, stood three deep, waiting for the next trolley at the Newport station. "There," said Mr. Opthof, a New Jersey Transit operations engineer. "See? Over there. Definite first-timer."

Lorraine Kirby had that look. "The dazed look," whispered Mr. Opthof, heading her way. Ms. Kirby's head was moving a lot.

"Regulars stare straight ahead," Mr. Opthof whispered.

She was eyeballing the automated ticket machine like it was a VCR that needed programming.

"Can I help?" Mr. Opthof asked.

Could anybody? Since Sept. 11, Ms. Kirby's hour and 20-minute commute has grown to as much as three hours, and so, after experimenting with combinations of car, PATH train, subway and ferry, she was adding the trolley. "Today we're trying this plus a ferry," she said.

Merrill Lynch, Deutsche Bank, Morgan Stanley, American Express and Lehman Brothers have opened replacement offices here, and ridership on the practically brand-new, still-unfinished trolley system, the Hudson-Bergen Light Rail, has doubled. About 5,000 new riders appeared overnight.

"The day after the attack, we thought this was coming," said Mr. Opthof, 43, whose job is to make the trolleys run on time. "But that next Monday, standing in the station, watching all those people come at you, it was still a shock."

Mr. Opthof has doubled the number of trolley cars while also trying to clear dangerously crowded platforms in seconds by shouting for everyone to get on without tickets. He rises for work at 3 a.m., gets home after 7 p.m., and even asleep, he can't rest, waking up to make a note on how to shave a minute off the Newport-to-Exchange Place run. "They're so upset about the trip taking longer," he said. "They're sensitive to every minute."

Though he lives in Princeton Crossing, he has patience for New Yorkers, who, by the time they reach the Newport platform, have typically taken a Manhattan subway and a PATH train under the Hudson River to New Jersey, and then rushed a block to the trolley. "They may start out smiling," he said, "but by the end of the commute they aren't smiling."

Efficiency experts break big problems into small ones to solve them, and so, by training, he noted the little detail. "I see a lot of people, with very startled looks," he said. "I see people, arms in casts, walking with limps,

fingers bandaged. You wonder: were they in the disaster? They're out here trying to get on with their lives."

His mild-mannered engineer's precision often defuses their anger. When Scott De Core complained about a door not opening, Mr. Opthof gave the man his e-mail address, his fax number and two other telephone numbers, and said: "Next time take down the train number, note the time and e-mail me. Our computers will tell me the exact problem."

During this adjustment period, anyone with a monthly PATH ticket has been allowed to ride the trolley free, until Oct. 31. Still, people are upset.

"What's the deal with charging after the 31st?" Christine Brooks said. "I complained. One of your guys yelled at me."

"It's true about charging," Mr. Opthof said, "but he shouldn't yell."

"Well, he didn't yell," Ms. Brooks said. "I'm exaggerating."

Normally, Mr. Opthof relies on marketing research to gauge rider patterns, but as Jeffrey A. Warsh, New Jersey Transit's executive director, said: "People are changing their commutes daily. Today's data is useless tomorrow."

So Mr. Opthof watches them, racing like ants, more than 100,000 every morn-

ing on this side, above and below ground, trying to find the best new trail now that their old routes—like the PATH line under the river to the World Trade Center—are closed off.

Where will they go? Sometimes you guess wrong. With the new car-pooling rules and restrictions into Manhattan, transit officials thought morning drivers would flood a park-and-ride lot at Giants Stadium.

"We thought we'd have up to 7,000 new riders," Mr. Warsh said. "We had 165."

Mr. Opthof said that he sees another trolley surge coming. With PATH's Exchange Place station damaged and closed, western New Jersey commuters have been getting off at PATH's Grove Street station and walking a mile to their offices at Exchange. Mr. Opthof has put out the people counters. "There are 5,000 people walking," he said.

When the weather changes, Mr. Opthof said, he believes that they will ride the PATH to Newport, then take the trolley. He is considering cutting the wait for a rush-hour trolley from six minutes to four.

"We'll need more operators, more cars—two minutes, it's a lot," he said.

Yesterday at 2 a.m., he woke up in Princeton Crossing and made a note that a four-minute interval was doable.

users. A new transportation choice will be present that at this time carries a superb public image and is attractive to individuals and beneficial to the community.

Light rail transit is eminently capable of providing comfortable and responsive service; it is a popular means of travel even in an automobile-obsessed society; and it can materially assist positive land development and redevelopment when combined with other programs of encouragement and guidance.

Bibliography

Cervero, Robert: "Light Rail Transit and Urban Development," *APA Journal,* spring 1984, pp. 133–147. A survey of all LRT efforts in the United States in the early 1980s and their potential development impact.

Gray, George E., and Lester A. Hoel (eds.): *Public Transportation* (2d ed.), Englewood Cliffs, NJ: Prentice Hall, 1992, 750 pp. The basic textbook on all public transportation modes.

"LRT Guide," *Mass Transit,* March 1987, pp. 8–41. A compendium of all services in operation worldwide, plus work in progress at that time.

Pattison, Tony (ed.): *Jane's Urban Transportation Systems* (18th ed.), Coulsdon, UK: Jane's Information Group Limited, 1999–2000, 660 pp. The encyclopedia of all cities with public transport services, including streetcars and LRT.

Transportation Research Board: *Light-Rail Transit: Planning and Technology,* Washington DC: National Academy of Sciences, Special Report 182, 1978, 172 pp. Overview with separate articles of the LRT industry in the early days.

Transportation Research Board: *Public Transit: Rail,* Washington, DC: National Academy Press, Transportation Research Record No. 1623, 1998. Various articles on current LRT topics and case studies.

Transportation Research Board: *Seventh National Conference on Light Rail Transit,* Washington, DC: National Research Council, American Public Transit Association, Conference Proceedings in Baltimore, 1995, two volumes, 341 and 201 pp. A series of presentations, primarily on specific systems and technological aspects.

U.S. Department of Transportation: *Light Rail Transit: Technology Sharing; State-of-the-Art Overview,* Washington, DC, May

1977. Summary of principles and practice in the early days of LRT.

Vuchic, Vukan R.: *Urban Public Transportation: Systems and Technology,* Englewood Cliffs, NJ: Prentice Hall, Inc., 1981, 673 pp. A very complete public transit review, with an engineering orientation, but becoming dated in its examples.

Monorails

Background

Monorails have sometimes been referred to as a solution in search of a problem. They are, indeed, interesting and workable devices, but it is not so easy to identify many situations where they would be a clearly superior choice, compared to other rapid transit possibilities. While they have been around for more than 100 years, actual operating systems are rare. They are a distinguishable variation of rapid guideway (rail) transit, and they enrich the total inventory of transit options, but they respond well only to specialized situations.

There are many variations among monorails, but their one common element, of course, is a single rail, beam, or channel that supports or carries the passenger container.[1] The vehicle may be large—comparable to a subway car—or small—a cabin for a few passengers. The principal difference is whether the passenger compartment hangs from an overhead beam or channel, representing a *suspended* monorail, or sits atop a single horizontal beam, representing a *straddling* monorail. In almost every other respect, a monorail behaves exactly like any other rail-supported

[1] Another definition of monorails is that the vehicles are wider than the supporting structure.

system, and service delivery and capacity are comparable to like-sized regular public transit modes.

Development History

It does not take particularly great imagination and engineering skill to visualize and build a transportation device in which a wheel runs atop an overhead rail or along a channel, from which goods or a passenger compartment can be suspended. Engravings from the nineteenth century show that such devices existed and were used in warehouses to move freight by hand, and such equipment is still in wide use today.

Despite the fact that maintaining stability with a single support is a much more intricate task than bracing a device on two legs or rows of wheels—or perhaps because of this challenge—inventors have attempted various ingenious ways to keep a straddling vehicle on top of a single track or to control the sway of a suspended moving compartment. Thus, records show that in 1825 an overhead track was built in England from which a string of gondolas was hung, pulled by a horse.[2] In 1833, at the Rouen Agricultural Exhibition, a single track, supported by a low triangular trestle, carried vehicles placed like saddlebags on both sides.[3]

The latter idea was used again several decades later for a demonstration project to carry passengers at the 1876 Centennial Exhibition in Philadelphia, which was notable for showing a number of new technological devices.[4] The monorail was the straddling kind, supported by a triangular truss, with a steam locomotive and passenger car running on top, but was also cleverly balanced by lower compartments. It shuttled back and forth over a length of 500 ft (150 m). Despite its popularity at the fair, it spawned only a few follow-up efforts. Such a device was in operation in Ireland from 1883 to 1924, and also briefly in Pennsylvania, where it was placed in operation in 1878, but closed the next year after a major disaster.

[2] The Cheshunt Railway was built for carrying bricks, but it accommodated passengers during the opening festivities.
[3] As described by S. H. Bingham, "High-Speed Mass Transport by an Overhead System," *Civil Engineering,* January 1961, pp. 61–64.
[4] See *Scientific American Supplement,* August 12, 1876.

As is well established, the second half of the nineteenth century, extending to World War I, was a fertile period of visualizing and attempting various transportation devices, certainly encompassing monorails. There was one—the Brennan system—that utilized large gyroscopes (!) to maintain balance, but apparently not too many were convinced of its reliability in regular public service. Another invention—the Kearney system—suggested a single support rail below the vehicle, but also a second rail on top to keep the car in place. There were also more down-to-earth attempts that explored the possibilities available with conventional technology and structures. During this period of experimentation, the patterns were set for the rest of the twentieth century, and while the monorail is one of the lesser branches on the modal evolution tree, it has a presence there nevertheless.

These late-nineteenth-century engineering efforts presumably led to the next—and fully workable—development. The first monorail system of the *suspended* type to serve passengers for an entire city was built more than 100 years ago (opened in 1901) in Wuppertal, Germany. This system is still in full operation today over a 13-km (8-mi) track with 19 stations, and it has been recently renovated with some completely rebuilt stations. Its official name is the *Schwebebahn,* alluding to its swinging operational character.[5] The system remains in place not only because of historical nostalgia—it works well enough to provide regular transit service, but it also undoubtedly gives the city a special and unique feature. It is slow by today's standards, it experiences problems with high winds, and the overhead structure is most visible (if not oppressive), but it has become an accepted part of the cityscape over the years.

There were no real followers of the suspended Wuppertal example for a long time, except for many small-scale or temporary operations on fairgrounds,

The oldest monorail in continuous service, in Wuppertal, Germany.

[5] *Schweben* in German means "to hang, to be suspended."

in amusement parks, and in similar recreational situations where the ride on the monorail became a part of the leisure time activity, and the device had a promotional role.

A recent development at the city scale, however, has taken place in Chiba, Japan, near Tokyo. Two lines utilizing suspended monorail technology have been built since 1988, and plans exist for an extended network of this mode.

The development path for the *straddling* monorail has been slightly different. Full-scale applications started later because the mechanics are more complicated: the vehicle not only has to be supported vertically to carry its weight; it also has to be braced laterally by horizontal wheels so that it does not tip off the support. This means that the "monorail" has to be a sizable beam with some depth to it, so that all these mechanical tasks can be accomplished.

A most curious aspect of monorails in the second half of the twentieth century is their very high name recognition and a general popular perception that these devices represent top-drawer advanced technology that communities should aspire to. This reputation can probably be explained by the persistent appearance of monorails in futuristic drawings of utopian cities for many decades. It is hardly possible to find Sunday supplement articles on visionary cities that do not include an illustration with a straddling monorail (and a flock of helicopters) on the scene. Both Disneyland and Disney World have had monorails as their principal means of access, and it can be assumed that Disney "imagineers" shrewdly capitalized on this futuristic image and, in turn, enhanced it considerably further. Anybody who has not actually taken a ride on one of them has certainly seen plenty of pictures of the colorful and clean Disney transportation environment, with or without Mickey Mouse.

In real life, straddling monorails became visible after World War II, when a number of manufacturers developed prototypes. Many of these found specialized applications again in service to or within recreational areas, in effect as components of private business developments. The principal promoters of the mode have been vehicle builders in Germany (Alweg) and Japan (Hitachi),[6]

---

[6] A Swedish investor developed monorail technology with a test track in Germany under the name *Alweg* after World War II. This corporation was the principal promoter of monorails until the 1960s, when the firm went out of business, but its basic designs were continued by Bombardier and Hitachi.

later joined by others, who embarked on a global campaign to find receptive locations for their products. These candidate places, after a longer or shorter evaluation of options, have almost without exception decided not to use them and turned to other rail-based modes or buses. Operational straddling monorails did, however, appear at world's fair sites, and Walt Disney, of course, deemed them particularly suitable for his new recreation ventures. No public service system was built in Europe, but the German hardware did find application in Seattle as a federally sponsored demonstration project in 1962, and it was even brought to Japan in the early years. The French (Safege) experimented with several suspended monorail concepts, but only limited opportunities for real service emerged. A temporary operation was an internal service loop at the 1964–1965 World's Fair in New York.

As a matter of historic fact, the Japanese borrowed and learned from the European technology (Alweg), until they developed their own technology and became major players in the market today (Hitachi). The Japanese found no municipal clients abroad at first, and gradually built monorail systems in several of their own cities. A number of them can be found in and around Tokyo (the first, opened in 1964, was actually the Alweg model), as well as in other places.

The peak period of attention toward monorails in American cities was undoubtedly the 1960s, when several projects were under construction, the mode received much publicity as a possible solution to many urban transportation problems in the popular press, the manufacturers had mounted active promotional campaigns, and many people looked at pure technology with great expectations. Miami, Orlando, Los Angeles, Houston, Dallas, Las Vegas, Boston, Chicago, Detroit, Cleveland, Philadelphia, Pittsburgh, Atlanta, San Francisco, Minneapolis, St. Paul, El Paso, San Diego, Newark, New Jersey, and a number of other cities seriously considered the mode as a possibility.

In the meantime, as various people movers and automated guideway transit (AGT) devices have been developed, several of the models have incorporated monorail support systems. Thus, a question emerges as to whether these examples should be classified as monorails or as AGTs. The decision is not easy, but not particularly critical either, because all of these examples are special and different from each other, and they do not lend themselves to simple classification systems. For example, although the

Reduced scale monorail for access to Disneyland.

Jacksonville, Florida, Skyway Express sits on a single thick beam with lateral wheels, it looks and behaves very much like a true AGT, and has to be included in that group (see Chap. 15). The Newark, New Jersey, airport Monorail, on the other hand, specifically carries that name, and, while it is fully automated, its size and appearance would place it in the monorail class.[7]

At the present time, although monorails appear quite frequently on lists of alternatives to be evaluated in corridor studies in American cities, there is no regular public-service system under construction in this country.[8] There are, however, a number of places where smaller-scale monorails operating over short distances have been developed as shuttles for special activity centers. Monorails have not lost their standing as a potential urban transit mode, at least conceptually, and there are active construction projects in several Asian cities.

Types of Monorail Operations

Rapid Transit Service with a Citywide Network

The monorail mode has not really succeeded at this level. The Wuppertal system is, after all, only a single line and has never been extended further, and other existing examples are limited in

---

[7] Even more confusion in terminology is generated by the currently ongoing construction of a transit system in Chonging, China, across two rivers. The vehicles will consist of manually operated passenger compartments that are suspended from large upside-down mechanical units on top with wheels running along aluminum tracks, which, in turn, are carried by suspension cables (monorail or aerial tramway?). The 2-km (1.2-mi) line, however, is referred to as *light rail* by the developers, while its trade name is *Aerobus*.

[8] There are proposals on the table in Las Vegas and in Seattle. They are referred to later in this chapter.

their scope. The city of Chiba, Japan, has plans to build a whole network of monorail lines, but so far there are only two lines, and it remains to be seen how well the next stages do and whether the concept catches on.

Shuttle Service Connecting Nodes Within a City

The best monorail examples are found in this group, albeit there are not too many of them, either. The line in Seattle connects a downtown retail cluster to the Seattle center, originally a world's fair site, over a distance of 1.2 mi (2 km). The Tokyo Monorail runs 10 mi (17 km) from a peripheral city node to the Haneda airport, which was originally the capital's principal international facility. The Tama Urban Monorail is under construction to link a new town 10 mi (16 km) away to a center in the Tokyo suburbs. A short line (1.2 mi; 2 km) connects a peripheral shopping district in Rio de Janeiro to a regional access node.

The Jacksonville Skyline Express—AGT in a monorail configuration—can also be considered. The original 1-km (0.6-mi) demonstration line (opened in 1989) has been reequipped and modified and is being extended, with further additions planned. If it proves successful (initial ridership has been low), it might become a service at the city scale. The Tampa airport service (opened in 1991) is similar, insomuch as it utilizes the same M-III Bombardier cars as the Jacksonville system.

Internal Connectors Within Controlled Areas

This refers to situations in which the transportation service is integral to the activities of a special district or project, with all the elements most often under the same ownership or control. Examples are the many monorail systems in amusement parks and recreation areas, starting with Disneyland.

This group also includes the Newark, New Jersey, airport facility, which—while it is publicly accessible—operates entirely in support of air passenger and airport worker needs, connecting the three terminals with each other as well as with several large parking lots. An extension to a station on the Northeast Corridor, the principal rail spine for the East Coast of the United States, has been opened recently. A peculiarity of this arrangement is that, due to institutional constraints, only rail passengers are allowed

Miniature monorail in an amusement park in San Antonio.

to use the monorail, not people from the surrounding communities. A somewhat analogous situation exists in Dortmund, Germany, where the H-Bahn (an automated suspended monorail) serves principally the university campus over a total length of several kilometers.

Private developers in Las Vegas have built and are operating a no-fare automated monorail that connects two major resort hotels and parking facilities. There is some thought that this facility could become the core of a larger system reaching the airport.

Reasons to Support Monorails

- *Monorails are nonpolluting, quiet, safe, controllable, and can be automated.*[9]   All these characteristics are approximately the same for any electrically powered transport system on a guideway or rails.

- *They have a limited visual impact.*   Of all the modes that employ overhead structures for support of the guideway, the elevated viaduct of the monorail beam or channel is smallest in cross-section. The monorail structure can be quite slender, but it will not be invisible.

- *A narrow right-of-way is needed.* Because of the limited width of the structure and vehicle, the alignment can be threaded through tight spaces. Frequently, the structure can be incorporated into buildings and constructed through the interiors of city blocks. It can be placed along street alignments, with the row of columns occupying surface areas of only a few square feet at extended intervals.

[9] See the home page of the Monorail Society.

- *Vehicles are not likely to derail.* It is practically impossible for the vehicle to leave the beam or channel, although other mechanical problems are not precluded. Suspended monorails claim to be weatherproof because rain and snow cannot enter the guideway channel, i.e., the longitudinal slot on the bottom. (High winds are another story.)

Service line penetrating the atrium of a Disney hotel.

- *Monorails carry a special image.* As previously discussed, monorails are associated in the public mind with technological advancement and visionary concepts. This may be a considerable positive force, possibly generating considerable public and civic support for implementation.

Reasons to Exercise Caution

- *Switching is cumbersome.* While vehicles can certainly be switched from one line to another, an entire section of the supporting beam has to be moved to accomplish each maneuver. This is a mechanical effort that is more intricate than, for example, moving two slender steel rails a couple of inches at one end.

- *Monorails can only operate in an elevated configuration.* The lines cannot be placed on the surface, because cross traffic cannot be accommodated on the same level. On long uninterrupted sections, the construction of the beam on the surface would be much more expensive than laying standard railroad track. If underground placement were to be considered, the tunnels would have to be considerably higher than for regular rail to accommodate the supporting mechanisms above or below the vehicle.

The elevated configuration allows minimally disruptive and fast construction, particularly with prefabricated assembly, and it precludes the collision of surface vehicles with the overhead transit cars, but it cannot necessarily avoid structural columns.

- *The vehicles are likely to be more expensive than regular rail cars.* While the passenger compartments can be identical to those of any other rail car, the suspension or straddling mechanisms are more complex than regular bogies or trucks under standard rail cars. All other things being equal, because of their rarity, each monorail manufacturing contract would be a special order, with a premium price. Experience in building and maintaining these vehicles is certainly limited.

- *Evacuation of a stalled or disabled train is a problem.* Since the slender beam or channel does not provide for any walkway, the safe accommodation of passengers along the elevated structure under emergency conditions will require special arrangements and catwalks.

Application Scenarios

Monorail systems can provide rapid transit service to cities and districts equal to other rail modes at the scale of full-size metros or as local connectors with small cabins. However, because this mode has a number of special features and characteristics, its suitability is not universal, and it responds best to the needs of certain situations, as compared to other modes. Generally, the selection of monorail technology would be driven primarily by a desire to achieve a high-profile civic image, accepting in the process some premium costs in implementation and operation. It would also have to be a place where elevated service structures are not seen as visual intrusions and where surface space constraints have to be overcome. In most cases, with full two-way operations, two largely parallel guideways would have to be built. If both beams were accommodated by the same support structure, the visual appearance, looking from the side, would not be too different from or less intrusive than that of any other elevated guideway with a deck on columns.

The ideal situation, however, would be the following:

- The service is needed along a *single corridor,* with no branches.

- The area is generally *built up,* with alignment space available only along existing street rights-of-way, but not on the surface because of existing overloads. Easements through private properties could possibly be utilized.

- Within this environment, *elevated structures* are visually and aesthetically acceptable to the public.

- Subsoil conditions or groundwater levels *preclude tunneling.*

Components of the Monorail System

Most elements of a monorail system (leaving aside the small operations in fairgrounds, recreation areas, and other controlled environments) are or can be identical to those of regular rail transit.[10] This encompasses power delivery and types of motors, signals or in-cab operational controls, the interiors of the passenger compartments, and user spaces in and around stations.

There are, however, significant differences with the following elements:

Guideway

For the straddling monorail, the longitudinal structural member has to provide a flat but relatively narrow surface for the main support (rubber tire) wheels in the center and side surfaces for the horizontal lateral wheels to keep vehicles upright; it must also have sufficient structural strength to span from one column to another (Fig. 12.1). It accommodates power, communication, and control lines as well, and it can be curved. The cross-sectional shape can be a vertical rectangle, an I, a T, or an inverted T. The guideway can be made of steel plates or structural shapes or as a

[10] The technical features and specifications are different for each manufacturer, and new systems can be designed according to specifications—thus, generalizations about sizes and physical arrangements have to be rather broad and approximate. Besides information that can be obtained directly from manufacturers, the Monorail Society Web site home page provides extensive technical summaries.

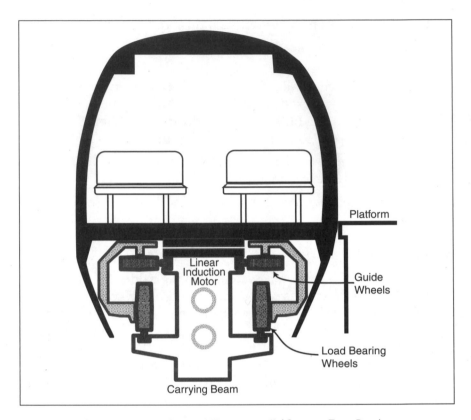

Figure 12.1   Support system of a straddling monorail. (*Source:* TransPort.)

truss, or it can be in the form of a precast concrete beam or a hollow box girder. An example of the latter is the Seattle (Alweg) system (Fig. 12.2), which utilizes 3- × 5-foot girders, spanning up to 100 ft (approximately 0.9 × 1.5 m, up to 30 m). The Disney system (Bombardier) relies on 26-in- (0.66-m)-wide prestressed girders that have a variable depth ranging from 48 to 80 in (1.22 to 2.03 m), for spans of up to 110 ft (33.5 m). The Hitachi systems have 33.5- × 59-in (0.85- × 1.5-m) solid beams. Any number of other, usually smaller, variations can be found.

For the suspended monorail, a channel has to provide support for a set of wheels in the form of a bogey or truck, from which the passenger compartments are hung (Fig. 12.3). In principle, the suspension arrangements are the same as those used for sliding doors in apartments, with the slot either on the bottom or along one side, or the guideway can be in the shape of an I beam.

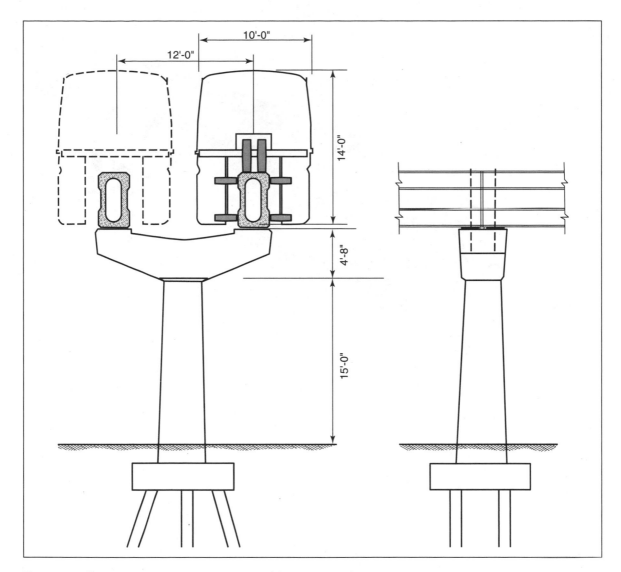

Figure 12.2 Typical straddling monorail structure. (*Source:* Alweg.)

Switching vehicles from one track to another requires that the entire beam or channel be moved to make the proper connection. This can be accomplished either by sliding laterally an entire section with a train on it or by "bending" the beam, which is either flexible or consists of pivoted segments. This operation takes anywhere from 7 seconds to a minute.

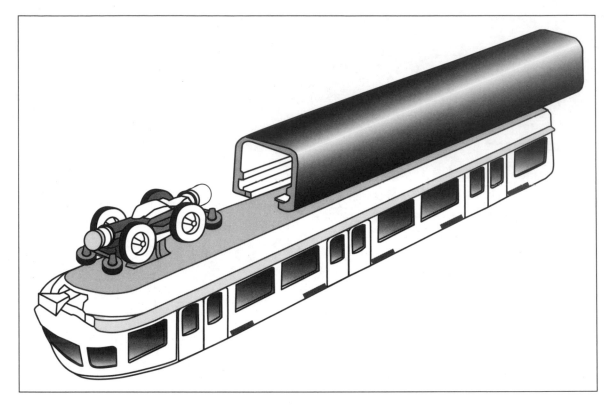

Figure 12.3   Suspended monorail. (*Source:* Safege.)

Stations

Since all monorail systems are elevated, the station design does not offer much room for variation. There may or may not be mezzanines in the intermediary level, and the layout may utilize side platforms or center platforms. The platform floor will have to be at the same level as the car floors. Escalators and elevators (not just for people with handicaps) should be provided. (See discussions of stations in the chapters on rail modes.)

Yard

As with any other transit system, monorail operations have to be supported by protected space that can accommodate the safe storage of cars and maintenance and repair facilities. The latter is a crucial consideration, since—unlike rail equipment—the monorail vehicles and guideway elements will most likely be unique in any given community, the vehicles cannot travel off-site, and all servicing will have to be accomplished at this facility. Another built-in complication will be the need for extensive switching in the yard and the constraints of moving personnel and other equipment across the parallel beams constituting the yard.

Rolling Stock

Every manufacturer offers a different vehicle, with a great variety of configurations and capacities. Undoubtedly, the manufacturers will be happy to design and provide cars and trains according to any reasonable specifications (for a price). In most cases, and again leaving aside the small people-mover systems, the passenger compartment will be quite similar to those of regular light or heavy rail transit cars. The principal difference is whether the placement of support and load-carrying wheels allow passage from one car to another and whether they visibly intrude on the compartment. The cars tend to have extremely streamlined front and back ends, largely to create an image of hypermodernity, and thus remain coupled in fixed train sets.

The standard Alweg train in Seattle, for example, consists of two vehicles, each with two 30-ft (9-m) sections, for a total length of 122 ft (37 m). It is 10.25 ft wide and 14 ft high (3.1 × 4.3 m). The train has seats for 124 passengers and can accommodate 326 standees, for a total of 450 patrons.

The Disney World Bombardier train, on the other hand, consists of six separate cars with a total length of 203.5 ft (62 m). It carries 360 passengers, of whom 120 are seated.

The Hitachi four-car train, which is comparable in its dimensions to the two previous examples, is designed for a larger passenger load and can accommodate 632 riders.

Capacity and Cost Considerations

Since it is possible to operate monorail trains at 2-minute intervals (on straight sections with no switching), utilizing already developed and tested technology, a carrying capacity of 20,000 passengers per hour in a single direction can be achieved. This exceeds the normal throughput assumed for light rail operations—which is not surprising, given the difference in running transit vehicles largely on street level versus on an elevated structure. Since it is entirely possible to design and build larger monorail cars and assemble them in longer trains, the capacities of rail rapid transit are approachable.

Suspended monorail linking a recreational area to rapid transit in Tokyo.

Likewise, it is easily possible to design systems with much lower throughput capacities—down to a few thousand per hour—by selecting smaller vehicles and/or running them at longer intervals. Experience with a great range of smaller systems can be found among the many operations in numerous recreation and leisure venues.

Cost comparisons of monorail versus conventional rail systems are not particularly meaningful. Monorails will certainly be less expensive than subways and more costly than light rail. The principal point is that they are elevated structures, with the same foundations, support columns, and other structural elements as any other such construction. The only appreciable difference may be that the longitudinal monorail girders *may* be marginally less expensive to produce and easier to construct than a deck carrying rail track—but not necessarily.

There are quite likely to be considerable savings in acquiring the necessary right-of-way, because relatively little space is needed, certainly not on the overloaded surface. Vehicles, as previously observed, will probably carry a premium price. Most of the system elements, particularly yards and maintenance facilities, will have to be special and exclusive, not to be shared with any other operations.

With these considerations in mind, some of the more recent projects have recorded capital costs around $160 million per mile ($100 million per kilometer), but with a rather wide amplitude. Bids for exploratory systems in American cities (Houston and Honolulu) were about $65 million per mile ($40 million per kilometer) a few years ago.

Possible Action Programs and Recent Experience

Monorail systems are currently under construction in Kuala Lumpur, Malaysia; Tokyo Disneyland; and Naha, Okinawa—all utilizing Hitachi technology. The Siemens H-Bahn system is also being constructed at the Düsseldorf airport in Germany. There are no active monorail projects in the United States, but there is heated debate in Seattle and Las Vegas about the possible extension of this mode. There is also a lingering proposal to connect northern Kentucky to Cincinnati across the Ohio River by utilizing a long-span suspension bridge–like structure carrying a Safege/Aerorail service. This is not a long list. (See Table 12.1 for a list of monorails in service.)

Since monorails are known to every transportation planner, the extant examples show that they are quite capable of providing service, and the general public regards them as the most advanced technology available, they have been mentioned as a possible alternative in just about every recent large-scale transportation and transit study conducted for American communities. The sponsoring agencies, specifically the U.S. Department of Transportation, have never suggested that monorails should a priori be given an inferior standing in the alternatives analysis process. Yet, almost without exception, they have not stood the test of comparison to other modes, particularly light rail lately.

Such an evaluation in Salt Lake City in the early 1990s concluded that "it would not be physically or politically feasible to put an elevated rail system in the downtown Salt Lake area. Monorail is best suited where the guideway has to be grade separated from all other traffic crossings, where low-cost right-of-way is unavailable and for intermediate capacity applications."[11] It was said that:

- The visual impact of elevated structures in general can be controversial.
- Monorails do not encourage activity at ground level.
- Emergency evacuation is more difficult for elevated rail structures than for at-grade applications.

Aside from the aforementioned suggestions by some groups that the small private monorail service now operating between several resort hotels in Las Vegas be extended into a citywide network for the convenience of visitors, the current situation in Seattle is of considerable interest. It brings to the forefront in a public forum all the considerations as to what an urban transit service should offer.

[11] See *I-15/State Street Corridor Alternatives Analysis and Environmental Study* (Utah Transit Authority, 1994), p. 10.

Table 12.1 Monorails in Urban Service, 2000

City	Name	Type	Manufacturer	Year Opened	Length, mi (km)	Stations	Daily Ridership	Cost, $/km	Remarks
Wuppertal, Germany	Schwebebahn	Suspended	Langen	1901	8.3 (13.3)	18	70,000		Built mostly over a river. Regular public transit.
Disneyland, CA		Straddling	Alweg	1959/61	2.3 (3.7)	2	100,000–200,000		Reduced scale; part of amusement park operation.
Seattle, WA		Straddling	Alweg	1962	1.2 (1.9)	2	2.5 million per year		CBD to World's Fair site.
Inuyama, Japan		Straddling	Alweg/Hitachi	1962	0.7 (1.1)	3			Rail station to recreation area.
Tokyo-Haneda		Straddling	Alweg/Hitachi	1964	10.5 (16.9)	10	200,000	15 million	Regional rail station to airport.
Mukogaoka, Japan		Straddling	Kawasaki/Nihon-Lockheed	1965	0.7 (1.1)	2			Rail station to amusement park.
Shonan, Japan		Suspended	Safege	1970	4.1 (6.6)	8	30,000		Rail station to coastal area.
Disney World, FL		Straddling	Bombardier	1971/82	14.7 (23.6)	6	150,000+		Connector between attractions.
Dortmund, Germany	H-Bahn	Suspended AGT	Siemens	1984	1.5 (2.4)	4	6,000		University service with link to metro.
Kitakyushu, Japan	City Monorail	Straddling	Hitachi	1985	5.5 (8.8)	13	60,000	62 million	Regular urban service.
Chiba, Japan	Townliner	Suspended	Safege/Mitsubishi	1988	9.6 (15.5)	18	40,000		Main railroad station to suburbs; expansion plans.

Sydney, Australia	Metro Monorail	Straddling	Von Roll	1988	2.2 (3.6)	8	30,000		Downtown to Darling Harbour loop.
Osaka, Japan*		Straddling	Hitachi	1990	4.1 (6.6)	16	50,000		Peripheral connector; expansion plans.
Tampa, FL	Airport	Straddling	Bombardier M-III	1991	0.6 (1.0)	8		12 million	Terminals to parking lots.
Newark, NJ	Airport	Straddling	Von Roll/Adtranz	1995	3.9 (6.3)	7			Terminals to parking lots; extension to rail station.
Las Vegas, NV	MGM & Bally Resorts	Straddling	Bombardier	1995	0.7 (1.2)	2	20,000	25 million	Privately built; expansion plans.
Rio de Janeiro, Brazil	Barra Shopping Center	Straddling	Intamin	1996	1.0 (1.6)	3			Parking lot to shopping center.
Jacksonville, FL	Skyline Express	Straddling AGT	Bombardier M-III	1997	4.3 (6.9)	6			Downtown service; plans for expansion.
Tama, Japan		Straddling	Hitachi	1998	10.0 (16.0)	19	43,000	144 million	East of Tokyo, with intermodal connections.

Source: Monorail Society home page; *Jane's Urban Transport Systems* (Jane's Information Group, 2001, 2002).

* Osaka has a vigorous monorail system expansion program, and its total length is continuously changing. Final length of 15 mi (24 km) is planned.

Monorail shuttle to the World's Fair site in Seattle.

Given Seattle's long and successful experience with the monorail to the fair site, a significant level of public support for this mode has been established. For this reason, and recognizing that the only other transit modes known locally are buses and trolleybuses, the citizens of Seattle approved a referendum in 1997 calling for the staged construction of a full 40-mi (65-km) monorail network, with an estimated eventual cost at the time of $1 billion.

One of the first considerations was that prior decisions about the technology and the mode are not acceptable if federal assistance is to be considered for technical studies and for actual construction—the selection has to emerge on its own merits in each specific case as a cost-effective choice among all other possibilities. (Similar concerns may also exist in Las Vegas, Birmingham, Fresno, and other places that are discussing the merits of a *monorail* system.)

The wishes of the electorate were again confirmed in November 2000, when another referendum instructed the municipal government to proceed with a monorail study and identify appropriate funding sources. The majority of the city council has opposed this approach and, on the advice of its technical staff, continues to support a more conventional metropolitan transportation plan relying on light rail and high-occupancy vehicle lane systems.

Conclusion

Monorails remain a possible urban transportation mode, but only in special situations where constraints preclude the use of more conventional technology, and a shuttle service or a loop are appropriate.

Bibliography

Most recent books on urban transportation barely mention monorails. At best, this is done in passing, without providing any technical or service information. Articles in the professional periodicals have been rare since the 1960s. Papers on monorails are included in the ASCE conference proceedings on automated guideway transit, such as Sproule et al. (1997). The only other available sources are manufacturers' literature, which does not particularly stress any possible limitations of the products, and the home page of the Monorail Society, which is very thorough in providing historical and technical information, but always with a strong advocacy orientation.

Botzow, Hermann S.D., Jr.: *Monorails,* Simmons-Boardman Publishing Corp., 1960, 104 pp. One of the early, but rare, full reviews of the monorail technology at that time.

Sproule, William J., et al. (eds.): *Automated People Movers VI: Creative Access for Major Activity Centers,* ASCE, Proceedings of the Sixth International Conference, Las Vegas NE, April 9–12, 1997.

CHAPTER 13

Heavy Rail Transit (Metro)

Background

Metro service—to use the international name derived from the original Metropolitan Line of London (1863)—is the mode capable of doing the greatest amount of useful work, carrying massive volumes of people, fast and efficiently, at the city scale. It is a means of transport that was quite a simple but effective railroad operation when first instituted a little more than 100 years ago, and the improvements in the intervening time have been gradual and largely marginal. It is only in the last few years that significant technical changes have been made, but those modifications do not really affect the basic character of service as experienced by users (except for air conditioning). The recent advances have addressed efficiency and safety issues primarily (power supply, braking, and control systems), moving toward automated operations. If construction of the Second Avenue subway along the East Side of Manhattan indeed gets started by 2004 and is eventually completed, its riders would have no difficulty also using the parallel line opened in 1904, had it remained in its original state. The cars of that time could be run on the new line.

The mode under discussion in this chapter carries many names: *rapid transit, heavy rail transit, subway, metro,* the *tubes* or the *underground,* the *U-Bahn* (*not* the S-Bahn), and the *T.* They all mean the same thing, but we will favor here *heavy rail transit,* which is the more precise technical designation, and *metro,* which is by far the most common label worldwide. The term *metro* has become somewhat generic in popular usage and is unfortunately being applied to many things. Some transit agencies that operate only buses use the designation in their official names.[1]

The term *subway* is largely associated with the New York system and those that have an affinity toward American practices and the New York model in particular. In British English, *subway* means any underground channel, and this meaning is also sometimes used in North America. The Newark Subway, for example, is a long-established streetcar service utilizing an old canal bed, and it runs underground in the center of the city. *Tube* and *underground* are British; *U-Bahn* is uniquely German; and the *T* (*tunnelbana*) is Swedish (and also Bostonian). *Rapid transit,* as is discussed elsewhere, can also consist of buses on an exclusive busway. (See App. A, Rail Transit Definitions.)

Thus, *metro* is defined as a passenger transport mode for urban use that is single-mindedly designed and built to operate rapidly and safely while carrying large loads of patrons in multicar trains. It is usually *rail-based* (steel wheels on steel rails), although some systems employ rubber tires; the rights-of-way are absolutely *exclusive,* precluding anything or anybody (except maintenance personnel) from entering the tracks; *high platforms* and *wide doors* are used to achieve safety and quick boarding and exiting; and the cars are *self-propelled* with *electric motors* to gain rapid acceleration and deceleration and to avoid air pollution in tunnels.

If any of the preceding characteristics are missing or are modified for any given system, there is a question as to whether it is true heavy rail transit. Since each metro system is somewhat different in order to respond to local needs and situations, judgment sometimes has to be applied to determine precise classification.[2]

---

[1] This also sometimes confuses the otherwise authoritative Jane's Information Group Inc.

[2] The Buffalo, New York (1985), system, for example, which is largely in tunnels but runs on the surface in the downtown section, is often placed in the metro class. However, its official designation is "light rail rapid transit," and it will be kept under LRT in this discussion. Likewise, Vancouver's SkyTrain and the systems in Lille and Toulouse, France, are large in size, but are actually automated guideway transit (AGT). There are a number of instances in which the differences between a metro service and regular commuter rail operations are not clear at all. This situation is discussed with respect to San Francisco's BART and Philadelphia's Lindenwold line in Chap. 14, but there are others as well. The JFK Airport AirTrain is another case.

This is not a serious problem, unless there is an overwhelming desire to maintain an orderly discussion structure.

Some systems have made compromises with respect to the previously listed features—usually to save resources—and a series of descriptive labels has been coined for such hybrid approaches: premetro, light metro, *Stadtbahn* (in Germany), and *sneltram* (in Holland).[3] Basically, light rail–type vehicles are usually used in these systems, but the lines are placed in tunnels as much as possible, certainly within the central districts. Platforms are usually low, but provisions are made for the upgrading of facilities.

There is no doubt that the principal asset of heavy rail is its unequalled ability to carry masses of passengers—40,000 and more per track per hour. The technology and operational procedures have been deliberately shaped over the years to achieve this capability—capacious trains that stop frequently but still move fast between points. Unfortunately, these benefits come at a considerable cost. That has been the case from the very beginning, and the situation has become more intense. The latest construction effort in Los Angeles has required capital expenditures exceeding $300 million per mile ($186 million per kilometer). Since any meaningful service cannot be just a mile or two in length, the total bill for any metro in any city in any country will amount to a considerable—and often enough unaffordable—sum.

The future of metro expansion and development will depend on the answers to a series of questions. Are there cities that badly need transit services with this magnitude of carrying capacity? There certainly are many in the developing world (and there will be more), but the answer is not too clearly affirmative in the countries of the industrialized world. Are there specific corridors in metropolitan areas that cry out for such service? Sure, but most of the obvious opportunities are already provided for in the West, and continuously decreasing overall development densities do not bode well for high-volume service in many places. Who can afford the price tag? Probably not any given city by itself, and external or national resources would have to be marshaled for any specific project. Should we depend religiously on cost-effectiveness calculations that appear to dominate decision making today, balancing the number of new riders on a system when it is first opened against the expended construction dollars? Fine,

---

[3] In Brussels, Frankfurt, Cologne, Amsterdam, Rotterdam, and Antwerp.

but let us also include some vision and some long-range expectations. Is there sufficient civic and political will and imagination to make a major investment in the basic citywide infrastructure for the future? Let us hope so.

As shown in Table 13.1, there have been periods of intensive metro building in major cities, and there has been a shift in the concentration of activity—from North America and Western Europe to the former Soviet Union, and now to Asia and Latin America.

One last cautionary note is in order before undertaking a detailed review of the heavy rail mode. While any rational person can, with some thought, define needs, visualize impacts, and make reasonable suggestions regarding some modes of transport services, decisions about metro operations have to be approached with considerably more care. The technical elements are complicated, and the consequences of various choices are not immediately obvious. This requires technical expertise—presented, of

Table 13.1   Summary of Global Metro System Development

	Time								
Region	**1910**	**Late 1930s**	**1970**		**1987**		**2000**	
United States and Canada	4	4	7	(2)	12	(2)	13	(1)
Latin America		1	2	(3)	9	(3)	11*	(4)
Europe (Western and Central)	5	9	18	(7)	28	(3)	33	(4)
USSR/newly independent states		1	5	(4)	11	(5)	14	(5)
Asia		2	4	(3)	15	(5)	25†	(8)
Africa						(1)	1	(2)
Australia							1‡	
Total	9	17	36	(19)	75	(19)	98	(24)

Source: Jane's Urban Transport Systems, various editions; V. Vuchic, *Urban Public Transportation* (Prentice-Hall, 1981).

Note: Numbers indicate systems in operation. Numbers in parentheses for 1970 onward indicate systems in design or under construction.

* Six of these are in Brazil.

† Nine of these are in Japan.

‡ The Sydney metro is an extension of the regular rail network.

course, in a format the public can understand. Above all, metro construction will involve the expenditure of massive amounts of public resources, and that responsibility cannot be taken lightly.

Development History

In the second half of the nineteenth century it became urgent to create high-volume means of mass transportation, because the large cities of the industrializing world were choking themselves with an overload of carts, wagons, carriages, and people on the streets. Any progress in production required efficient movement of goods, workers, and consumers, but complete immobility threatened instead. Public services along surface streets had begun earlier, but they had modest capabilities,[4] and they were not able to cope with the explosive growth and the intensity of demand as the major cities became production centers. Origin and destination points of daily travel moved farther apart, beyond the range of human legs and horse carts.

It did not require an engineering genius to identify a solution beyond the limited range, capacity, and speed of the horse— steam power had to be brought into the dense city to move wagons carrying people.[5] Since the early steam engines were clumsy, cumbersome, and unsafe devices, they could be kept in a protected building, pulling a cable to which cars were attached. This eventually proved to be an ineffective solution. Various unusual power sources—such as compressed air, stored steam, wound-up springs, and flywheels—were brought into play, but they quickly were discarded as being wasteful or simply unworkable.

---

[4] See Chaps. 8 and 11.

[5] Sources on the history of rapid transit are many and varied. Any urban history, general or specific, for cities with metro systems provides some coverage of this subject. Discussions with an urban transportation perspective are found in various chapters of G. E. Gray and L. A. Hoel, *Public Transportation* (Prentice Hall, 1992, 750 pp.) and V. Vuchic, *Urban Public Transportation* (Prentice Hall, 1981, 673 pp.). See in particular B. J. Cudahy, *A Century of Service: The Story of Public Transportation in North America* (American Public Transit Association, 1982, 80 pp.); B. Bobrick, *Labyrinths of Iron: A History of the World's Subways* (Newsweek Books, 1981, 352 pp.); C. W. Cheape, *Moving the Masses: Urban Public Transportation in New York, Boston, and Philadelphia, 1880–1912* (Harvard University Press, 1980, 285 pp.), and various editions of American Public Transportation Association (APTA), *Public Transportation Fact Book* and *Transit Fact Book*.

Since there was no way to accommodate steam locomotives directly on city streets, nothing else could be done but to place them on elevated structures atop wide enough streets, pulling passenger coaches. This was not a technological breakthrough; it was simply a logical application of available hardware elements to a new problem. Thus, rapid transit was born, and, since much experience had been accumulated in running trains by the 1860s, there were no serious difficulties in operating the service. It was not so comfortable for the riders, however; and everybody on the sidewalks certainly suffered from falling ash and hot cinders—soot and lack of sunlight established a gloomy streetscape below. Noise and horse manure added to the urban misery within this environment. But people had to be moved, in large masses every day.

The first elevated line (*el* or *L*) was built along Greenwich Street in New York City between 1867 and 1870,[6] and this private project can be regarded as the original rapid transit effort in the New World. The route was extended along Ninth Avenue soon afterwards, and gradually a whole network of els provided service in the center of the city (built in the 1870s and 1880s). As underground subways were constructed some decades later, the old els were progressively removed, but many of the early subway lines were also placed on elevated structures on their outer ends, in the undeveloped sections of the city. They are still there above many miles of Brooklyn, Queens, and the Bronx.

Chicago, too, opted for the early elevated railway mode, and construction of a network operated with steam locomotives began in 1892. This city, however, was the first to introduce self-propelled, electric multiple-unit rolling stock at the earliest opportunity, without going through a stage of using electric locomotives.

Painting by John Sloan of a New York elevated line. (*Six O'Clock, Winter;* courtesy of the Phillips Collection, Washington, D.C.)

[6] See "Elevated Railways" entry in the *Encyclopedia of New York City,* (K. Jackson, ed., Yale University Press, 1995).

It is interesting to note that European cities, even those with the greatest transport demands, did not accept the intrusion of elevated structures into the urban environment, at least not in the high-density centers. London took the heroic first step in 1863 by building the Metropolitan Line below street level while steam locomotives still reigned. This was a shallow-draft project, left as open as possible for ventilation, but for years afterward the smoke below provided much copy material for journalists and writers concerned with the city's ills. An article in the London *Times* expressed these apprehensions as follows:

> Who of his own accord would choose to travel in tunnels buried beyond the reach of light or life in passages inhabited by rats, soaked with sewer drippings and poisoned by the escape of gas mains. . . .

The original 3.75-mi (6-km) route connected two major railroad stations, and the alignment is still there, incorporated into one of the largest mass transit systems in the world, which grew decade after decade. The London lines were built either near the surface with openings on top or, when electric power became available, in deep tunnels or tubes (thence the name) of a rather small diameter (12.5 ft; 3.8 m) that could penetrate any dense district, fitting within the confines of even narrow streets. In the early days, electric locomotives powered from a third rail pulled the cars.[7]

Other metro efforts followed in Europe, preceding similar systems in the United States. These were built in Liverpool in 1886,[8] Budapest in 1896, Glasgow in 1897 (distinguished by the smallest diameter tubes ever, the cars being practically miniatures), Paris in 1900, and Berlin in 1902 (which defined the *Untergrund* or *U-Bahn* concept).

There were no admirers of the New York and Chicago els, except a few graphic artists depicting the contemporary urban scene; patrons, nearby residents, landlords, and people walking below complained all the time. A tunnel solution was long awaited, but there were some initial obstacles. Electric traction had to be developed and tunneling methods had to be perfected, particularly since owners of adjoining property feared damage

---

[7] A protected and insulated rail parallel to the track, somewhat elevated, along which slides a "shoe" to feed power to the vehicle.

[8] Actually a steam railroad tunnel under the River Mersey that was later converted to transit operations.

General W. B. Parsons at the groundbreaking for the first New York City subway. (Courtesy of Parsons Brinckerhoff.)

due to the tight clearances. The riding public was apprehensive about being subjected to unprecedented speeds in an "unnatural" underground environment. A major delay, however, was caused by a typical New York problem of the time: the presence of the notorious Tweed Ring and railroad and real estate barons who tried to protect lucrative monopolies associated with their elevated railway franchises.[9] In 1891, however, a Rapid Transit Commission was organized to decide on future plans, and it selected the tunneling approach. Work started in 1900, and the first 9.1-mi (14.5-km) line was opened in 1904. It was accomplished as a private project, supported by municipal backing and guarantees, with William B. Parsons as the chief engineer and John A. McDonald as the builder. The principal mover behind the early subway efforts in New York, however, was August Belmont as the financier and manager. As president of the growing transit enterprise, he even had his own lavishly appointed private subway car, like any other railroad president of the time.

The line started at City Hall, ran northward along Manhattan's East Side, crossed westward along 42nd Street, and continued north on the West Side. It provided local and express service from the very beginning, which set the pattern for the rest of the network as it grew. This still remains a rare arrangement among metro systems.

In the meantime, Boston had constructed the first transit tunnel by 1897, which accommodated streetcar lines in the very center at Tremont Street. Boston's Green Line can be regarded as the original light rail transit (LRT) operation, but additions to the system are of the heavy rail type, with the first such tunnel section

[9] Every history of New York in the nineteenth century deals with this situation; see in particular the chapter "Manhattan, Inc." in E. G. Burrows and M. Wallace, *Gotham* (Oxford University Press, 1999, 1383 pp.).

opened in 1908. Philadelphia also built a central tunnel under Market Street in 1907, which accommodated streetcar and rapid transit operations, with the latter placed on elevated structures further out.

The first two decades of the twentieth century saw much construction of metro lines, not so much in opening new systems (except for Buenos Aires and Hamburg) as in building additions to existing networks. The large heavy rail systems—in New York[10], London, Paris, and Berlin—took the basic shapes that we see today. The 1920s and 1930s suffered from severe economic problems, but rapid transit work continued. The first systems appeared in Asia (Tokyo and Osaka), and Moscow entered the field in 1935. The latter was built with deliberately ornate stations under harsh working conditions in order to present an image to the population of what the Socialist political system could achieve:[11] "This is how everything will look if you just be patient for a while." Moscow was the most dramatic, but not the last, example of a metro seen as an element of national and civic pride. In any case, the Moscow system was the last one opened before the start of World War II, and its stations served as air raid shelters and military command posts. So did the deep stations in Germany and Great Britain, and a memory of that experience can still be seen in some places as semihidden blast doors that can seal off passenger spaces.

In the meantime, New York had built the Independent system as a municipal project, in contrast to the private efforts that had earlier created the other two systems—the Interborough Rapid Transit Company (IRT) and the Brooklyn–Manhattan Transit Corporation (BMT)—under the so-called dual contracts, between 1913 and 1931. This basically concluded the network

---

[10] The history of New York's rapid transit is particularly well documented in such thorough works as *Interborough Rapid Transit, New York Subway: Its Construction and Equipment* (IRT, 1904, and facsimile edition by Arno Press, 154 pp.); J. B. Walker, *Fifty Years of Rapid Transit: 1864–1917* (Arno Press, reprint edition 1970, 291 pp.); J. B. Freeman, *In Transit* (Oxford University Press, 1989, 434 pp.); C. Hood, *722 Miles: The Building of Subway and How They Transformed New York* (Simon and Schuster, 1993, 336 pp.); S. Fischler, *The Subway* (H&M Productions II Inc., 1997, 256 pp.); P. Derrick, *Tunneling to the Future* (New York University Press, 2001, 442 pp.); and many others of a more popular nature.

[11] S. Grava, "The Metro of Moscow," *Traffic Quarterly,* April 1976, pp. 241–267.

development, and, except for some short additions and changes, it is what we have today. It is not really a coherent and balanced *system,* because the three enterprises saw themselves in competition for access to the highest-demand districts and had little interest in servicing lower-density areas. Fortunately, there has been enough demand to fill just about all routes, but the streamlining and rationalization of the network is not yet accomplished. Cars designed for some of the original networks cannot be operated on the others. (However, things are even more complicated in Boston, where each of the four lines have special rolling stock not compatible with the others. In Philadelphia, the two lines were built with different track gauges.)

Another issue that affects us today in New York, as well as in other American cities, is that subways were seen as purely city-level services under a single municipal government. Thus, the old lines did not extend across the boundaries of the central city, and only decades later, when actual urban development had spread across the metropolitan landscape, did system development take on a larger dimension. Today agreements and cooperative arrangements are negotiated among counties and other local political entities as partners in regional service networks, usually under regional transportation authorities.

The early period saw relatively little change in technology or service quality. In the United States, public concerns about urban transportation were largely directed toward maintenance of a politically acceptable low fare and the management of franchises by private companies that served the public at the municipal level. None of these were events that would make the services more attractive or necessarily encourage higher ridership, yet—as shown by Fig. 13.1 later in this section—patronage levels continued to grow in the United States, even with the severe impact of the Depression, reaching record volumes during World War II.

Basic rail technology was employed, with third-rail power pickup and wayside signals. The operations were robust—not a bad characteristic—but not particularly sophisticated or comfort oriented either. While there were 11 systems in operation worldwide at the time of World War I, the total inventory had grown to 17 by the late 1930s. No new systems, however, were opened in North America during the 1920s and 1930s. The high costs of construction had become a real concern, particularly in the United States, with its low urban densities and the powerful emergence of the private automobile—a condition which also started to affect decision making by municipalities and public agencies.

After World War II and a period of recovery, rapid transit construction resumed with considerable vigor, particularly in Western Europe and the Soviet Union. During the 1950s and 1960s, new systems opened in Brussels, Frankfurt, Rome, Rotterdam, Oslo, Lisbon, and Stockholm, and other projects were initiated. These large and dense cities really had no choice but to build heavy rapid rail to maintain their viability in the new urban situation. Stockholm's plan of dispersed satellite new towns depended on the availability of rapid transit service.

The USSR adopted a national policy of equipping any city that reached the 1 million population mark with a metro. This was practiced religiously; while some of the efforts consisted of only one line, starting with Leningrad in 1955, by the end of that political regime, 14 systems had been opened. Other projects were under way in the late 1980s, one of which was stopped due to local opposition (Riga);[12] others dropped out or became delayed due to resource shortages, but one was completed a few years ago.

In the New World, work also resumed in the early period after World War II. Cleveland built a rather conventional heavy rail connection to the airport (1955); Toronto (1954) and Montreal (1966), on the other hand, embarked on extensive metro development programs that now place them in the top ranks of rapid transit communities. Mexico City did likewise (1969). The Montreal and Mexico City systems are distinguished not only by the architectural attention devoted to the passenger environment and stations, but also by the use of rubber-tired rolling stock. The latter choice was due to the substantial participation of French engineers in the design, who had developed such equipment earlier for a new line in Paris.

In the United States, the early postwar period showed a grim picture, with ridership dropping precipitously. It appeared that the heavy rail systems in the few old cities could hold on only to the patrons who had no other commuting choices. Something had to be done.

The watershed year was 1965. That was the year that Congress authorized the construction of the Washington, D.C., metro as the most visible expression of confidence in and support for rapid tran-

---

[12] S. Grava, "The Planned Metro of Riga," *Transportation Quarterly,* July 1989, pp. 451–472.

sit nationally. Public concern had coalesced into a widespread demand for action, and actual efforts had begun. The vigorous programs in Canada were most encouraging, national support legislation had just been passed, and construction of a new system propelled by local civic energy and resources was underway in San Francisco–Oakland. All the old systems had expansion or modernization plans, and more than a dozen other cities were examining metro possibilities. U.S. Steel built and displayed an attractive transit car model; General Electric promoted high-density development along rail lines, expecting that the residents would need many household appliances. While the nadir in ridership still lay ahead (the early 1970s), a reversal of trends was accomplished.

A breakthrough in heavy rail planning and design was made in the building of the San Francisco—Bay Area Rapid Transit (BART), which is considered the first modern system of the current period. It was planned during the 1960s and opened in 1972, and it introduced advanced electronics technology throughout the system—in train control, car operations, information systems, maintenance management, etc. There were serious debugging problems with the new and untried equipment at the beginning, with most participants suing each other at the end, but corrections were made and solutions were found. BART set a number of precedents, and it is a point of reference for all subsequent efforts. The network is a regional system extending over several municipalities and counties, which had the option of joining the system or staying out. As a consequence, unfortunately, the system is not geographically balanced; some logical routes are missing in several directions because local communities deemed the associated costs too high for the service that they would obtain. Perhaps the most important feature of BART was the fact that the original system was built without federal assistance. This has not happened again.

The building of the BART system was a bold step—not only had general transit usage in the United States continued

Station in San Francisco's BART system—the first modern rapid transit system.

on a downward course after the peak war years, as previously mentioned; most public transportation agencies were unable to offer fully adequate service by the 1960s. The steps that were taken in this desperate situation basically amounted to two major types of action. The first of these was of an institutional nature: cities were to take over bankrupt operations because service had to be maintained, and—more important—the various small separate operating agencies were to be consolidated into large regional units. BART and MTA in New York are examples of metropolitan agencies with state charters that were expected to apply large-scale efficiencies and managerial overview to the vital task of keeping public services in operation. Today, just about all large American cities are the core units of regional public transportation organizations. Still, while mobility services and facilities may be coordinated—though not always perfectly—at the logical metropolitan level, such capability is lacking in the land use and activity distribution sector. This is how American communities want it, even if short-term inefficiencies and long-term imbalances are frequently quite obvious.

The second action to cope with urban transportation needs was to develop programs of financial assistance, since the cost burdens had become too onerous for any given locality if fares were to be kept at levels most residents could afford. In the large metropolitan areas, the heavy rail networks remained the largest and most crucial components of the overall regional systems, with a quite apparent internal conflict between the needs and capabilities of the old central city and the new suburban communities.

A favorable public attitude toward transit, even if it was not accompanied by actual increases in ridership, was moved forward by growing public understanding of what responsible transportation solutions should be in cities that were visibly suffering under intensifying pollution and surface congestion. Grade-separated transit, with high-volume service capability, was indeed seen as an attractive response for large cities. This public attitude and support became embodied in official policies, programs, and legislation, notably the federal urban mass transit assistance acts of 1964 and 1970, which allowed and promised 80 percent grants toward capital construction costs. After the opening of BART, which undoubtedly burnished the image of rail transit through the extensive media coverage that it received, the United States experienced much metro construction (Table 13.2).

Table 13.2 Heavy Rail Transit Systems in North America, 1999

City and Operating Agency	Open*	Riders, million	Total Length, km	Total Lines	Total Stations	Max. Gradient, %	Min. Curve Radius, m	Power V dc[†]
Atlanta MARTA	1979	78.4 (1997)	62.9	3	36	3	230	750
Baltimore Maryland MTA	1983	12.8 (1998)	23.7	1	12	4		700
Boston MBTA	1908	107.6 (1996)	125	4	84			600
Chicago Chicago Transit Authority	1892	84 (1996)	173	7	140			600
Cleveland Greater Cleveland RTA	1955	6 (1996)	30.7	1	18			600[†]
Los Angeles MTA	1993	12 (1998)	28	1	16			750
Miami Miami-Dade Transit	1984	14 (1997)	33	1	21			700
Montreal STCUM	1966	197 (1997)	65	4	65	6.5	140	750
New York MTA NYC subways	1868	1132 (1997)	371	25	468			625
New York MTA Staten Island Railway	1884	4.8 (1997)	23	1	22			600
New York–Jersey City PATH	1908	67 (1999)	22.2	4	13	4.8	27.4	650
Philadelphia SEPTA	1907	48.2 (1998)	41	4	82	5	32	625
Philadelphia–Lindenwold PATCO	1969	10.7 (1996)	23.3	1	13	5	61	685
San Francisco BART	1972	76 (1998)	153	5	39	4	120	1000
San Juan[‡] DoT & Public Works, PR	2002	30	17.2	1	16			750
Toronto Toronto Transit Commission	1954	142.1 (1996)	56.4	2	61			600
Washington WMATA	1976	194 (1996)	166	5	83	4	213	750

Sources: Jane's Urban Transport Systems, Jane's Information Group, 2001–2002; periodicals, particularly *Mass Transit;* data from agencies Web sites.

Note: ABS = automatic block signaling; ATC = automatic train control; ATO = automatic train operation; ATP = automatic train protection; ATS = automatic train supervision.

Peak Service Interval, min	Daily Hours	Signaling	Total Cars	Car Manufacturers
8	20.5	ATP, ATO	240	Franco-Belge Hitachi/Itoh
8	19	ATO	100	Transit America
4.5	19.5	Driver's control	602	various
3–8	24 and less	Cab signals & ABS	1192	Budd, Boeing, Morrison Knudsen
6	21	Three-aspect lights; cab signals	60	Tokyu
5	19	Central control room	60	Breda
6	18.5	Partial auto. control	136	Transit America
3–5	19.5	ATC, ATO, cab signals	750	
2–10	24	Wayside, train controls	5799	Various
5	24	Light signals	64	St. Louis Car Co.
Frequent	24	Block signals	335	St. Louis Car Co.
2–3	24		411	Budd, Kawasaki, Adtranz
3–4	24	Cab signals	121	Budd, Vickers Canada
2.5–5	20		669	Rohr, Alstom, Amerail
			74	
2.5	20	ABS, wayside	806	Hawker Siddeley, UTDC, Bombardier
3–6	18.5	ATS, ATO, ATP	764	Rohr, Breda

* For New York and Chicago, the opening date refers to the elevated lines serviced by steam locomotives, prior to underground electrical operations.
† Overhead supply for Cleveland; third rail for all others.
‡ Under construction; opening date and ridership estimated.

Single-line systems were built in Baltimore (1983) and Miami (1984); full metropolitan networks were created in Atlanta (1979) and Washington, D.C. (1986). The Washington effort, serving the capital of a global superpower, received generous support from national resources, and it qualifies as an exemplary endeavor that attracts high patronage and is instrumental in shaping development patterns. The subway construction and station location plans were accompanied by constructive land use controls and development-fostering programs. The Miami effort, on the other hand, has received much negative publicity since ridership levels have consistently fallen far below forecasted volumes. A coordinated policy toward all forms of transportation appears to have been missing.

The Los Angeles heavy rail project was begun after much debate by breaking ground for the Red Line in 1986, and the adopted plans envisioned a 400-mi network (including all types of rail modes) that would reverse the trends in this, the quintessential automobile-dominated community. The first phase opened in 1993, but unanticipated construction difficulties and severe cost increases gained much local and national attention during the implementation period. The Los Angeles and Miami events have given much ammunition to opponents of heavy rail transit in North America, and place under a question mark the affordability and viability of capital-intensive systems in low- and medium-density urban environments. A formal decision has been made through a public referendum to halt any further subway building in the Los Angeles region.

At this time, the only active new-start rapid transit project in construction under United States jurisdiction is the *Tren Urbano* in San Juan, Puerto Rico. While many cities have explored the feasibility of the heavy rail transit option over the last few decades (including Honolulu, Dallas, Houston, Denver, Pittsburgh, Orlando, Minneapolis, Seattle, Kansas City, St. Louis, and a few others), none of those concepts has survived, or the decision has been in favor of light rail transit. There are not too many active plans for additions to existing systems, either. Washington continues to build based on its current strengths, but Atlanta has no large future plans; in San Francisco, a major link to the airport is under construction. In New York, a very expensive but crucial line interchange node has been completed in Queens, but an actual go-ahead decision on the Second Avenue subway in Manhattan, which has

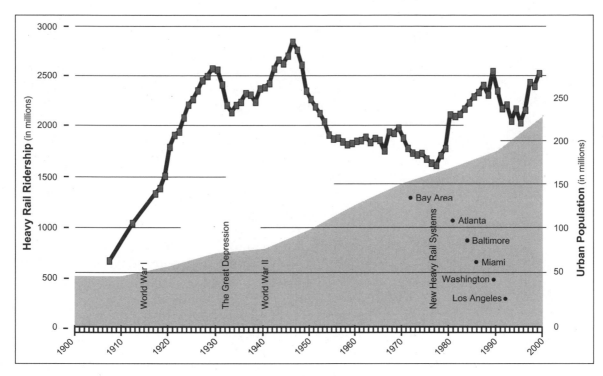

Figure 13.1   Heavy rail ridership in the United States compared to urban population. (*Source:* American Public Transportation Association and U.S. Census data.)

been anticipated for more than half a century,[13] has not yet been made, except that preliminary engineering design is underway.

A close examination of the ridership curve for the last three decades in Fig. 13.1 is instructive. Increases in total national patronage are always associated with new systems coming on line. A few years later, drops are again seen, until a new opening occurs. Clearly, this cannot be repeated endlessly with massive inputs of resources, and new approaches toward gaining and holding rapid transit users will have to be developed.

Some serious obstacles face American communities that may consider the heavy rail choice, besides the dominance of the private automobile and the shortage of corridors with intensive development that would cry out for high-capacity public service. The cost factor looms particularly large. While the federal govern-

---

[13] S. Grava, "Is the Second Avenue Subway Dead on Its Tracks?" *New York Affairs,* no. 3, 1980, pp. 32–41.

ment has changed its approach toward grants from a largely discretionary base to formula allocation, giving each city more freedom to advance its own preferred programs, there are stricter expectations for documenting the feasibility of major projects. For example, it is not just a question of total expected ridership, but of whether the patrons will be *new* transit users—i.e., not simply diverted from existing bus operations. That is difficult to achieve under normal conditions in American communities.

Taxpayers are increasingly concerned about continuing operations and maintenance expenses. It is one thing to assemble the funds to build a system; it is another matter entirely to pay for annual deficits with escalating price tags for personnel, materials, and power year after year. In the 1970s and 1980s, federal and some state operational assistance programs were developed, but they have since been cut back if not completely eliminated. To take long-term advantage of rail transit's community-building and land use distribution-shaping capability remains an uncertain aspiration, because each political jurisdiction within a metropolitan area retains and protects its exclusive responsibility over development controls. To shape an entire region by planning major transport services and activity locations in concert is still a dream in the United States.

In Europe, too, while several second-rank cities are exploring metro possibilities, activity in building new heavy rail systems shows a slowdown. The construction costs are equally high. In the developing world, the needs are greater and much more obvious. Many cities face operational strangulation if metro systems are not developed, or they may break up and spin off activities to dispersed locations. A number of metro projects at various levels of likelihood have been envisioned and even programmed, but, again, the question of cost becomes the critical concern. None of these cities is in a position to sponsor subway build-

Hollywood station in the most recent American metro, Los Angeles.

ing through its own resources; national commitments at the expense of other programs have to be made, and international financial sources need to be engaged.

Figure 13.1 shows the good news that ridership on heavy rail systems, as well as on transit in general, has been on an upswing over the past few years, and volumes keep increasing. We may be back where we were in the early 1990s, when the last downward slide of a rather serious nature started. These are very encouraging signs; perhaps the residents of the larger cities of the United States have recognized the advantages of rail transit and more of them will use it. Much has certainly been done since the 1970s— the low point in utilization rates—to make the services attractive and responsive. Ridership may even get back to the levels of the late 1940s.

Yet, a sense of reality has to be maintained, and an overly optimistic view, unfortunately, cannot be justified. The simple fact is that the country in the early 2000s is not what it was in the 1940s or even the 1980s. Since the end of World War II, the urban population has more than doubled, but transit is struggling to keep what it has within a much larger reservoir of potential customers. Granted, much of this growth has been in low-density suburban districts, but that is not an excuse for stagnant transit use; it is rather the explanation of why no real or proportional progress has been made. More than 60 percent of the heavy rail ridership is carried by the New York subways, and, if that component were to be excluded, the national picture would be dismal indeed. The several new systems that have been opened in the last few decades, ranging from the San Francisco Bay Area to Los Angeles, have undoubtedly made a difference, but they have not turned the situation around, either. There are currently no successor programs in heavy rail construction on the continent.

Types of Heavy Rail Operations

There really is only one type of heavy rail transit, although there are multiple variations regarding the separate elements. Indeed, the physical structure of the networks of all forms of rail transit— heavy, light, or hybrid—follows the same patterns. (See "Components of Heavy Rail Systems" later in this chapter.) The networks take many forms, and each of them has been generated specifically for any given city, usually incrementally over many years, to

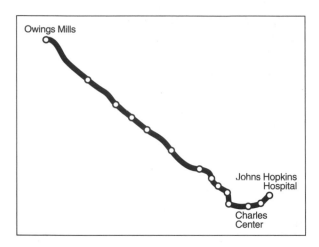

Figure 13.2   Single line (Baltimore).

respond to the needs and capabilities of its service area. Nevertheless, some general structural concepts can be identified.

- *Single line.* The single-line structure characterizes the smaller systems, and there are quite a number of them. In most instances they are seen as the first phase of a larger network, but they have to be *operable*—i.e., be able to provide useful service as they are at any time, even at the beginning.[14] By definition, they constitute the local public transportation spine, with feeder and distributor links directed to the stations. Systems today have to include park-and-ride and kiss-and-ride (drop-off/pickup) automobile access at the noncentral stations.

The line usually runs from one edge of the built-up city through the center to the other edge (Fig. 13.2). The operations are thus very simple—basic shuttle service back and forth. It may branch out at either or both ends, which usually requires grade-separated guideways for each movement. If the line terminates in the center (as in Istanbul, for example), this is in most cases a temporary situation, with the expectation that construction will continue on the other legs.

- *Radial network.* This is the most common type of structure. The system is usually developed step by step over decades by continuing to add lines oriented to the traditional business core. They may all come together at a single node to facilitate transfers (as in Atlanta, for example; see Fig. 13.3) or the intersections may be spread apart within the central district to ensure wider geographic coverage (as in St. Petersburg, Kiev, or Prague, for example, and even in Washington, D.C.).

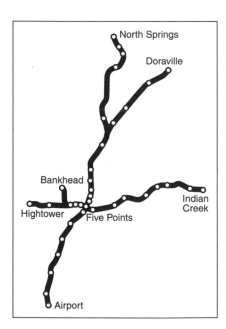

Figure 13.3   Radial network (Atlanta).

[14] These systems are defined in federal programs as *minimum operable segments*—the smallest units that can receive assistance.

All the other feeder and access service needs mentioned previously apply as well, of course.

- *Grid.* A grid system with multiple (approximately) parallel lines crossing each other at many points is a theoretically advantageous pattern since it provides good access to many districts, and efficient transfers can be made without everybody going through the center (Fig. 13.4). This, however, is predicated on the existence of a large network and a distribution of major destination points over a large area. Many of the very large systems can claim to approach or actually have achieved this structure—Paris, Berlin, Madrid, Mexico City, Osaka. Any station may be reached with no more than one transfer.

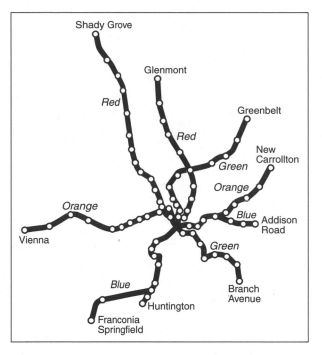

Figure 13.4   Grid (Washington, D.C.).

- *Circle line.* This structure distributes patrons efficiently toward their destinations without causing them to be delayed in the center, serves a large business core with internal linkage, and interconnects long-distance terminals (Fig. 13.5). London, Moscow, Beijing, Chicago, and Glasgow have such an arrangement, and it works well in high-intensity situations.

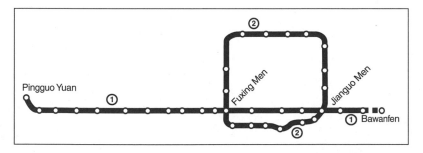

Figure 13.5   Circle line (Beijing).

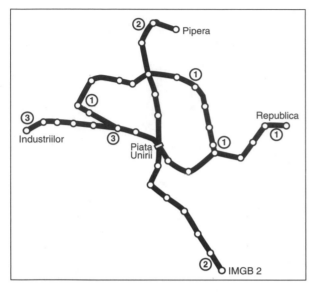

Figure 13.6   Peripheral loops (Bucharest).

- *Peripheral loops.* In this structure, segments of a circle, or lines running on a partial ring around the center and intercepting radial lines, provide a distribution and transfer function, but this layout is not very common. New York, for example, has only one such segment (the G Line), since the overwhelming majority of transit trips are still center-oriented. Such arrangements, or at least partial examples, can be found in Lisbon, Hamburg, Seoul, Singapore, Barcelona, Bucharest, and some other cities (Fig. 13.6).

- *Parallel lines.* This structure is appropriate in instances where major demand is concentrated in a broad corridor, and a single line would become overloaded. This is the case in Caracas, which is basically a linear city, and where a second line is under construction next to an older one (Fig. 13.7). Similar arrangements can be identified in Athens, Milan, Montreal, and Toronto. The north-south spine of Manhattan is also equipped with a series of subway lines, all running in the same direction.

Service for any given rider can be expedited only by not stopping at stations that are not the rider's origin or destination points[15]—which is patently impossible, since each train may carry

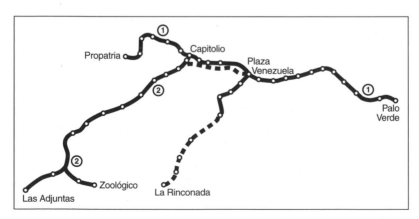

Figure 13.7   Parallel lines (Caracas).

[15] The concept is not quite absurd, and one of the objectives of automated guideway transit is to achieve exactly this type of individualized service on a mass transit mode. As is shown in the following chapter, it has not yet been accomplished with even the most advanced technology.

thousands of patrons. Instead, skip-stop operations may be employed, in which alternating trains do not stop at every second less-busy station, as it is done, for example, in Chicago. The riders, of course, must select the proper train before boarding. Or, trackage may be doubled and local and express trains run in parallel, with the latter stopping only at major stations, as is done on most central lines in New York.

Vestibule between platforms in a Beijing metro station.

While there is no rule against it, heavy rail is not particularly suitable for short trips, primarily because of the effort involved in climbing staircases up and down in stations or spending time riding long escalators. Thus, in high-density areas, subways do not entirely displace regular surface services.

Reasons to Support Heavy Rail

The strengths of the metro stem largely from its ability to capitalize on long-tested, robust railroad technology.

High Capacity and Low Space Utilization

The operations are compact, riders are concentrated into cars, and they move fast in trains consisting of many cars, one after another. No other service can transport as many passengers on a single track or lane during an hour or a day. Basic elements of the service have been honed over many years to achieve this unequalled working capacity. (See "Scheduling and Capacity" later in this chapter.) The high-density operations also ensure that the use of surface urban space will be minimal—the rights-of-way are relatively narrow, or the track does not occupy surface space, being placed above or below grade level. In most cases, no scarce street-level land is preempted at all, except for access to stations.

As shown in Fig. 13.8, there is a reasonably wide range in possible station configuration to respond to various demand and loading situations.

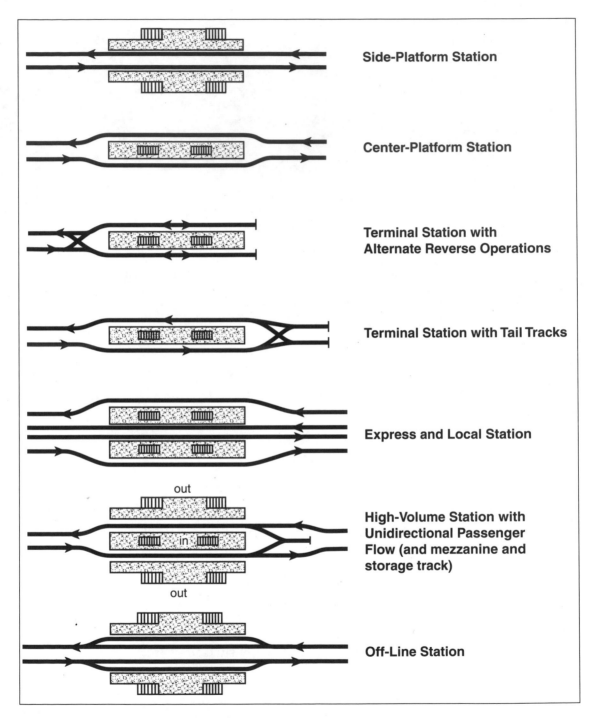

Figure 13.8   Possible basic station configurations.

Efficiency of Urban Patterns

Heavy rail, again more than any other mode, has the ability to influence and shape land use and activity locations. A metro station is a major point of access, enhancing the utility of all properties around it. It makes a difference in the real estate market. Thus, development is usually attracted to the vicinity, and it tends to be high-density commercial development to best take advantage of the transportation service. Building such a transit system is a commitment toward a concentrated urban environment with strong nodes and emphasized corridors. Yet, evidence suggests that that effect is not automatic. This may not happen if other development forces dominate (or concentrating forces are absent), even though the policy intent is clearly to build compact districts.

Significant developments, for example, are not easily identified along the Independent lines constructed in New York in the 1930s, and the stations in the peripheral districts did not attract much beyond local shops. On the other hand, Toronto's metropolitan structure includes very pronounced major building nodes around many stations. Likewise, strong new urban concentrations have been created in Washington, D.C., through effective multidimensional programs. It is a curious and somewhat disturbing fact that edge cities (major retail and office concentrations), vital activity cores in the contemporary suburban landscape, have neither sought nor depended on mass transit access. This may change in the future, at least in some instances.

Avoidance of Surface Congestion

Because of complete grade separation and exclusive rights-of-way, metro operations stay clear of traffic jams on streets and highways, and trains can move unimpeded by the usual urban constraints. Heavy rail has a reasonable chance of competing with the private car in terms of travel time, at least during rush hours in congested corridors. By removing sizable loads from the surface system and placing them on the exclusive channels, transit does have a material effect of reducing overall congestion—as long as the newly freed capacity is not preempted by new motorists taking advantage of the opportunity to move on previously saturated roadways.

A small illustration of the difference in service speed between underground and surface public transit can be seen in the results of a race conducted on November 22, 2001 (a Wednesday, the day before Thanksgiving) in Manhattan.

Travel by subway from Grand Central Terminal (by the Shuttle, with a transfer to the Number 1 line) to Broadway and West 110th Street station took 25 minutes.

Starting at the same time (1:15 P.M.) from Madison Avenue and 43rd Street (one short block from Grand Central Terminal), the M4 bus took exactly 60 minutes to reach the same Broadway and West 110th Street intersection above the subway station.

Mechanical Efficiency and Energy Conservation

A steel wheel rolling on a steel track requires very little energy to maintain motion, since friction is at a minimal level. (Rubber tires on a smooth concrete guideway are also quite efficient.) Thus, the consumption of energy to move weight over distance is as low as it is possible to get with normal technology. Efficiency, however, should not be measured against total weight, but rather against *payload*—i.e., the number and weight of passengers actually carried, not the weight of the vehicle itself. This means that an empty heavy rail car has no efficiency at all, and a lightly loaded one would show poor performance no matter how fast it travels drawing even a limited amount of power.

Speed and Quality of Ride

On an open and well-maintained track, electric trains can achieve speeds in excess of 100 mph (160 kph)[16]—not that this is recommended or that it should be attempted with heavy transit. Nevertheless, 60 mph (97 kph) is a reasonable possibility, provided that there is enough distance between stations to accelerate to that level and then slow down again. Even with repeated stops at stations, the overall running speed can remain in the 20 to 30

[16] This refers to *maximum speed*—the top velocity that can be achieved with a full load after the acceleration phase. There is also the more meaningful *overall running speed* (sometimes called the *platform speed*) that accounts for station stops—the distance between any two points on a line divided by the actual time taken to travel between those points. In describing operational performance of trains a *schedule speed* is also sometimes used, which includes layover and schedule recovery times on round trips.

mph (32 to 48 kph) range. (BART, for example, runs at 70 to 80 mph [112 to 128 kph] on open track and can maintain an overall speed of 30 to 40 mph [48 to 64 kph], thanks to relatively long station intervals.)

Modern cars have well-engineered suspension systems, which, on a good track, results in a smooth ride for the passengers. Some swaying and shaking will be present, but this can be kept to very acceptable levels. Reading and solving the crossword puzzle can be comfortably done, although trying to write much could result in scribbles. Acceleration and deceleration rates are not limited by the power of the motors, but rather by human tolerance thresholds—standees want to maintain their balance without excessive effort, and sitting passengers want to remain upright.[17]

Environmental Quality

Since all metro operations use electrical power, they emit no air pollution. Power generation by burning various fuels may cause some problems, but this will ordinarily be done at a large-scale remote facility where proper control measures can be applied. High-voltage feeder lines will be normally placed underground. Dust and abraded metal particles can be controlled with proper sweeping and cleaning of platforms and track. Noise, except with very badly deteriorated rolling stock and misaligned track, usually remains within acceptable limits.

Safety and Reliability

More so than any other rail-based system, metro service is contained within its own exclusive network of trackage, over which complete control can be maintained. This does not mean that mishaps cannot occur, but the chances are minimized, and operations are not much affected by external forces. Split-second timing and reliability can be maintained, and should be attainable even in old systems. Automation is increasingly introduced, thereby reducing chances for human error and upgrading control reliability.

[17] These rates have to remain below 3.0 to 3.5 mi/h·s (4.8 to 5.6 km/h·s). See John D. Edwards (ed.), *Transportation Planning Handbook* (Institute of Transportation Engineers/Prentice-Hall, 1992), pp. 140–143.

Heavy rail operations should be immune to weather problems, since trains run on rails and frequently under cover, in contrast to surface modes. It takes a major weather event to stop rail service, although heavy snowfall does achieve this from time to time.

Durability

The systems and their elements have to be robust, and they are usually built to last. Anything that is fragile is likely to be quickly damaged and destroyed under the stresses of rapid railroad operations. Therefore, cars with adequate maintenance will stay in service 30 years and more, and tunnels will remain in place forever (barring massive catastrophes, such as the World Trade Center attack).

Record of Experience

Usable experience with heavy rail transit, not to mention railroads in general, has been accumulated over 100 years. A new system can be built without engaging in any research or development at all, and cars can be bought from manufacturers' catalogs. This is not necessarily always done, however, since every metro system has unique demands and has to respond to particular local requirements. But even then accumulated experience with most elements is useful, keeping in mind that any pioneering effort may have to face the considerable tasks of breaking in and debugging new components (as was the case, for example, with BART).

Automation

Because of the "closed" and self-contained nature of metro service, automation in the control and operation of trains and many other tasks is readily possible. Much is currently being done in this direction, and the level of automation is perhaps the prime factor that separates the systems opened before the 1970s and the more recent ones in the most developed countries. Major consequences of this trend are the reduction in total labor requirements, but a growing need for very skilled personnel in maintenance and supervision. There are systems now in operation in which the riders may not see any transit agency people at all, except those who are there for security reasons and to give the riders assurance that human beings are still in control.

Civic Image

There is no doubt that many see a metro system as a true symbol of a city that has reached the top ranks of urbanity, sharing something in common with Paris, London, and New York. A reputable opera or a major-league sports team is not quite equal in generating that perception. This is a most significant factor in local and even national politics, as a part of the decision-making process regarding public transportation (within reason). It has perhaps

Elaborate decorations in an early Moscow metro station.

led to a few misguided efforts, but by and large it is to be applauded. Under the proper leadership, it can lead to long-range upgrading of the urban structure, since capital infrastructure development is otherwise likely to languish within the time horizon of elected officials who are subject to term limitations.

The former Soviet Union deemed it appropriate to use national resources to build a metro for every city that reached a population of 1 million, which may not have been a cost-effective action then or now, but generated much favorable internal and external publicity. The implementation of the Washington, D.C., system, which is a successful and entirely defensible program in its own right, was expedited by the fact that the national capital was the showcase. A number of cities in the developing world, badly in need of high-capacity transit, have benefited and will benefit from nationally sponsored attention.

Artwork on the platform of the Atlanta system.

Reasons to Exercise Caution

The splendid characteristics of heavy rail transit have to be weighed against one crucial consideration—costs—with all other potentially negative factors assuming a secondary role.

Capital Costs

As has been mentioned repeatedly in this discussion, exclusive channels in tunnels have required and continue to require major capital investments. There is no way around this fact (barring some devoutly hoped-for breakthrough in tunneling techniques), since very few labor-saving opportunities exist and heavy machinery and skilled labor have to be employed. In many instances, an impasse has been reached today.

Any metro project has to be of a relatively large scale, because a route of less than 10 km, or even 10 mi, will not be a meaningful and particularly useful effort. Thus, significant amounts of financial resources have to be marshaled and committed, starting with at least $1 billion (most likely, $2 billion).

Any healthy metropolitan area with a viable economic base (not any central city alone) should theoretically be able to assemble the resources needed to build a metro system by itself (as BART did). However, given the availability of federal assistance programs that make the task of implementing any rail system much more feasible, it is doubtful that this will happen.

Operations and Maintenance (O&M) Costs

If a heavy rail system were to be used intensively around the clock, labor costs of all kinds and expenditures for materials and energy would be very favorable on a passengers-carried basis. The productivity, given high load factors, is very advantageous. If full loads occur only during peak hours, the annual budget will show major deficits. This has been shown time and again, with perhaps the only exceptions being the Hong Kong and Singapore operations. Those systems are in good use throughout the day and on weekends. Otherwise, the large fixed costs and semi-idle staff, rolling stock, and plant will quickly eat up the budget.

Long and Difficult Implementation Period

Leaving aside the approval period, which in countries with extensive environmental and community-protection safeguards will consume considerable amounts of time (10+ years?), construction is an elaborate and slow process, extending over years. Normal life along the corridor will be dislocated, even if deep tunneling methods that involve a limited number of construction shafts and controlled spoils removal are employed. If much care is not taken, some recent evaluations suggest that those temporary disruptions may be so large that the later service benefits might not balance them over a long period, if ever.[18]

Good management of these processes is a key requirement; there are too many examples of projects where the complexities have overwhelmed the available administrative skills.

Passenger Comfort and Convenience

Generally speaking, riders will be satisfied with metro service, but there are aspects that require attention. One of these is the fact that with a grade-separated system, patrons will usually have to change levels up or down when entering or leaving a station. Escalators may be highly desirable. Also, there are strict requirements regarding accessibility by people with disabilities (using wheelchairs) in many countries.[19] These will require elevators to and from platforms.

New and well-maintained systems should be quiet, but in older ones the rattling of cars, noise of wheels and motors, and screeching along rails may make conversation in vehicles or on platforms impossible. Vibration and shaking may also be excessive, if proper mitigation and maintenance steps are not taken.

The confined spaces call for proper ventilation and fire protection, and climate control is now mandatory. All these are issues that can be dealt with, but they do increase capital and O&M costs on the final budget sheet. (See "Passenger Amenities and Environment" in subsequent section "Components of Heavy Rail Systems.")

Inflexibility of the Network

Once a tunnel is in place, it obviously cannot be moved; changing the location of a heavy overhead structure also calls for a major

---

[18] D. H. Pickrell, "Estimates of Rail Transit Construction Costs," *Transportation Research Record No. 1006,* 1989, pp. 54–60.

[19] See Americans with Disabilities Act (ADA), passed by U.S. Congress in 1990, and a number of follow-up manuals by various agencies.

effort. Thus, once a metro system is constructed it will remain in place whatever may happen to land use distribution and activity locations on the surface. Experience in some older American cities has shown that even closing heavily underutilized stations, justified because most people have moved away, is not an easy task, since the remaining patrons and the neighborhood will fiercely oppose such action. The principal question then is whether new activities can be deliberately placed at such locations to make effective use of a sunk investment, not whether major service adjustments can be implemented.

Technical Constraints

While anything can be built anywhere, given sufficient engineering skill and resource investment, there are situations in which the effort may become unreasonable by any measure. These would include hostile subsurface conditions, proximity to fragile but crucial landmark buildings, and long crossings under bodies of water. There are examples of projects in which any one of these conditions has been overcome, but at a price.

There are limitations on the steepness of grades that guideway systems (even with rubber tires) can utilize; therefore, in hilly terrain alignments may not be able to follow surface topography very well, requiring deep tunnels and high structures in some places.

Technical Vulnerabilities

The large investment in a heavy rail network and the safety of the riders have to be protected, and there are areas of concern. For example, power failures may paralyze the transit system of a city entirely. Fires in tunnels can be particularly destructive and dangerous, requiring not only normal safeguards and continuous cleaning out of debris, but also elaborate escape and safety spaces for passengers. Any accident in the underground environment will be much more difficult to deal with than if it were to occur in the open. New metro designs must satisfy very stringent fire code requirements.

Finally, all tunnels leak water, by definition, and elaborate systems have to be built in to cope with this situation.

Personal Security

The amount of criminal activity in a subway is usually not any worse than above ground, but the perception of personal danger on the part of the riding public is considerably heightened. Much of this fear is due to being in confined spaces, where people may be apprehensive about having no way to escape; much of it may be fostered by media attention, which will not ignore any significant negative event in heavily used metro systems.

Elevated station in the Atlanta system.

Whether this perception is supported by the facts or not is quite immaterial; metro systems demand better-than-average security systems, in the form of either police presence, technical devices, or special design features.

Social Status of Heavy Rail Transit

There are people in New York who proudly announce that they would never use the subway and have not stepped into one for years. Remembering the condition of the system in the early 1980s, when it was on the verge of collapse, this attitude may be understandable but not excusable. The situation has improved considerably, although everybody knows that more should be done.

The rail transit operations in the older cities of the United States suffer from an unattractive public image. At best they are regarded as egalitarian services that provide basic mobility to every city resident and visitor regardless of economic or social standing; at worst they are seen as the means of transportation for only those who have no other choice. Their basic purpose is, indeed, to serve everybody, and the highlighting of class distinctions is an unfortunate by-product of society in very large cities. The relatively few but dramatic incidents of crime that have occurred on subways have received most extensive coverage in the national media, and the presence of homeless people, hawkers, unlicensed musicians, and plain panhandlers is quite visible since this public environment suits their purposes. There have been tourist guidebooks that have specifically warned their readers to stay away from the old subways in American cities, and

European tourists still regard a short trip on the New York system a major adventure worth writing home about. The media always devote much attention to whether the mayor of New York and the commissioners use the subway on a regular basis.

However, the newer systems in the United States (San Francisco Bay Area, Washington, and Atlanta) do not carry this historical baggage of negative experiences accumulated during periods of neglected maintenance, and people take civic pride in these systems. Ridership profiles show an equitable distribution of user characteristics, and taking the metro for social and discretionary trips is a largely accepted practice. The lesson appears to be that heavy rail transit, as a massive investment and an addition to the public infrastructure that people contact directly, has the force to define its own image and attractiveness. Quality of service and popular approval do require continuous attention to prevent any slippage, however.

It is argued—with some justification—that crowded systems are associated with some psychological tension, which may become distress if the congestion is severe and riders are thrown into physical contact with each other. Our expectations of personal sanitation may be violated, and our personal dignity may be invaded. These are serious concerns, and most people in the prosperous countries will not tolerate such experiences in their daily lives; they will go to considerable lengths and expense to avoid them when they travel. There have been systems abroad, which shall remain nameless, that have deliberately set fare levels high and have excluded basic amenities (such as toilets) to keep unwanted patrons out as much as possible.

While we have classes of service with corresponding charges on planes and trains, and there are clear class distinctions among automobiles, that would be unthinkable today on basic public transit. Paris no longer has first-class metro cars, either. In New York, the thought of instituting a special executive subway route with premium fares (let us say from Grand Central Terminal to Wall Street) is mentioned quietly from time to time and dismissed immediately. Yet, regional express and subscription buses are certainly understood to be of a special class. It might be appropriate for urban rail transit to strive for the status and comfort of commuter rail.

Recapturing public confidence and bringing riders back voluntarily to transit service, even on the old systems, is thus a major task that goes beyond purchasing sleek cars and refurbishing stations. In the new systems in cities that have no urban rail history, the challenge would be to maintain the original image of attractiveness, efficiency, and modernity. Perhaps the most effective invention in the rail transit field in the latter half of the twentieth century was the term *light rail transit,* because it is not associated in the popular mind with either the old subways or streetcars.

Mexico City metro operations, utilizing cars with rubber tires.

The problem goes deeper yet. There are documented instances in this country in which communities and neighborhoods have deliberately opposed building new rail transit facilities because they might bring in "undesirable" people, either as residents or as opportunity-seeking visitors. Just about every system, old or new, that has considered expansion into new regional territories has encountered at least some such opposition.

Application Scenarios

Heavy rail transit is obviously a mode for large places and not for very short trips. The principal operative questions that have been on the table for decades are: what is the smallest city size for which a metro service can be reasonably considered, and at what population level do such operations become almost mandatory to maintain adequate mobility in a large city? These are not particularly astute questions because many factors are involved, but they can be the base for some investigation nevertheless.

The principal issue is that a large volume of riders must be attracted and assembled every day to load the trains sufficiently to justify the large investment. This is not even a matter of true amortization by generated revenues, because as a matter of social policy fares have to be kept at an affordable level for almost the entire population. The question is how large a subsidy a society or community is willing to allocate for this purpose. It has become increasingly difficult in North America to gain acceptance for any large investments, especially if they are not for profit-making ventures, and rapid transit proposals in particular will have to show respectable and confidence-inspiring ridership estimates in order to gain support.

It is theoretically possible to envision a city that would be modest in size but have all its trip-generating activities concentrated in a single corridor. However, even Honolulu (1990 population of 0.4 million), which is as linear in its shape as we are likely to find, has considered rail transit several times, but opted not to take that step.

In the early post–World War II period, the suggestion in American professional circles was that a metropolitan area[20] with a population of at least 1 million and a central city of 0.5 million that contains a central business district of 25 million ft^2 is the *minimum* size for rapid transit.[21] Doubling all those figures would represent a *desirable* condition. Although these figures agree with the Soviet standard of the period, it can be noted that neither then or now could any heavy rail operations be found in any North American metropolitan area of that modest size. In terms of central cities, however, in 1998 the population of Cleveland was 0.50 million; Washington, D.C., 0.52 million; Boston, 0.56 million; Baltimore, 0.65 million; and San Francisco, 0.75 million.[22] This observation strongly suggests that the determining factor is concentration of the urban fabric in nodes and corridors; the total size of the entire service area matters, not the anachronistic political boundaries in the center.

---

[20] In most of the world, the term *city* is a fair approximation of the total urban settlement; however, the urban development of the larger cities frequently does extend beyond the political boundaries of the central city. In those cases, the appropriate term would be *urban agglomeration,* as used by the United Nations. In the United States, the "city" is only the central historical municipality, which may be surrounded by several dozen or more contiguous urbanized political units that together constitute a *metropolitan area*—the true urban unit in operational terms. The U.S. Census assembles demographic data into *consolidated metropolitan statistical areas* for particularly large conurbations.

[21] Wilbur Smith and Associates, *Transportation and Parking for Tomorrow's Cities* (Automobile Manufacturers Association, 1966), p. 208.

[22] The respective consolidated metropolitan area (CMA) sizes of these five cities in 1998 were 2.9, 7.3, 5.6, 7.3, and 6.8 million. (Washington and Baltimore are a single CMA).

A major 1982 urban transit analysis, often cited subsequently, suggested that a base density of 200 families per acre in a corridor is required before rail systems can be considered.[23] Other criteria were that 15,000 daily passenger miles per mile of route[24] have to be generated to justify rail service and that 20,000 to 42,000 daily passenger miles per mile of route are necessary to support tunnel construction.

The situation is different in developing countries, and it has been suggested that a population of 5 million is the threshold size for rail transit, and that 700,000 or more person trips per day have to be generated in a corridor for the heavy rail mode to be feasible.[25]

A look at actual demographic statistics and metropolitan area sizes in North America shows a different picture. The second largest urban agglomeration in the United States—Los Angeles–Riverside–Orange County (15.8 million)[26]—recently reached a decision to stop further rapid rail construction, and the largest urban unit without such a system or plans for same is Detroit–Ann Arbor–Flint (5.5 million). The next four largest areas without heavy rapid transit are Dallas–Fort Worth (4.8 million), Houston–Galveston (4.4 million), Seattle–Tacoma (3.4 million), and Phoenix–Mesa (2.9 million). The smallest metropolitan area with at least one rapid transit line is Cleveland–Akron (2.9 million), built in 1955.

Therefore, the magic threshold for North America appears to be a total of 3 million people, yet several metropolitan areas in that range have looked at the option and decided not to proceed at this time.

In the rest of the world, high-density corridors do exist, and rapid transit has been built in some surprisingly small cities.[27] These include Amsterdam (0.7 million); Antwerp, with premetro (0.5 million); Bilboa (0.4 million); Bratislava (0.4 million);

[23] B. Pushkarev and J. Zupan, *Urban Rail in America: An Exploration of Criteria for Fixed-Guideway Transit* (Indiana University Press, 1982, 289 pp.).

[24] A measure of the density of loading, taking into account how many passengers enter the system and how far they travel on each average segment.

[25] Halcrow Fox and Associates, *Study of Mass Rapid Transit in Developing Countries* (Transport and Road Research Laboratory, United Kingdom, 1989).

[26] U.S. Census estimates for 1998.

[27] The figures have been rounded off since they refer to various recent years, as reported in *Columbia Gazetteer of the World* (Columbia University Press, 1998).

Copenhagen (0.5 million); Frankfurt, with *Stadbahn* (0.7 million); Glasgow (0.7 million); Helsinki (0.5 million); Lisbon (0.7 million); Lyons (0.4 million); Marseille (0.8 million); Newcastle (0.3 million); Nuremburg (0.5 million); Oslo (0.5 million); Rotterdam (0.6 million); Stockholm (0.7 million); Toulouse (0.4 million); and Valencia (0.8 million). All of these are Western and Central European cities with high densities; the respective urban regions are often considerably larger than the central city, but in some cases they do not quite reach the million level, either.

In Asia, Africa, and Latin America, the situation is again different. There are many large metropolitan areas above the 5 million mark that undoubtedly would benefit greatly from heavy rail transit—not to mention those above 10 million—yet lack resources to undertake such projects. For example, of the 15 largest urban agglomerations in the world today, none are in Europe, 11 are found in the Third World,[28] and the trend toward the latter group is unstoppable. Of these 15 megacities, only 4 have extensive rapid transit systems that are being actively expanded (Tokyo, with a population of 26.4 million; Mexico City, 18.1 million; São Paulo, 17.8 million; and Osaka, 11.0 million); 1 started a network, but further extension has been officially terminated (Los Angeles, 13.1 million); 2 have old systems that are receiving some additions and improvements (New York, 16.6 million; and Buenos Aires, 12.6 million); 1 has a major expansion program underway after a late start (Shanghai, 12.9 million); 1 has a single line that may be languishing (Calcutta, 12.9 million); 1 has started construction (New Delhi, 11.7 million) and another is making plans (Jakarta, 11.0 million); but 4 have not shown any visible progress toward rapid transit at all (Mumbai/Bombay, 18.1 million; Lagos, 13.4 million; Dhaka, 12.3 million; and Karachi, 11.8 million).

Whatever the feasibility situation may be at the metropolitan level, the justification of actual construction is still a local matter: *corridors with sufficient density* and *major destination nodes* that can fill many trains in rapid operation, assisted by feeder services. The rule of thumb supported by American transportation planners at one time was that the total daily ridership on a line should be at least 30,000, with the peak hour demand in one

---

[28] 1999 U.N. estimates, as reported in most almanacs.

direction exceeding 8000. Another suggestion was that at least 5000 riders per hour in a single direction are needed to *consider* heavy rail transit, and that with 10,000 patrons such service is definitely supportable.[29]

Given recently skyrocketing costs, those norms are no longer viable. Such loads can be accommodated by light rail transit and even bus rapid transit. Thus, the base patronage demand projection in American communities would have to be placed in the range of 15,000 to 20,000 passengers per peak hour in a single direction. Unless, of course, special conditions exist, and there is a targeted long-range plan that moves deliberately toward a high-density pattern with synergetic land use and transportation policies.

The Manhattan West Side Subways

To confirm the performance characteristics described in this chapter, surveys of the local and express subway routes along the West Side of Manhattan were done in June 2001. This covered the Number 1 and 9 locals and the Number 2 and 3 expresses,* which run along the same alignment. Before the current designations were established, this service was known for decades as the IRT Seventh Avenue Local and Express. The upper leg, north of 42nd Street, was a part of the original NYC subway route opened in 1904; the lower leg, south of 42nd Street, was added in 1918. In the 1950s, peak hour frequency was often less than 2 minutes; today the scheduled interval, not always attained, is 4 minutes. The loading was much heavier then, judging from the author's personal experience when he commuted to college on this route.

The northern terminal of the local route is 242nd Street/Van Cortlandt Park in the Bronx, near the municipal boundary with Yonkers. The line ran almost straight southward to the South Ferry station at the tip of Manhattan at the time of the survey. Here the trains used to turn around on a loop track, but the old curved platform is so short that the doors of the last 5 cars of the 10-car trains could not be opened. The express service overlaps the local service in the middle portion between the 96th Street and Chambers Street stations, where four tracks are in operation. There are also multiple tracks to the north, but they are used only for storing trains and emergency bypass.

The surveys were done during regular weekdays, specifically during off-peak hours, to record unconstrained operations. No special events occurred in the city during the survey periods; student and school children usage during

[29] Wilbur Smith, op. cit.

the summer months is light. The local trains filled all seats with some standees in the middle portion of the route; the express trains always carried standees, but with no crowding and no special delays at the doors.

The local route is 15.1 mi (39 km) long, with 38 stations. The total travel time was about 60 minutes southbound and 55 minutes northbound, give or take 2 minutes, because the last southbound section almost always encountered delays awaiting departure of the previous train at the South Ferry station. This results in an overall running speed of 16 or 16.5 mph (26 kph) without the southbound delay.

Dwell times (seconds from the time when a train stops at the station to when it starts moving again) were 13 to 20 seconds at local stations, with 7 seconds recorded on a few occasions. At transfer and express stations they were about 25 to 30 seconds, or more if the local train was held back to meet an express, to adjust schedule, or for some other operational reason.

The express segment of the route is 5.8 mi (9.3 km) long, with 6 stations; the average running time is 14 to 15 minutes. This gives an overall running speed of approximately 24 mph (39 kph). The dwell times were 15 to 40 seconds under normal circumstances, or longer as appropriate. During rush periods, the dwell times may of course be considerably longer, largely unpredictable, with many passengers forcing their way in or out and bags and clothing preventing the doors from closing.

On the local routes, trains were in motion about 80 percent of the time; on the express routes, 85 percent.

* After the September 11, 2001, attack, all service between Chambers Street and South Ferry was stopped, and other adjustments were made. The No. 2 runs as a local on this part of the route, the No. 9 has been eliminated, and the No. 1 terminates at Chambers Street.

The speed results from the Manhattan West Side subway survey are somewhat disconcerting, because the overall running speed on the Seventh Avenue express is only 24 mph (39 kph), while many transit planning references give the high range of this parameter as well above 30 mph (49 kph) and even as high as 50 mph (80 kph). It is hard to see how any transit system could operate faster than the Seventh Avenue express, with modern equipment serving stations about a mile apart, and achieve higher overall running speeds. To test this concern, similar time measurements were undertaken a few weeks later in Caracas, Venezuela, a city that is served by a well-designed and well-built metro system that opened in 1983.

The Caracas Metro

The system consists of a principal line that runs along the spine of the largely linear city. There are two branches to the south, and a new parallel line is under construction along the central corridor.

The surveys were done in July 2001 during regular weekdays, off-hours, on the principal line (Linea 1). It is in intensive use, with many standees in the trains even in the middle of the day (except for the two end segments extending over two or three stations). The trains are full but not excessively crowded. The deceleration and acceleration of trains are not particularly agile, while all other elements operate at a high level.

The dwell times at stations are mostly in the 18- to 24-second range. In some instances they are as low as 15 seconds. The longest dwell times during off hours reach the 40-second range, primarily at the several transfer stations where much passenger interchange takes place. The total length of Linea 1 is 13 mi (21 km); the running time is normally 33 to 35 minutes, which results in an overall running speed of 23 mph (37 kph).

The stations are spaced rather closely on the Caracas metro; thus, on any given run, the train is in motion about 77 percent of the time, stationary 23 percent.

Components of Heavy Rail Systems

Since heavy rail transit is basically a railroad, the physical components are not too different from those of commuter rail and light rail elements, except that they have to be more robust to stand up under very intensive use, and more advanced and refined to respond quickly to rapid service needs. Heavy rail is the *extreme* railroad.

Rights-of-Way and Track

Since total grade separation has to be maintained (except for hybrid systems), the guideway is to be placed underground or on an elevated structure. Surface trackage can only be considered if lateral intrusions can be prevented absolutely by impenetrable fences, buffer strips, under- or overpasses, and similar devices. Highway medians, existing railroad corridors, and land behind large industrial or institutional properties are possible candidates for the surface placement of parts of lines. (See Fig. 13.9.) There are four choices for track placement:

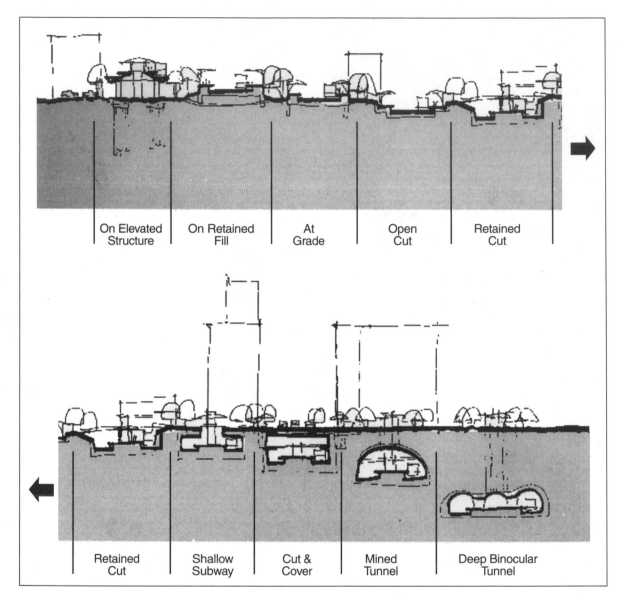

Figure 13.9 Functional placement of rail stations. (*Source:* H. A. Kivett and K. Peterson, *Rail Station Compendium,* Parsons Brinckerhoff, 1995. Used with permission.)

OPEN CUT (OR SHALLOW DRAFT)

Depressing the track one level below the street surface is a relatively low-cost solution, provided that the right-of-way can be acquired without too much difficulty, since a strip of land will be preempted for metro use. Ventilation will be provided naturally, and cross streets can be accommodated by direct overpasses. Reliable fencing along the edges is mandatory. All open depressed alignments face drainage and dewatering problems. Placement of buildings and other activities spanning the cut is entirely possible.

ELEVATED STRUCTURES

If the community can tolerate the visual intrusion and some possible noise impacts, raising the track above the surface satisfies all operational requirements and saves construction expenses. Strong columns about 100 feet apart will carry a deck (usually prefabricated today) that accommodates two tracks. The concerns are the exact placement of the alignment and securing a right-of-way. If placement is to be on top of streets, movement obstructions and shutting off of light and ventilation below are likely to be issues. In the United States, the memory of the old East Coast els is still strong, and this scenario quite likely will not be acceptable in high-density districts. If the alignment is to be placed in the rear of properties and in the middle of blocks, acquisition of the right-of-way and relocation of activities are likely to be costly and time-consuming issues.

CUT AND COVER

This method has been in use for a long time and is particularly suitable if the subway is to be placed below existing streets—as close to the surface as practicable. The construction process follows a certain sequence (Fig. 13.10): the entire street width is first covered with a temporary floor of heavy beams, which allows surface traffic to move after a short interruption and allows adjoining buildings to remain in normal use. Excavation is done under the deck until the necessary volume is emptied out. Utility lines are a particular problem; they are usually found in a great maze under the street pavement, and they have to be suspended from the deck so that service can continue. Usually, the base slab is poured first, then the side walls in lifts, and finally the roof slab. The finishing steps are equipping the tunnel with track and power and control systems, refilling the space above the tunnel, and repaving the street. The advantages of this method are not only the tolerable construction costs, but also the fact that station

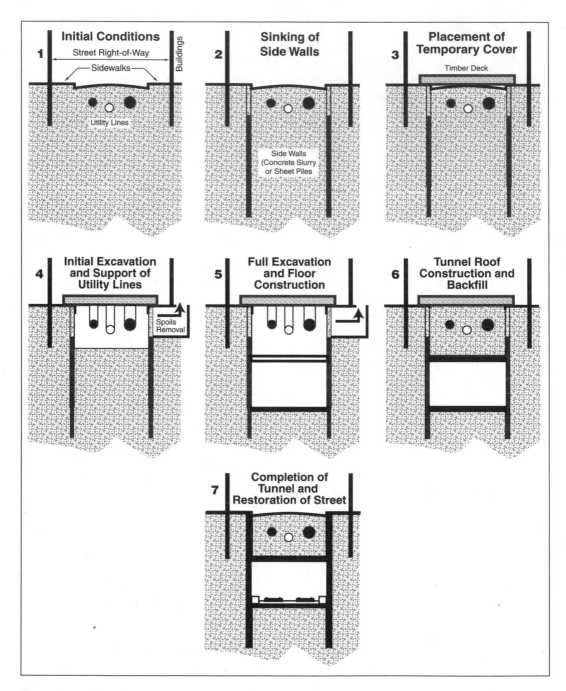

Figure 13.10   Cut-and-cover method of subway construction.

platforms will be quite near the surface, thus minimizing vertical distances for passengers.

A variation on this method is *cover and cut:* building the side walls first by filling narrow but deep slit trenches with concrete slurry, then erecting the tunnel roof and replacing street pavement, and finally completing all the work below, starting with excavation.

TRUE TUNNELS

This is a construction method that proceeds horizontally without touching the surface, except for the presence of construction shafts at regular intervals. These are necessary for the removal of the excavated material and for lowering construction materials, machinery, and crews or workers. They may later become part of the ventilation system.

In rock or hard material, the work can be done by the traditional drilling and blasting approach, removing the broken spoils after each phase and building the structural shell immediately behind the blast face; or by tunneling machines that, once assembled underground, grind their way through the rock or soil layers continuously, transport the excavated material to the rear, and provide for the placement of the tunnel lining. To shorten the construction period, tunneling may be done from several places at once or in opposite directions from each end.[30] Sometimes, if there is no resale market for the used machinery after a major job, it may be left below in a side hole.

Tunneling in soft soils is more intricate because of groundwater presence, potential cave-ins, and soil subsidence. Shields that protect the crews and keep the shape of the bore are pushed forward gradually. The material is recovered manually or by machines. If the soil is waterlogged, the difficulties increase considerably. Compressed air chambers are usually used at the face of the bore, and other methods are available as well.

Tunnels can be done either as a single large bore that can accommodate two tracks side by side and all ventilation ducts, or as two smaller diameter bores, one for each track. Stations, of

---

[30] The old joke among tunnelers is that if the crews working from each end miss each other, you get *two* tunnels. They usually do meet, however, within inches of perfect alignment.

course, require large spaces and surface connections that have to be carved out in addition to the rail tunnels.

Ventilation and drainage require special attention in the enclosed underground environment (as do passenger access, evacuation, and safety features, as discussed in the next subsection).

The unit costs of tunneling are closely tied to the quality of the tunneling medium and the diameter of the bore in any given situation. The larger tunnels are not only more difficult to excavate, with more massive machinery; they also require much stronger and more elaborate shells to keep the surrounding soil in place. The Japanese are experimenting with linear induction transit car motors,[31] which allow the vehicle floors to be lowered. This results in a lower height for the vehicles, so they can fit into smaller tunnels (diameters of 17.5 ft [5.3 m] instead of 24 ft [7.3 m]). It is not certain that this approach will make subway construction affordable, particularly because linear induction motors are not yet in wide use, but any progress in reducing tunneling costs is welcome.

Heavy railroad rail will be used for the trackage, supported by ties or a concrete deck. Normal railroad gauge (4 ft 8.5 in; 1.435 m) will usually be used, although there are exceptions. The first line in Philadelphia used a 5-ft 2.5-in gauge; BART has 5 ft 6 in to enhance train stability.

The interior dimensions of a tunnel or the longitudinal channel along any guideway are determined by the *kinetic envelope* of a train (see Fig. 13.11 and Table 13.3). This is calculated from the external dimensions of the cars; plus additional clearances required by the swaying, leaning, and bouncing of the vehicles and the cantilevered protrusions of the ends or middle of vehicles when going around tight turns; plus a safety factor.

One size of tunnel does not fit all conditions; therefore, the cross section shown in Fig. 13.11 refers to the Red Line in Boston (Massachusetts Bay Transportation Authority [MBTA]).

These basic dimensions translate into the following physical width requirements for a two-track alignment (rights-of-way or

---

[31] Basically, these amount to open electric motors. Electrical elements are placed along the centerline of the guideway floor. These pull the vehicle forward magnetically. The system is used on the Vancouver SkyTrain system and several AGTs.

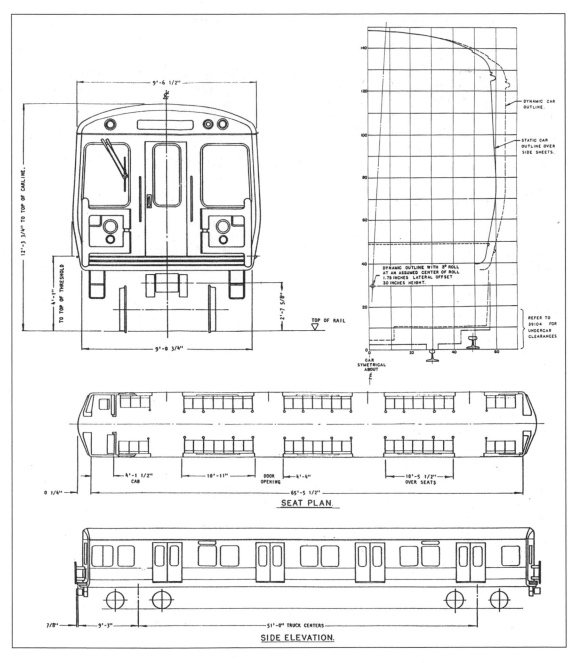

Figure 13.11 Dynamic envelope of heavy rail car. The dynamic envelope is different for each model of vehicle. (*Source:* Massachusetts Bay Transportation Authority, Boston.)

Table 13.3   Physical Characteristics of Heavy Rail Transit Vehicles and Infrastructure

Length of car	50–85 ft (15.2–2 m)
Width of car	8.2–10.5 ft (2.5–3.2 m)
Height of car (from top of rail)	9.5–12 ft (2.9–3.7 m)
Number of seats in car	40–75
Number of standees	Up to 240*
Total regular capacity of car	140–280 passengers
Number of cars in train	2–11
Maximum running speed	80 mph (130 kph)
Usual operating speed	15–30 mph (24–48 kph)
Minimum radius of horizontal curves	
On main line	350 ft (107 m) in old systems
	1000 feet (300 m) or more in new systems
In yards	25 ft (7.6 m) in old systems
	50 ft (15 m) in new systems
Maximum Acceptable Grade	
On main line	4%
Desirable maximum	3%
With rubber tires	6%
Standard track gauge	4 ft 8.5 in (1.435 m)
Length of platform	
For 8-car train	600 ft (182 m)
For 10-car train	750 ft (230 m)
Minimum Width of Platform	12 ft (3.7 m)
Distance between Stations	1500–10,000 ft (460–3000 m)

Source: Various operating agencies.

* Depends on the amount of floor space available for standing and the assumed space per average passenger.

easements; not including stations, which have their own space needs):

At grade (assuming no embankments
or heavy retaining walls)   32 ft (9.8 m)

On an elevated structure
(including side parapets)   26 ft (7.9 m)

In tunnels (assuming no unusual
space requirements along sides)   40 ft (12.2 m)

At the end of every line, unless there is a yard, tail tracks with crossovers have to be provided behind the last station (sometimes before the station). These must be long enough to store and reposition trains at the platforms for the return journey. Crossovers and switches have to be provided along the entire alignment to cope with emergencies and offer flexibility in operations.

Stations

Customers usually judge the quality of a metro system by the quality of its stations, where they spend many minutes with nothing much to do. The quality of construction and level of maintenance of a rail transit system can be gauged quickly by how well and how closely the door sill of the car lines up with the edge of the platform to minimize the gap that passengers have to cross.

There are two basic types of stations, distinguished by the configuration of the platforms;[32] they may also be distinguished by whether they have a mezzanine. (See Fig. 13.12.)

CENTER PLATFORMS

With this station arrangement, trains stop on both sides, accommodating the passenger circulation and waiting space in the middle. Such stations are more compact in their arrangements, and therefore are somewhat cheaper to build, even though the track alignment has to be spread out. Center platforms are just about mandatory if transfers are to be expedited across the platform for trains going in the same or opposite directions.

Center platforms are not particularly suitable if large volumes of passengers move on and off during peak hours, since they will get in each other's way, and the streams cannot be easily separated. Also, if the station is close to the surface, access from the sidewalks to the middle-of-the-street alignment may be cumbersome and costly to achieve.

SIDE PLATFORMS

In central district locations with high passenger volumes, this type of station arrangement allows the best response to internal circulation needs. The streams of passengers can be reasonably provided with direct and wide enough paths when adequate

---

[32] This discussion does not differentiate between underground and elevated stations. The functional configuration is basically the same, except that they are stacked in reverse order.

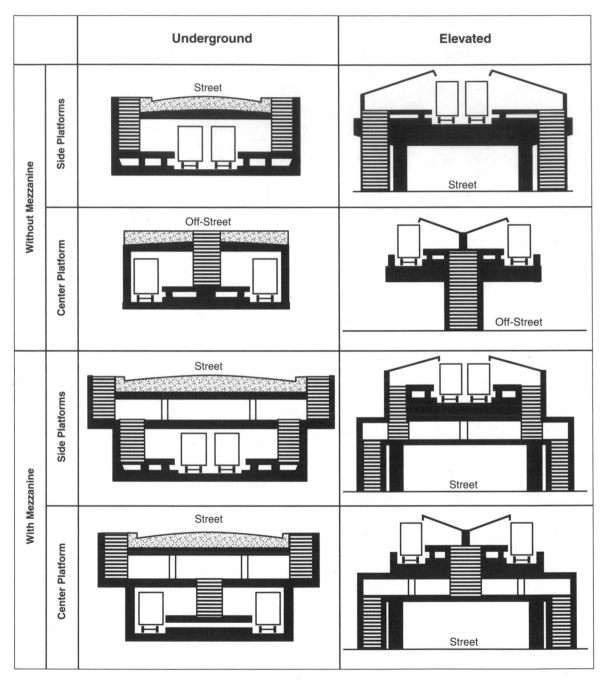

Figure 13.12   Rail transit station types. [Based on V. Vuchic, *Urban Public Transportation* (Prentice-Hall, 1981), p. 242.]

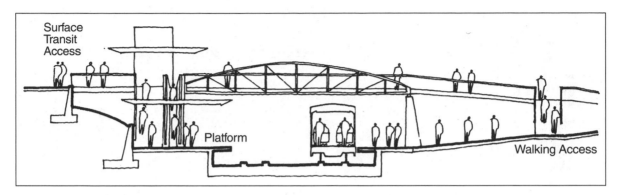

Figure 13.13 Open-cut station. (*Source:* H. A. Kivett and K. Peterson, *Rail Station Compendium,* Parsons Brinckerhoff, 1995. Used with permission.)

staircases or escalators are available along the sides, in proximity to street sidewalks. (See Fig. 13.13.)

A recommended feature of the longitudinal grade profile through a station is a "hump": the grade ascends as the tracks enter a station, thereby assisting the train in slowing down without unbalancing the passengers; and the grade descends as the tracks leave the station, to assist the train in accelerating by gravity.

The design of the passenger spaces of stations is largely an exercise in accommodating flows from the street in and out of trains, even though waiting times are involved and patrons may make some purchases along the way. (Precise methods for calculating all interior space requirements are available.[33]) These paths have to be as direct as possible and of sufficient width. The users must receive clear information to remain oriented in an otherwise rather stressful environment. One feature that particularly assists in attaining these objectives is a mezzanine level between the platform and the street. Reservoir space is made available, patrons can be guided to and from the proper staircases up and down, and overall orientation is facilitated.

The presence of mezzanines can also assist in the development of an underground pedestrian environment (shops and services) extending well beyond the stations. Direct connections to base-

[33] See G. Benz, "Transit Platform Analysis Using the Time-Space Concept," *Transportation Research Record No. 1152,* 1987, pp. 1–10; also see Chap. 2.

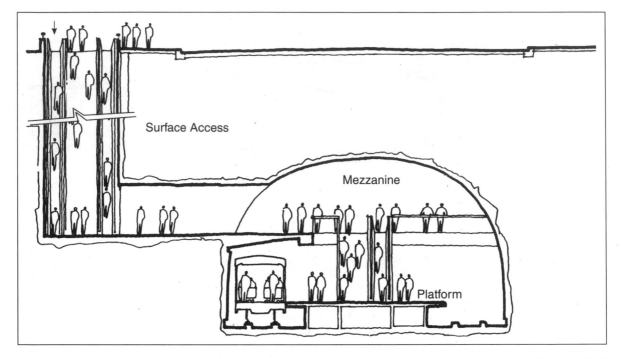

Figure 13.14   Mined station with mezzanine. (*Source:* H. A. Kivett and K. Peterson, *Rail Station Compendium,* Parsons Brinckerhoff, 1995. Used with permission.)

ment levels of stores and office buildings can be made, and the network of concourses and passages in tunnels for weather-protected circulation can be created, as has been done in Montreal and Toronto on a major scale. (See Fig. 13.14.)

Platforms have to be of sufficient width to hold the maximum number of passengers expected to accumulate between trains, while providing space for exiting passengers to get off a train and move away. There have to be enough staircases so that all exiting passengers can leave the platform before the next train pulls in. It also must be recognized that moving and waiting passengers will not be distributed evenly along the entire length of the platform; therefore, the central sections should usually be wider. The platform length, of course, is a function of the number of cars in a train times the car length, plus a short overrun. It would seem to be a wise step to build them longer than loads in the early phases would call for, since lengthening them later is a very expensive undertaking, especially underground.

A particular challenge is the design of stations and platforms near facilities that discharge large masses of people all at once—sports stadiums, for example. The platforms can be made very wide to hold large volumes of waiting passengers, but this may create serious safety problems due to pushing from the rear, because everybody will not be able to get on the next train. A better approach is to retard the passenger flow through holding areas, thereby "metering in" only the volumes that can be safely processed. Although not found on U.S. systems, such platform controls have often been implemented abroad (for example, in Mexico City and Seoul). Accommodation of emergency evacuation requirements would be a challenge under U.S. regulations.

At the street level, entryways and staircases have frequently been placed within overcrowded sidewalk spaces, thus constraining general pedestrian circulation. This condition highlights the fact that station design should not stop with the staircase to the surface. Entry/exit gateways should be set within building lines if at all possible,[34] and paths leading to and from such gateways in high-usage areas should be adequately wide and properly laid out.

Station spacing is a particularly important consideration in planning a rapid transit system.[35] If the stations are close together, walking access will be encouraged, but overall running speeds will drop and service will move slowly. In the old systems, such as the New York City subways, stations were frequently placed only 6 blocks apart (1560 ft; 475 m), which may be appropriate in very high density districts, allowing access on foot for everybody within a corridor 0.5 mi wide. In low-density districts, this is not workable if any rapidity is to be maintained. On the Moscow metro, the spacing may be 2 or

Major bus connections to metro station in São Paulo.

[34] New York City zoning ordinances give developers of new buildings bonus floor space for off-sidewalk entries.
[35] See Edwards, op. cit., pp. 140–147.

Long escalators leading to deep stations of the London tubes.

3 km (1.2 or 1.8 mi) on the outside parts of routes, which results in many rather unpleasant long walks in the Russian winter. On the more recent systems in North America, the intervals are also a mile or more on the peripheral sections, but access is largely provided by feeder services and automobile facilities, where not much walk-in patronage is expected.

Passenger Amenities and Environment

There is a long list of considerations that affect passenger well-being and the perception of a good trip, which for all practical purposes can be regarded as mandatory if customers are to be retained on public transport services in prosperous communities.

VENTILATION

Air circulation has to be maintained, since air quality can deteriorate quickly in the confined underground environment.[36] In most instances the so-called piston effect—the trains push air through the tight tunnels—is sufficient. Yet exhaust vents have to be available, and mechanical assistance by fans is mandatory for removal of smoke and the introduction of fresh air. This is a particularly serious issue in deep tunnels, where major ventilation shafts will most likely be needed. (See Fig. 13.15.)

On elevated platforms, the problem may be the opposite condition in inclement weather, and windscreens often have to be provided to protect waiting patrons.

HEATING AND AIR CONDITIONING

Full climate control is to be expected with heavy rail transit. This means adequate, but not excessive, heating during winter within stations as well as inside cars. It should be kept in mind

[36] Particularly useful reviews of ventilation and heating issues are contained in Papers 4 and 5 of the Institution of Civil Engineering proceedings *Urban Railways and the Civil Engineer.* There is also a *Subway Environmental Design Handbook* (Parsons Brinckerhoff, 1975) that provides a design tool with model simulations.

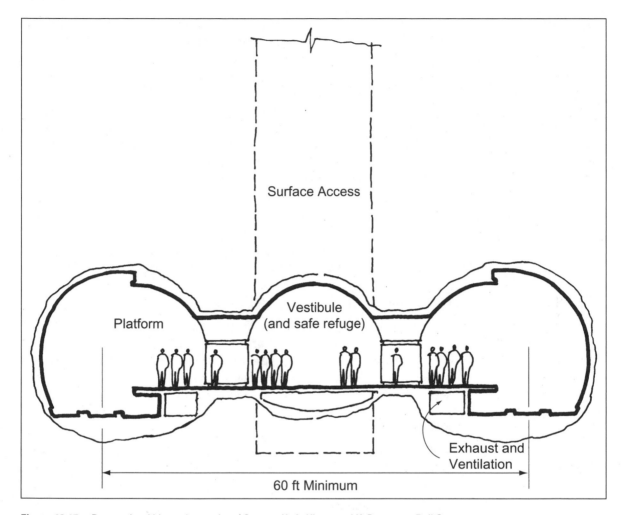

Figure 13.15 Deep-mined binocular station. (*Source:* H. A. Kivett and K. Peterson, *Rail Station Compendium,* Parsons Brinckerhoff, 1995. Used with permission.)

that human bodies and train operations generate considerable amounts of heat—particularly if old-fashioned braking is used, which dissipates wasted energy through rheostats. Even the flexing of rubber tires generates heat, as was experienced in the early days of the Montreal system, calling for emergency corrective measures.

Expectation of comfortable conditions on all systems in North America is now a matter of course, and the debate concentrates largely on how to retrofit the older systems. A basic question is how extensive climate control coverage should be. Ideally, it should encompass the stations, platforms, and the cars—the entire passenger environment. Since there is no point in cooling the tunnels, air conditioning stations is a very difficult task; many openings for cool air to escape are present.

Some systems (St. Petersburg and Singapore) have placed screens along the edges of platforms with doors that line up with the doors of the cars, thereby containing the air volume within stations, at least on the platform level.

The standard response is to purchase rolling stock with complete onboard climate control. These systems now operate quite reliably, after considerable startup difficulties in developing sufficiently robust equipment. But stations remain in a "natural" state, under the assumption that passengers will not have to spend too much time waiting on the platform.

LIGHTING

Adequate levels of illumination are important so that passengers gain a sense of security and are able to read while on the system. Purely transitional spaces can be dimmer. The only exception to these expectations for brightness is the Washington, D.C., metro, which maintains a twilight atmosphere on the platforms.

NOISE CONTROL

Rubber-tired systems have been advocated as particularly effective in achieving low noise levels. They have been implemented on four lines in Paris, and there are entire such systems in Lyons, Montreal, Mexico City, Santiago, and Sapporo. The noise and vibration reduction claims are basically true, except that incoming trains in Montreal stations are not particularly more quiet than ordinary steel-wheel vehicles, although the noise is different in sound quality. Vibration that may affect nearby buildings is certainly precluded; however, rubber tires do pose a fire and smoke hazard. Well-maintained regular steel-on-steel hardware should not produce excessive noise or vibration inside or outside the cars. For vehicles, this means no loose parts, tight mechanical connections, perfectly round wheels,[37] well-lubricated bearings,

---

[37] If a wheel slides when brakes are applied, it will develop a small flat spot, which thereafter will generate a click with each turn. This can be corrected by grinding down the wheels in a special machine to restore a smooth and continuous running surface (wheel trueing).

and tuned motors. Wheels can be made resilient by sandwich construction, using inserts of an elastic material.

Within stations, sound-absorbing acoustic material can be placed on walls and ceilings, but particularly should be used to semienclose the lower parts of cars where trucks and motors are located. Platform edge screens will keep train noises away from platforms, as well as dust, drafts, and smoke and heat that may be generated in the tunnels.

The track requires attention as well, since a poorly maintained guideway may be the principal noise generator. Welded rails with no joints will eliminate one source of noise; fastening the rail tightly to the ties or floor deck will help considerably; resilient pads can be inserted below rails, or floating slabs can be used. The top surface of rails should be kept absolutely smooth;[38] the screeching sound produced by wheels going around tight turns (which does occur on the older systems) can be mitigated by lubricating the rails.

If a guideway is placed on an elevated structure near buildings, particularly residences and other sensitive uses, sound-absorbent parapets can be placed along the edges, or the track can be enclosed entirely (a tunnel in the air), as has been done on some sections in Toronto and Hong Kong.

As always, playing radios without earphones and using cell phones should not be allowed.

SEATS

Since waiting times for trains may be extensive, benches and seats on the platform are desirable. It has been argued in some cases (Moscow, for example) that their frequency of trains is high enough so that sitting down is hardly possible. There are other instances, in cities with large homeless populations (New York and Paris, for example) where considerable pains are taken to design benches so that they are unsuitable for sleeping.

TOILETS

It can be argued that public toilets should be found in all public environments for the sake of basic human comfort. Yet, this creates major maintenance and security tasks, and transit providers—stressing that they are not social service agencies—will

---

[38] Rails tend to corrugate under heavy use; this cause of noise can be removed by grinding down the surface periodically.

often deliberately not provide such facilities, particularly because they are said to attract undesirable customers onto the system.

VENDORS

Kiosks selling reading material and basic refreshments, as well as vending machines, are a normal part of the metro environment. There have been efforts, however, to exclude them from all stations in some systems to minimize cleaning and waste collection tasks. One of the principal reasons why Singapore has banned chewing gum throughout the entire city-state is to keep the shiny tile floors of the metro stations free of stains. (The other side of the coin is the revenue received from the concessions.)

ORIENTATION AND INFORMATION SYSTEMS

Except for commuters who make the same trip every day, most riders need useful information before they enter the station, as they pass the payment barrier, when they are on platforms, and when they ride the trains. This is particularly the case with large complex networks. Such elements of interest encompass systemwide maps, onboard announcements, and graphic material showing the sequence of stations along any given route, transfer options, and principal attractions found around any station. Symbols, color codes, and schematic drawings can be effectively used for this purpose, and there are specialists who are skilled in this type of design. A number of transit systems have done well in this field recently; among the old systems, Boston particularly stands out.

Another desirable feature that is now quite common on platforms is the strategic placement of variable message signs that give the time and information on the next trains. They do not shorten the waiting time, but give the patrons useful information and an assurance that things are under control. (The Moscow metro has a clock at the end of the platform that counts the seconds since the previous train left the station. This would appear to be useless information for riders on the platform, unless it is known that trains will maintain their scheduled intervals precisely.)

VISUAL QUALITY AND ARTWORK

As a matter of civic pride and consumer satisfaction, the transit environment should be agreeable to the eye as well. Given the resources committed to the construction of metro systems and the great amount of time that customers spend in this environment, nothing less should be expected.[39] The excesses of the Stalin-era metro decorations in the former Soviet Union are not called for, but it can be noted with some satisfaction that practically every new system built today endeavors to display good design and artful elements. These can range from the old terra-cotta plaques in restored New York City subway stations to complete museum-quality environments such as the Stockholm Blue Line.

The Stockholm Blue Line, with extraordinary attention to visual design.

Efforts in São Paulo, Caracas, Los Angeles, and Montreal are exemplary, among a number of others. In some instances, a certain percentage of the total construction cost is allocated for the acquisition of artwork and quality decorative elements. There are even some efforts to generate a kinetic visual experience as a fast-moving train passes elements along the trackside. This might open up an entirely new branch of applied arts and the creation of visual excitement on infrastructure facilities that traditionally have tried only to keep the walls reasonably clean.

Safety Features

As has been noted previously, many transit users are concerned about their personal safety from criminal acts, and a fire or train accident in a tunnel can cause much damage. The latter concern is to be addressed primarily through good maintenance of equipment and track, having emergency equipment available and in good order, reliable housekeeping within the system, proper train-

[39] These aspects have captured the attention of several authors, and the following references can be mentioned: M. Strom, *Métro-Art dans les Metropoles* (Jacques Damase [ed.], 1990, 184 pp.) and F. A. Cerver, *The Architecture of Stations and Terminals* (Arcol, 1997). Many more references on railroad stations are available, including M. Thorne (ed.), *Modern Trains and Splendid Stations* (Merrell/Art Institute of Chicago, 2001, 160 pp.).

ing of personnel, and the establishment of communications links with regular emergency response forces.

A significant element in rail transit design is the mandated need for safe and quick evacuation of patrons, particularly in the event of fire. Tunnels can be filled with smoke, and the high-voltage power lines are always a danger to people. Catwalks along the sides of tunnels have to be present, and ventilation systems have to be able to cope with smoke.

Station design is subject to strict regulations requiring that platforms have sufficient exit lanes to evacuate a full load of patrons in 4 minutes or less, that no point on the platform be more than 300 ft from an exit, that unobstructed paths be available, and that the people be able to reach a "point of safety" within 6 minutes from the most remote location on the platform. A *point of safety* is any at-grade location outside the station structure, an enclosed and protected fire exit leading to a safe location, or another space that affords adequate protection against fire and smoke.[40] For example, the mezzanine may be such a place if the stairs to the platforms can be sealed off by fire doors.

The safety concern on platforms—besides fire and smoke—is primarily patrons falling inadvertently or by deliberate action in front of incoming trains.[41] This is one of the principal justifications for providing platform-edge barriers (full or partial screens)—as mentioned previously in connection with passenger amenities—but that requires precise positioning of trains every time they stop, and other problems may occur (passengers may be caught between the doors). In most instances, a satisfactory safety feature is to mark a wide, very visible strip along the platform's edge, within which patrons should not stand when trains are in operation. It must be textured so that blind riders can sense its presence; the Washington, D.C., system adds a row of lights that flash when a train pulls in.

Criminal activity against subway patrons is a threat, and vandalism and general rowdiness cannot be tolerated either. Pickpockets and graffiti writers find the transit environment particularly attractive. There are several ways to combat such

---

[40] National Fire Protection Association, Inc., *NFPA 130 Standard for Fixed Guideway Transit and Passenger Rail Systems* (2000 ed., 43 pp.).
[41] People with mental disorders have been known to push other riders, and suicides are not unheard of.

unfortunate situations, and it is a never-ending battle. For example, after special paints that allow easy cleaning of graffiti were developed, vandals in the New York subways started to express themselves by scratching window panes, which are much more difficult to protect.

The presence of uniformed police is the most effective but the costliest response. This is being done in a number of instances, particularly in stations and on trains that have a bad record and during vulnerable times (at night or when a local high school ends its classes). Closed-circuit TV monitoring is currently the most often used security system, and it is effective—provided that ready two-way communications links can be maintained and that personnel can reach an incident site quickly. A standard design practice, employed today in every new system, is to ensure that there are no secluded or unobservable nooks and spaces that would create opportunities for criminal or undesirable actions.

The preceding lengthy discussion of safety and security issues in the metro environment should not be interpreted as indicating that major dangers are endemic. As a matter of fact, heavy rail transit is the safest and most reliable mode of all. It is just that any incident becomes magnified in reality and in popular perception when it occurs in the closed environment of a metro system, and care must be exercised always.

Control Systems

Some of the older metros (New York, for example) still depend on nineteenth-century wayside signal systems that require the person in the front cab to be always alert. From time to time human errors occur, even with mechanical brake-tripping devices when a red signal is passed. Keeping these systems in place is not a question of a lack of responsibility, but rather one of cost, because a change to advanced control systems would involve billions of dollars and serious modification of ingrained labor practices. Operations that depend on the old devices remain reasonably safe but not particularly efficient, because prudent managers maintain longer intervals between trains than would be possible otherwise. Changes will have to be made in steps, one route at a time, to eventually achieve an affordable conversion.

The old standard controls usually employ the *block signal* or *automated block signal* (ABS) concept, which governs train separation by ensuring that the next "block" (i.e., a length of track) is free of any previous trains, and that the following train may safely enter it.

Major advances in control systems were made with the construction of BART, the first fully automated system, and there has been a series of subsequent improvements. The systems depend on precise methods to locate the position of trains at all times and provide that information to a communications-based train control center. The principal tasks are to maintain proper spacing between trains and to control their speeds. Accurate and reliable controls allow safe reduction of headway, thereby gaining significant increases in capacity.

All new networks or routes (such as the Jubilee Line extension in London and the Météor Line in Paris) employ automated control technology through sensors that monitor each train as to its precise location. Command centers govern all operations (speed, spacing, and schedules of trains), and instantaneous instructions to the cab of each train control its performance (possibly including opening and closing the doors). The driver[42] in the front of the train has only a supervisory responsibility, with the ability to override any automated controls. However, to preclude complete monotony, the driver may be given some functions (such as monitoring trackway conditions at stations, making special announcements, operating the doors, etc.).

The overall term for such complete systems that govern all operational aspects is *automatic train operation* (ATO). Within that, there is *automatic train control* (ATC), which controls movements with safety as the uppermost concern; *automatic train protection* (ATP), which is a fail-safe program against collisions, excessive speed, and other hazards; and *automatic train supervision* (ATS), which monitors trains, controls schedule, and manages route selection. On any given route or network, these programs may be used fully or in a modified form, at various levels of completeness.

Note: Train controls, much more so than any other subsystem of heavy rail operations, are complex and critical elements. A general discussion may provide some information and understand-

---

[42] The traditional term was *motorman,* but it is deemed no longer politically correct.

ing, but any decisions about the components and their operation have to be left to qualified specialists.

Fare Collection

As is the case with other transit modes, the method used to ensure that all passengers have paid the required fare has substantial impact on the efficiency of operations. The best procedure from that point of view would be free and unobstructed entry and exit. Considerable resources would be saved in eliminating the cumbersome fare collection mechanisms. There would be no fare-box recovery of costs, however, and all O&M funds would have to be allocated from general public budgets. Alas, we are not ready to do that, and there are no free heavy rail systems, except on opening day and during some celebrations.

The *proof-of-payment* or *honor* system—requiring that all patrons pay an appropriate fare, but having them prove it only on request by inspectors—is very workable because movement is not delayed, and the only physical device needed is a well-marked line that notifies everybody about crossing from the free space outside to the paid space inside. If the fare structure is complicated by zones and time of day, nonregular customers may need assistance in acquiring the proper ticket. The level of fines and intensity of enforcement (i.e., the number of inspectors deployed) should reflect the prevalence of fare evasion in any community and the policy regarding a defined level of tolerance toward revenue losses. The adoption of a proof-of-payment fare collection system can have a profound effect on station design. If the original configuration is on such a basis, with open access, it may be impractical to revert to a barrier-type concept later.

Mechanical turnstiles have become museum items in the industrialized world, although not for a very long time in some cases. Magnetically encoded cards provide ample flexibility to charge fares by distance, time of day, characteristics of user (student, elderly person, pregnant woman, handicapped person), and the need to swipe the card through a reader or carry it past the proximity sensor on entry and sometimes on exit is not a difficult or time-consuming chore.

In all instances the possession and use of a prepaid pass, good for a defined time period, expedites all tasks and encourages mass transit use, since any additional trips taken beyond the base number are in effect free. It is not surprising that holders of passes depend less on the private automobile. This could conceivably lead to a recommendation that all city residents be encouraged, and perhaps required, to acquire a transit pass (on favorable terms), thereby building a larger clientele for public transportation.

Yards

Heavy rail systems, like all other railroads, have to have yards where rolling stock is stored when not in use, cars are cleaned, maintenance is performed, and crews report for work. With a system of any size, these facilities will not be shared with other operations (such as buses or regular rail), since transit vehicles have special characteristics and their own housekeeping needs. There is also likely to be institutional separation of responsibilities among agencies.

The yard should be directly on the operating network so that deadheading of trains at the start and end of runs can be minimized. Such facilities also occupy significant parcels of reasonably level land and present an industrial appearance—finding an appropriate and acceptable site is therefore not always an easy task. As a rule of thumb, about 2.5 acres (1 hectare) of land is required for each 5 trains to be stored.

Power Supply

All heavy rail transit operations depend on electric power, and the standard supply arrangement is the third rail, with the insulated running rails providing ground return for current. Each self-propelled car or multiple unit (MU)[43] has shoes that slide along this rail and pick up power. The voltage is 600 to 750 V dc and even higher (Hong Kong uses 1500 V dc), and thus is extremely dangerous to humans. Third-rail systems can obviously be employed only if the right-of-way is completely protected against intrusion by any nonauthorized personnel. (Short gaps can be tolerated because each train picks up power at many points along its length.)

---

[43] Multiple units (MUs) are two or three cars semipermanently joined by a drawbar, all controlled from a single cab. They usually share undercar equipment. A three-car MU may include a trailer car without motors. A train of any length (a *consist*) can be made up of several MUs or of all fully equipped single cars (or some MUs and some single cars).

When such exclusivity is not the case (as, for example, with the several premetro and *Stadtbahn* systems), power has to be supplied by overhead wires, out of reach of people and other vehicles. This has an impact in tunnels because the pantograph on the roof of the vehicle takes up space (even in a collapsed configuration), and the tunnels therefore have to have a higher vertical dimension than would otherwise be required.

Rolling Stock

All metro vehicles have the same general configuration, yet—because they tend to be manufactured under special orders by various transit agencies—they are usually somewhat different from one place to another. The passenger compartment is supported by two trucks (bogies) with two axles each. There are also two-car units mounted on three trucks with an open through-passage; sometimes two (a *married pair*) or three vehicles are permanently coupled together, thereby saving somewhat in control systems and motors. The motors are in the trucks, and all the vehicles are electrically powered single cars coupled together or multiple units. The significant differences are in the overall dimensions (see Table 13.3), the number and location of the doors, whether longitudinal passage from one car to another is possible, and the number of seats provided and their arrangement. (See Fig. 13.16.)

Car capacity refers to the number of seats plus standees, all within the interior floor area of the car. Transit cars range in length from 50 to 85 ft (15.2 to 26 m), with 75 ft (23 m) being common. Car width ranges from 9.5 to 10.5 ft (2.9 to 3.2 m), with 10.2 ft (3.1 m) being a norm. The allocation of floor space to seats, standees, vestibules, wheelchairs, and luggage space will be decided by each operating agency, reflecting their perception and policy toward expected service demands. Fewer seats and more standing room will be the likely choice on systems with high ridership volumes and stations placed close together.

A seat should take up at least 3.2 ft^2 (0.3 m^2) of floor space; a standee can be placed on 2.3 ft^2 (0.2 m^2). These are not at all comfortable standards, and for noncrush loading might be increased to 3.75 and 3.0 ft^2 (0.5 and 0.28 m^2), respectively. Reviewing the literature and norms followed by various

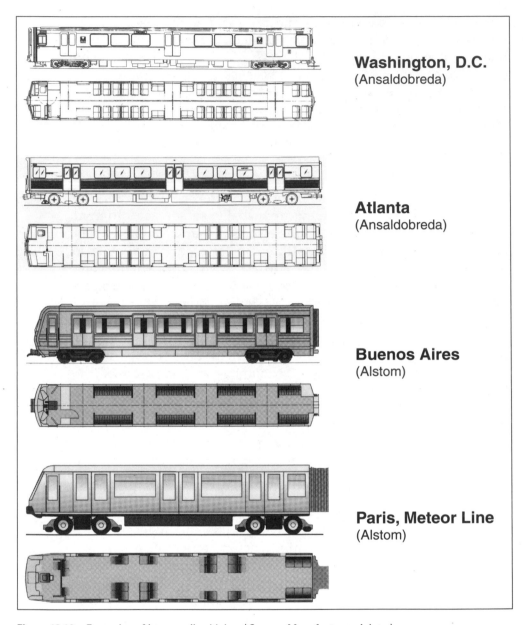

Washington, D.C.
(Ansaldobreda)

Atlanta
(Ansaldobreda)

Buenos Aires
(Alstom)

Paris, Meteor Line
(Alstom)

Figure 13.16   Examples of heavy rail vehicles. (*Source:* Manufacturers' data.)

transit agencies,[44] the ranges for floor space allocations are the following:

Floor space of car

Range	Per Seat	Per Standee
Smallest	2.9 ft² (0.27 m²)	1.6 ft² (0.15 m²)
Largest	5.7 ft² (0.53 m²)	4.0 ft² (0.37 m²)

Thus, if a 75- × 10.2-ft (23- × 3.1-m) car has 74 seats, it can accommodate 225 passengers. A more civilized loading would be 180 passengers. If the number of seats in the same car were reduced to 40, the capacity would be 240 to 186 for the various loading conditions. Thus, particularly because the dimensions of the car can also be changed, there is a range of car capacities.[45] For preliminary assessments, the figure of 200 passengers per car is a reasonable approximation.

Car design from the user's perspective becomes an increasingly critical concern in light of lingering memories of discomfort with older systems and growing environmental expectations with a higher quality of life. People in prosperous countries simply refuse to accept unpleasant conditions on public systems today. In transit cars, this attitude is not determined by the question of square footage per rider alone, although it is the first concern. Many of the basic issues have been outlined previously, largely as elements of station design; inside cars, people are even more sensitive to noise, vibration, illumination, temperature, and humidity levels because of the space constraints.

In no strict order, the following items come up repeatedly in user evaluations:

- *Seats* that are reasonably comfortable and vandalproof. Upholstering may not be practical, and neither are indentations on benches, because all bottoms are not the same width.

- *Sanitation arrangements,* with litter receptacles available and debris easily removable.

- *Cantilevered seats* that allow easy cleaning of floors and provide space for packages.

[44] See Edwards, op. cit., p. 134.
[45] BART uses 69-ft cars with 2 doors per side, providing about 70 seats; MARTA has 75-ft cars with 3 doors per side and about 74 seats.

- *Public address systems,* with understandable announcements.

- *Maps and other information* readily visible.

- *Handholds* and stanchions that are also useful to short people and do not obstruct movement on board.

- *Wide doors* that expedite boarding and exiting. Closing mechanisms that do not injure people, but minimize delays.

- *Nonslip floors* that do not show dirt.

- *Cleanable surfaces* from which graffiti and scratches can be removed (including windows).

- *Wheelchair positions* that are secure and nonobstructing. Space for bicycles can also be considered.

- *Access to transit personnel* in case of emergency.

- *Security screens* that increase security (at doors, for example, to prevent purse snatching as the doors close).

From the engineering point of view, transit cars have a number of features that are best left to specialists. These relate to the precision of control systems, the efficiency of motors, the structural strength and weight of bodies and trucks, and surface finishes. The aims in hardware development have been to reduce power consumption (largely by making the cars lighter) and improve the agility of trains by upgrading acceleration and deceleration capabilities. As noted previously, changes in speed coming into or leaving stations have to be governed by human comfort levels, particularly by keeping the *jerk rate* (rate of change of acceleration) in check.

The development of *regenerative braking* systems is particularly notable. Instead of wasting the energy that is released in braking by discharging it through rheostats as heat (and also thereby heating up the tunnels), the energy is used to drive the motors, which then act as generators, feeding power back into the system and helping to brake the car.

The recent history of rail car manufacturing in the United States is not an inspiring story. American manufacturers have deemed the transit market small, and the volume is uneven from year to year. Delivery schedules have been tight with strict deadlines. Inflation has been high and unexpected on some orders; specifications have changed rapidly, particularly with respect to advanced controls and reduced weight. When operating agencies demanded strict guarantees from the prime contractors, American manufacturers, who basically assembled components produced by others, could not respond.[46] The aerospace companies, which at one time considered making a serious entry in this field, have not been successful either. There are considerable differences in how a space rocket and a transit car are built. For example, the Urban Mass Transportation Administration sponsored the development of a state-of-the-art model rapid transit car (by Boeing-Vertol Division and St. Louis Car Division), but the results did not gain practical application. It has taken some years to develop reliable hardware from fragile, sophisticated, and expensive components, replacing the simple and heavy but robust vehicles of a previous era.

In the early 1960s, three American companies produced cars—the St. Louis Car Division of General Steel Industries, the Pullman-Standard Division of Pullman Inc., and the Budd Company. Twenty years later, only Budd remained in business, but it had been acquired by the Thyssen Group of Germany. Rohr, Boeing, and General Electric entered the field for short periods. Thereafter, bids on new and replacement fleets came from Japan, France–Belgium, Germany, Italy, and Canada. The rule that the value of a government sponsored purchases has to be at least 51% American-made remained in effect, and, consequently, some foreign manufacturers established US assembly plants, while others made arrangements with existing prime contractors.

The newer Washington, D.C., cars are Italian (Breda); Cleveland and the older systems rely on Japanese and Canadian models (Kawasaki, Kinki Sharyo, Tokyu, Hawker Siddeley, Vickers, and Bombardier); Atlanta uses French, Japanese, and Italian cars (Franco-Belge, Hitachi/Itoh, and Breda), and Baltimore and Miami look toward Canada (Transit America). In recent years, another major player had emerged—Adtranz, with Swedish origins, a German base, and plants at many locations around the world—but, it has since merged again, with Bombardier. It is now apparent that the market acts at a global scale, and the pattern of mergers and acquisitions is by no means completed.

---

[46] A sobering experience was the occasion in 1980 when the New York Transit Authority successfully sued Pullman because it delivered R-46 cars that developed multiple cracks in the undercarriages, which had been produced by Rockwell.

Scheduling and Capacity

Repeated experience has shown that rapid transit trains can be operated at a 90-second headway. But it is not always assured that such short intervals can be maintained over extensive periods, because all operational capabilities are stretched to their limits.[47] If 40 trains can be moved per hour, with a consist of 10 cars, each carrying 250 passengers, the single track capacity would be 100,000 riders. This figure has to be left in the realm of theory, because it cannot be relied on under real-life conditions, nor should so many passengers be packed into each train. Yet, 80,000 per hour has been attained on the Hong Kong system, carrying 2700 passengers per train at a 2-minute headway,[48] and the West Line currently under construction expects to accommodate 100,000 passengers in peak periods at a 105-second headway in 9-car trains (at least 33 trains per hour).

With 30 ten-car trains per hour, each car carrying a reasonable peak hour load of 200, the capacity of 60,000 riders per track is achieved. There is nothing impossible about this, since several routes of New York subway are able to do exactly that—or at least they did it routinely some years ago, when the system operated with shorter intervals between trains.

A modest scenario but one that still does justice to the heavy rail mode would be a 5-minute headway for 8-car trains carrying an average of 180 passengers per car. This will result in a capacity of 17,280 passengers per hour per track. Any system can obviously be operated with shorter trains and less frequent schedules, even in the peak hours, but then there is less, or hardly any, reason to consider heavy rail transit in the first place.

The basic method for calculating a fleet requirement for heavy rail transit is identical to that discussed under "Bus Scheduling Example" in Chap. 8. The total round-trip length in minutes is divided by the train headway in minutes. It should be noted that the round-trip time for any given train has to include layover and

---

[47] In some instances it has been claimed that 80-second intervals or less can be achieved (45 trains per hour), but it would be very difficult to maintain such frequency, certainly not at speed and not with line-end turnbacks or close central business district station spacing.

[48] Edwards, op. cit., p. 438.

recovery periods at both ends and any other expected regular delays along the way. To this number of active trains in the fleet, a reserve component should be added because all scheduled train sets may not be available at any given time. Some will be in for repair or maintenance, and others may develop problems at the last moment, leaving a potential gap in the schedule. Thus, a 10 percent reserve fleet should be the minimum, with 25 percent a much more confidence-inspiring number, but not always attained. New systems usually plan on 15 percent spares, to balance new-car reliability and new-car risks.

Cost Considerations

Metro systems are built at irregular intervals, under widely diverse demand and site conditions, in different parts of the world, and during changing economic situations. Historical records of actual experience are therefore of limited utility, but we do not have much else to go by in making first estimates. There have been quite a number of preliminary estimates associated with early rapid transit feasibility studies in American cities. Regrettably, the follow-up record shows that they do not carry much reliability. It has been suggested that the low cost numbers reflect a tendency to put the pending proposals in the best possible light. When basic planning and engineering have been done for any given project, careful construction cost estimates can be prepared, but even then with a healthy allocation for contingencies and unexpected events. The underground environment can generate many great surprises, unless very thorough (and costly) subsurface exploration is done.

The construction costs are broken down in the following groups, each having its own conditions and concerns:

- Right-of-way
- Track, guideway, and channel
- Stations
- Motive and control systems
- Support facilities, such as yards
- Rolling stock

The first consideration is the acquisition of the right-of-way, and no general cost guidances can be offered here. This may be a small effort if existing public properties are to be used—streets, for example—or it may amount to a massive program if properties are to be bought through eminent domain in high-value districts. Clearly, a continuous strip of land is necessary, and even one holdout can dislocate implementation schedules if court action has to be taken. That possibility, as well as relocating residences, small businesses, and institutions, is to be guarded against. To be economically correct, a "free" right-of-way should also be evaluated as to its *opportunity costs*—the value of the land if it were to be used for some other reasonable purpose, which opportunity is now foregone.

If deep tunnels are to be used, the land acquisition problem might be minimized, but not avoided entirely. Openings for construction shafts will be needed, stations and their entrances will almost certainly require some surface properties, and easements will also have to be obtained or purchased if the tunnels run under private properties.[49] If the tunnel is so deep that it could not affect any conceivable use of that space by the owner and will not threaten any buildings, the cost of the easement will not be excessive. An elevated transit structure, while it touches the surface only at points spaced far apart, will preclude any reasonable use of the land below, particularly if railroad maintenance needs have to be accommodated; therefore, outright acquisition of a continuous right-of-way will usually be involved.

Without doubt, the lowest construction costs will be achieved if the transit line is placed on the surface. Comparative analyses suggest certain ratios between this option and all the others. As a very general guide only, an elevated alignment will usually cost 2 times as much as a surface solution (leaving aside all right-of-way expenses).[50] (See Fig. 13.17.) A tunnel will be 2 or 3 times the cost of the elevated guideway, but only under normal conditions. If the subsurface conditions are difficult, the ratio may escalate considerably.

A program much cherished by planners is the *value capture* effort. Since the location of a transit station increases property values in the immediate vicinity because of increased accessibil-

---

[49] An *easement* is not the acquisition of full title to a property, but only of some specified rights (such as entering a parcel for a defined purpose or running a line above or below it not affecting anything else).

[50] A summary is found in Edwards, op. cit., p. 166.

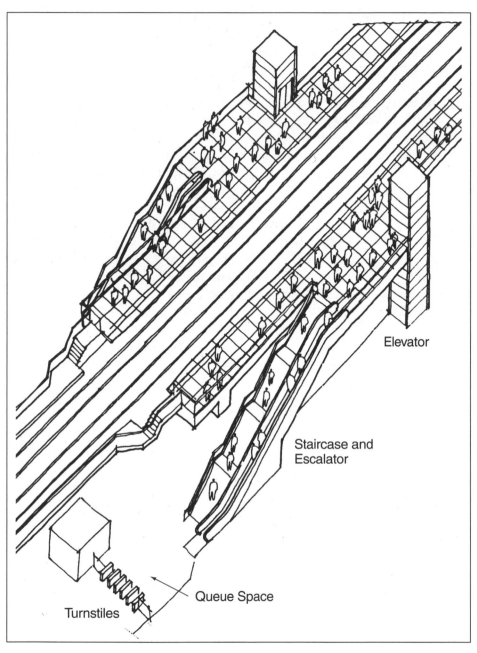

Elevator

Staircase and
Escalator

Queue Space

Turnstiles

Figure 13.17 Elevated station with side platforms. (*Source:* H. A. Kivett and K. Peterson, *Rail Station Compendium,* Parsons Brinckerhoff, 1995. Used with permission.)

ity, that increment—created by public expenditure—should bene-fit the public, rather than provide a windfall to private owners. This can be done by acquiring all the affected parcels before con-struction starts at the then-prevailing market price and then reselling them under the new market conditions. The theory is fine, but there are rules in the United States against "excess con-demnation," and transit agencies are not particularly skilled play-ers in the real estate business. There are usually no up-front resources for the purchases, and risks are certainly present. Besides maintaining prudence in the expenditure of public funds, it can also be argued that the value is recaptured eventually any-way through higher real estate taxes on more valuable property. Air rights over transit properties can be sold, and some recovery of investment may thereby be achieved.

Short of significant real estate ventures, there is the concept of *joint development:* a plan is adopted that utilizes private and pub-lic investments to achieve the most effective concentration of new buildings and activities around stations. The government contri-bution, besides the infrastructure components, may be as little as constructive zoning changes. Further steps can be taken, as, for example, the development of public spaces, walkways, and local transit access. In Mexico City, station construction has been extended upward into multistory buildings that have rentable commercial space.

Station plaza in the Mexico City metro system.

The construction of the track and guideway, including tunnels and elevated struc-tures as appropriate, is a fairly routine process, with not too much of a range in expendi-tures. It is usually expressed as a sum of dollars per mile for a two-track alignment, including stations. In the early postwar period, it hovered between $20 million and $40 million for elevated guideways and just below $100 million for tunnels. To the surprise of most, the steep inflation of the 1970s and early 1980s pushed the price through the $100 million mark. A comprehensive review in 1989 (utilizing 1983 dollars) gave the following summary:[51]

In tunnel with stations	$137 million per mile
In tunnel excluding stations	103 million
On elevated structure with stations	55 million
On elevated structure excluding stations	39 million
At grade with stations	31 million
At grade excluding stations	22 million

That was before the Los Angeles Red Line project started. After it ended (i.e., was terminated), there was another surprise: the total price tag for the 17.4 mi (28 km) entirely in tunnel was over $4.5 billion. This averages out to well over $250 million per mile. In other words, in little more than a decade, the construction costs of heavy rail had doubled. Granted, this is a single case, and arguably not representative of what most new systems can or should be, but the trend is dramatic, nevertheless.

The experience in Hong Kong is no different. The 19-mi (30.5-km) West Rail Line under construction at this writing is a $6.5 billion project.[52] While the construction conditions have been quite difficult and high standards are being maintained, only about a third of the alignment is in a tunnel, and the average cost is still $342 million per mile. Twenty years ago, the cost on a previous metro project was about $130 million per mile.

Station design and construction are special efforts with various options open. First, large cavities for platforms, mezzanines, and access paths have to be carved out underground, and much of this is manual work. Elevated stations involve elaborate and large structures. Second, the stations can be rather spartan in arrangement and appointments (as was done in Lindenwold and Baltimore, for example), or they can be very spacious, with wide platforms, generous mezzanines and passenger corridors, and extensive surface plazas and entryways (as was done in the BART

[51] Pickrell, op. cit.
[52] "Hong Kong's West Rail Line Quietly Setting New Standards," *Engineering News Record,* October 2000, pp. 40–43.

and Los Angeles cases, and in a few other places). The costs are large and variable, and they are separate cost items. The previously mentioned study identified $40 million for underground stations, $23 million for elevated, and $10 million for at-grade facilities.[53] By 2001, $100 million for an urban subway station would be a reasonable first approximation.

Distributed over the dual trackage and within each station are the various subsystems that power the trains, control their movements, permit vital communications, and collect the fares. The costs of all of these taken together depend heavily on the choices made regarding levels of reliability and technical advancement, station amenities, safety provisions, and fare collection methods. Otherwise, the costs relate not only to the length of line, but also the complexity of routes, number of stations, and level of service to be provided. There is a range for the subsystems between a bare-bones and a gold-plated approach, with most transit agencies opting for a position between those extremes. Election of minimal-cost systems and components will eventually impact the O&M expenses and likely require upgrading as demand grows (at even higher costs). These costs will usually be separated on a per-mile basis as follows:

- Traction power supply and distribution
- Train control systems
- Communications systems
- Fare collection systems

(These costs are included in the summary estimates given previously.)

The expenses of building yards are primarily associated with finding and purchasing the necessary real estate and deciding about the extent and quality of the maintenance facilities and equipment. The per-mile cost summaries usually include these improvements.

The price of a heavy rail transit car is not only a function of the size of the vehicle; it is substantially affected by the amount and sophistication of included components, the size of the order, the current workloads of competing manufacturers, the cost of delivering the vehicles to the site, the urgency of implementation sched-

---

[53] Pickrell, op. cit.

ules, the warranties required, and the financing arrangements. In the early 1970s, such a car could be purchased for $273,000 (New York R-46 cars). The price passed the $1 million mark some years ago, and today it will be at least $1.5 million to $2 million.

Operational expenses of heavy rail transit have justifiably received much scrutiny lately. This is particularly triggered by the fact that construction may be sponsored to a large extent by external sources, but O&M is increasingly a local responsibility.

The design approach in São Paulo, with open stations.

Metro systems have the theoretical advantage of minimizing labor input on a per-passenger basis due to the heaviness and large capacity of the equipment and infrastructure. With increasing automation, further steps can be taken. However, as is discussed under other transit modes, the time span encompassing the two daily peak periods is more than a regular working day, and paying overtime every day to many employees is an expensive proposition. The responses are such arrangements as split shifts (with very low pay for the intervening hours), entirely unpaid breaks, or the use of part-time labor on an hourly basis. All these arrangements usually mean modifications to traditional labor practices and will require negotiations with and acceptance by labor unions.

Possible Action Programs

In the case of North American metropolitan areas, possible action programs with proposed metro systems—even substantial extensions to existing networks—will have to be approached with extreme care. Nothing will be accepted by the public, elected officials, or sponsoring agencies on faith or good expectations. The mass media have widely publicized the recent cost problems, and the general tenor is to search for the most economical and noncapital-intensive programs in any community. The burden of proof to show unequivocally that metro service is not only necessary but also affordable therefore rests heavily on the shoulders of advocates.

A thorough planning process is thus indicated that concentrates, first, on patronage estimates showing in an understandable and most reliable way the expected ridership. Almost certainly, this will have to be accompanied by a comprehensive restructuring of the circulation and feeder systems in broad corridors. It will also have to be shown that no other mode can cope with the anticipated loads. This will not be difficult to do in many of the megacities in the developing world; it will be a major challenge in North America.

Second, very thorough financial analyses will be expected that show how the necessary massive funds are to be assembled. The sources may be federal, state, or local, but the review cannot be just theoretical—the reality of capabilities and expectations over a considerable time period will have to be documented. Private participation may play a significant role, ranging from some peripheral tasks to complete design, building, operation, and management arrangements. Communities are becoming increasingly aware of and concerned about continuing operations and maintenance needs, particularly since government assistance programs in this sector are waning. If the fares have to be subsidized—which they certainly will have to be—where will the money come from?

The hopes of planners remain that metro systems may be seen as powerful tools to structure and achieve a better and more efficient regional development pattern. No other transportation mode can quite do this, but, as noted previously, we appear to lack the necessary institutional structure and political will to go boldly where a number of European cities have gone. Regrettably, such possibilities may

New housing districts in Singapore connected by rapid transit.

not sway public referendums and financial organizations at the point when decisions have to be made, looking ahead to an immediate and difficult implementation period.

Conclusion

There is no doubt that heavy rail systems are the most effective mode available to cope with large public transportation demands and to serve large urban agglomerations. The problem is that we have developed an urban pattern in the new

Access plaza to the Los Angeles Red Line.

American metropolis that is not able to support this option or does not actually need it in most instances. Low density does not lead to concentrated passenger loads. Our grandchildren may find some decades in the future that this was not a wise path to take because major operational inefficiencies will have been built into this spread-out metropolitan structure, but that is what we mostly have now. The trend continues in spite of a general recognition of the problems associated with sprawl and the "nonsustainable" city. These conditions may be corrected, and they should be, but it will take heroic political and community effort to do so. The metro is a transportation device that can help achieve it.

In the meantime, it is still a large world out there, and urbanization trends continue unabated on other continents. The existing and emerging megacities can survive, if not flourish, only with internal means of high-capacity transportation. For those cities there is hardly anything to consider except heavy rail transit.

Bibliography

Edwards, John D. (ed.): *Transportation Planning Handbook*, Institute of Transportation Engineers/Prentice-Hall, 1992, 526 pp. In particular, Chap. 5, "Urban Mass Transit Systems," by Herbert S. Levinson summarizes most of the planning and engineering features.

Garbutt, Paul: *World Metro Systems* (2d ed.), Capital Transport, United Kingdom, 1997, 136 pp. Profusely illustrated review of all aspects, with references to most systems, taking an advocacy position.

Gray, George E., and Lester A. Hoel (eds.): *Public Transportation* (2d ed.), Prentice-Hall, 1992, 750 pp. Comprehensive coverage of all modes; heavy rail is discussed specifically in Chaps. 4, 5, and 11.

Institution of Civil Engineers, *Urban Railways and the Civil Engineers,* Proceedings of a Conference in London, 1987, 257 pp. Various papers on different aspects of rail modes, mainly covering European experience in planning, design, construction, and maintenance.

Transportation Research Board, *Transportation Research Records,* National Academy Press. The following recent issues provide specialized articles on heavy rail: no. 760, 1980; no. 1451, 1994; no. 1503, 1995; no. 1571, 1997; no. 1604, 1997; no. 1623, 1998; no. 1669, 1999; and no. 1735, 2000.

Vuchic, Vukan R.: *Urban Public Transportation: Systems and Technology,* Prentice-Hall, 1981, 673 pp. Includes full coverage of the heavy rail mode, stressing engineering aspects.

Commuter Rail

Background

The traditional rail mode is still the most efficient way to move large volumes of people over many miles at a reasonable speed. When these systems operate at the metropolitan scale, they take the form of commuter[1] or regional rail service. Indeed, the operations discussed in this chapter can only exist in large conurbations with distinct employment centers and population concentrations in corridors, since the stations have to be relatively far apart, and they must attract sufficient numbers of riders to warrant stopping a train. This is, after all, the mode with the heaviest rolling stock and the most extensive infrastructure. Commuter rail has been the principal means to allow metropolitan areas to happen historically and to hold the larger ones together even today. This mode works effectively where movement demands are on a massive scale, and it is certain to have a role in the very large metropolitan areas of the future.

Because rail lines and trains in America have a long history, and because significant evolutionary changes have taken place in

---

[1] The dictionary definition of *commute* is to travel back and forth on a regular basis. It is said that this designation stems from the early days of railroading, when operating companies, in order to attract steady customers through already existing suburban stations, "commuted," or reduced, the fare.

the last half century, there is an inventory of common preconceptions and attitudes regarding this mode. One, for example, is that regional rail can provide "cheap" transportation. That is not really the case, even though existing rights-of-way may be used, because modern attractive service does require quality rolling stock, a reliable infrastructure, and responsive management. Passenger rail is seen as having less agile, slower, and lower-volume operations than heavy rail transit (metro). That is basically true, because traditional rail service was not originally designed to provide quick movement for a massive ridership: the intention was to offer comfort and reliability for a public that was not always in a hurry. Another widely held idea is that commuter rail is a mode for the wealthy. To the extent that late twentieth century residential distribution patterns prevail, with higher-income families located in suburban areas, there is some statistical truth in this. But this condition is not caused by the mode itself. Indeed, there are many reasons to correct this imbalance, and reformed rail service may be a means to help do that.

The general label for this type of service should be *regional rail*—passenger service at the metropolitan level in self-propelled trains or trains pulled by locomotives. Strictly speaking, *commuter rail* should refer only to operations at the beginning and end of the workday;[2] however, the latter term is in general use for all nonmetro rail operations even if they accommodate shoppers, students, visitors, etc. In this work, both terms are employed, leaning toward *commuter rail,* thus respecting common, if somewhat imprecise, usage. Europeans frequently use the term *pendulum service;* sometimes *suburban rail* appears as a name and a characterization.

There is practically no instance where a commuter rail route has been built in North America entirely for the sole purpose of serving commuter trains, at least not in the last hundred years. New services on existing track, however, have been implemented many times in response to opportunity and need. The first such full-scale operations happened in the late nineteenth century because prosperous families discovered that they could build a spacious house on a large lot in a remote village with a railroad station and because they could afford the daily round-trip ticket for the breadwinner who worked in the city. They invented the suburb and made it

[2] The precise distinction is maintained by V. Vuchic, *Urban Public Transportation,* (Prentice-Hall, 1981), pp. 647 and 649.

work.[3] Today, commuter lines operate successfully, and new ones are started as the urban development spreads out, but only along old railroad alignments and existing rights-of-way.

There are distinct differences regarding this mode in different parts of the world. In Western Europe, with its dense network of track, frequent service, and intensive general usage, it is often not easy to separate commuter rail as an operation distinct from regular trains, except when special systems have been

Small commuter rail station on the Long Island Rail Road.

built, such as the S-Bahns of Germany and the RER of Paris. Also, trains operating at the metropolitan scale usually make many stops within the city, thus being not particularly different from metro or heavy rail transit.

In the developing world, the situations are diverse, but frequently rail lines, if they exist, are brought into almost emergency commuting service to gain some means of mobility in metropolitan areas where explosive growth has badly outpaced all service capabilities.

In North American communities the distinction is clear, however, because hardly any intercity passenger service still exists. Regional passenger rail service has become a special activity, much closer to transit in its role. It is usually separate in its management from long-distance rail operations, which continue to experience their own difficulties.

Only a handful of large cities enjoy the presence of regional rail, and those services almost exclusively connect suburban nodes to the central business district, with very few stops within

[3] John R. Stilgoe, *Metropolitan Corridor* (Yale University Press, 1983, 397 pp.), describes the formative period from 1880 to 1930, as documented in contemporary literature and popular publications. See also *Borderlands: Origins of the American Suburb, 1820–1939* (Yale University Press, 1988, 353 pp.) by the same author. A somewhat different suburb for the middle and working classes came with streetcars—see S. Warner Jr., *The Streetcar Suburbs* (Harvard University Press, 1962).

the old parts of the city. Even if the instances of commuter rail operations are few (13 places in the entire United States[4] and 3 in Canada; see Table 14.1) the total volume of users is quite respectable. In terms of passenger miles, the traffic carried by regional rail is only somewhat less than that by metro (heavy rail), as shown in Fig. 14.1. But, because commuter rail accommodates longer trips than the other forms of transit, of all the transit trips (or number of boardings) made nationally only 4.4 percent are on commuter rail, while the share for metro is 27.4 percent.[5] Commuter rail accounts for 25 percent of the value of capital investment in transit and 12 percent of the annual operating expenses; on the other hand, the sizes of the vehicle fleets are not as different as the other proportions—4907 commuter rail cars versus 10,301 heavy rail transit cars (in 1998).[6]

Since commuter rail systems operate over the trackage of regular railroad systems, frequently mixed in with freight and long-distance passenger traffic, they have to respect the overall standards and practices as defined by the Federal Railroad Administration of the U.S. Department of Transportation and the Association of American Railroads. This fact makes the construction and operations parameters substantially different from those of the other types of rail transit.

The lore of trains is an indelible part of American culture, certainly until the last half century. They are seen as symbols and practical devices in literature; they provide material for Hollywood, ranging from *The Great Train Robbery* (Edison, 1903) to *Throw Momma from the Train* (Orion, 1987). It is not an exaggeration to say that railroads made the country and its cities what they are—before the automobile came. That history of development explains what we have inherited today.

Development History

Commuter rail service started and continued as an intrinsic part of general rail operations. Only in the second half of the twentieth century has it become a readily separable service, more akin to transit than long-distance trains. Therefore, at the risk of going off on a sidetrack, but in order to place this effort in its proper

[4] See Tables 14.1 and 14.2.
[5] American Public Transportation Association (APTA), *Public Transportation Fact Book 2000,* p. 66.
[6] Ibid, p. 83.

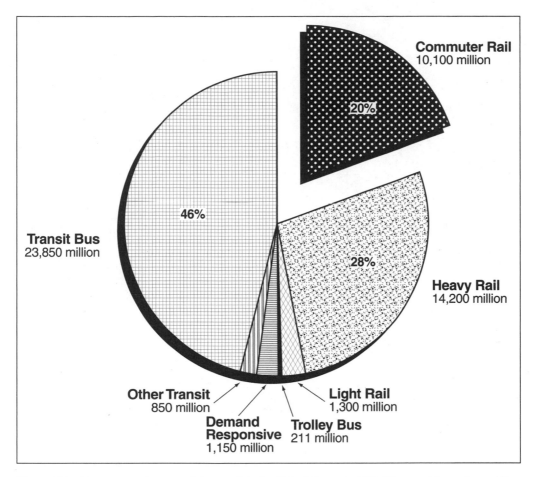

Commuter Rail
10,100 million

20%

46%

Transit Bus
23,850 million

28%

Heavy Rail
14,200 million

Other Transit
850 million

Light Rail
1,300 million

Demand Responsive
1,150 million

Trolley Bus
211 million

Figure 14.1 Passenger miles on public transit in the United States in 2000. (*Source: Metro, Annual Fact Book,* 2000, p. 11.)

context, we will outline the United States rail history in its entirety, but as briefly as possible in spite of its richness.

The history of railroads in the United States[7] is not just a matter of a new transportation mode being introduced and placed in

[7] A simple bibliography of U.S. railroad history would extend for many pages. There are thousands of publications, ranging from scholarly treatises on policy matters to detailed monographs for every line that has ever been constructed. It seems that every good or bad event in the railroad industry has generated at least one book. Some that may be useful in expanding the information base for the discussion here are: John F. Stover, *American Railroads* (University of Chicago Press, 1961, 302 pp.); John F. Stover, *The Life and Decline of the American Railroad* (Oxford University Press, 1970, 324 pp.); John R. Stilgoe, *Metropolitan Corridors: Railroads and the American Scene* (Yale University Press, 1983, 397 pp.).

Table 14.1 Commuter Rail Systems in North America (1997)

Central City/ Responsible Agency	Number of Routes	Total Length	Number of Stations	Diesel Locomotives	Electric Locomotives	EMUs	Coaches	Passengers per Year, millions
Baltimore, MD Maryland Commuter Rail (MARC) train service	3	—	40	25	4	—	101	4.7
Boston, MA Massachusetts Bay Transportation Authority	13	328 mi (528 km)	116	55	—	—	287	29
Chicago, IL Metra, Northern Indiana Commuter Transportation District, Burlington Northern Santa Fe	11	334 mi (537 km)	250	130	—	223	861	75.2
Dallas–Fort Worth, TX Trinity Railway Express	1	10 mi (16 km)	—	—	—	—	13	—
Los Angeles, CA Southern California Regional Rail Authority	6	416 mi (669 km)	45	33	—	—	119	6.2
Miami, FL Tri-Rail	1	71 mi (114 km)	19	10	—	—	30	2.5
Montreal, Canada L'agence Métropolitain de Transport	3	73 mi (118 km)	37	13	—	29	122	8.4
Newark, NJ–New York City New Jersey Transit	12	390 mi (627 km)	162	60	32	300	514	52.1
New York City Metropolitan Transportation Authority/Long Island Rail Road	11	323 mi (520 km)	124	67	—	916	191	75.8
Metropolitan Transportation Authority/MetroNorth	—	267 mi (429 km)	118	39	7 dual	750	78	62.2

Philadelphia, PA Southeast Pennsylvania Transportation Authority	13	304 mi (489 km)	176	—	8	304	35	22.5
Providence, RI Rhode Island Public Transit Authority	1	43 mi (69 km)	—	—	—	—	—	—
San Diego, CA North County Transportation District	1	43 mi (70 km)	9	5	—	—	16	—
San Francisco, CA Caltrain	1	83 mi (133 km)	34	20	—	—	93	8.1
Santa Clara, CA San Joaquin Regional Rail Commission	1	—	—	—	—	—	—	—
Toronto, Canada GO Transit	6	224 mi (361 km)	—	46	—	—	315	27.2
Vancouver, Canada BC Transit	1	40 mi (65 km)	8	—	—	—	—	1.5
Washington, DC MARC Train Service	3	191 mi (308 km)	38	27	4	—	94	4.7
Virginia Railway Express	2	87 mi (140 km)	18	14	—	—	65	1.9

Sources: *Jane's Urban Transportation Systems, 1999–2000*; operating agencies.

service; the events are not solely associated with the technology of the Industrial Revolution. The development of railroads is the development of the country. They opened the continent, they propelled population and business expansion forward and westward, and they built the economic might of the United States in the nineteenth century. While the principal attention was devoted to long-distance linkages, nodes had to be formed that became crucial concentrations of activities. Railroads located cities, and this crucial access made a number of them prosperous and important. Even the settlements that had existed before the railroad era were profoundly transformed. Railroads also shaped the inner configuration of cities by defining points and corridors of best accessibility. The tangible legacy that has been left—fundamental to the review of current and potential commuter rail service—are the extensive spiderwebs of reserved rights-of-way that still penetrate communities and metropolitan areas.

Experiments with steam engines on track started in the United States in 1825. The pioneering efforts with operating lines centered on Baltimore, Charleston, Boston, and Albany. The expansion picked up steam immediately, and by 1840 twenty-two states had some rail service in place. It was clear that this means of transportation responded exactly to the pent-up demand for mobility and access on a national scale: sources of raw material had to be tapped, production facilities supplied, and markets and ports reached. Railroads could do that cheaply, efficiently, and reliably. People also found the service most useful, giving passengers unprecedented ability to travel easily to remote destinations.

In the very early days rail lines entered cities along existing streets, even if this frightened horses, demolished some carts, killed a few people by exploding engines, and played havoc with the urban environment. The unnatural speed of 20 mph (32 kph) was difficult to accept. A solution in some places was to have a man with a red flag ride ahead of the train. Eventually municipal ordinances were passed that tried to address the safety issues, but that also triggered a continuing battle among the private railroad companies to secure their own rights-of-way into and through the principal cities. Part of the game was to block rival companies in the highly competitive situation that soon developed. The result was that in a number of communities more land was reserved for railroads than was strictly necessary.

It appears that in American communities, terminals, yards, and stations have penetrated deep into the urban fabric, and principal activity centers have been established in the very core. In European cities the large terminals seem to be somewhat further out—placed originally at the edges of the then contiguous development. Lines certainly could not cross the dense medieval centers, and frequently a ring of stations characterize the larger cities under the transformations of the Industrial Revolution. Actually, the basic events were the same in American cities. The difference is that there was much subsequent development here of downtown districts that extend beyond the sites of the first railroad stations, thus encompassing them in the high-density fabric.

There was a period of almost 100 years in North America during which railroads reigned supreme, having no effective competition from other modes. Continuous expansion of the network took place that was only interrupted now and then by wars, economic depressions, or financial panics. At the start of the Civil War, there were more than 30,000 mi in operation. Lines were pushed westward to penetrate beyond the Atlantic states; construction of track had even started in California. Chicago emerged as a focus for several important routes.

Technological improvements had to be made since the systems were still rather fragile and prone to mishaps. More suitable passenger coaches were built, and the steam whistle was perfected, but hardware and equipment remained at the simple stage. Most of the attention was devoted to raising money and organizing companies to build increasingly more mileage. Much was obtained through public subscriptions and the sale of stock, but state governments also supported construction through direct financial participation and—most important—by awarding powers of eminent domain to railroad companies. Land grant programs also appeared before the Civil War.

That major conflict proved conclusively the logistical value of rail transportation, and after the war expansion programs resumed with unprecedented vigor. A most notable event was spanning the continent by track, which was achieved in 1869, but that was only the most visible evidence of rail progress. Within the decade, five lines were completed that connected the East to the West, and the national system grew to 53,000 mi (85,000 km) by 1870. The principal concern in building lines was speed—to overcome as quickly as possible the large space that became the continental United States. The record was 10 mi (16 km) in a single day, but the inescapable consequence of this practice was that the quality of the roadway, bridges, track, and all other infrastructure elements suffered greatly. The country was left with a very large system that worked, but required continuous maintenance and upgrading. We are still suffering from that legacy. The railroad gangs, as they raced westward, left in their wake construction camps, which sometimes turned into settlements that became cities, but in most instances faded from the scene.

The most striking feature of the railroad efforts in the half century that preceded World War I was the prevailing business practices and ethics (or more accurately, the absence thereof). The railroad industry emerged as big business of tremendous economic and financial importance, and almost all the men who led it to that position were complete scoundrels. This colorful and greed-motivated situation has been described in every detail in many publications and does not particularly concern the discussion here, except to understand how the networks that we have today were created. Hardly anything associated with labor, organizational, financial, and administrative railroad practices then would be tolerated today, and toward the end of the century public displeasure was sufficient to achieve control legislation and supervision. Principal among those efforts was the establishment in 1887 of the Interstate Commerce Commission, which was fully abolished only in 1995.

Nevertheless, a remarkable system was built that gave the country a new basic infrastructure and changed its economic and social life forever. An interdependent economy with a high productive capacity was established. A web of railroad lines held it together and reached wide domestic as well as international markets. The methods used to accomplish this—if some rationalization is needed—only followed the accepted processes and attitudes of that time. The overall network, unlike those in other industrialized countries, may be held by private corporations, but, nevertheless, it became a national system with a national purpose. The rail rights-of-way crossing cities have to be seen as resources serving, or able to serve, public interests, even if ownership issues are somewhat complex. If nothing else, it can be noted that the federal government gave land grants to finance construction amounting to 131 million acres. In 1900, the total network was 193,000 mi (310,000 km); it reached its peak of 254,000 mi (409,000 km) in 1916.

By the end of the nineteenth century, further technological improvements had been made, including powerful and fast steam locomotives,[8] heavy steel rail, reliable (air) brakes, improved and safe couplers, strong bridges, and block-signal systems—all of which achieved much higher safety, and also improved efficiency. Perhaps even more important toward structuring a national system were agreements on a standard gauge for track (4 ft 8.5 in; 1.435 m) and standard times among the various companies. Integration of service with interchange of cars and the use of high-capacity and special freight cars streamlined the movement of goods and materials.

The turn of the century was a period of restructuring and consolidation of many individual enterprises into large corporations. In 1900, the giants of the industry were the New York Central, Chicago & North Western, Pennsylvania, Baltimore & Ohio, Chesapeake & Ohio, Erie, Southern, Missouri Pacific, Rock Island, Great Northern, North Pacific, Burlington, Union Pacific, Southern Pacific, and Illinois Central. These were household names at the national level and certainly within their own territories.

Also at this time, while commuting operations started to become a factor, there was a perceptible shift in attention by the operators toward freight, which was easier to accommodate and certainly more profitable. Passenger business grew in volume, but not proportionally, generating only one-fifth of total railroad revenues. Nevertheless, in this period of prosperity, the major corporations deemed it appropriate to build a series of monumental passenger stations to exhibit their might and create a public image. The more visible examples were St. Louis Union Station (1894), South Station in Boston (1898), Union Station in Washington (1907), and Pennsylvania Station (1910) and Grand Central Terminal (1913) in New York City. These and many others are still with us, trying to cope with new roles and diverse missions under completely different demand situations.[9]

---

[8] Locomotive No. 999 of New York Central exceeded 100 mph (160 kph) in May 1893 near Batavia, New York.

[9] Penn Station has been pushed underground, having lost its monumental aboveground structure; many of these buildings, particularly in smaller cities, have been converted to other uses. A number of old stations are left with very few trains each day.

The importance of rail transportation, which had no competition in passenger service at this time, is illustrated by the development of luxurious dinning cars and sleepers (Pullmans since the 1870s). Even the P.T. Barnum circus became a rail-based operation in 1872. Railroad companies acquired large land holdings in central locations, which also became a significant urban development factor in subsequent decades. In the 1920s, more than 90 percent of all intercity national passenger volume was carried by rail.

After World War I, during which the federal government assumed responsibility for all rail operations, conditions changed, and a downward trend began. While some technical improvements continued to be made, the railroad industry was unprepared for the emerging competition from automobiles, motor trucks, buses, and eventually airplanes. While excuses can be found in the strict regulations and controls under which railroads had to operate, and in the fact that companies had to maintain their own infrastructure (whereas the use of roads and highways was essentially free) and had to pay real estate taxes, there appeared to be little managerial energy and initiative to cope with the new situation.

The response was to cut back service, reduce mileage, and struggle with featherbedding practices by labor.[10] Smaller lines were abandoned, and the total mileage spiraled down to 171,000 (275,000 km) in 1939 and 100,000 (160,000 km) in 1959. Soon, all less-than-carload business, as well as most perishable and fragile or valuable freight, was lost.

The losses of passenger business were particularly severe. Starting with 1929, which incidentally was the year when the Greyhound bus system was organized, almost all rail passenger operations claimed losses continuously. Smaller stations were closed, low-density lines were abandoned, and frequency of service even on major lines was reduced at a steady pace. As service quality deteriorated, more customers were lost. One very visible effort by the rail industry to counteract the downward trend was to institute and promote streamlined "name" trains, such as the Zephyr, Super Chief, Owl, and Congressional, not to mention the Broadway Limited and the Twentieth Century Limited. This did not turn the tide, but became a further opportunity to lose money. Commuter operations were a particular drain. By 1959, only one-

---

[10] For example, diesel locomotives had to carry a fireman even though there were no fires to tend, and a 100-mi (160-km) run represented a day's paid work.

tenth of railroad income came from passenger traffic—after all, the companies have always seen their primary mission as moving freight—and the expenses continued to mount. The obvious goal of the operators was to shed this burden entirely.

After World War II, during which the railroad industry performed admirably under emergency conditions, there was a period of reorganization and mergers, frequently under desperate pressure. The nadir was around 1970. There were complex and continuous changes, and, to cut a long story short, out of all that federal and state programs emerged that relieved railroad companies of the passenger "burden." Since it was believed as a national policy that a basic intercity railroad network was required, the National Rail Passenger Corporation (Amtrak) was established in 1971. This is a federally supported effort to maintain at least skeleton passenger linkages between major population centers, usually utilizing the existing rail infrastructure under lease. It has lost money steadily, and at the time of this writing it is under a congressional mandate to become self-supporting by the end of 2002. Evidence is building that this will not be achieved. What will happen then, nobody knows, but local agencies may have to take over in the few metropolitan areas where Amtrak operates regional service, and it is expected that some of the busier intercity links may be operated by state agencies or private enterprises. There are great differences between the Northeast Corridor—where, without doubt, sufficient demand exists to maintain good rail service—and the rest of the country.

The other major thrust has been for regional authorities to assume responsibility for commuter service to the extent that it is reasonably supportable, beginning in the late 1960s. Commuter volume had also dwindled in 1959 to about a half of what it was in 1929 nationally, and it continued to drop. The railroad companies complained bitterly that they had been prevented from increasing ticket prices while the operations had become highly inefficient 20-hour weeks—two 2-hour periods of demand during 5 days, with very limited income during the off-hours, with expensive equipment standing by idly. Commuter passenger volume was one-fifth of the total carried, but generated only one-seventh of the revenues nationally. Private enterprise cannot handle that for long.

Today, commuter service under public agencies is subsidized from various sources because these operations are deemed vital for the economic and social well-being of large metropolitan areas. That is a policy, based on fact, which argues not only for the maintenance of this mode where it still exists, but also for the seeking of further opportunities, as discussed in the rest of this chapter. Fare box recovery[11] is frequently in the range of 30 percent (in a few cases up to 50 percent), but we consider the need for subsidies an acceptable situation in terms of social policy.

While all these institutional events took place, there were also several technological developments that affect service today. The most important of those was the change in the means of train propulsion. Electric motors had become practical for transit use in the 1880s, and experiments with electric locomotives started in the 1890s, with the first actual operation in 1895. The further development of this type of propulsion occurred primarily with heavy rail transit (metro), since regular railroads had no compelling reason to incur the considerable expense of electrifying their lines (providing power lines and a distribution grid). Slow expansion of this submode did take place, however, particularly where rail service had to go underground (such as at the two main terminals in New York City). By 1894, 6000 mi (9650 km) of route were so equipped.

Diesel power had a much greater impact on the rail industry. While the basic engine had been invented much earlier, the first practical switching locomotive was developed by 1925. Thereafter, a veritable revolution took place as diesel locomotives were placed in service during the 1930s, and by the end of the 1950s there were no steam locomotives to be seen anywhere, except in railroad museums. While diesel locomotives were quite expensive to purchase, their fuel efficiency was three times better than that of steam locomotives, they needed no water supply and daily inspection, they were simple to operate and maintain, they needed no start-up time at the beginning of any run, and they lacked the pounding force of steam engines that damaged equipment and track.

It is much to the credit of the railroad industry that, after the grim 1960s and 1970s, it was able to recover quite well, at least in the freight sector. Reorganization helped, but much was due to improved management practices and proper responses to market

[11] The portion of operating and maintenance budget (not including capital investment) covered by revenue from ticket sales.

demands. New and efficient rolling stock was acquired and new types of operations were instituted (such as piggyback, trailers on flatcars, containers, and unit trains). All that is beyond the scope of this discussion related to passenger modes, except to note that growing freight traffic has filled up the capacity of several lines so that the level of infrastructure underutilization is not as extreme as it was not so long ago.

During more than a century and a half, while many elements in the transportation picture have changed, one fact has remained constant: rolling a steel wheel over a steel track consumes the least amount of energy to move people or goods forward, compared to all other possible choices (except the bicycle). That fact alone should be sufficient to retain a role for commuter rail in metropolitan systems.

Types of Commuter Rail Operation

Commuter rail routes in North America are basically remnants of once massive systems of train operations with frequent schedules and a long reach. The current lines fit physically into the historical network. Alignments are usually utilized that were in place 100 years ago, and station locations have not moved either (with a few exceptions). There may be a few instances where a new station has been established in the outlying districts, and certainly a number of the old ones have been closed because of shifts in demand concentrations, but by and large the old sequences of stops stay in place. New track may have been laid along some stretches, and certainly much rehabilitation of the infrastructure has been done, but no new route alignments have been created and no new rights-of-way have been acquired. Much has been abandoned. The operations are simple enough—running trains on schedule along set routes. The issues are how various parts of the system are being used, to what extent, whether more can be recovered, and whether new systems should be built.

It can be argued that the San Francisco–Bay Area system under BART, the first new heavy rail project in the United States after World War II, has many of the distinct characteristics of commuter rail. However, it is usually listed under urban transit, and we will retain that classification here. The Lindenwold line operated by PATCO out of Philadelphia is of the same type.

Beginning of Roald Dahl's short story centered on rail commuting in London.

Galloping Foxley

1953

Five days a week, for thirty-six years, I have travelled the eight-twelve train to the City. It is never unduly crowded, and it takes me right in to Cannon Street Station, only an eleven and a half minute walk from the door of my office in Austin Friars.

I have always liked the process of commuting; every phase of the little journey is a pleasure to me. There is a regularity about it that is agreeable and comforting to a person of habit, and in addition, it serves as a sort of slipway along which I am gently but firmly launched into the waters of daily business routine.

Ours is a smallish country station and only nineteen or twenty people gather there to catch the eight-twelve. We are a group that rarely changes, and when occasionally a new face appears on the platform it causes a certain disclamatory, protestant ripple, like a new bird in a cage of canaries.

But normally, when I arrive in the morning with my usual four minutes to spare, there they all are, these good, solid, steadfast people, standing in their right places with their right umbrellas and hats and ties and faces and their newspapers under their arms, as unchanged and unchangeable through the years as the furniture in my own living-room. I like that.

I like also my corner seat by the window and reading *The Times* to the noise and motion of the train. This part of it lasts thirty-two minutes and it seems to soothe both my brain and my fretful old body like a good long massage. Believe me, there's nothing like routine and regularity for preserving one's peace of mind. I have now made this morning journey nearly ten thousand times in all, and I enjoy it more and more every day. Also (irrelevant, but interesting), I have become a sort of clock. I can tell at once if we are running two, three, or four minutes late, and I never have to look up to know which station we are stopped at.

The walk at the other end from Cannon Street to my office is neither too long nor too short—a healthy little perambulation along streets crowded with fellow commuters all proceeding to their places of work on the same orderly schedule as myself. It gives me a sense of assurance to be moving among these dependable, dignified people who stick to their jobs and don't go gadding about all over the world. Their lives, like my own, are regulated nicely by the minute hand of an accurate watch, and very often our paths cross at the same times and places on the street each day.

For example, as I turn the corner into St. Swithin's Lane, I invariably come head on with a genteel middle-aged lady who wears silver pince-nez and carries a black briefcase in her hand—a first-rate accountant, I should say, or possibly an executive in the textile industry. When I cross over Threadneedle Street by the traffic lights, nine times out of ten I pass a gentleman who wears a different garden flower in his button-hole each day. He dresses in black trousers and grey spats and is clearly a punctual and meticulous person, probably a banker, or perhaps a solicitor like myself; and several times in the last twenty-five years, as we have hurried past one another across the street, our eyes have met in a fleeting glance of mutual approval and respect.

Stations

Commuter rail routes, with few exceptions, start at the old established downtown railroad stations, run outward along old radial
alignments, and make stops at the old suburban stations.[12] All
this does not represent a bold and innovative approach in creating urban transportation systems, but it is a most commendable
practice in making good use of major assets that would be discarded otherwise (or, in the case of station buildings, be converted to quaint arts and crafts shops and restaurants).

There is an associated dimension to this situation that does not
affect transportation system development as such, but is important in the culture of cities—the adaptive reuse of historical landmarks. Since a great many of the old stations are of that quality,
and they were located deliberately on highly visible sites, the matter deserves attention.[13] Appropriately, protected status has been
given to many of these buildings, but the problems of conversion
remain. Grand spaces that were designed to accommodate long-
distance travelers with much luggage and that were provided with
comfortable waiting rooms and respected restaurants have to
cope today with commuters who rush through the building twice
a day.

After the deplorable cultural vandalism that was committed in
1962 by razing Pennsylvania Station in New York City, attitudes
have certainly changed. Grand Central Terminal, for example, has
recently been refurbished, and much effort has been devoted to
the task of making a facility built to serve a few thousand passengers on some 150 trains each day respond to the needs of
about 500 commuter trains and 500,000 people who enter and
leave the building. The list of similar accomplishments continues
to grow, and includes the monumental stations of Boston, Washington, Cincinnati, St. Louis, Los Angeles, and a few others as the
larger examples.

At the suburban ends of commuter routes, the challenges are
similar, but the needs are somewhat different. Since service is

[12] To understand the seminal role and the development process of railroad stations within American communities, see Stilgoe, op. cit., chapter on depots.
[13] Specific literature is available on this subject, such as Carroll L. V. Meeks, *The
Railroad Station: An Architectural History* (Dover Publications, 1956, 203 pp.),
and Lawrence Grow, *Waiting for the 5:05: Terminal, Station and Depot in America* (Universe Books, 1977, 128 pp.).

provided by trains at relatively long intervals, weather-protected waiting space is mandatory, opportunities to buy newspapers and some basic supplies are desirable, and purchase of tickets should be possible. The results with respect to this sector are most diverse. There are splendidly restored and well-equipped old station houses, but there are also instances in which a prefabricated metal box and vending machines are expected to suffice. The latter may be the high-tech, efficient solution for the future, but it would seem that a sensible regard for human amenities is called for to attract and keep customers.

To maintain adequate service levels while respecting the slower acceleration and deceleration characteristics of trains pulled by locomotives, station spacing closer than 1 mi (1.6 km) is rarely to be suggested. The location of stations would depend on the presence of well-defined nodes where access modes can be effectively concentrated, and distances of 3 mi (4.8 km) apart or more are the norm.

A principal issue at suburban stations is the means of access from the residential districts. Effective local feeder services, such as buses, paratransit, and taxis are essential, since walk-in patrons will be few at the home end. All of the feeders should touch the station as closely as possible, with loading bays near the rail platform. However, given the prevailing mobility practices in low-density American communities—the single-minded reliance on the automobile—transit access rarely works by itself, even in the best of circumstances. Therefore, the next priority has to be given to *kiss-and-ride* operations (a driver, usually a spouse, dropping off or picking up a rail passenger and continuing on his or her way). Convenient access lanes and some waiting space until the train arrives are important in this case as well.

Repeated experience, however, shows that the critical demand is for park-and-ride facilities. It is almost axiomatic today that the success of commuter service will depend on

New station on the Los Angeles rail network (Riverside).

such availability. For example, it is quite clear that further growth in patronage on the many regional routes in the New York metropolitan area hinges on the creation of adequate parking. The potential riders have reached such a level of prosperity that a car can be acquired and maintained for the sole purpose of using it twice each workday for a few minutes, allowing it to stand idle for about 99 percent of the time during the week. Be that as it may, if a station is located in the center of an old village—as many are—very little open space can be found for a parking lot. If the automobiles start to inundate the surrounding streets, local residents and businesses will show great unhappiness and countermeasures will be taken. The construction of multistory garages still appears to be out of scale in terms of expense and purpose, but has become necessary in some instances.

There are a few possible solutions to this dilemma, short of more extensive local transit use, that can be advocated vigorously. These responses, however, may not resonate very well with American commuters if current attitudes are maintained. The most obvious is the use of *bicycles,* as discussed in Chap. 3. The access distances are short, and parking should not be a space problem. They are seen at most railroad stations in American communities, but the volume is far short of what it could be or what is experienced in Europe under similar circumstances.

The other response could be *station cars*. These are very small automobiles, communally owned, with a distinct appearance, that can accommodate perhaps two passengers and some parcels. They are parked at a commuter station, and anybody can pick them up upon payment of a fee or utilizing a magnetic card and drive them to a local destination. They are kept at the house overnight and driven back to the station the next morning and left there. There are obviously some problems in making such a system work, particularly if there is no previous experience and trust, but experiments in several communities in Europe, as well as in the United States, have shown promising results. (See Chap. 5.)

Operating Schedules

The rail systems that serve large volumes of patrons, including shoppers and other travelers besides commuters, will provide service during the entire day (with lesser frequency in the middle of the day and no service at night). This includes the services within the New York/New Jersey region, Miami, and Los Angeles. There are some that operate only during the peak periods, e.g., in Washington, D.C., and on the West Coast. Some systems have certain routes with daylong service, as well as others with rush service only, e.g., Baltimore, Chicago, and Toronto.

There may also be service distinctions in terms of stations served. Express operations will bypass stations with low volumes to reduce the total trip time for most passengers; zone arrangements will provide service to groups of stations, but skip others (every train will stop at key stations).

Routes

Most of the commuter rail systems in North America consist of separate routes that connect some of the denser and older suburbs to the central core. They run invariably on existing rail rights-of-way. There are examples where branching at the outside ends takes place, but this too is normal railroad practice creating no special problems.

A major operational issue in some instances is the presence of freight traffic on the same track or within the same right-of-way. While, generally speaking, conflicts have disappeared or become minimal because industrial and warehouse activity has mostly relocated out of the central districts, there are significant safety and priority concerns. It is not infrequent that different agencies are responsible for different types of traffic on the same right-of-way, and therefore clear operational procedures have to be defined. The best approach, of course, is the designation of separate track for each purpose, with only a few crossing points that are carefully controlled. If the number of trains of either kind is not particularly high, it is possible to run them on the same track as has always been done in regular rail operations. This is acceptable because commuter passenger coaches are built strong enough with sturdy frames to carry an adequate crashworthiness rating (an unfortunate term) to hold their own in disaster situations, i.e., not crumple upon impact. If the volume of commuter trains is high, and they are given a priority status, in several instances freight trains are operated only during nighttime hours.

An issue that is being debated at this time is whether it is acceptable and advisable to operate light rail vehicles, which are more vulnerable, on the same alignments used by freight trains. Current American railroad rules bar such practice, but there are instances in Europe where this is being done. Karlsruhe, Germany, has such a system in full operation, and its experience is watched by everybody.

A basic issue related to the use of existing rail alignments is their placement. They were usually established more than 100 years ago to serve a completely different city configuration and respond to the needs of that time. They are not necessarily central to the current corridors of residential and commercial activity. To the extent that the early passenger lines generated villages and suburban clusters around their stations, the fit is still fine, but that is not the case with alignments that were intended primarily to serve industrial districts and carry mostly freight trains. To a large extent this is a moot question because nobody in the United States has proposed carving out new commuter rail alignments through built-up districts in a long time. The dominant operative policy at this time is to use (wisely) what we have—cost effectiveness governs politically correct thinking at this time.

Another issue associated with the use of the existing network is the dominant orientation of the lines to the historical center. Since the current metropolitan development trend emphasizes the establishment of major centers of activity within the urban field, there may be a limit to the now prevailing programs of commuter rail expansion. The old city business and cultural cores—at least not the major ones—are not expected to decline in absolute terms, and there are still many opportunities and needs to augment their heavy rail accessibility; however, the provision of similar services to the new peripheral centers is a new issue that remains problematic. Several studies have been made of reactivating circumferential links, but feasibility could not yet be shown. Reaching dispersed employment centers from the inner city remains a significant challenge for American communities today. It is clear that the key to any success with commuter rail is to ensure that a sufficiently large volume of commuters is eager to avail itself of this service, in both directions if at all possible.

Returning to the question of network configuration, it has always been obvious that branch lines terminating at the center have a serious drawback: there is no place to store the trains that arrive in the morning and have to wait for the afternoon rush back. Even if there are old railroad yards, they usually represent valuable space useful for a higher-value function. Deadheading[14] of trains twice each day is a wasteful practice. The logical response would be to provide for "through-running," with the trains moved in service from one edge of the metropolitan area to the other through the center. Most rail stations, however, because the lines penetrated into the built-up city historically, are "stub-end" terminals—in and out operations with trains reversing directions and negotiating switching arrangements among tracks.

A through system was constructed in Munich, Germany, with all the S-Bahn routes placed and "bundled" in a single tunnel that runs under the entire downtown with a number of stations in series where easy transfers are possible (see Fig. 14.2). This was an expensive solution, not repeated elsewhere, except in Philadelphia, which built a short tunnel in the center (1984) connecting the former Reading and Pennsylvania systems. In Tokyo, the regional rail system has been built up over the years so that it defies description, but it does attempt to provide linkages among routes and subsystems (see Fig. 14.3).

In New York City the situation is most complicated as well, with three networks entering, but all in effect stub-ending in the center (Long Island Rail Road, MetroNorth, and New Jersey Transit). It is not so much that too many passengers would want to travel from one end of the region to another, but the daily operations are highly constrained by this arrangement. Long-distance Amtrak service moves through this network as well. Even with the construction of new access by the Long Island Rail Road to Grand Central Terminal (in addition to Penn Station), the systems will touch at more points, but they will not be integrated. The respective agencies appear to have no urge to combine efforts, pointing to state lines, established labor practices, and the undeniable fact that the original builders of the separate networks went to great pains to make sure that the systems would not be compatible and that takeovers or encroachments would be precluded. The governing dimensions of the cars are similar but not exactly the same; most important, the power supply arrangements are distinctly different:

- *Long Island Rail Road.* Mostly multiple units (MUs); dual-powered locomotives (third-rail 760-V dc/diesel); conventional diesel locomotives

---

[14] Trains not in revenue service, i.e., running without any paying passengers.

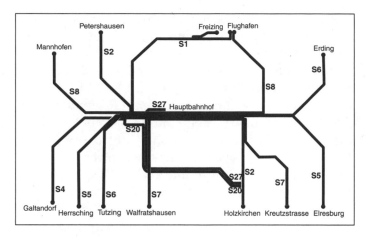

Figure 14.2 Schematic map of the Munich S-Bahn system. (Does not include the U-Bahn system.)

- *MetroNorth.* MUs with under-running shoes for third rail; dual-powered locomotives; dual AC catenary and diesel locomotives

- *New Jersey Transit.* AC catenary (12,000-V) locomotives and MUs; conventional diesel locomotives

- *Amtrak.* AC catenary; conventional locomotives

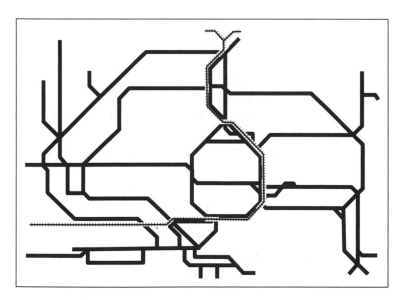

Figure 14.3 Schematic map of the Tokyo commuter rail system.

Nevertheless, the principal obstacles toward service integration are not technical (multimodal locomotives could be used), but rather institutional.

The most significant accomplishment in the development of regional rail systems in the second half of the twentieth century is undoubtedly the *Réseaux Express Régionale* (RER) of Paris (Fig. 14.4). It has never been called commuter rail in popular usage, but neither is it a conventional metro service. The five lines (A through E) reach out very far toward the edges of the region, and the stations are quite far apart. The new routes go directly through the center of Paris, and because the immediate levels below street surface are quite crowded already, RER is placed deeper yet. The stations are spacious and well designed, and they connect with principal metro nodes.

Purpose and Quality of Service

Commuter rail has always been distinct from the other modes of public transportation because of a certain aura. It started as suburban service for families prosperous enough to build large homes in remote locations, and this condition has largely persisted—even though reverse commuting is receiving increasingly more attention. These customers expected good quality, they were in a position to demand it and to pay for it, and they have not changed this attitude over generations. It would be politically and socially unacceptable to call regional rail a premium service, but it is as close to that level as we are likely to get in transit. Proper ventilation, comfortable seats, adequate lighting, air conditioning, and safety are expected and are (mostly) provided. This is seen as a justifiable public policy to maintain ridership because the clientele is in a position to use automobiles or even seek employment beyond the rail corridor.

All this is not idle speculation, but an accepted set of criteria for commuter rail. These considerations can be either seen as desirable conditions that point the way for all other transit services or can be kept out of public review as much as possible. The latter situation is not, however, always achieved, as shown by recent controversies in Los Angeles and New York. Consumer and city transit advocates have taken or have threatened legal action to have fare levels, comparing regional rail and city transit (buses and subways), reflect not only the difference in service quality, but also the ability to pay by the respective groups of passengers.

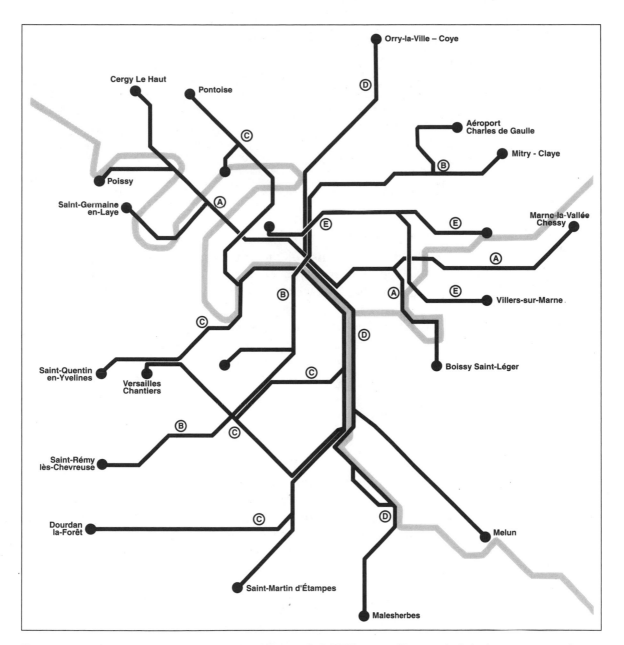

Figure 14.4 Schematic map of the Paris Regional Express Rail (RER) system. (Does not include the metro system.)

Current Extent of and Future Plans for Commuter Rail in the United States

In operation	3248 mi	(5230 km)
Under construction*	104 mi	(167 km)
In design	168 mi	(270 km)
Under planning	1611 mi	(2590 km)
Proposed	1600 mi	(2575 km)

* In Burlington, Dallas, New York, Seattle, and Washington, DC.
Source: APTA 2000, *Public Transportation Fact Book*, p. 26.

The challenge for commuter rail today and in the near future, besides the concerns just touched upon, is *reverse commuting,* i.e., to accommodate not only passengers who commute to the center, but also inner-city residents who could have jobs in the outlying districts. This is not just a matter of running the trains back on a frequent enough schedule, but would involve some changes in established operating practices, particularly related to turning trains around quickly and finding space for temporary storage at various locations. The major difficulty is providing for adequate means of distribution at the suburban end to scattered destination points.

Cases in the United States

In 1987, there were only seven metropolitan areas in the United States that had commuter rail service. As shown in Table 14.2, not only has the number of operations grown, but ridership on existing systems has increased substantially. Only two small systems have been abandoned (Pittsburgh and Detroit), and a marginal decrease in total patronage has been registered in another case (Philadelphia).

While commuter rail appears to operate in the easily recognizable and traditional railroad mode (except perhaps for some advanced equipment), there are significant differences from one place to another. These exist largely because of the specific historical development of the systems in each region, as each locality has responded to its own particular needs and opportunities. This situation also illustrates some of the possible variety in institutional arrangements under which commuter rail service can be provided in this country.

Table 14.2 Recent Developments in Commuter Rail Operations in the United States

	In 1987		In 1997*	
	Number of Routes	Passengers per Year, millions	Number of Routes	Passengers per Year, millions
Baltimore	—	—	3	4.7
Boston	5	14.3	13	29.0
Chicago	11	66.5	12	75.2
Detroit	1	—	—	—
Los Angeles	—	—	6	6.2
Miami	—	—	1	2.5
Dallas–Fort Worth	—	—	1	—
New York/New Jersey	Several systems	179.0	3 systems	190.1
Philadelphia	13	24.1	13	22.5
Pittsburgh	1	0.2	—	—
San Francisco Bay Area	1	5.4	2	8.1
Washington, D.C.	3	1.8	5	6.6
Total		291.3		344.9

* Table 14.1 shows the situation in 1997 as well. A few more operations have been added in the last few years, serving 13 places in the United States and 3 in Canada. More are on the way.

Brief summaries are given of three of the larger systems in the United States: the Long Island Rail Road, operating in and out of New York City; the extensive "federation" of operations around Chicago; and the suburban rail services centered on Boston.

Long Island, New York

The Long Island Rail Road (LIRR) is the oldest Class I railroad in the United States still operating under its original name and charter (1834). It is also the largest commuter rail operation in the country, carrying each weekday more than 290,000 riders on 735 trains.

The railroad was built originally to provide a link to Boston (via ferry); when that need was satisfied more directly by a land route through Connecticut, LIRR entered a period of minimal activity. This changed gradually as development started to spread eastward from New York City. A number of competing lines were also built, but they were quickly absorbed by the LIRR system. Its territory is a cul-de-sac, with the only connection to the mainland through New York City and Manhattan; industrial development here has not been particularly heavy to demand extensive rail

The LIRR ran past the Corona Dump before it was converted in 1939 into Flushing Meadows Park, which accommodated two world's fairs.

About half way between West Egg and New York the motor-road hastily joins the railroad and runs beside it for a quarter of a mile so as to shrink away from a certain desolate area of land. This is a valley of ashes—a fantastic farm where ashes grow like wheat into ridges and hills and grotesque gardens, where ashes take the forms of houses and chimneys and rising smoke and finally, with a transcendent effort, of men who move dimly and already crumbling through the powdery air. Occasionally a line of grey cars crawls along an invisible track, gives out a ghastly creak and comes to rest, and immediately the ash-grey men swarm up with leaden spades and stir up an impenetrable cloud which screens their obscure operations from your sight.

But above the grey land and the spasms of bleak dust which drift endlessly over it, you perceive, after a moment, the eyes of Doctor T. J. Eckleburg. The eyes of Doctor T. J. Eckleburg are blue and gigantic—their retinas are one yard high. They look out of no face but, instead, from a pair of enormous yellow spectacles which pass over a nonexistent nose. Evidently some wild wag of an oculist set them there to fatten his practice in the borough of Queens and then sank down himself into eternal blindness or forgot them and moved away. But his eyes, dimmed a little by many paintless days under sun and rain, brood on over the solemn dumping ground.

The valley of ashes is bounded on one side by a small foul river, and when the drawbridge is up to let barges through, the passengers on waiting trains can stare at the dismal scene for as long as half an hour. There is always a halt there of at least a minute and it was because of this that I first met Tom Buchanan's mistress.

access. At this time, only consumer supplies move in, and those few freight trains can be accommodated during night hours.

The system went through various reorganization efforts, eventually coming under the ownership of the Pennsylvania Railroad as a means to provide access from New Jersey to Manhattan. As that corporation found it increasingly more difficult to cope with the passenger service demand generally and on Long Island specifically, the State of New York acquired the property in 1965 and placed it under the newly organized Metropolitan Transportation Authority (MTA) largely intact. Thus, MTA/LIRR owns the network and the rolling stock and is fully responsible for all operations. There have been periods of user discontent when service quality suffered due to resource shortages and managerial inattention, but at this time the situation is well under control with satisfactory performance. That required the infusion of considerable public resources ($6.8 billion during the 1980s and 1990s).

Nine of ten branch lines in the east converge on a single node—the Jamaica Station, which is a major transfer point and center of operations at a massive scale (see Fig. 14.5). The branches fanning out from Jamaica to the east carry different size loads, but they cover the territory quite well. This system has been the framework within which the dramatic development of Long Island has taken place, from the 1930s to today. Highways alone, which arrived during the suburbanization period, could not possibly cope with

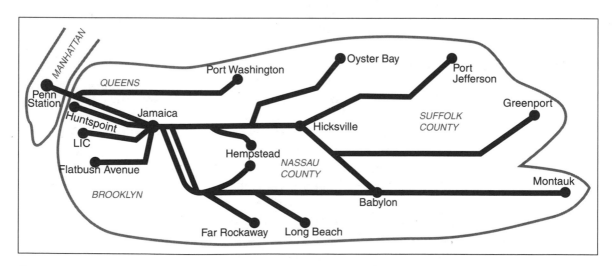

Figure 14.5 Schematic map of the Long Island Rail Road system. (Does not include the subway system.)

the transportation demands, and the levels of roadway congestion, particularly on the Long Island Expressway, are legendary.

From Jamaica westward, some trains reach the Atlantic Terminal in Brooklyn, but this connection has ebbed in importance, reflecting the trends in this part of the city. It is possible, to envision a greater role for this link within the larger system, but only with considerable capital investment. Most peak period trains from the 10 eastern branches move via Long Island City and Woodside, past the large Sunnyside Yards, into Manhattan through the tunnels to Pennsylvania Station. This is one of the largest commuter centers anywhere in the world, but it is located somewhat to the west of the principal concentration of Midtown destinations. To remedy that situation, a project is currently under way (East Side Access) to bring LIRR trains through the 63rd Street tunnel, which was completed in 1973, into Grand Central Terminal (GCT).

While Penn Station is connected by tunnels under the Hudson River to New Jersey, there is no through running of trains because the other side is the territory of another operator—New Jersey Transit (NJT). However, arrangements have been made to lay over NJT trains during the day in the Sunnyside Yards to the east, while LIRR trains have a large yard on Manhattan to the west, near the Hudson River. Most of the LIRR system is electrified, but not the eastern parts of four of the longer lines. The current capital investment programs, while they include track renovation, are mostly geared toward the acquisition of advanced rolling stock and dual-mode locomotives. A proposal to build another tunnel to New Jersey and possibly to connect Penn Station directly to GCT has also been under discussion for some time. It is not only the expense of such a project that delays progress; the constraints are to a significant degree the complex institutional politics among a number of very large transportation agencies.

Penn Station is slated to receive a new terminal building one block to the west in place of the long demolished grand original station. Another pending long-range issue is whether commuter service can be created to Lower Manhattan, which suffers significantly from the lack of direct access by managers and workers from the larger region. The September 11, 2001, disaster has diverted attention to more urgent issues, but the question of accessibility by patrons at all income levels from all sectors of the tristate metropolitan area remains unanswered.

Chicago, Illinois

Chicago, once the undisputed premier railroad node in the county, is still located in the center of an extensive network of tracks. It certainly does not retain all of its rail services, but there is an effective commuter system still in place (see Fig. 14.6). The total of Chicago's commuter rail activity is only a small amount below that of LIRR (see Table 14.3).

As in other large metropolitan areas of the United States, most Chicago public transit services are operated today under the over-

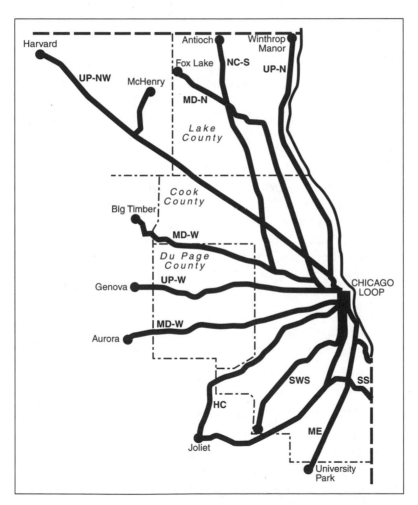

Figure 14.6 Schematic map of the Chicago Commuter Rail system (Metra). (Does not include the rapid transit system.)

Table 14.3   Ridership on the Principal Commuter Rail Systems*

	Annual Ridership, 2000	Average Weekday Ridership, 2001, first quarter
New York City (LIRR)	104.7 million	378,000
Chicago (Metra)	72.4 million	293,600
New York City (MetroNorth)	71.7 million	246,900
Newark, NJ/New York City (New Jersey Transit)	61.9 million	224,400
Boston (MBTA)	36.7 million	131,100
Philadelphia (SEPTA)	28.4 million	105,900

* APTA, *Transit Ridership Report* (unlinked passenger trips).

all purview of a single agency—the Regional Transportation Authority (RTA), chartered in 1974. It supervises three operating agencies, one of which, Metra, has been responsible for commuter rail operations since 1984. The original arrangement under RTA was to sign purchase of service contracts with the separate then operating railroad companies to provide service, while the public body set standards and handled the budget (including subsidies). With the creation of Metra, this agency has also assumed operating responsibility for almost all trains, but the routes are still referred to by the names of the original private rail companies, even though the arrangements have been officially changed. In each instance, specific arrangements have been made with these companies as to what role they have in operations and how much ownership they retain of trackage and infrastructure.

The components of the system are:

- Chicago and North Western—three lines out of North Western Station, operated by Union Pacific

- Milwaukee District—two lines out of Union Station, operated by Metra

- Rock Island District—one line out of LaSalle Street Station, operated by Metra

- Burlington Northern—one line out of Union Station, operated by Burlington Northern Santa Fe

- South West Service—one line out of Union Station, operated by Metra

- Heritage Corridor—one line out of Union Station, operated by Metra

- Electric District—one line out of Randolph Street Station with branches, operated by Metra

- North Central Service—one line out of Union Station, operated by Metra (newest line, opened in 1996 utilizing trackage of Wisconsin Central Railroad)

- South Shore Line—one line out of Randolph Street Station, operated by Northern Indiana Commuter Transportation District

Different lines from different central stations provide service to different corridors. While this is not an integrated system in a true sense, individual customers, who usually are not too concerned with the connectivity of the network but primarily commute back and forth to the downtown district, receive basic service within their own corridors.

The ridership numbers have shown steady increases in recent years, reaching about 280,000 each working day. After considerable public demand for service improvements and renovated infrastructure and equipment, positive results are visible. Metra continues a vigorous program of promotion and keeps exploring further expansion possibilities.

A chronic problem remains the availability of parking at suburban stations, and Metra has developed procedures in this area that are effective in securing local cooperation. It acquires land and constructs a parking facility, but then turns over the operating responsibility to each municipality, which also collects the fees.

Metra has taken the unusual step of examining two possible circumferential routes—one running from Waukegan in the north through Elgin and Joliet to the Indiana state line and the other connecting O'Hare and Midway Airports. The first proposal covers a distance of 105 mi (169 km) and may cost as much as $1.3 billion, depending on how much double track will be needed; the second is a 22-mi (35-km) run that is likely to require a $350 million investment.

A number of other lines are being considered for reactivation, several extending well into Wisconsin and being sponsored by the planning agency of that state. Similar arrangements and extension plans are being examined in Indiana.

Boston, Massachusetts

Boston, having the oldest public transportation system in the country, also has a regional overarching agency—the Massachusetts Bay Transportation Authority (MBTA), established in 1964, which extends over 175 cities and towns (see Fig. 14.7). In 1965, the agency acquired and reserved for its use the entire regional network and rights-of-way of the New Haven Railroad. The MBTA, however, does not operate commuter services itself; Amtrak, under a management contract, assumes that responsibility.

The 12-route system consists logically of two parts, each oriented to its own station—the South Station or the North Station in the downtown area. The North Side lines are Rockport, Ipswich Branch, Haverhill, Lowell, and Fitchburg; the South Side lines are

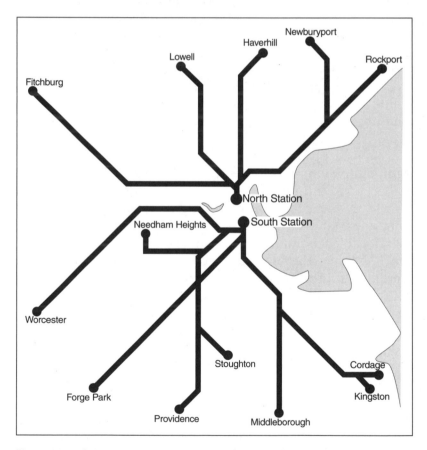

Figure 14.7 Schematic map of the Boston Commuter rail system (MBTA). (Does not include the rapid transit system.)

Attleboro/Stoughton, Framingham/Worcester, Needham, Franklin, Fairmont, Middleborough/Lakeville, and Plymouth/Kingston. The total daily ridership was 126,800 patrons in 2000. None of the commuter lines are electrified, and therefore diesel locomotives operate throughout the system.

Recent history has been quite encouraging in the commuter sector. The Old Colony Line, which started in 1845 but died in 1959, had two branch lines reopened in 1997, with a third in the planning stage. This service has been a success, providing access to seaside communities, with parking spaces at the stations in great demand. Park-and-ride lots have been expanded at several locations; new stations have been added (Grafton) or are being programmed; and service has been extended as far as Worcester recently. The idea of connecting North Station to South Station with a tunnel for through-running remains, however, a distant possibility.

Development Programs in Other Cities

Commuter rail development and expansion programs are to be found at a number of locations in the United States currently.[15] As a matter of fact, there is hardly any sizable metropolitan area where some examination of such options has not taken place. The places where early explorations have resulted in further progress are outlined in this section.

The Dallas Trinity Railway Express has reached the center of Fort Worth; a line has been reactivated in Burlington, Vermont; a new branch line has opened on the MARC system serving Washington, DC; further significant additions are planned for the Boston network beyond those recently inaugurated (to Portland, Maine, for example); SEPTA (Philadelphia) has been considering linkages to Delaware; Virginia Railway Express is considering multiple line extensions and new stations; Chicago is evaluating additional opportunities; New Jersey Transit has completed and has under construction several major interchanges and transfer points that will considerably ease travel to Manhattan; and a number of other central cities are looking at possible routes with

[15] See "Commuter Rail Update 2001" in the March/April and May 2001 issues of *Mass Transit,* pp. 38–48 and 30–37, respectively, and the special issues of *Passenger Transport* (APTA weekly newspaper) devoted to commuter rail conferences, such as April 2, 2001, April 10, 2000, and others.

various degrees of attention and urgency. Efforts that were or are quite visible are to be found in Atlanta, Cleveland, Akron, Durham–Raleigh, Harrisburg, Hartford, New Orleans, Salt Lake City, St. Louis, and Tampa.

Peak period commuter service has been started between Seattle and Tacoma by Sound Transit; further improvements have been programmed for that system. When all extensions are completed that are now under development, the network will extend over 82 mi (132 km). The Miami system (Tri-Rail) is double-tracking its 71-mi (114-km) route, and, instead of painting its cars, is wrapping them in vinyl film with printed on advertising. Minneapolis is building a 70-mi (113-km) line to St. Cloud for peak hour service. The situation around San Francisco is quite complex, with new proposals and counteractions, but progress is being made by strengthening the system now in place. The Coaster in Southern California is another new effort.

An interesting aspect regarding commuter rail systems is the element of flexibility as to operational responsibility. In most instances, the regional transit agencies that now dominate the public transportation field in the United States are also the operators of service (they employ the crews) and own the equipment. The track and other fixed infrastructure components, however, may still belong to others (municipalities or railroad companies). The responsibility for operations may be contracted out, either to railroad companies or Amtrak. The national agency, for example, runs the trains in Boston, Baltimore, Virginia, Connecticut, San Diego, Los Angeles, San Francisco, and Seattle. If Amtrak becomes reorganized or curtailed under a national mandate in the near future, these arrangements will have to be adjusted.

Reasons to Support Commuter Rail

The strengths and weaknesses of commuter rail are outlined here in the context of utilizing existing rail alignments, not attempting to create new rights-of-way.

Efficient operations are the principal mechanical advantage of all rail-based transport. Once trains are in motion, it takes very little power to maintain forward progress. Thus, in terms of energy consumption per passenger, heavy rail scores very high, as long as the cars are reasonably occupied. An empty coach does no useful work, and therefore this calculation is only meaningful if good loading ratios can be maintained.

Fast and comfortable service is offered, characteristic again of all trains if they are properly operated. The overall average running speed in commuter service is not so much a matter of the maximum speed that the equipment can attain, which is well over 100 mph (160 kph), but rather station spacing and the power of engines that govern acceleration and deceleration rates. As a rule, stations on commuter lines are reasonably far apart (at least 1 mi), which makes auxiliary access modes mandatory for most patrons. Since very few riders live and work within walking range of railroad stations, the fast and comfortable rail journey may be only a part of an otherwise long and tedious commuting trip.

Reliability and safety on rail systems under normal circumstances should be appreciably higher than on other modes because of the separation of the operations and overall internal control. The trains and passengers are largely isolated from external impacts, the vehicles are sturdy, and weather has little influence. Thus, as long as at-grade crossings are few and maintenance of all components is satisfactory, precise schedules can be kept and mishaps precluded. Statistics show that the death rate associated with automobile travel is 0.91 per 100 million passenger miles, while that for intercity and commuter rail is 0.06.[16] (Transit buses are better yet at 0.01.)

Use of existing resources is a major factor in the evaluation of urban transportation options. This refers to track, stations, infrastructure, and most other physical elements inherited from the golden age of railroading, but now frequently subject to neglect. Perhaps the most valuable asset is the right-of-way—a long and narrow channel of land that can accommodate not only transportation systems but also utility and communication lines. If nothing else, pedestrian and bike trails can be built along the rights-of-way, provided that compatibility and acceptance issues are resolved. There are many regrettable cases around the country where these potential channels have been sold and thereby lost permanently for the public service sector. On the other hand, there is Rails to Trails, an active program supported by federal assistance, which has considerable capability to upgrade the recreational inventory in many communities.

Service can be implemented quickly, as compared to other high-capacity transit projects, if existing rights-of-way are being used. New infrastructure does not have to be built as construction

---

[16] 1996–1998 averages, as assembled by the National Safety Council, 2000.

projects of long duration, the approval process should be easy and direct since a new use and activity is not being created, and no significant condemnation of properties is likely to be necessary.

Good public image characterizes commuter rail today. There have been periods when the service has dwindled down to only a few examples, and quality on existing routes has deteriorated badly, but a remarkable recovery has been made. Financial resources have been invested to upgrade the fleet and facilities, to buy new equipment, and to generally refurbish the image of this mode. Recently, a number of routes have been reactivated, and plans exist for further expansion of service along existing routes and in the creation of new ones (on existing rights-of-way).

Reasons to Exercise Caution

Locational constraint is present to the extent that existing rights-of-way are to be used. Clearly, there is no flexibility in the placement of the routes, and the only question is whether strong enough service demands can be generated along any given corridor. None of this applies if the right-of-way is considered for line haul only, i.e., connecting two mutually dependent districts via a long intervening track.

Space conflicts will exist if the alignment carries other types of traffic (such as freight) and an accommodation has to be made between the new and the previous users of the channel. Various approaches toward a compromise are possible, as outlined previously.

Implementation can be expensive if the available infrastructure has to be rebuilt extensively or created anew, even if the right-of-way can be obtained at small or for no cost. Rolling stock, including quite likely expensive locomotives, will have to be acquired. If ridership volumes are not going to be particularly high, unit operations

Underground station of the Paris RER system.

and maintenance costs on a per capita basis may be unacceptably high. Railroad operations are efficient, but large capital investments are involved that can only be justified in economic terms under intensive use—unless the infrastructure is available already and regarded as sunk costs not to be recovered. Even then, the marginal costs to bring the system to acceptable standards and maintain quality service may be excessive in light of expectable benefits, most of which are not specifically recoverable anyway (air quality, development patterns, civic image).

Environmental considerations generally favor rail operations greatly, certainly as compared to street traffic carrying equivalent passenger volumes. However, there are areas of concern suggesting the need for proper care in the implementation and operation of rail-based service. With the use of diesel locomotives, air quality issues may appear unless much attention is devoted to maintenance and the use of proper fuels as well as the acquisition of proper equipment. The size and speed of train operations may expose nearby buildings to considerable noise and vibration. These problems can be mitigated quite well with good design and continuous maintenance of rolling stock and track. However, there may be an impact on property values immediately adjacent to the line, no matter what measures are taken. This would also include visual impacts because a full-size train is at a considerably different scale than low-density neighborhoods. Grade crossings with barriers will also frequently cause queues of motorcars with idling engines.

There may be real or perceived safety issues, particularly if at-grade crossings are present and there are possibilities for persons, particularly children, to wander onto the right-of-way. Unauthorized trespassing is frequently a cause for concern. If the lines are electrified at high voltage, the possibility of blundering into dangerous conditions may become significant.

In many cases the right-of-way is held by private corporations that are wary about possible intrusions into and curtailment of their freight operations. As owners, they can dictate terms, delay implementation, or refuse joint efforts entirely. Eminent domain (compulsory purchase) is theoretically possible, but it is an expensive and time-consuming effort to achieve passenger service.

Not-in-my-backyard (NIMBY) concerns are frequently present when new commuter rail operations are under consideration. Neighborhood residents next to the line may feel threatened by

Bilevel commuter cars on the Chicago system.

heavy operations with large equipment that probably has no direct utility to the local area. This includes safety concerns related to surface crossings and entry on the right-of-way, noise (whistles, locomotives, and fast movement), aesthetics of elements beyond a neighborhood scale, possible derailments, and other characteristics of rail operations. Property values may be affected, and there is likely to be a fear of the unknown caused by a new presence, particularly if there have been no trains running on the line in question for some time. All of these impacts can be mitigated or eliminated, but some effort will be required to overcome the possible local objections.

Application Scenarios

First, it can be noted that none of the North American metropolitan areas that have any commuter service have less than 2.8 million total population (San Diego),[17] and that the largest U.S. metropolitan area without such service is Detroit–Ann Arbor–Flint (5.5 million). Total size is not a governing factor, since separate routes serve separate corridors, but it may serve as a preliminary indicator of feasibility. Beyond that, several characteristics appear to be relevant in the review of commuter rail suitability in any given locality:

- Availability of underutilized railroad rights-of-way or those with completely abandoned operations, but still in single ownership; expected willingness of the railroad companies or transit agency(ies) controlling the right-of-way to accommodate commuter train operations.

[17] Providence and Santa Clara are actually smaller, but they are the outlying terminals for service operating from a larger center.

- Location of such channel approximately along a corridor where regional travel demand will concentrate.

- Presence of a strong central business district that will maintain its viability in the foreseeable future. It probably should not contain less than 100 million ft² of commercial space.

- Presence of broad corridors of residential neighborhoods developed at medium densities (contiguous urban-type districts, not exclusively with single-family homes), i.e., districts that are likely to generate travelers to the central location at reasonable levels.

Protected grade crossing on Long Island.

- Access to major facilities that may operate intermittently but attract large volumes of patrons (sports stadiums, for example).

- Persistent traffic congestion on the street and highway networks, making commuting and travel at the regional scale a significant constraint.

- High parking costs in the destination districts or severe parking shortages at convenient locations.

- If implementation costs are not particularly high, rush hour service only may be a viable option to serve large employment centers.

Components of Commuter Rail Systems

Rolling Stock

Rail vehicles associated with passenger service can be classified in the following groups:[18]

[18] V. Vuchic, op. cit., pp. 314–316, provides a set of more detailed definitions.

- *Locomotives.* Powered units with large traction capability able to pull or push trains, carrying no passengers themselves. Since steam engines are not used anymore, the basic options are electric or diesel locomotives.[19] The former may receive power from overhead wires or a third rail along the side of the track. Available as well are dual-mode locomotives that are able to operate on electrified and regular (nonelectrified) track.

- *Coaches or Trailers.* Nonpowered vehicles that are towed or pushed by locomotives or by other powered units. They provide passenger accommodations only. The principal variations are regular coaches with 2 & 2 or 2 & 3 seating in rows with a central aisle, or bilevel coaches that accommodate seats on two levels (see Fig. 14.8). The latter are either of the "gallery" type, with elevated rows of seats or vehicles with two full floors and intermediate decks. Their popularity is currently on an upswing and increasingly more agencies are phasing them into service. Large coaches reduce space requirements throughout the system, but, because of their height, may encounter restrictions in tunnels and underpasses.

- *Powered Cars.* Vehicles with electric motors in the trucks[20] below and direct power pickup through overhead wires or a third rail. If this vehicle has all train controls and can operate alone, it is called single unit (SU). Much more common are vehicles designed to operate in train *consists*[21] with a single driver or engineer up front. These are multiple units (MUs), sometimes referred to as "emus" (electric multiple units).

- *Railbus or Diesel Multiple Units* (DMUs). Passenger-carrying vehicles operating on regular track, but propelled by a diesel engine. They may run singly or in consists, towing one or more trailers. They have not found much favor in North

[19] "Diesel" locomotives are actually diesel-electric locomotives. The prime mover is a diesel engine connected directly to a generator, which in turn feeds power to electric motors that are linked through gears to drive axles.

[20] Or *bogies,* which constitute a unit with two axles and four wheels. A passenger compartment is usually carried by two trucks.

[21] A *consist* is an operating unit with a number of vehicles acting together. A *train set* is about the same, except that the cars are coupled together semipermanently.

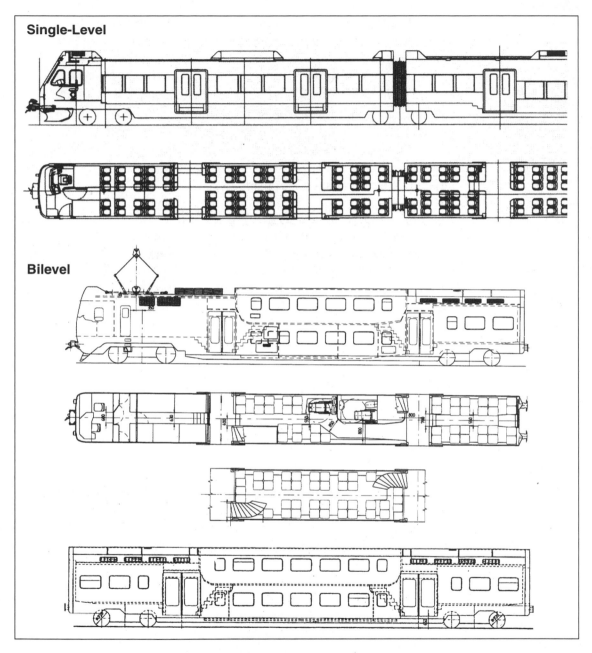

Single-Level

Bilevel

Figure 14.8 Examples of commuter rail cars. (*Source:* Ansaldobreda.)

America, but several systems exist in Europe, and the Dallas–Fort Worth system employs such vehicles currently. Attitudes may see a change, with Ottawa starting a railbus line in 2001 and the Camden-Trenton system now under design also deciding on this technology.

Beyond these basic types, any number of variations can be found or envisioned. For example, there are "married pairs" that can only operate together because they share components; there may be similar arrangements for three cars. Regular commuter coaches allow passage from one car to another, but this may not be always the case; articulated units (i.e., with flexible joints in the body) may be employed; and any other conceivable variation has probably been built by somebody at some time.

Cars can be built with doors only on one side, which would mandate that they are to run only in one direction and that all station platforms within the system are to be located on that side. This is not a recommended practice to save marginally on purchase price, because requirements may change, and flexibility is always desirable.

Right-of-Way and Track

Since the reuse or new adaptation of existing rail alignments is under discussion here, it can be assumed that the geometrics, grades, clearances, and other physical parameters will be satisfactory for regular commuter service. The principal requirement, as mentioned previously, is to achieve safety—not only at previous levels, but better, because intensive passenger operations and today's higher expectations are in play. Trains move very fast, and the vehicles, particularly locomotives, are heavy. Under such momentum, nothing much will survive a collision with a train. There have been instances in which a train has hit people on the track, and the engineers in the cab have not even noticed the impact.

The physical dimensions horizontally and vertically of movement space on top of the track are determined by the characteristic of a "design vehicle"—the largest unit of rolling stock that will be operating on any line. However, because the vehicle sways and bounces up and down, and its edges protrude when going around curves, the clearances are defined as a "dynamic outline," giving the envelope of the longitudinal space.[22] (The basic dimensions are given in Fig. 14.9.) It is mandatory that an emergency evacuation and maintenance walkway be provided along each line outside the clearance envelope.

Separation in time and space as much as possible from freight operations is a basic requirement. This is best achieved by an exclusive right-of-way, with minimal possible intrusions by other vehicles, people, animals, or objects. This means grade separation throughout, if at all possible. It is fortunate that in the earlier period of railroad prosperity resources were available to accomplish much of this inside cities because to build even a few rail under- or overpasses today would involve considerable expense, probably beyond the capabilities of a retrofit budget. If some at-grade-intersections remain, every measure will have to be taken that they are properly protected. That, of course, also means that surface motor traffic will be regularly delayed crossing the line.

If commuter trains are run in tunnels, as they are in some of the larger cities, there is no choice based on modern standards but to utilize electric locomotives or emus, with the accompanying electrification of the track along the entire length of the route (or with the use of dual-mode locomotives or provisions for changing engines). It is possible to conceive a diesel-propelled system with extensive tunnel ventilation and enclosing platform screens, but such efforts do not appear to be warranted in light of reality.

Stations

Principal design requirements of stations,[23] besides the provision and location of passenger amenities (waiting rooms, concourses, food sales, newsstands, information boards, ticketing facilities, rest rooms), are concerned with passenger movements, i.e., the ease and safety of negotiating the space between trains and external access as a pattern of flows.[24]

---

[22] The detailed parameters can be found in all engineering reference books, notably American Railway Engineering Association, *Manual for Railroad Engineering*.

[23] See under "Types of Commuter Rail Operation" for general discussion of station configuration and functional variations.

[24] American Railway Engineering Association, *Manual for Railway Engineering*, pp. 6, 8, 10, and 11.

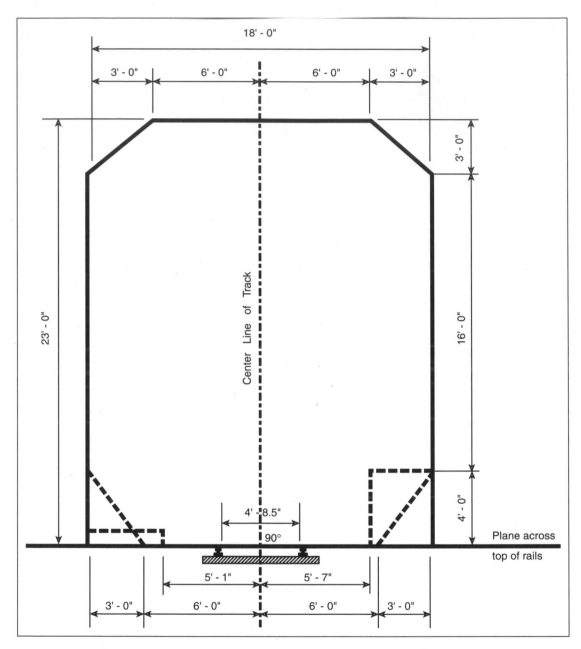

Figure 14.9   Clearance outline for passenger trains on straight sections. On curved track, the lateral clearances have to be increased. On superelevations, the centerline remains perpendicular to the plane across the top of the rails. (*Source:* American Railway Engineering Association.)

Central (island) or side platforms may be used; the principal question is whether it is acceptable to have the patrons cross the tracks at grade when they walk to and from the platforms. This was done frequently with the old systems, but currently is regarded as very inferior practice. With low train volumes, this might be permissible in exceptional cases, as long as ample warning of incoming trains can be ensured and no physical obstacles are located on the path (such as steps). Otherwise, grade-separated over- or underpasses will be required. This is a sizable capital expenditure, particularly because wheelchair accessibility with elevators or the use of 1:12 gradient ramps with landings is now mandatory. It also represents a permanent security concern, requiring direct or closed-circuit TV monitoring of all internal sections, with associated maintenance costs.

Another significant issue is the question of low or high platforms. Traditionally, trains have operated with low platforms that require several high steps to reach the car floor. Assistance by conductors and porters is no longer to be expected, certainly not on commuter lines. In the case of commuter trains, this is not so much a concern with dwell times as it is with the Americans with Disabilities Act. For all these reasons, high platforms have to be looked upon favorably, particularly where close headways have to be maintained and the rail service is expected to provide some of the desirable characteristics of transit. High platforms require considerable precision in the construction of all elements so that the gap

between the edge of the platform and the sill of the coach door is acceptably narrow. The use of low-floor cars, which are now available with at least partial low floors, will ease considerably the step-up problems presented by low platforms.

It should be noted that these close tolerances (edge of platform) restrict the dynamic envelope within which cars move. This will place limitations on the speed and performance of freight cars or block them entirely if they use the

Renovated Union Station in Los Angeles.

same track. A response in such cases may have to be the provision of fully or partially offset tracks through the station.

Signaling and Control Systems

The time is long gone when a locomotive engineer could see a damsel tied to the tracks and stop the train in time. Such line of sight systems have been replaced at least by manual or automatic block arrangements that ensure that two trains are precluded from occupying the same stretch of track. Since commuter trains operate on the trackage of regular railroad properties, they have to conform to the needs and capabilities of modern control systems as they exist throughout the network within a given metropolitan area. Those are likely to be—but not necessarily (because all lines have not been upgraded)—automatic train protection systems, which are most reliable, but will not go as far as the advanced automatic train control and operation systems that are now common on new heavy rail/metro lines, which have to operate at very high densities.

Many complex considerations and possible responses are involved in this sector. There is, for example, the option of having centralized control arrangements covering large territories vs. local towers.[25] Trains operating at 80 mph (130 kph) or more require supplemental cab signaling, which is a considerable added expense. If freight and passenger trains operate on the same track, optimal signal systems are not the same for both modes because of differing operating patterns and safe stopping distances.

Control systems are a crucial subject, and the reader has to be warned that this is certainly an area where a little knowledge is a dangerous thing. The matter has to be left to experts who have the responsibility of maintaining the highest reasonable safety on the networks. The principal concerns are prevention of collisions between trains, assuming that human errors will occur from time to time; ensuring no derailments and fires within the right-of-way; and making it very difficult for other vehicles, persons, or animals to enter the track. The options in selecting and implementing any given control system revolve around the trade-offs of enhanced

[25] Control nodes or rooms that still are referred to by that name because historically they were housed in such structures.

safety vs. maximum capacity in terms of achievable train movements per hour (see "Capacity and Cost Considerations").

Fare Collection

Fare collection on most commuter routes still follows the traditional practice: tickets or passes are obtained before boarding, and they are all checked by a conductor who can also sell tickets at a slightly higher price. To cut costs and expedite operations, some agencies have started to shift to transit-type arrangements with turnstile barriers and magnetic passes. Passes may be available on a weekly, monthly, or annual basis. Automatic fare collection does reduce the need for staff at stations and on trains. These procedures are becoming accepted, utilizing inspectors who conduct spot checks and collect fines immediately (and sell no tickets on the trains). If a high percentage of riders are regular commuters with monthly passes, the enforcement procedures require little resources and lost revenue due to fare beaters is minimal. The automated approach may encounter opposition by organized labor and violations of inherited work rules as they have been defined historically for railroad operation.

Yards

All railroad operations require yards for storage and maintenance of rolling stock. Since the discussion here addresses largely the use of existing infrastructure, the assumption is that such facilities would also be available. In some cases, former freight yards can be reactivated for commuter rail use. The specific purposes of yards include the storage of equipment overnight and often during midday; routine repair, cleaning, painting, and refurbishing; and overhaul (which may also be accomplished off site at larger facilities in joint use). Federal Rail Administration (FRA)-mandated inspections will also take place in yards.

The creation of a new railroad yard would be a major effort, since considerable acreage of suitable land would have to be found. Local zoning issues may come into play. Theoretically, the best location of a yard would be at the end of a line or where several lines cross to minimize deadheading, but such options are rarely available. In most cases, arrangements will be made with already operating regular rail systems to accommodate commuter rail equipment.

Power Supply

Besides diesel power, electric locomotives are frequently employed. The original systems depended on 11,000-V ac, 25-Hz current supplied by overhead catenaries. Modern power supply utilizes 25,000 V ac, 60 Hz. Some commuter rail systems rely on metro-like arrangements—600 to 650 V dc drawn from a third rail.

Capacity and Cost Considerations

The capacity of commuter rail operations—as is the case with all other public transit modes—is primarily a function of the passenger holding capacity of each vehicle, the number of such vehicles operated together in a train, and the number of such trains moved past a given point. However, other elements and possible constraints may be in play as well. For example, long dwell times at stations may reduce possible frequency, but this is not likely to happen with commuter rail because the headways are usually quite long compared to dwell times. There may be, however, operational delays caused by switching complications entering and leaving terminals and yards. During peak periods with high-intensity operations there may be problems of marshalling a sufficient number of trains at the loading end to feed them quickly to platforms, and passenger overloads may occur.

The throughput element on the line, measured as possible train movements per hour on a single track (tph), is largely a matter of the signal systems in place. For example, conventional arrangements with 3-mi blocks will allow only 4 to 6 tph. Very advanced systems, rare on commuter rail lines and involving special rolling stock, may achieve short headways, resulting in up to 25 or 30 tph.

Unlike most other public transportation modes, commuter service is predicated on the assumption that every passenger will have a seat even in the rush period. Thus, if bilevel coaches are used, with a seating capacity of 140, and they are coupled in eight-car consists, each train will be able to carry 1120 passengers. If it is possible to run trains 4 min apart, and as long as sufficient platforms are available at the origin and destination ends where the longest dwell times are to be expected, and there are reservoir space and tail tracks to absorb the empty trains, a very respectable amount of work can be achieved. In that case, with

Physical Characteristics of Commuter Rail Vehicles and Infrastructure

Length of a coach	65 to 85 ft (20 to 26 m)
Width of a coach	10 to 10.5 ft (3.05 to 3.2 m)
Height of a single-level coach	14 ft (4 m)
Height of a bilevel coach	16 ft (5 m)
Number of seats in regular coach	Up to 128
Number of seats in bilevel coach	Up to 175
Capacity with standees	360
Number of cars in a train	1 to 12
Maximum running speed*	80 mph (130 kph)
Usual average operating speed	18 to 50 mph (30 to 75 kph)
Maximum curvature	
On main line	570-ft radius (174 m)
In yards and terminals	300-ft radius (91 m)
Maximum acceptable grade	
On main line	3 percent
On main line with mixed freight	1 percent
Desirable maximum	2 percent
Standard track gauge	4 ft 8½ in (1.435 m)
Minimum width of reserved envelope	13 to 15.5 ft (4.0 to 4.75 m)
Minimum height of reserved envelope	17 ft 10 in (5.4 m)
With double-stack freight cars	22 to 23 ft (6.7 to 7.0 m)
Distance between center lines of track	15 ft (4.6 m)
Distance from center line to side wall in tunnel	9 ft (2.7 m)
Desirable center line offset to new track parallel to existing main track	25 ft (7.6 m)

* With wayside signals. Higher speeds are acceptable with cab signals or automatic train stop devices.

15 trains each hour, the throughput capacity would be 16,800 passengers per track per hour, comparable to general rail transit capabilities.

While there may be a few corridors where such demand levels could be found or be envisioned, that number is unrealistic in day-to-day operations. A more reasonable scenario would be 100 passengers in a coach and six-car trains that run every 10 min. That results in 3600 passengers per track per hour—still a respectable figure, not exceeded in actual demand on almost all corridors, but considerably lower than expectations with heavy rail transit, light rail transit, or even bus rapid transit. Most commuter rail service in North America operates at 20-, 30-, or 60-min intervals.

The prices of rail vehicles continue to escalate. General numbers are not particularly reliable because each situation is different, and the purchase orders do not come at a steady rate, nor are they consistent. Much depends on the size of the order and the general business situation at any given time. The best that can be done is to record that in 2001 the following approximate prices prevailed:[26]

Passenger coach (not powered)	$1.3 million
Bilevel passenger coach	$2.3 million
Electric multiple unit	$2.5 million (or more)
Diesel locomotive	$4.0 to 5.0 million
Electric locomotive	$5.0 to 6.0 million

Possible Action Programs

The normal planning process for transportation systems proceeds rationally from the estimation of future demand, to the identification of modes that are able to respond, to an evaluation of the positive and negative consequences of each mode to select the most suitable choice. This is a long and iterative process. Planning for commuter rail service, on the other hand, may be best done somewhat in reverse. Under the contemporary situation in large American cities, there is little chance that the acquisition of a new right-of-way in a desirable location, with its associated costs and community disruption; the building of a most heavy infrastructure; and the purchase of expensive and technically advanced rolling stock could be justified to provide convenient and comfortable service to a few thousand suburban commuters. The idea only makes sense, in a practical world, if an already existing alignment is reasonably available, many of the supporting structures are still in place, and rolling stock can be borrowed or obtained cheaply—that is, if the capital start-up costs can be reduced to a minimum. There will be enough problems generating sufficient income to cover regular operations and maintenance costs in any case, at

---

[26] *Metro Annual Fact Book* for those years, and cost estimators in consulting firms.

least in the context of the contemporary urban situation. Clearly, the suggestion here is to cut the coat after the cloth.

This statement may be seen by some as too harsh and limiting—why not consider building brand-new rail lines for metropolitan commuting? The fact is that this has not happened ever in North America, and examples abroad are not too many either. The RER of Paris could be regarded as such an instance, and the only place today anywhere in the world where a new passenger rail line is under construction is Caracas, Venezuela.[27] A commuter line from the end of the metro, through hilly terrain, to new districts is expected to make a badly needed urban expansion space accessible.

Fortunately, the idea of reusing available rights-of-way is not naive at all because many underutilized rail lines still exist within American cities. The planning process, therefore, can be somewhat as follows:

1. Identify all or specific rail rights-of-way as a resource inventory.

2. Evaluate each such alignment as to potential technical problems or fatal flaws with respect to the operations of commuter rail service.

3. Envision various operational and service alternatives; do very preliminary cost estimates for required improvements to initiate operations.

4. Undertake patronage estimates for the various alternatives and different access scenarios for regular and special services.

5. Equilibrate (i.e., repeat the process) several times to achieve a reasonably balanced package of recommendations.

6. Compare possible demand levels to all impacts, i.e., costs and benefits (quantifiable and qualitative), and calculate cost-benefit ratios or evaluate against policy considerations in a public forum.

7. Review likely sources of financing.

8. If the findings are reasonably positive, repeat everything with fewer alternatives in greater detail.

---

[27] To the best knowledge of the author.

9. If the conclusions are favorable, prepare engineering and operations plans. Arrange for financing.

10. Advertise for construction and supplier bids, select contractors, build and monitor progress.

11. Open the system and congratulate everybody concerned.

12. Monitor performance and usage levels.

Commuter rail services are eligible for all the federal assistance programs in public transportation provided under current legislation (Intermodal Surface Transportation Efficiency Act of 1991 and Transportation Equity Act for the 21st Century of 1996).

Conclusion

There is no question that rail lines and stations were major land use generators, and that they structured the metropolitan patterns in all the older cities of North America. The principal nodes were nailed down, and that structure prevailed until the ascendancy of the automobile. The threatened extinction of rail service did not happen, but there certainly was a period (the 1960s and 1970s) when the best that that the remaining services could do was to plead for help and battle a negative image. That situation no longer prevails, but neither can it be asserted that rail nodes are again places where new vigorous activities concentrate. Such instances can certainly be found with new developments clustering in the vicinity of stations, but they are not as dominant as edge cities relying on highway access. Nevertheless, the potential of rail as an urban growth generator is again a reasonable consideration under the right circumstances.

Because commuter rail is a mode very much fixed in place and likely to take advantage of the radial track networks built in the early periods of city development in North America, its role is very much tied to the fortunes of the central cores of

Bilevel commuter cars in Toronto.

metropolitan areas. If the centers are to be preserved and enhanced with an array of constructive policies and action programs, then commuter rail should certainly be a key component of such efforts. Indeed, it may be a critical, as well as an affordable, element in success. For low-density, sprawling developments, rail service will do little, unless extensive and well-integrated feeder/distribution services are planned and implemented jointly. High-intensity employment or commercial centers may benefit from rail access, provided a reasonable service corridor or network can be structured.

If a commuter rail service were to be built from scratch, the balance sheet would not look particularly good. The difficulties of acquiring the right-of-way and the extraordinary expense of building the infrastructure and buying rolling stock would have to be contrasted with the potential demand in patronage. Whatever those figures might be, the investment would be fully employed only a few hours each weekday, which is most likely to create an untenable economic situation. Commuter rail thus becomes only a modal option under today's conditions if a substantial portion of the up-front costs can be foregone—securing the right-of-way at minimal expense and being able to adapt the existing infrastructure to current needs. The rolling stock will almost certainly have to purchased new, since the expected service quality is not likely to be achieved with worn-out and obsolete equipment. Mitigating factors to all of this may be the existence of a rather strong market in used but refurbished rolling stock and the fact that peak period demands may be sufficient to justify the implementation of commuter rail services.

Given the vast network of rails that was created in North America during the previous centuries, such opportunities do exist. The presence of underutilized rights-of-way has to be regarded as an asset for which we have not yet discovered a fully effective use. Perhaps some constructive ideas will emerge before these rights-of-way gradually disappear or are completely committed again to freight operations.

Bibliography

The two standard reference books on public transportation—Vuchic, and Gray and Hoel—contain coverage of commuter rail as well, as does Edwards.

American Railroad Engineering Association: *Manual for Railroad Engineering,* 1994. This 6-in-thick manual contains all possible technical and construction data and procedures related to the implementation, operation, and maintenance of railroad systems.

Armstrong, John H.: *The Railroad: What It Is, What It Does: The Introduction to Railroading,* Simmons-Boardman, 1977, 240 pp. Basic railroad engineering and operations as of the 1970s.

Edwards, J. D., Jr. (ed.): *Transportation Planning Handbook,* Institute of Transportation Engineers/Prentice-Hall, 1992, 525 pp.

Gray, George E., and Lester A. Hoel (eds.): *Public Transportation* (2nd ed.), Prentice-Hall, 1992, 750 pp.

Hay, William W.: *Railroad Engineering,* John Wiley & Sons (2d ed.), 1982, 758 pp. The definitive reference work, leaving no stone unturned regarding technical detail.

Vuchic, Vukan R.: *Urban Public Transportation: Systems and Technology,* Prentice-Hall, 1981, 673 pp.

Automated Guideway Transit

Background

"If we can send a man to the moon, we should be able to solve traffic problems in our cities," was said quite frequently and with conviction in the late 1960s.[1] The implication was that advanced technology can tackle any task and achieve spectacular results. Various means of propulsion became available; the capabilities of computers were seen as boundless, including the control and management of urban transportation services. This national attitude opened the door for many conceptual and applied explorations of automated transportation systems and sophisticated equipment.

The preoccupation with the perceived potential of "gee whiz" technology was further encouraged by two factors. One was the desire on the part of the high-tech defense industries to diversify into civilian markets; the second was the inauguration of a federally sponsored research and development program called Tomorrow's Transportation, envisioned as a catalyst to encourage the application of advanced technology to the solving of urban transport problems. The proposals that emerged were less a result of

---

[1] President Richard Nixon used almost this exact phrase in his 1972 budget message to Congress. The landing on the moon took place in 1969.

America's industrial giants developing new concepts than of backyard inventors and garage tinkerers doing their traditional innovative work.

Under these encouragements, during the 1960s and 1970s, many automated transportation devices were envisioned, designed, and engineered; some reached the pilot project stage, and a few were placed in service in the 1970s and 1980s. Work continues, but with reduced intensity and more realistic expectations.

These means of urban transportation first carried the name *people movers,* but since that term was too all-encompassing, the eventual formal designation became *automated guideway transit* (AGT).[2] "People movers" remains the popular designation. Several dozen such systems have been developed by various manufacturers, and even more have been proposed by a multitude of inventors. They are quite different from each other, offering a range of capabilities at different levels of practicability. Precise definitions are difficult,[3] but a working consensus has emerged, which describes AGT as a fixed guideway transportation mode, on an exclusive right-of-way, which operates automatically under central control (no drivers), either as individual vehicles or in trains, running either on a fixed schedule or activated upon demand.

It is convenient to subdivide automated guideway transit into the following categories:

- *Personal Rapid Transit* (PRT). Small vehicles for one or a few persons requested by a call button and directed individually to a particular station without intervening stops.

- *Shuttle-Loop Transit* (SLT). Vehicles or short trains operating back and forth on a single line between two stations without

[2] Some other acronyms that have appeared are automated people mover (APM) and advanced rapid transit (ART).

[3] As discussed throughout this chapter, there are automated transit systems that could be included under AGT, but are not because their other characteristics dominate. Such examples include the monorail system of Newark, New Jersey, Airport, the advanced rail network of Docklands in London, and the automated light rail lines under construction currently in JFK International Airport, as well as several other recent monorails in Japan and Australia. Likewise, the H-Bahn of Dortmund, Germany, the light metro of Ankara, Turkey, and the suspended cable-drawn service to Mud Island in Memphis, Tennessee, are deemed to be outside the AGT scope.

intermediate stops or on a one-way or two-way continuous loop, usually at fixed intervals.

- *Group Rapid Transit* (GRT). Larger vehicles (up to 100 passengers) or trains operating on networks with station stops, usually on a schedule, although demand activation is also possible.

There are some semantic problems with these terms, but they are in general use and therefore have to be accepted. For example, the guideway is not automated, the system operations are. Most of the systems are not really rapid transit because of the small-scale, localized service that they provide. And true PRT simply does not exist, as will be discussed later.

AGT has often been described as being a "horizontal elevator," which is fair enough, although there are some basic differences, such as the fact that AGT vehicles run on different paths and are able to switch between lines, which vertical elevators cannot do. Many of the systems in operation today also have similarities to conventional light and heavy rail transit; some could actually be classified as monorails and cable cars; a few even aspire to the service quality offered by taxis and jitneys. The lack of a clear-cut definition in the real world occurs in part because AGT was developed to take advantage of high-technology devices, which at their best should be able to provide a more direct and immediate response to user needs than conventional transit, to achieve greater safety by precluding human error, to utilize equipment at maximum efficiency, and to minimize labor costs. These results are not always obvious to the casual observer.

Some of these goals have been attained by operating AGT systems, but the original premise of creating an all-purpose transportation system that would constitute the basic service for any city has not been reached. After 30 years, there are only four systems in the United States operating in the public realm—Morgantown, West Virginia; Miami, Florida; Detroit, Michigan; and Jacksonville, Florida. There are several more in private or institutional ownership, a number of which operate in airports. The aggregate length of all of those is not much above 20 mi (32 km), and total annual ridership is too small to appear on national passenger summaries. Nevertheless, advanced people movers have a specific niche in the transportation service spectrum, being well able to provide mobility at the local scale within defined areas that encompass distances beyond walking range. As regular fixed guideway systems have become more automated, the distinctions between a modern metro and an AGT have started to blur, and this trend of convergence is certainly continuing.

Development History

Leaving aside the imagination of science fiction writers and the visions of utopian engineers that enriched popular and technical publications during the first two-thirds of the twentieth century, realistic possibilities for a new transportation mode were brought into being by the computer. Automated control systems are the principal revolutionary characteristic and defining feature of AGT. All the other elements, such as propulsion systems, means of vehicle support, materials used, and passenger amenities, may be employed at a very advanced level, but they are not new concepts. The package that constitutes AGT contains more familiar but upgraded parts than revolutionary ones.

From the time of Buck Rogers, and perhaps earlier, inventors and writers have dreamed about individual pods that would be almost magically transported from any point in the city to any other. There were early efforts to at least visualize systems whereby automobiles, coupled into trains, could run on guideways. As the urban transportation problem became worse, there were increasingly more attempts by technologists to conceptualize systems that would achieve high throughput capacities while maintaining the privacy and attractiveness of the private car, Such systems would also reduce labor costs associated with conventional transit and would eliminate the chore of driving a separate vehicle. Unfortunately, while many of these ideas were imaginative and inspiring, few were accompanied by a sense of engineering feasibility and realistic awareness of costs.

As we entered into the 1950s, swift advances in computer science opened previously unheard-of possibilities, and automated transportation controls became conceivable. During the 1960s, several interesting experiments and applications were attempted. For example, in 1961 the 0.4-mi (0.6-km) Times Square Shuttle under 42nd Street in New York City was equipped with devices that allowed it to run back and forth without a driver (known as Shuttle Automatic Motorman, or SAM). Labor unions protested vigorously; the train was allowed to operate only after it was agreed that a crew would be aboard at all times. A fire of unknown origin soon destroyed the installations. The New York City transit system has resumed explorations of automated controls only now after four decades.

There were tests of an automated Presidents' Conference Car (PCC) streetcar in Erie, Pennsylvania, in 1962 to 1963. A major success of this era was the Bay Area Rapid Transit (BART) metro system, which was designed and built during the same decade as a completely automated system (for its time). As with any truly pioneering effort, BART encountered many problems due to the imperfection and early capriciousness of the equipment. Yet, these difficulties were solved, even if some parts had to be completely rebuilt and lawsuits had to be settled. Today, BART is regarded as one of the best of the "new-generation" rapid transit systems. The Lindenwold line (Port Authority Transit Corporation [PATCO], 1969), connecting Philadelphia to its New Jersey suburbs, is also an early example of automated train operations. Like BART, PATCO continues to function reliably and efficiently. Other automated rapid transit systems in the United States operate in Washington, D.C., and Atlanta.

A very visible demonstration project, built in Pittsburgh's South Park, was a 1-mi (1.6-km) test track on which a special automated vehicle—the Transit Expressway or Skybus—was put through its paces during 1965 to 1967. Single vehicles (50 to 70 passengers) or short trains were placed on an elevated structure; they could reach 55 mph and could operate at 2-min headways. The system worked, and Westinghouse Electric Corporation, which was responsible for this as well as many of the other early projects, emerged as the leader in the automatic vehicle control field. After Skybus there was no longer any doubt that technical feasibility had been established, and the snags encountered along the way were seen as challenges rather than real obstacles. Not only public confidence but also

Early experimental vehicle—Pittsburgh's Skybus.

expectations mounted regarding the capabilities of advanced technology. Those were the signs of the times. The late 1960s and early 1970s were an exciting period, promising real and effective solutions to urban transportation problems. Most journalists, citizens, public officials, and businessmen believed that technology could solve urban transportation problems. It is sad to think that such confidence in a bright and efficient transportation future may not be experienced again in our communities for some time.

As was to be expected, many system developers, large and small, jumped into the field to compete for the prestige of having the most effective hardware and gaining the expected lucrative contracts for system implementation. A major factor in the expansion of the automated transportation field in the United States was the contemporaneous change in federal policy toward aerospace and defense contractors. It became apparent that space exploration programs would not expand indefinitely. At the same time, criticisms mounted toward expensive military programs that brought no tangible benefits to the population at large. Therefore, it became appropriate for these large aerospace and defense companies to seek other (civilian) markets and to become "socially relevant."[4] An obvious marketing target was advanced urban transportation systems, where the unquestionable technological capabilities of space and military vehicle builders were expected to serve well. A number of these enterprises, with the full encouragement of the government, marshaled their skills and resources and developed prototype AGT systems. Many were actually built and applied, but the anticipated markets did not develop. Also, it was discovered only later by the transit newcomers that it is one thing to manufacture a rocket or missile for a single shot under the auspices of a single agency, with a practically unlimited budget; it is something entirely different to build a public means of transportation that has to operate continuously, day after day, within an urban environment under political and media scrutiny—almost always with modest financial resources, but with great expectations of cost effectiveness.

Some of the early systems are quite interesting, if not in terms of actual practical results, then certainly in their aspirations

[4] A. G. Meyer, unpublished master's thesis, *The Development and Urban Deployment of AGT,* 1980, Columbia University.

regarding service character and quality.[5] Almost all concentrated in the early days on the personal—not group—rapid transit concept, pushing the frontiers of applicability.

One of the first of these systems was designed by Edward O. Halton, a promoter of monorails, who in 1953 envisioned small six-passenger vehicles suspended from an overhead guideway, which he called *Monocabs*. The idea was sold to Vero, Inc., which built a test track in 1969, and was sold again to the Rohr Corporation, a major space contractor that had built the original BART cars. The concept underwent a full high-technology metamorphosis by placing the cabs on magnetic levitation tracks and equipping them with linear induction motors. A test track was built at Rohr's Chula Vista, California, facility, and the device was demonstrated at the Transpo72 exhibition in Washington, DC. (More about that event later.) The system was selected for installation in Las Vegas, but the economic recession in 1974 brought further progress to a halt. The patents were sold once more, this time to the Boeing Company, which participated in the Urban Mass Transportation Administration's (UMTA's) Advanced Group Rapid Transit program (more about that later as well) until it was terminated in the mid-1980s.

Another development effort was initiated by a group at the General Motors Research Laboratory starting in the late 1950s. Its device, called Hovair, relied on air suspension instead of wheels and also used a linear induction motor for propulsion in the early models. Since GM was under antitrust regulations and was barred from entering the urban transit field, a separate enterprise was formed, called Transportation Technology Incorporated (TTI). It too conducted full-scale testing in Detroit during 1969, but was acquired by the Otis Elevator Company in 1971. After a demonstration at Transpo72, the project moved to Denver with a new test track. The system was actually implemented at Duke University with larger cars than were originally envisioned. It was found along the way that the linear induction motors of the time were not particularly efficient, and, therefore, Otis—fittingly enough—developed and sold several models pulled by a cable.

---

[5] The early history is documented quite well by J. Edward Anderson in various publications, including "Some Lessons from the History of Personal Rapid Transit," jeanderson@taxi2000.com. Anderson remains a principal advocate of PRT, and promotes this mode against all nonbelievers.

The Alden StaRRcar, an attempt to achieve personal rapid transit.

Another entrepreneur, William Alden, developed the Alden staRRcar in the 1960s. He started with a more radical idea—the dual mode concept: small cars that could be operated individually on regular city streets, but also driven onto ramps that would lead to a network of automated guideways that would cover the community. Alden formed a company, built a test track in Bedford, Massachusetts, in 1968, developed a fully operational small-scale model, and demonstrated the vehicle on numerous public occasions. Alden won the contract for the original Morgantown, West Virginia, system, but his concept was displaced by other technology.

One more early example was a device patented by Sam Dashew, who called it, not surprisingly, the Dashaveyor. This concept was acquired by the Bendix Aerospace Corporation as its entry in the people mover field. It was tested extensively by UMTA, used at the Toronto Zoo, and eventually shown as one of the official entries in Transpo72. However, in less than a decade Bendix was out of this business as well.

There were some dozen other efforts by firms, individuals, and institutions to develop their own versions of PRT concepts. Most of these remained as theoretical studies. At least one was actually built—the Jet Rail for Braniff airlines at Love Field in Dallas, suspended from an overhead beam—but it was later abandoned. Another exploration—the Uniflo device—was a small container enclosed in a tube and supported and propelled by track-side air jets. Cornell Aeronautic Laboratories did studies to see whether headways of less than a second were feasible (Urbmobile); the Massachusetts Institute of Technology (MIT) published a comprehensive review of AGT possibilities (Project Metran). The Aerospace Company (a nonprofit group associated with the U.S. Air Force) did theoretical studies in the early 1970s, and estimated that headways in fractions of seconds were feasible, but was also concerned about the visual impacts of overhead structures.

In the meantime, Westinghouse was joined by AEG Transportation Systems, which enabled this group to establish and maintain leadership in the AGT field. The companies did not concern themselves very much with PRT, but concentrated on larger vehicles and higher-volume systems, i.e., group rapid transit (GRT). They built and opened in 1971 the first of a series of shuttles that connected the central terminal building of the Tampa Airport with several satellite units. This was followed by work in most of the other (but by no means all) U.S. airports, and in other significant applications.

Considerable momentum toward AGT development was generated at the federal level as a national policy in the 1960s. A clause (Section 6) was inserted in the Urban Mass Transportation Act of 1964 that specifically mandated research and development of "new systems of urban transportation [that] take into account the most advanced available technologies and materials." The first result of this were studies of all the systems then being developed or proposed, with the findings published in 1968.[6] These reports, with their optimistic tone, can be credited with building support for AGT and further activity in this field not only in the United States, but also in a number of other technologically advanced countries.

It has to be observed, however, that the early efforts abroad were minor or derivative of U.S. examples. An exception was the German *Cabinentaxi,* developed by a consortium of firms beginning in 1970 and continuing with the construction of prototype models and test tracks. Despite much promotional effort at home and in North America, no new working systems of this type were built. As could be expected, the Japanese also entered the field, and while their first effort—Computer Controlled Vehicle System (CVS)—was not successful and no markets could be identified, they recovered lost ground eventually.

In France, scattered efforts were taken over in 1970 by the aerospace firm Engins Matra, which attempted a very advanced

---

[6] These are known as the 17 *HUD Reports,* since the Urban Mass Transportation Administration at that time was a unit of the U.S. Department of Housing and Urban Development (HUD). They were summarized in an influential publication by W. Merritt, *Tomorrow's Transportation,* that was widely distributed and read.

and sophisticated concept (*Aramis*): fast-moving small vehicles in closely spaced platoons, with individual units entering and leaving the central track at speed. The control systems, however, were not able to achieve the necessary precision and reliability, and the concept was never implemented. Matra, however, did become a major player in the field a decade later with a more conventional transit approach. The Canadians also started with a PRT vision in the late 1960s, and formed the Urban Transportation Development Corporation (UTDC) in 1973 to promote research and lead to the manufacture of advanced technology systems. It developed workable systems with GRT characteristics in several cities, notably Vancouver and Scarborough (Toronto).

The real start of AGT systems as a functioning public transit mode was the development of the automated connector between several units of the University of West Virginia in Morgantown.[7] The initiative for this system was taken by an engineering professor in the late 1960s, and he soon gained support both from the university and the local municipal government. UMTA designated this effort as a demonstration project (1970), and it became a full participant with financial and managerial involvement. There was also a national political agenda in play, with pressure on UMTA to show results before the 1972 election, which resulted in an accelerated but not well-advised implementation schedule. The hardware choice shifted from the Alden staRRcar to models of the Boeing and Bendix Companies, which had to invent new elements and systems under extreme time constraints. The civil engineering infrastructure was designed and built before the characteristics of the vehicles were known. The results of all these problems were fourfold cost increases, a number of clumsy elements, serious delays, and loss of general confidence. Even after completion, the university was hesitant to contribute the necessary financial support to maintain the system on a permanent basis. At one point it was seriously considered whether the U.S. Army Corps of Engineers should not be called in to demolish the half-built structures. Ultimately, the university agreed to help fund the system, and the demolition plans were cancelled.

[7] The Morgantown system has been reviewed in countless articles. A concise summary is "People Movers Head into Costly Future," *Engineering News-Record,* July 19, 1979, pp. 37–38.

The Morgantown system was completed and opened for operations in 1975. The original concept was to achieve a demand-responsive service with off-line stations and circulation loops, all of which increased the costs and operational complexity. Despite its problems, Morgantown has been working successfully for almost 30 years; it has been inspected by just about every transportation planner and municipal official contemplating new transit systems for their communities. It

The first full AGT system in North America (Morgantown, West Virginia).

is also referred to as an example of how *not* to manage and structure projects, particularly when they are of a pioneering nature. (While the Morgantown system serves a university, because of the massive government assistance in construction and open usage, it can be classified as a public service.)

Along the way, UMTA organized an international show of transportation equipment at Dulles International Airport called Transpo72. Many vehicles and devices were displayed, but the key attractions were full-scale working guideways and vehicles by several AGT manufacturers (the Bendix Aerospace Corporation, Ford Motor Company, Otis Elevator Company, and Rohr Corporation). As a means of providing useful information with tangible examples, the show was most successful, and those attending were suitably impressed. Much further interest was generated, but not necessarily matched with commitments of local public funds for immediate implementation. The pattern had been set in the United States of communities expecting grants from

Air-supported vehicle on display at Transpo72.

the national treasury for research and development of any new system and sponsorship of actual implementation, not just for pilot projects.

The early 1970s was a time when construction grants for rail transit systems became available, and a number of cities started building or had serious plans in progress. The Clean Air Act of 1970 mandated programs that would reduce pollution caused by automobiles. Much work was also being done to create major new airports. Yet, there were no tangible efforts by cities to undertake AGT projects, and the large aerospace firms that had developed hardware systems at considerable expense became most concerned.[8] General pessimism set in, which became worse with the recession of 1974. The automated systems that had opened by that time (BART, Morgantown, and Dallas/Fort Worth Airport) all had serious startup problems with unreliable equipment that received much attention in the national media. There was a catch-22 at the federal level: government attitudes had overstimulated expectations, but, when it came to actual implementation, UMTA insisted on cost effectiveness. Unfortunately, cost effectiveness will start off low if R&D costs have to be absorbed by the first system. Also, to insist that systems sponsored by public funds should have a proven track record does not work with pioneering efforts.

Influential senators and congressmen stepped in and instructed UMTA and the Office of Technology Assessment (created in 1972) to review the existing situation and the future potential of AGT. This—as frequently happens—was largely an examination by the industry itself, and questions did not address so much the service capabilities and suitability of automated people movers as the reasons why a market had not emerged. The conclusion was direct and somewhat predictable: American technology deserved support, its value had to be shown through pilot projects, and those were only possible with federal sponsorship. No locality was going to risk its own resources in deploying new technology.

In an effort to remove or at least alleviate the risks inherent in being the first with a new technology, UMTA instituted a new

[8] Summaries of the state of AGT technology and expected utility at that point in time are provided by F. L. Schell, "Peoplemovers: Yesterday, Today, and Tomorrow," *Traffic Quarterly,* January 1974, pp. 5–20; and H. W. Demoro, "People Movers," *Mass Transit,* July/August 1977, pp. 53–58.

program in 1976 for the development of Downtown People Movers (DPMs) as demonstration programs of prototypes. Care was to be exercised by concentrating on projects that took a direct and simple approach, and only those equipment vendors who had working models were listed as eligible. There were nine of them at that time, and the expectation was that as many distinct systems as possible would be created.

The response was surprisingly large. Sixty-five cities submitted letters of intent, presumably to take advantage of an open grant situation, not so much following a careful analysis of needs and suitability. From these, a group of 11 finalists was selected, including the first demonstration sites: St. Paul, Los Angeles, Cleveland, and Houston. (It is instructive to observe, of course, that none have actually built an AGT system.) Later, UMTA reclassified the applicants as First Tier cities (Detroit, Houston, Los Angeles, Miami, and St. Paul), which were allocated construction grants, and Second Tier cities (Baltimore, Indianapolis, Jacksonville, Norfolk, and St. Louis), which received only study grants. Most of these cities dropped out for any number of reasons, but three systems were built—the Miami Metromover, Detroit DPM, and Jacksonville Skyway.[9] The results were mixed, and they continue to be debated even today.

Administrations and policies continue to change in Washington, and the DPM program was terminated in the mid-1980s. There has been no active program or serious plan for new public service AGT systems in any United States city since that time. Further efforts, however, can be seen in the private sector and at airports. Work also continues abroad, particularly in Japan and in other countries that tend to purchase Japanese equipment. Perhaps the major beneficial result of all the AGT efforts has been the development and testing of automated system control technology for regular transit. Experience and confidence have been gained, and the range of applications has broadened. Such control systems are now utilized routinely with new rail transit projects, and projects have started to convert a number of conventional light and heavy rail lines to automated operation. The current state of activity with AGT and automated

[9] K. Myint, "Where Have All the People Movers Gone?" *Center City Reports,* November 1983, pp. 6–8.

systems is outlined subsequently under "Application Scenarios."[10]

Types of AGTs and Their Operation

Each AGT system now in operation differs at least in some respect from all the others; therefore, many ways of grouping them could be employed. The first approach would be to use the conventional classification system introduced at the beginning of this chapter, which distinguishes among the general types of service that can be provided. The problem with these subdivisions is that they are neither mutually exclusive nor fully expressive of the actual forms of AGT operations and service.

Shuttle-Loop Systems

This group is intended to encompass services that are limited in scale, both in route length and in passenger volumes. The cleanest examples would be simple connectors between an activity center and a parking facility or mass transit station, or linkages between separate airport terminals. Usually only two terminal stations are involved, although there may also be other stops along the way. The loops may be short routes that either follow a somewhat circular path, have separate alignments for the two directions, or are basically shuttles with vehicle turnarounds at the ends (pinched loops). Considerable density of use can be achieved in controlled environments, and the image of the horizontal elevator is most appropriate. It is difficult to say where this type stops and the next one begins. Examples include most airport shuttles, such as those in Atlanta, Denver, Orlando, and Tampa.

Group Rapid Transit

These are systems that carry many passengers in single vehicles, which may run in trains, between stations on a fixed schedule. They are thus basically similar to any other exclusive right-of-way mass transit system, except that they are fully automated and the

[10] See also W. D. Middleton, "Automated Guideway Transit Systems Come of Age," *Transit Connections,* March 1994, pp. 12–20; "People-Movers Around the World," *Mass Transit,* May/June 1995, pp. 22–24; C. Hanke, "Driverless Metros Win Fans throughout World," *Metro* magazine, December 1998, pp. 36–37.

vehicles tend to be smaller than regular subway or light transit cars. The bulk of AGT systems now in existence are of this type, being able to operate with some efficiency. Examples include the Vancouver SkyTrain and the Toronto Scarborough line.

Personal Rapid Transit

This is the original high-technology dream that inspired most early efforts. Each rider would be able to reach a boarding place very conveniently—if the guideway does not touch the origin house or apartment building directly, then an entrance may be found on the next major street. An empty vehicle would be summoned by a call button, and it would be small enough to accommodate only one or a few persons, requiring no sharing of space with strangers. Another push button would be used to select the destination of the trip; the pod or vehicle would automatically merge into the main traffic stream on the guideways and be routed through multiple switches and merge/diverge arrangements until it reaches the desired building, or at least stops within one or two blocks of the final destination.

It is a beautiful scenario, but no such system has been built, nor can it be built with currently available or reasonably foreseeable technology. A short digression is necessary at this point to put this concept to rest. It certainly would not be wise to say that the required extreme precision in vehicle control can never be accomplished, but, despite valiant and persistent design efforts that continue even today in some quarters, the capabilities are still not there. The required reliability would come at a high price.

Conceptual design of what a PRT might do in Los Angeles.

Networks

Returning to the realities of AGT, an alternate means of classification would be based on network configuration rather than service type. Thus, different networks could apply both to shuttle-loop and group

The Mirage of Personal Rapid Transit

PRT of the type described here would require the introduction of a dense network of overhead guideways on most major streets in our cities. Conceptual drawings show very slender structures and beams, but they have been drawn with considerable artistic license by illustrators who are not structural engineers. The multiple second-level entry/exit points would require some sort of a station with staircases if not escalators at many locations along the line. If communities are concerned about the visual intrusion of trolleybus wires, the reaction toward these facilities can be easily predicted. The concerns of Americans with Disabilities Act have not even appeared in the debate yet.

The real problem with PRT is even more basic and insurmountable. A staggering capital investment would be called for, generating a very low volume of service.[11] The use of private containers, even at split-second intervals (which are currently not attainable with any kind of precision, safety, or reliability), would mean, in effect, a moving line of individual passengers on any guideway. That is not sufficient density under most reasonable transit demands. How would thousands of users enter or leave a large office building in the space of an hour with such a system? The entry and exit of people cannot be automated or accomplished instantaneously.[12] For off-hours casual travel we already have a PRT in full operation: private automobiles.

A case in point is the French effort to develop a PRT, with significant inputs of resources and national prestige between 1969 and 1987. It carried the name *Agencement en Rames Automatisées de Modules Independants dans les Stations* (*Aramis*), which gives some idea of the extreme technological aspirations behind the basic concept. It never worked, but the project is well documented;[13] the experience is used today as an example of misguided expectations, and it continues to cause some merriment among the skeptics of technology. A carefully researched evaluation in 1999[14] concluded that PRTs would face some serious constraints if they were to be implemented:

- Single-platform stations would not be able to accommodate the numerous small vehicles that would converge during peak periods.

- Many entry and exit ramps/channels would be required at every station, thus creating a complex facility that would cover much space.

- Acceleration and deceleration could not be done on the principal guideway because of close headways, thus increasing the number and extent of the ramps at stations.

- The large number of vehicles in operation would result in some expectable breakdowns, which would paralyze at least parts of the network and be difficult to remove.

- Because of all of this, station spacing cannot be very close, thus precluding door-to-door accessibility.

Currently available control systems have not yet proven themselves able to accomplish safe and quick merging of individual vehicles into a moving traffic stream. If this were to be achieved at some future date (and there is no reason to doubt that it will be possible), the remaining questions are whether any reasonable capacity in terms of transit loading can be attained and whether the capital investment can be justified.

---

[11] This argument has been advanced and maintained most strongly by Vukan Vuchic in various publications and presentations (including pp. 497 in *Urban Public Transportation,* and several short articles in *The Urban Transportation Monitor,* November and December 1996).

[12] We are all aware of the imagined transporter capabilities seen in just about every episode of *Star Trek.* But we are certainly generations away from being able to say "Beam me up, Scotty," and expecting to get somewhere in one piece.

[13] A book has been translated into English by Bruno Latour: *Aramis, or the Love of Technology,* (Harvard University Press, 1996, 314 pp.). The story is often told that when Prime Minister Jacques Chirac was briefed on the "nonmaterial couplings" of the cars, he could only claim lack of personal experience with such a concept.

[14] M. A. Sulkin, "Personal Rapid Transit Déjà Vu," *Transportation Research Record 1677,* 1999, pp. 58–63.

The dream of personal rapid transit is not likely ever to disappear, not even to the *New York Times Magazine.*

The Train You're Never Late for

Peter Richmond

Doug Malewicki envisions a future full of unusual things. Like a way of liberating American cities from their clogged-artery highways using a lacework of elevated rails that would carry thousands of commuters in individual pods coasting silently above the sidewalks at 100 miles per hour. Like a 40-foot-high robotic dinosaur that eats cars and spits fire to thrill crowds of carnival spectators. Malewicki has already designed the latter, but while his Robosaurus will be chewing Chevy Novas throughout the land this summer, the only way to see his magnetic-levitation-powered SkyTran is to pore over the snappy retrofuturistic illustrations on its Web site (www.skytran.net). Malewicki and the rest of his team at Skytran Inc.—a half-dozen staff members scattered from Raleigh, N.C., to Seattle—are fully convinced that once a full-size prototype is up and running, it's only a theoretical hop, step and jump to the day when the beleaguered American worker will walk to his local pod station (at a shopping mall on Main Street), wave a personal ID in front of a sensor and be whisked away in his fiberglass cocoon. Picture the offspring of a Newark Airport monorail and a Killington chairlift, and you can envision SkyTran's commuting machine: no straphanging with strangers, no carpooling with verbose neighbors. Just a personal zoom with a view.

"There are a lot of people saying, 'You're pie in the sky,' " says Malewicki, 61, who has a master's degree in aeronautical and astronautical engineering from Stanford University. "No. This is just engineering. People are good at saying, 'When the time is right.' Well, the time is right."

It may be right, but it hasn't arrived, and with no money yet in place, there's no assurance that it ever will. It's one thing to have Ralph Nader touting SkyTran on the Green Party campaign trail, or to have transportation officials from Seoul to El Paso exploring SkyTran's potential to relieve their urban woes. It's quite another to get the thing off the ground, so to speak—especially considering the inventor's belief that the financial backing ought to come from the private sector. Malewicki's last stint of gainful employment—as "a senior technical specialist in advanced composites manufacturing" on the B-2-bomber project—soured him on governmental red tape.

On paper, SkyTran does take on a certain fanciful patina—like, say, that of a musty back issue of Popular Mechanics with "the Diesel-Powered Flying Bicycles of the Future" on the cover. But SkyTran makes use of solid physics: specifically magnetic levitation, which has been in development for years for trains. Engineers have been betting that a train with magnets built into its carriage can float along a hollow guideway. But where most mag-lev systems need electromagnets or superconductors powered by an outside source, Malewicki's SkyTran relies on a pioneering, pollution-free mag-lev system called Inductrack. Developed by a Lawrence Livermore Laboratories scientist named Richard F. Post, Inductrack needs nothing but simple magnets, and in Malewicki's

scheme they'd be mounted in the carriage of the people pods. As the carriage passed above two kinds of wire coils embedded in the guideway below, the resultant electric fields would create a levitation force that would keep the thing aloft and a propulsion force that would keep it moving forward.

A small jolt of electricity would be required to start each pod, which would be obediently awaiting its passenger in the pod station down on a second guideway beneath the main "superhighway" guideway. Assuming the cost of gas at $1.50 a gallon and a kilowatt-hour of electricity at 10 cents, SkyTran's chief marketer, Robert Cotter, claims that SkyTran could move thousands of pods along a rail designed to hold six lanes' worth of automobile volume at 300-miles-per-gallon efficiency—in theory, anyway.

Cotter, for one, has believed in Malewicki's theories for years. The two first crossed paths in 1979, when Cotter was the vice president of the International Human Powered Vehicle Association and Malewicki entered his Mini Micro Missile, driven by his 8-year-old daughter, in an I.H.P.V.A. competition. She averaged 29 miles per hour that day, the world's fastest self-propelled kid.

In the end, even if SkyTran suffers the fate of another of Malewicki's inventions, the manned flying kite cycle, give the SkyTran team its due: while beltways continue to bigfoot our last few acres of exurbia and choke our cities into submission, SkyTran is looking up, and ahead.

riding services. The utopian PRT network would be a fine-grained, interconnected grid or multiple loops with unconstrained movement possibilities in all directions. Real-world systems fall into one of the following categories:

1. Single line with one vehicle shuttling back and forth between terminals (see Fig. 15.1).

2. Single line with two vehicles operating simultaneously with a double track in the middle to allow bypassing of the cars (see Fig. 15.2).

3. Double line shuttle between two terminals with one or both lines in operation at the same time (see Fig. 15.3). The terminals may be stub-ends with the vehicles returning in reverse, or there may be a turnaround loop. Branches may be added. The Morgantown and Jacksonville systems are examples of this type.

4. Single one-direction loop with a series of stations (see Fig. 15.4). The loop should be relatively small, because otherwise movement between two nearby stations may require a long trip if the destination point is in the reverse direction. The Detroit DPM system is a single loop with 13 stations; a complete round trip takes about 15 min.

5. Double loop with two-directional movement and any number of stations (see Fig. 15.5). This system operates like regular fixed guideway service. The Miami Metromover is the best example, although it has added both a north and a south branch.

6. A combination of any of these arrangements is possible (see Fig. 15.6), as would be a grid network for a larger system. The Seattle-Tacoma Airport, for example, combines loops and a shuttle with transfer connections.

The first three classes are the simplest possible systems as they are found in various airports, shopping centers, recreation areas, between parking lots and destination points, and on institutional campuses (see Table 15.1). It is possible to have several independent shuttle lines within the same development (see Fig. 15.7), as is the case in the Tampa Airport.

Another way to classify AGT systems is by their ownership

Figure 15.1 Single line (one vehicle).

Figure 15.2 Single line (two vehicles).

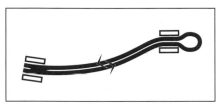

Figure 15.3 Double-line shuttle.

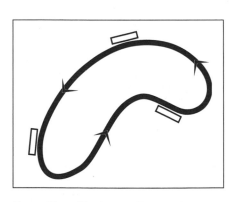

Figure 15.4 Single one-direction loop.

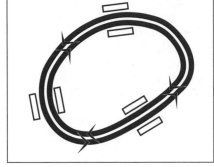

Figure 15.5 Double loop.

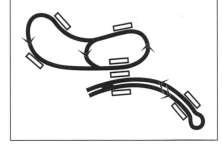

Figure 15.6 Combination system.

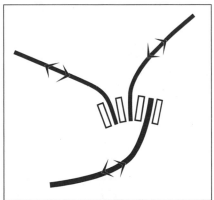

Figure 15.7 Independent shuttles.

Table 15.1 Automated Guideway Systems in North America, 2001

System/Operating Agency	Year of Opening	Passenger Boardings/ Year (Daily Ridership)	Total Guideway Length, mi (km)	Type of Network	Guide-way Place-ment	Number of Stations	Power Subsystem	Support Subsystem
Public Service								
Morgantown, WV/ University of West Virginia*	1975	2.2 million in 1993 (14,000)	8.7 (14.0)	2-lane shuttle, off-line stations	Elevated/at grade	5	575 V ac	Rubber tires
Detroit DPM/Detroit Transportation Corporation	1985	(7,000)	2.9 (4.7)	1-lane loop	Elevated	13		Steel wheels
Miami Metromover/Miami-Dade Transit Agency	1985	3.6 million (12,000)	4.0 (6.4)	2-lane loop with branches	Elevated	21		Rubber tires
Scarborough, Canada/ Toronto Transit Commission	1985	(72,000)	4.3 (6.9)	2-lane alignment	Elevated	6	Linear induction	Steel wheels
Vancouver SkyTrain/ BC Rapid Transit Company	1986	20 million (120,000)	13.3 (21.4)	2-lane alignment	Elevated	15	Linear induction	Steel wheels
Jacksonville Skyway/ Jacksonville Transportation Authority	1989	0.3 million	1.4 (2.3)	2-lane Y-shaped network	Elevated	3		Rubber tires
Private/Institutional Sponsorship								
Fairlane Town Center	1976		0.5 (0.8)	1-lane shuttle	Elevated	2		
Honolulu/Pearlridge	1978		0.2 (0.4)	1-lane shuttle	Elevated	2		
Durham, NC/ Duke University	1980		0.6 (0.9)	1- and 2-lane shuttle	Various	3	Linear induction	Air cushion
Las Colinas, TX	1989		0.7 (1.2)	2-lane shuttle	Elevated	4	Linear induction	Rubber tires
Tampa/Harbor Island	1991		0.5 (0.8)	1-lane shuttle	Elevated	7	Cable	Air cushion
Washington, DC/U.S. Senate	1995		0.7 (1.1)	Shuttle	Underground	3		
Airport Service								
Tampa Airport	1971		1.9 (0.8)	2 lanes, 5 shuttles	Elevated	12		
Seattle-Tacoma Airport	1973		1.7 (2.7)	2 loops, 1 shuttle	Underground	8		
Dallas/Fort Worth Airport Airtrans	1974		13.0 (21.0)	1 lane, multiple loops	Elevated/ at grade	28	480 V dc	Rubber tires
Atlanta Hartsfield Airport	1980		2.3 (3.7)	Loop	Underground	10		
Miami Airport	1980		0.5 (0.8)	2-lane shuttle	Elevated	2		
Houston Airport	1981		1.4 (2.2)	1-lane loop	Underground	5		
Orlando Airport	1981		2.2 (3.6)	2 lanes, 3 shuttles	Elevated	6		
Las Vegas McCarran Airport	1985		0.5 (0.8)	2-lane shuttle	Elevated	2		
Chicago O'Hare Airport	1991		5.5 (8.0)	Loop	Elevated/ at grade	5		
Pittsburgh Airport	1992		0.5 (0.7)	2-lane shuttle				
Denver Airport	1993		1.9 (3.0)	Loop	Underground	4		
Newark Airport[†]	1994		4.4 (7.1)	Loop	Elevated	7		

The many examples of AGTs in amusement or recreation areas are not listed because of the large variety of features. See "Types of AGTs and Their Operation."
* The Morgantown system was a demonstration project sponsored by UMTA, the municipal government, and the University of West Virginia.
[†] UTDC has been absorbed by Bombardier in 1992, and AEG/ Westinghouse had become a part of Adtranz. In 2001, Adtranz also became part of Bombardier.
[‡] Replacing the original Rohr Monotrains (1972).
[§] Newark Airport operates an automatic monorail with six-car trains. It is discussed in Chap. 12.

Sources: Data from agency Web sites; U.S. Department of Transportation, FTA, *Characteristics of Urban Transportation Systems,* September 1992; *Jane's Urban Transport Systems,* 2001–2002.

Peak Service Interval	Hours of Operation	Number of Vehicles	Vehicle Length × Width, ft (m)	Vehicle Capacity		Vehicle Speed, mph (kph)		System Manufacturer[†]
				Seated	Standing	Maximum Speed	Operating Speed	
Public Service								
	76 per week	73	15.4 × 6.6 (4.73 × 1.83)	8	13	30 (48)	15 (24)	Boeing
2–3 min	89 per week	12		34	66	30 (48)	12 (19)	UTDC
90 s	18.5 daily	12		14	82	27 (43)	8 (13)	AEG-Westinghouse
		24						UTDC
		130						UTDC
3 min	13 daily	2		12	80	41 (66)	16 (26)	MATRA Bombardier
Private/Institutional Sponsorship								
	12 daily	2	24.7 × 6.67 (7.53 × 2.03)	10	14	30 (48)	19 (31)	Ford Motor Company
	69 per week	4		323	393	8 (13)	4 (6)	Rohr Industries
	24 daily	4		4	14	28 (45)	14	Otis
	On demand	4		12	33	30 (48)	18 (29)	AEG-Westinghouse
	24 daily	6		8	9	12 (19)	4 (6)	Transportation Group
		12		9	3	14 (23)		Transportation Group
Airport Service								
	24 daily	10	36.3 × 9.33 (11.05 × 2.84)	0	50	30 (48)	9 (14)	Westinghouse
	20 daily	24	37.0 × 9.3 (11.28 × 2.84)	12	45	26 (42)	9 (14)	Westinghouse
	24 daily	51	21.25 × 7.3 (6.48 × 2.24)	16	12	17 (27)	10 (16)	LTV/Vought
	20.5 daily	17		16	34	27 (43)	10 (16)	AEG-Westinghouse
	24 daily	6		2	50	30 (48)	10 (16)	AEG-Westinghouse
	20.5 daily	18		6	6	15 (24)	6 (10)	Walt Disney Company[‡]
	21.5 daily	18		0	51	30 (48)	9 (14)	AEG-Westinghouse
	24 daily	4		4	48	25 (40)	17 (27)	AEG-Westinghouse
	24 daily	13		8	49	50 (80)	23 (37)	MATRA
								AEG-Westinghouse
	24 daily	16		14	42	32 (51)	11 (18)	AEG-Westinghouse
	24 daily	72		122	542	28 (45)	14 (23)	Von Roll

and specific service role. AGT is a mode with wide utility in private development efforts as a component of those projects, and only a few of the systems are operated as true public transit.[15]

Central Business District Public Transit (Downtown People Movers)

The purpose of these systems is to provide a convenient means of transportation to workers and visitors in high-density districts for short trips—to carry them from access points (rail stations and large garages, for example) to destination places, to provide for internal linkages among activity nodes, and to give tourists an orientation tour. They serve the same purpose and are expected to achieve the same results as are any number of conventional downtown circulator systems (buses and trolleys) that operate on surface streets—except with an expected greater effectiveness and an aura of technical sophistication. The three major United States public systems (Miami, Detroit, and Jacksonville) fall into this group.

Citywide Transit

If AGT routes are extended and larger vehicles are utilized, network coverage can be increased, and the systems can operate as any other regular fixed guideway, exclusive right-of-way transit. Feeder services, station design, and passenger facilities are very much similar to regular rail transit elements; the hardware, rolling stock, and control systems are, of course, quite different.

The best-known such system is the *véhicule automatique leger* or *Villeneuve d'Ascq-Lille* (VAL) network in Lille, France, which has been operating since 1983. This 40-mi (64-km) network of two lines services 45 stations and is rapid transit for all practical purposes. The only large-scale AGT operation in North America is the SkyTrain of Vancouver, Canada.

Support System for Private Development or Institution

Land development projects and campuses that are in private or institutional ownership, managed by a single agency, but dispersed horizontally, can benefit from a responsive internal means of transportation. AGTs in their role as horizontal elevators are

[15] D. Shen et al., *Automated People Mover Applications: A Worldwide Review* (National Urban Transit Institute/U.S. DOT, 1995).

particularly suitable for this task, and have been implemented in a number of instances. These include shopping centers with remote parking (Pearlridge, Fairlane), campuses with scattered major buildings (Duke University),[16] office districts with separate large buildings (Las Colinas),[17] and large real estate developments (Kobe and Osaka). While these are commercial and institutional environments under private control, entry into the vehicles is free and unrestricted, although limitations can be imposed (as would be the case with regular building elevators).

Airport Service

In a manner very similar to the previous group, AGTs continue to flourish and operate effectively within a number of airports. They are particularly useful in connecting various terminals and parking clusters for the convenience of air travelers, visitors, and airport workers. Again, access to these systems is unrestricted. The facilities are usually owned by the airport agency, and the major difference is that the vehicles have to be able to accommodate luggage carried by the travelers. Since it is crucial that airplane connections not be missed, backup systems are necessary in case of even a temporary delay on the people mover network. Since the distances between gates in the larger airports have become exceedingly long, and transfer time allocations are frequently short in the major hub airports where large volumes of travelers have to reach the proper gate, mechanical movement devices become almost mandatory, and AGT fills this role very well.[18] Some dozen such systems are found today in American airports, as well as in Germany, Italy, France, Great Britain, Hong Kong, and Singapore.

Service for Amusement/Recreational/Cultural Facilities

Many of these installations attract large volumes of customers, may have remote entry points, and are spread over large territories. They include a great variety of AGT applications, ranging from a full internal loop in the Bronx Zoo atop the Asian plains to

---

[16] L. J. Fabian, "People Movers: the Emergence of Semi-Public Transit," *Traffic Quarterly,* October 1981, pp. 557–568.

[17] D. Dillon, "Las Colinas," *Planning,* December 1989, pp. 6–11.

[18] T. Austin, "Traversing the Terminals," *Civil Engineering,* September 1993, pp. 40–43.

observe the animals, to a shuttle up the hill from the parking garage to the Getty Museum complex in Los Angeles. The hardware too is found in various configurations and levels of technological advancement, ranging from full AGT capability to rather simple mechanical devices.

Applications include amusement parks and recreational facilities, among which the better known examples are Disneyland and Disney World, Busch Gardens and Kings Dominion in Virginia, Hersheypark in Pennsylvania, and the California State Fair in Sacramento; several are in major zoos (the Bronx, Minnesota, and San Diego); others are special services in urban environments, such as the Getty Museum in Los Angeles, Mud Island in Memphis, and between certain hotels in Las Vegas.

Reasons to Support AGT

The AGT concept has received much publicity over the last three decades, and vocal advocates have lost no opportunity to highlight the systems capabilities. Avoiding the overly enthusiastic claims, some definite advantages can be identified. Those factors that are common to all transit systems (the absence of air pollution, for example) are not included here.

- *Customer responsiveness.* While the PRT concept, which would offer an ideal ability to satisfy each user's specific travel desires, has not yet become feasible, the relatively small scale of the existing AGT systems, the frequent headways, and the agile operations expedite movements and minimize waiting times. Service quality can be, and is expected to be, at the top level.

- *Safety.* The record has been extremely good, with serious operational accidents not yet encountered. Personal safety has also been exceptional, which may be due to the fact that many systems operate in controlled environments, that platform edge barriers (screens) are almost always present, and that extensive surveillance programs by TV monitors and safety personnel are in place.

- *Low labor input.* Since there are no drivers or conductors on the vehicles, and passengers may not see any employee of the operating agency at all, there should be considerable savings on the personnel side of the ledger. However, some caution has to be exercised here because the sophisticated technical features throughout the systems may require continued oversight and maintenance by skilled and well-trained specialists. Some systems also have a practice of placing very visible safety personnel within the passenger spaces and on vehicles.

- *Suitability for constrained spaces.*   Since all the dimensions of AGT elements are measurably smaller than those of conventional transit, and little noise or vibration is generated, there are better opportunities to thread lines through intensely developed districts, and even buildings can be penetrated. Since the guideways have to be elevated (or placed in tunnels), there is no interference with traffic on already overloaded surface streets.

- *Catalyst for development.*   Concentrated capital investment in infrastructure systems may and should materially encourage real estate development in their vicinity; locations near AGT stations undoubtedly have enhanced accessibility. Actual results, however, are difficult to isolate as being primarily due to AGT because in all instances a number of complex forces have been in play. Positive effects appear to have occurred in Miami and Jacksonville; the DPM in Detroit has not been able by itself to bring back urban activity.

- *Advanced technology image.*   AGT is the most advanced and sophisticated transportation technology currently available for general use. There is considerable prestige and publicity value associated with the deployment of such an innovative system in any community. It is seen as an indicator of the progressive nature of any host locality, and it is an object of civic pride (at least once the startup problems are overcome). This observation does not discount the fact that some of the recent heavy rail transit systems or lines are as advanced in their technology as any AGT effort.

Reasons to Exercise Caution

In the last decade, much of the original excitement and fascination with AGT systems has ebbed, and a much more careful, if not critical, attitude has been adopted by communities and private developers. Experience with actual operations has shown that theoretical expectations are not always achieved in the field.

- *Fragility of system.* AGTs are characterized by highly advanced technology with components that can be somewhat delicate. Under normal, civilized use, there should not be significant problems, particularly if good maintenance programs are in effect. However, the vehicles are not as robust and resilient as, for example, those of a regular subway or streetcar. Holding doors open, overloading a vehicle, and throwing debris on the guideway may create problems with equipment and control systems. Repairs are intricate and costly; spare parts are expensive, particularly because many of them have to be custom-made or adapted.

- *Visual impacts.* The need for a completely exclusive right-of-way gives little choice but to build overhead guideways within existing districts. While some sections can be channeled through buildings, most of the guideway and station structures will have to be placed above sidewalks and streets. The guideways are not nearly as heavy and oppressive as the old steel elevated lines, but they are noticeable nevertheless, which may be regarded as a serious problem in some situations. If an AGT system is built at the same time as the buildings to be serviced, greater opportunities for functional integration and even the construction of sublevel channels are, of course, present.

- *Cost factors.* Due to the availability of only a relatively few similar cases, reliable cost comparisons and definitive estimates cannot be made. However, capital investments will be considerable because a completely new exclusive guideway has to be created, advanced-technology vehicles have to be acquired, and sophisticated maintenance facilities have to be made available. Given the relatively low throughput capability of AGT, these costs are acceptable if loading factors are high over many hours of the day; they may be excessive if only peak periods generate reasonable ridership.

Service to commercial development in Honolulu.

Application Scenarios

Experience so far shows that the ideal environment for AGT application is a large airport with scattered terminals and parking garages, i.e., high-activity nodes separated by distances beyond a convenient walking range (particularly if luggage has to be carried). This situation means that large volumes of riders will move in all directions throughout the day and much of the night. The best examples are the airports in Atlanta, Dallas–Fort Worth, Seattle-Tacoma, Houston, Changi (Singapore), and Gatwick (London). On the other hand, the record is not good with smaller facilities. Love Field in Dallas, Bradley Field of Hartford, and the Birmingham Airport in the U.K. all had automated people mover operations that could not be justified and were not continued. In Birmingham, the replacement shuttle bus does sufficiently well.

At the other end of the scale are the citywide systems providing service over extended distances. Besides Lille, which is adding a second VAL line, similar systems are under development in the French cities of Toulouse, Rennes, and Lyons. These projects continue the trend toward making large-scale AGT systems difficult to distinguish from automated heavy rail operations. The newly opened (1998) *Météor* (*métro est-oest rapide*) line of the Paris metro looks like, and is classified as, a heavy rail transit line, but it carries most of the characteristic features of AGT.[19] There are also plans to gradually automate the other 13 older metro lines.

Likewise, progressive modernization of the New York subway system is expected, starting with the Canarsie line in Brooklyn.

In New York, an extensive fully automated system (the Air-Train) is under construction that will connect all the terminals of the John F. Kennedy International Airport in a loop

[19] C. Henke, "Paris Meteor Shows World the Way to the Future," *Metro*, July/August 1999, pp. 46–48.

VAL system in Lille, France, employing platform edge barrier.

and also will lead to remote parking lots and the Long Island Rail Road station in Jamaica, Queens. The equipment vendor is Bombardier. While the entire guideway is elevated, placed mostly on top of highways and roadways, it is still known as a light rail system, not an AGT.

These examples are illustrations of the blurring of the modal designations, which, not so long ago, were clear-cut and pleased the orderly professional, who could easily label each system by type. It is not such a direct exercise any more. For example, while VAL technology was originally slated for true AGT application, it ended up being a metro for the city of Lille. VAL was selected in large part because of its ability to be fitted with a small–cross section vehicle and to propel cars up steep grades. In method of service, as far as passengers are concerned, the system is about the same as the metro of Paris or the subway of New York. Similarly, the designation of the JFK Airport application is primarily driven by the use of linear propulsion motors chosen by the vendor. Otherwise, the service will respond not very differently from other urban rail transit.

In Los Angeles, the heavy rail transit Red Line is already under automatic control, and programs are structured to retrofit the light rail Green Line. This is causing some local opposition because of cost factors.

Japan is also continuing along this same path toward systems with full automation. There are well over a dozen systems recently completed or still under development that qualify for the AGT designation (several are of a monorail configuration). The two principal examples connect offshore new towns to major access nodes—Kobe Portliner (Kawasaki, 1981) and Osaka New Transit (Niigata, 1981).

In Canada, there are two extensive automated systems. The original 18-mile SkyTrain of Vancouver (1986) is expected to remain as the spine of the transit network, but it will be complemented by an additional

Network of AGT lines in Scarborough, linking to Toronto's rapid transit.

trunk line currently under construction. Light rail transit was the leading candidate for the new addition, but the decision to select a second automated line was in part prompted by the promise of a new manufacturing facility in British Columbia. In Scarborough, outside Toronto, the AGT line opened in 1985 remains as a single unique addition to the extensive regional transit network.

Systems that provide for group riding at a rapid transit level are also under development in other Asian cities. Kuala Lumpur and Taipei have plans for automated service, but the largest program is starting in Singapore. This city-state has a record of undertaking major transportation programs of an advanced nature and carrying them through to a successful completion without undue delays. At this time at least three separate AGT feeder lines are being designed as adjuncts of the overall rail transit system, and a new metro line ($2.9 billion) promises to be not only fully automated but also unusually fast. The equipment vendors are Adtranz and Mitsubishi for the AGT systems and Alstom for the heavy rail line. Retrofitting of the entire metro network is a longer-range program, and it is entirely likely that in a few years Singapore will have the most extensive set of automated transit services in the world.

The conclusion as to AGT suitability for CBD service is not quite clear. Initially, this was a major federal initiative to help rejuvenate old business centers by equipping them with better services; it was also a program that was terminated abruptly. No U.S. city has taken any self-generated initiative to plan or even examine such possibility since those days, with the exception of Las Vegas. There, private sponsorship is undertaking construction of a Disney-type monorail system, which is expected to become the spine of a future urban system. The local transit agency is required by law to ensure compatibility with any publicly funded extension it may undertake in the future.

The Detroit DPM is a very lightly used service, and clearly out of scale with actual demands. Neither the district nor the city in general has shown any real signs of renewed activity, and no spontaneous substantial commercial regeneration has taken place. The DPM has not made any apparent difference, but this is probably too much to expect given the serious nature and the depth of the regional decentralization trends. The Miami Metromover is an effective distributor from the main Metrorail (heavy rail) system. It is also a great device for tourists and visitors to see the center of the

Elevated AGT loop in downtown Miami.

city. Some claims are being made that it has brought new business activity near the stations. However, given the still dominant presence of the automobile in this city, despite the availability of heavy rail transit service, it is hard to assign any specific results as being caused by separate development forces.

Even in Las Colinas, which is a pure office and commercial complex, it is difficult to say that trips by office workers between buildings during the business day, going to lunch places, and moving to and from garages justify fully a capital-intensive means of internal transport. At this time, operations have been suspended for an unknown period.

In Jacksonville, on the other hand, a major addition to and rebuilding of the original AGT system has been opened recently. While ridership on the earlier short segment was very low, it is expected that the new line serving an array of destination points will draw additional customers.

In instances where AGT systems have been targeted to specific and simple needs, the results have been positive and the services work effectively. This encompasses such tasks as carrying people to otherwise difficult-to-reach but attractive sites and connecting to garages and other entry points. Even then, it may be well to ask whether a streamlined bus service could not do the job with better cost effectiveness. But buses would certainly not create the same promotional civic image, and they are likely to suffer from service irregularities due to surface street congestion.

One last interesting example of AGT effort, even though it was terminated in 2000, was the recent program in Chicago under the Northeast Illinois Regional Transportation Authority to build a 3.2-mi (5.1-km) PRT system in Rosemont, Illinois. The service would have connected a cluster of offices, hotels, and a convention center to O'Hare Airport. Egg-shaped small cars carrying up to four passengers would have used a network of slender overhead guideways. The hardware design was by Raytheon, and a test track had been built.[20] However, as a publicly supported project, it could not generate enough confidence among government

[20] J. Rattenburg, "The PRTs are Coming! The PRTs are Coming!" *Progressive Railroading,* May 1994, pp. 92–95.

officials to convince them to allocate the necessary resources. It was believed that the $130 million construction cost could not be justified for an untested system that, at best, could accommodate 1000 to 3000 riders per hour.

Therefore, the conclusions at this time in North America regarding the applicability of AGT systems and concepts would have to be the following:

- At the city scale, AGT projects have been an excellent means of exploring advanced technological concepts. The lessons learned and devices developed can be applied to more conventional and robust modes (such as metro and light rail transit systems) quite effectively. Much work in the near future can be expected in this field.

- As special connectors and linkages between nodes located at relatively short distances from each other, AGT has proven itself to be a responsive, attractive, and effective mode, provided that usage remains at high and steady levels.

- Personal rapid transit remains a utopian dream.

Components of the AGT System

A systematic review of AGT components cannot be very specific due to the variety of existing examples (some operating, some engineered but not yet built). Many possibilities exist; the field has not converged on a few hardware choices. Nevertheless, some manufacturers have been more successful than others in responding to the market.

Most of the basic concepts related to fixed guideway transit components have been outlined in Chaps. 11 and 13. They will not be repeated here to the extent that they are also applicable to AGT. We will concentrate in this chapter only on the special features of technologically advanced modes. A summary table of physical characteristics would also not be particularly useful with this mode because of the prevailing variety. Reference should be made to Table 15.1.

Right-of-Way and Track

A major requirement for AGT is a completely exclusive and grade-separated right-of-way, since intrusion by other vehicles or people on the guideway cannot be tolerated. The guideway itself is most frequently a flat U-shaped channel with the sidewalls providing

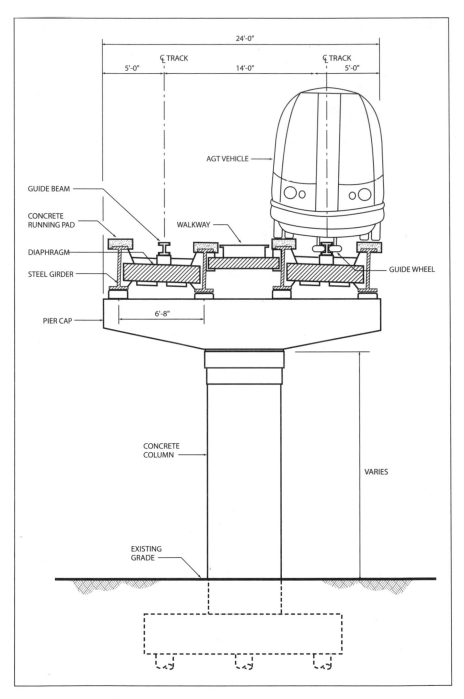

Figure 15.8   AGT elevated structure with double track. (*Source:* Jacksonville Transportation Authority and Parsons Brinckerhoff.)

lateral guidance for the vehicles and accommodating live power supply bands (see Fig. 15.8). Or a guiding beam may be placed along the centerline, and there would be no parapets along the sides. Since most AGT systems utilize vehicles with rubber tires, the bottom of the channel provides the supporting surface for the wheels, or there are two smooth strips along the sides. The semi-enclosed channel shape (a trough) may cause problems with snow accumulation and icing, which is difficult to cope with except by heating the carrying surfaces.

Because grade crossings cannot be accommodated, guideways have to be elevated or (for example, where guideways are being built at the same time as an airport terminal) placed under the main activity level in tunnels. The horizontal overhead beams are not particularly heavy, but they certainly are visible, and space has to be found for the supporting columns.

If the guideway accommodates rubber-tired vehicles running on concrete channels, switching (i.e., following the left or the right branch when tracks diverge) requires special attention. This is accomplished by steering the vehicle so that it remains in contact (via lateral guidance wheels) with either side of the channel, or by mechanical attachments that pull the vehicle in one or another direction, or by moving a central guide beam.

Rubber tires offer good adhesion to the concrete running surface, and therefore steep grades of up to 10 percent and even more are possible. (Both Detroit and Miami have some 10 percent sections.) Because of the relatively slow speeds of the vehicles, horizontal curvature can also be quite sharp. A radius of 100 ft (30 m) is usually the minimum used, but some smaller systems allow turns as tight as 30 ft (9 m).

All these general considerations also apply if the vehicles are supported by air cushions or are pulled by cables. In systems that utilize steel wheels on steel rails, normal railroad engineering standards would apply, except that the scale may be smaller.

All these physical features and the relatively small scale of true AGT systems limit achievable speeds to the modest range. They rarely exceed 30 mph (48 kph), and higher speeds are only possible with more robust equipment.

Stations

An interesting feature of AGT stations is that they may be placed "off-line," i.e., the main movement channel continues past the station, and only vehicles carrying passengers who wish to get off at a specific station pull away on a separate guideway section (or ramp) that leads to the platform. They reenter the main stream via another ramp after loading. Empty vehicles may also be directed to waiting passengers. The great advantage of this concept is that most travelers need not be delayed by the on and off movements of other passengers, but can continue rapidly to their destination. The problem with this arrangement is that extremely precise control systems are necessary to accommodate smoothly and safely, with split-second timing, the diverging vehicles, and particularly those that have to find a safe opening to rejoin the main flow.

This capability of off-line loading has not been accomplished efficiently and safely in actual practice with short headways. In fact, only the early Morgantown and Dallas/Fort Worth Airport systems were envisioned and built for that type of operation. Today, Morgantown and D/FW operate mostly on a regular schedule with sufficient headways to provide safe merging and diverging on the various ramps. It is also obvious that such operational capability would have to be supported by an extensive network of auxiliary channels and ramps and multiple switching points.

In all other respects, AGT stations tend to be scaled-down versions of regular transit stations recognizing the smaller size of the vehicles (see Fig. 15.9). A common feature, however, is a platform edge wall or other restraining barrier to prevent patrons from falling onto the guideway and to protect blind persons. This is possible because the accuracy of operational controls ensures precise placement of entering vehicles at the door openings.

Passenger Amenities and Environment

As a part of the image and service quality of AGT systems, all projects have maintained the best possible standards in passenger comfort. The improve-

Station platform of the Detroit's downtown people mover.

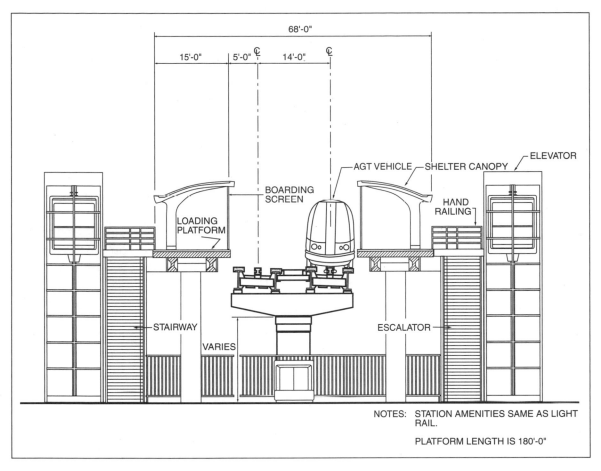

Figure 15.9 AGT station with side platforms. Station amenities are the same as for light rail; platform length is 180 ft (55 m). (*Source:* Jacksonville Transportation Authority and Parsons Brinckerhoff.)

ments have been treated basically as extensions of buildings, with the same attention and care in design. It has also been important for the public systems, since they have been principally demonstration projects, to achieve environmental quality and service levels beyond those usually experienced with regular transit. The riding public has responded quite favorably, suggesting that potential transit users in North America expect such characteristics and comfort features on any system that wishes to maintain good ridership.

All AGT systems have level boarding, i.e., the platform and the car floor are at the same level. It has been suggested that the horizontal gap should not be more than 1 in (2.5 cm) wide and the vertical difference no more than 0.5 in. Such fine tolerances may be somewhat excessive, but they are not too different from those now attempted on all modern transit systems.

Safety Features

AGT operating systems incorporate fail-safe features to minimize any possibility of hardware accidents, collisions, or fires. TV monitors are often placed throughout the passenger spaces, backed up with trained staff and appropriate equipment to deal with emergencies if they occur. Some systems, particularly in those places where incidents of vandalism, personal misbehavior, or outright crime have been observed, deploy security personnel, often quite visibly. As has been mentioned before, barriers and screens to shield the guideway from passengers are frequently provided, as are other safety devices.

Control Systems

All AGT systems operate through a unified control arrangement based on sensors along the guideway, response devices in vehicles, communications linkages, and a central computer. There is, of course, a human supervisor, but the operations are automatic, i.e., service is provided following preset protocols, or it is adjusted as information on passenger loading and vehicle location is received from the field. These are the defining characteristics of *automated* guideway transit. Vehicles are usually "smart" themselves, able to react to conditions encountered along the way and in stations.

Most of the control systems and programs have been developed by equipment vendors, and they remain largely proprietary. The implementation of an AGT project is basically a matter of purchasing the hardware and software (or making some other arrangement to place it on site), which encompasses vehicles and equipment from a specific manufacturer, together with the corresponding control systems. The civil infrastructure (guideway, stations) is designed in response to the selected equipment and built under separate contracts.

The proprietary nature of such systems can be a deterrent to implementation because, once the system is purchased, the owner is locked into the service programs and parts pricing of the initial supplier. Where agreement between agency owner vendor/supplier could not be maintained, the results have been quite dislocating. The Bradley Field system, for example, was dismantled, and the Jacksonville system was completely rebuilt with a different technology.

Fare Collection

Basically, there should be no differences in the fare policies and collection practices between AGT and regular transit, as related to public services in any community. AGT usage can be integrated with all other service modes, having the same joint tickets and passes, transfer arrangements, and special discounts. The exception may be that the AGT service may carry a lower fare if it is deemed appropriate to institute a special promotional policy.

On privately sponsored systems there may be no fare at all if the service is seen as basic support of the primary activity of the development—for example, at a shopping center. In cases where the transport link is internal to a facility that charges admission (for example, an amusement park or cultural institution), the use of the AGT may be a part of the total experience.

Yards

AGT systems certainly need storage and maintenance facilities like any other transit fleet. Given the often unusual and sophisticated nature of the equipment, these have to be exclusive installations. Besides the capability of dealing with mechanical problems and tasks, the ability to cope with any number of electronic control and communications components is vital. This calls for highly skilled personnel and advanced diagnostic and testing equipment.

Propulsion, Supply, and Support

Most AGT vehicles are propelled by electric motors under the floor that draw power via brushes that pick up current from power rail along the sides of the channel. This is rather conventional technology, but at one time or another almost any other conceivable power system (save steam) has been explored. Linear induction motors are in use on several systems; there are some air cushion vehicles pushed forward by air jets. Cables, reverting to very ancient technology, pull some of the simpler shuttle cars.

The more ingenious approaches include powered rollers in the guideway that carry forward an inert box/passenger container (the Carveyor). There is a short track in Porto Alegre, Brazil, and a longer route in Jakarta with a box beam that supports cars running on top (Aeromovel). Each car has an arm extending downward through a slot with a plate that fits inside the beam and acts as a sail when air pressure is applied to move the vehicle. Sponsors can be found apparently for almost any device.

Not all AGT vehicles roll on the guideway; some hover, levitate, or slide. The more common support systems are rubber tires on concrete channels or steel wheels on steel rails, but that does not exhaust the possibilities. Some systems utilize the hovercraft or air cushion concept, where compressed air on the bottom surface elevates the vehicle and eliminates almost all restraint to forward motion. A problem here is the escaping air that has to be controlled. Magnetic levitation—suspending the vehicle by a magnetic force—also eliminates all friction, and has been tried in some instances (the M-Bahn of Germany). This technology appears to need some further development before it becomes fully competitive.

Vehicles

The AGT vehicles themselves are often considerably smaller than regular transit cars, ranging from 5 or 6 passenger spaces to about 100. Those providing service over short distances (within airports, for example) may have very few seats to gain maximum standee and luggage capacity. The dimensions of the vehicles and the width of the guideway are kept within the usual range of transit lanes or are made smaller, so that challenges in creating a right-of-way are minimized.

Some systems are designed to operate single vehicles; others couple them into trains semipermanently. The ideal situation would be automatic and dynamic coupling of as many

Suspended shuttle service to recreation area in Memphis.

vehicles as required by demand at any given time for efficient line haul.

In virtually every case, an AGT system is developed and marketed as a package including the vehicles and propulsion and control technology. As with conventional systems, the vehicle becomes the symbol of the entire concept. Manufacturers have made no effort to attempt interchangeability of vehicle types. As a result, less than a handful of equipment vendors are successful in the AGT field.

Ever since the early days, and even today, there have been small inventors and system developers who have attempted to enter the market with their vehicles. Some of these have been acquired by large corporations, which have been able to engage in real research and development, to produce prototypes, and to vigorously promote their product. Foremost among those has been Westinghouse Electrical Company, together with AEG Transportation Systems. These companies have been able to sell their hardware and software to a number of agencies (see Table 15.1). Their C-100 vehicle has been a standard for the industry, but other models have been available as well. This enterprise was taken over by Adtranz, an international conglomerate, under Daimler-Chrysler Railsystems, with operations in many countries and producing a wide range of transit vehicles. DaimlerChrysler's current AGT models are the CX-100, which is used to replace earlier vehicles on some existing systems, and the new INNOVIA vehicle.

In Canada, pioneering AGT vehicle development work was done by the Urban Transport Development Corporation (UTDC), which, with government support, produced the Mark I and II ART models that were placed on several systems. This enterprise was taken over in 1992 by Bombardier, which has also become a very large manufacturer of transit vehicles with a large product line, including the current AGT model UM III. In mid-2001, an agreement was signed under which Adtranz has become a component of the Bombardier Consortium, which is now the dominant supplier worldwide.

Thus, except for a number of small manufacturers that are ready to sell AGT systems that are not always fully tested in the field, there is in effect only one manufacturer of AGT equipment in North America at this time. Any competition could come only from Matra, which is owned by Siemens, or several Japanese firms (Kawasaki, Niigata, Nippon Sharyo).

Scheduling and Capacity of AGT

One of the great expected advantages of AGT would be its ability to operate on demand for each rider. Regrettably, as has been noted earlier, this is still a somewhat theoretical concept that has been attempted in only a few network systems with unconvincing results. An elevator is the model where any user can push a button on any floor to summon a vehicle and direct it to any other floor, while also accommodating the demands of other users. But an elevator is a simple shuttle; once different paths have to be chosen, the problem becomes much more intricate. AGT systems of the shuttle type can and do have call buttons. The larger AGT systems operate like any other transit service on fixed schedules, since even with smaller vehicles the potentially conflicting trip destination instructions by many passengers would overwhelm the response capability of any system.

The capacity of any AGT system, as is the case with all transit operations, is a function of vehicle size and service intervals between units. Discussion continues about 1-second (and even shorter) headways, which may be possible, albeit theoretically, with off-line stations. Otherwise, as always, the frequency of service is limited by the time needed to stop at stations along the track (slowing down, opening the doors, letting passengers off and on, closing the doors, and picking up speed). Even with relatively small vehicles, anything less than 1 minute is difficult to envision.

Thus, four-car trains, with 100 passenger spaces in each car, running at 1-minute intervals, could achieve a throughput of 24,000 passengers per hour on a single lane. That is certainly pushing the boundary of operational feasibility, but the system in Lille is able to achieve that level (with slightly larger cars). A more typical AGT scenario would be two-car trains at 2 minutes, resulting in 6000 passengers per hour. The operations in Toulouse, running with 60- to 90-second headways, provide 8000 passenger spaces per hour. The rather extensive service in Kobe, Japan, operating six-vehicle trains with 53 passenger spaces in each car, achieves a capacity of 7600 passengers per hour in each direction. The Dallas/Fort Worth Airport system, with all its intricate loops, claims 9000 per hour.

With smaller cars, which is the common practice in AGT operations, the offered capacity would be proportionally lower (with the same headways). For example, 30-passenger cars operating singly at 1-minute intervals would offer a capacity of 1800 passengers per hour. It is therefore reasonable to place the general AGT capacity in the rather wide range of 1500 to 15,000 riders per hour per lane.

It is to be expected that on-demand (passenger-actuated) services, if they become truly operational, will have lower capacities (all other features remaining the same) because various potentially interfering movements will have to be accommodated. It is unlikely that any system would be able to operate in the on-demand mode during peak loading periods.

Finally, just for the sake of the argument, if a true PRT system were to be developed and implemented with two-person pods operating at 5-second intervals (and off-line stations), the throughput on the main channel would be only 1440 riders per hour. Let us leave in abeyance for the time being the scenario of six-passenger vehicles at 0.5-second intervals.

Cost Considerations

Little can be said about AGT capital costs that is definitive and convincing. All existing systems are different, built under very diverse conditions; many were experimental, implemented on a pilot project basis. The best that can be done is to give some examples of actual expenditures on recent projects, but there are problems in attempting any comparisons. For example, the Morgantown system has 8.6 mi (13.8 km) of guideway, but the total length that a passenger can travel from one end to the other is only 3.4 mi (5.5 km) because many auxiliary guideway sections had to be built to accommodate the off-line stations. Since the total construction cost in the early 1970s was about $130 million, the average was $15 million per mile of guideway or $38 million per mile of route.

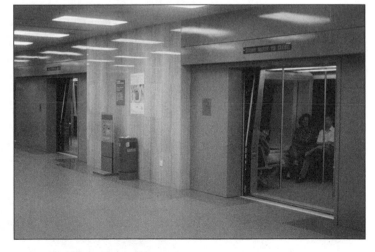

AGT within the Changi Airport of Singapore.

A larger complication is the fact that this construction effort was a first of its kind, with multiple errors and corrections along the way. Not least of these problems was the fact that the guideway was constructed prior to understanding all of the parameters of the selected technology. Thus, the guideway was overdesigned rather than being tailored to a specific system. The diplomatic way to account for that situation is to say that 40 percent of the cost should be attributed to research and development.

Since AGTs had generated much interest in the 1970s, a comprehensive survey of the cost experience was prepared in 1978.[21] The survey showed large differences among the 10 systems then open, with wide ranges under each cost item. The lowest cost per lane mile of guideway was $0.5 million; the highest $7.2 million. The lowest-priced vehicle was $47,000; the most expensive $413,000.[22]

Another similar comparative summary was prepared in 1992.[23] By that time, 23 systems could be examined, but the capital cost numbers still showed wide ranges. The most expensive American public system (corrected for inflation to 1990) was the Detroit DPM at $231.4 million for the 3-mi one-way loop. The airport service with the largest price tag was the Dallas/Fort Worth network at $125.5 million.

Since there has been no construction experience with AGTs during the last 10 years in the United States, the actual capital costs of the larger existing systems can be corrected for inflation to the year 2000 and averaged out as per-mile costs of the double guideway (not just a lane). Table 15.2 shows the results.

The conclusion from Table 15.2 is that a mostly elevated AGT system can be constructed at less that $100 million per mile. If it has to be placed underground, even if the tunnels are a part of a larger construction effort, the costs will run from $100 million to well above $170 million per route mile. Yet, the Tokyo Teleport Town 7.5-mi (12-km) AGT connector, opened in 1995, had construction costs that amounted to $228 million per mile. Granted, it had difficult site and foundation conditions at the edge of Tokyo

[21] N.D. Lea & Associates, Inc., *Summary of Capital and Operating and Maintenance Cost Experience of Automated Guideway Transit Systems,* June 1978.
[22] The low numbers were associated with the Jet Rail system in Love Field, Dallas; the high ones with the Seattle-Tacoma Airport underground system.
[23] U.S. Department of Transportation, *Characteristics of Urban Transportation Systems,* revised edition, September 1992.

Table 15.2 Actual Capital Costs of AGT Systems*

System	Total Capital Cost, Millions of 2000 Dollars	System Length, mi	(km)	Capital Cost per Mile, Millions of 2000 Dollars
Atlanta Airport (with tunnel)	$ 120.2	1.14	(1.8)	$105.4
Chicago O'Hare Airport (elevated)	216.0	2.67	(4.3)	80.8
Denver Airport (with tunnel)	159.4	0.93	(1.5)	171.0
Miami Metromover (with heavy structures)	241.7	1.86	(3)	129.6
Orlando Airport (elevated shuttle)	49.7	0.74	(1.2)	67.1
San Francisco Airport (elevated)	401.5	2.66	(4.3)	150.9
Seattle-Tacoma Airport (with tunnel)	110.2	0.85	(1.4)	129.6
Vancouver Skytrain (elevated)	$1106.1 (U.S. $)	13.1	(21)	84.8

* This summary was prepared by Parsons Brinckerhoff in 2000.

Bay, but otherwise it is a rather regular fixed automated guideway system. The JFK International Airport connector in Queens, New York, currently (2001) under construction, is expected to cost about $120 million per mile. It is fully elevated on a heavy structure atop major highways and roadways, and it can be classified either as a large AGT or as a fully automated light rail transit system.

Thus, while some of the civil engineering and structural elements of AGTs can be lighter and smaller than for regular transit systems because of the characteristics of the vehicles, the various elements constituting the systems are more complex, with technically advanced components. Therefore, AGT systems are not necessarily less expensive on an aggregate basis than other fixed guideway transit systems with comparable service capabilities and placement of alignment.

Likewise, the cost of the larger AGT vehicles, despite their limited capacity of not much more than 100 passengers, may not be lower than that for regular transit cars because of the large amount of advanced equipment that AGT vehicles carry. There is also no regular and steady market that would allow manufacturers to achieve efficiencies of scale; for all practical purposes, most vehicles are hand-built. Comparative data analysis of the systems in operation in 1992[24] showed that the vehicles in the smaller capacity range of 20 to 30 passengers cost $380,000 to

[24] U.S. DOT, *Characteristics of Urban Transportation Systems*, op. cit.

$480,000 each (in 1990 dollars); those in the midrange of 45 to 60 passengers $1.0 million to $1.6 million; and those in the high range of 90 to 100 passengers $2.3 million to $3.0 million. If these numbers are inflated to likely year 2000 prices, the following results are obtained:

- Low capacity $0.5 million (Morgantown)
                  to $0.6 million (Dallas/Fort Worth)
- Mid capacity $1.3 million (Seattle-Tacoma)
                  to $2.1 million (Denver)
- High capacity $3.0 million (Detroit)
                  to $3.9 million (Jacksonville)

The last major cost items are annual operation and maintenance (O&M) expenditures, where the uncertainties are even greater and precise breakdowns are often simply not available. Information assembled in early 1990s showed that the Miami Metromover, for example, had expenses of $2.72 per passenger mile. At that time, that ratio was six times higher than the same parameter for the local bus operations. Obviously, the principal remedy would be to attract more riders. On the other hand, the Vancouver SkyTrain system was operating at $0.11 per passenger mile, or at one-third the O&M expense of local bus service.

An even more extreme picture is presented by fare box recovery percentages. In Miami, only 5 percent of the costs of operating Metromover were covered by ticket revenues. It is suggested that this extraordinarily low number is due to the fact that many AGT riders utilize free transfers from the metro. Also, if the elevated loop serves as an attraction for tourists to see the sights, then there can be a justification to keep the fares low and make the experience enjoyable. In contrast, the VAL system in Lille claims a fare box recovery rate of 120 percent. Thus, the city's heaviest fixed guideway system is able to cross-subsidize other transit operations. The explanation for this sterling performance is the fact that in this city labor costs represent only 42 percent of the annual O&M budget, while in most transit operations two-thirds to three-quarters of expenditures fall in that category.

Possible Action Programs

AGT systems in the form of horizontal elevators will undoubtedly continue to be built by private and institutional developers as internal support services for large-scale projects. These certainly include airports, shopping centers, and amusement parks. This effort becomes a part of the site planning and architectural design process, depending on patronage estimates that anticipate the demand by several thousand users in any direction during each hour over extended time periods. Considerable weight in making decisions in favor of this mode will be carried by the high-tech and progressive image that AGT lends to any contemporary enterprise. The internal transportation tasks may be doable by conventional means (such as shuttle buses), but there is marketing value in a broad-based perception that an advanced system is in place.

In the public realm, the current prospects for AGT implementation are not very bright. The systems are not yet at a stage where costs and service reliability are completely predictable. A certain amount of risk is involved, and the current atmosphere in government decision making mandates a very careful approach. It would have to be a special situation where an American municipal government or a public transportation agency would engage in such a bold endeavor (such as in Las Vegas). This is not inconceivable, but the program most likely would have to receive considerable outside financial assistance and burnish the image of the locality in a wide forum—perhaps the organization of a world's fair that becomes a permanent park, or hosting a major sports festival. Clearwater, Florida, announced somewhat casually in 2001 an intent to explore AGT feasibility connecting its downtown to beaches and

providing a foundation for a possible bid to be selected for the 2012 Olympics.

The need to serve a central business district or to provide internal linkages in any other high-intensity district will probably not lead to the selection of AGT, because a number of other conventional transit modes can accomplish such tasks quite well with considerably lower risk.

There is one field of definitely positive application for AGT technology that may be

Interior of an AGT vehicle (Miami).

more important than anything else: the introduction of advanced concepts into the design and operation of regular transit systems. The AGT experiments and pilot projects have constituted a fruitful field from which much practical experience has been gained and tests of various components have been accomplished.

Conclusion

Most transportation modes emerged in response to some urgent and pervasive transportation needs at certain times. Streetcars, metro trains, and automobiles appeared first as fascinating concepts, but once it was clear that they had real utility, they were shaped and developed into devices that were able to capture mass usage within a competitive market by virtue of their service capabilities. Has this been the case with AGTs? Not exactly. These concepts also appeared rather suddenly on the scene, generated much early excitement, and were judged prematurely to have a much greater potential than they were actually able to deliver. Urban transportation solutions were to be kicked deliberately to a higher technical level and attain great effectiveness. In fact, AGT has not solved and cannot solve the basic transportation problems of our cities, and that lesson had to be learned the hard way. The expectations were high, but a rational evaluation by the market was blurred by substantial promotional assistance from the United States government.

All that is not the fault of the mode itself or of its basic technology. Automated systems are becoming more and more common. They create new capabilities and utility, and greater safety and reliability are being achieved, but not exactly in the way things were envisioned originally. AGTs are finding their proper niche, and their overall impact is significant.

Assuming that true AGT systems are not going to be particularly suitable for heavy-demand rapid transit use, the question remains as to what directions in the application of advanced technology should be explored further. If residents of American communities were to be surveyed, they would probably indicate a desire to see dual-mode options: small cars that can be individually operated on regular streets, but also carried automatically on guideways for line-haul trips. Or they may ask to be left alone since they have personal vehicles already.

Bibliography

American Society of Civil Engineers: Proceedings of Conferences on People Movers. There have been six such conferences, with the latest in San Francisco in 1997, and the published proceedings contain a range of articles covering various aspects of the AGT field.

Gray, George E., and Lester A. Hoel (eds.): *Public Transportation* (2d edition), Prentice-Hall, 1992, 750 pp. Chapter 24 by Clark Henderson (pp. 660–681), "New Public Transportation Technology."

Vuchic, Vukan R: *Urban Public Transportation: Systems and Technology,* Prentice-Hall, 1981, 673 pp. Chapter 6 (pp. 476–515), "New Concepts and Proposed Modes."

CHAPTER 16

Waterborne Modes

Background

Most cities are located on a bay or lake, and they are crossed by rivers and canals. Their original location was most likely associated with the presence of water as places where trade routes converged and crossings had to be made, or goods had to be moved between marine vessels and land vehicles (break-of-bulk points). There are some cities that sit on islands (Venice, Stockholm, Hong Kong, and New York), some that are next to a large system of bays and inlets (San Francisco, Sydney, Seattle, and Boston), and others that deliberately created their own water-based systems in earlier periods (Amsterdam and Bangkok). It is not "natural" to build communities in a water-dominated environment because of the difficulties that this presents to land creatures, but it has been done often enough for reasons of commerce and defense. On the other hand, it has always been of great advantage to locate settlements in reasonable proximity to waterways for long-distance transport and linkage to other centers.

Bodies of water are obstacles in the daily operation of communities (blocking movement between land-based activities) and in the orderly, contiguous expansion of development. Thus, they have to be overcome by bridges and tunnels, if at all possible. At the same time, water is also a connector, allowing travel by rafts, boats, and ferries. After all, the cheapest way to move between two points is to float down a river—provided that some sort of vessel is available, that it does not matter how long the trip will take, that the weather is good, and that the waterway is clear of obstructions (whatever they may be).

The result of all these elementary considerations is that the history of cities is rich with examples of urban waterborne transportation as a direct means of crossing and connection. Most of these activities disappeared when vehicular bridges and tunnels were built, but there are opportunities and perhaps needs today to expand waterborne services again. If nothing else, in many cities the only open uncluttered spaces left are the water bodies in the center, as the busy maritime cargo operations have declined or long since moved to peripheral and unencumbered locations. The best evidence of this condition is provided by a simple comparison of photographs of the harbor areas at the beginnings of the twentieth and the twenty-first centuries in New York, Boston, Baltimore, and a number of other old cities.

There are cities in the continental interior of North America that have rather minuscule waterways, but even some of those resources can be used for excursion and sightseeing purposes; most cities have significant bodies of water (albeit not always in the most advantageous configuration). Yet, the official list of waterborne passenger services in the United States today includes relatively few major operations in urban areas. As shown in Table 16.1, most are rather limited and specialized local services for commuters and sightseers. The 1990 U.S. Census recorded ferry use for commuting to work, but the total volume was not large enough to report, even as a fraction of a percent. (The actual use may be larger, as one part of the total commuting trip for many, but not reported as the *principal* mode, and thus overlooked in the official statistics.)

(Note that Table 16.1 does not include *Canadian* ferry oper-

South Ferry terminals in Lower Manhattan (before September 11, 2001).

Table 16.1 Ferryboat Transit Operations in the United States

Primary City	Transit Agency	Vessels	Annual Passenger Miles, million	Annual Unlinked Trips, million	Notes
Alameda, CA	Harbor Island Ferry				
Balboa, CA	Balboa Island Ferry				
Baytown, TX	Harris County Lynchburg Ferry				
Boston, MA	Massachusetts Bay Transportation Authority				
Bremerton, WA	Kitsap Transit	4	0.4	0.3	
Chicago, IL	Vendella RiverBus				
Cincinnati, OH	Anderson Ferry Boat				
Corpus Christi, TX	Corpus Christi Regional Transportation Authority				
Galveston, TX	Texas Department of Transportation				
Hartford, CT	Connecticut Department of Transportation				
Jacksonville, FL	Florida Department of Transportation				
Long Beach, CA	Long Beach Public Transportation Company				
New Orleans, LA	Louisiana Department of Transportation and Development	5	1.5	3.0	Cross-river
New York, NY	New York City Department of Transportation	7	99	19	Staten Island service
	Port Authority of New York & New Jersey*	5	4.1	2.4	Private operations
Norfolk, VA	Transportation District Commission of Hampton Roads	4	0.7	0.4	Cross-river
Oakland, CA	Alameda–Oakland Ferry Service				
Philadelphia, PA	Delaware River Port Authority RiverLink				
Port Huron, MI	Blue Water Area Transportation Commission				

(Continues)

Table 16.1   Ferryboat Transit Operations in the United States (*Continued*)

Primary City	Transit Agency	Vessels	Annual Passenger Miles, million	Annual Unlinked Trips, million	Notes
Port Townsend, WA	Washington State Department of Transportation				
Portland, ME	Casco Bay Island Transit District	5	3.1	0.9	Links to islands
Providence, RI	Rhode Island Public Transit Authority				
San Diego, CA	Coronado Ferry				
San Francisco, CA	Angel Island–Tiburon Ferry Company				
	Golden Gate Bridge	5	21	1.9	
	Highway & Transportation				
	Red & White Fleet				
	Vallejo Baylink Ferry				
San Juan, PR	Puerto Rico Ports Authority	9	1.8	1.1	
Seattle, WA	Washington State Department of Transportation	28	131	15	Largest in U.S.; 10 routes
Tacoma, WA	Pierce County Ferry Operation	9	1.4	0.2	
	Washington State Department of Transportation				Included under Seattle

Note: Not all are in urban service. Some are seasonal; others have special weekend schedules. Only the larger ferryboat operators are required to submit information to the National Transit Database. Private systems with less than 10 vessels, not accepting federal funds, and operating in nonurban areas below 50,000 population are exempt. They are listed in the table by name only.
* While the Port Authority is the agency with overall responsibility for most shoreside facilities, the actual service is provided by private operators, such as NY Waterway, Seastreak, and others. Furthermore, the statistics reported officially do not include many of the 15 routes that operate in New York harbor (besides the Staten Island ferry). Thus, before September 11, 2001, the total annual ridership on the private waterborne services was 9 million; since the attack, it has grown to 15 million at this writing. However, it is not yet known whether the upward trend will continue or ridership will revert back to earlier levels when rail services are fully restored some years in the future.
Source: American Public Transportation Association and respective Web sites.

ations, which carried approximately 39.2 million passengers and 15.3 million vehicles in 1999. Operations are concentrated in the West, but ferries operate also in the Maritime region, providing primarily nonurban service and accommodating vehicles [with the exception of Vancouver, which is discussed further later].)

Development History

The use of ferries for transportation is an obvious and just about unavoidable response in the proper places. When travelers encounter a waterway that is too deep to be forded, they seek a dry crossing. Nor does it take much business skill by local entre-preneurs to sell such a service to others who need to get to the other side. Any floating vessel, even a raft, can be poled, rowed, sailed, or pulled across by a rope using plain muscle-power. All a ferry needs is a flat deck that can be boarded by people, animals, and wagons (or vehicles), and a way for them to disembark with some safety and convenience. Such ferries have always been present, and they can still be found in remote locations, even in the United States.

This mode of transportation is thus truly ancient, with no basic changes in its operation for centuries. The only substantial developments over the last century and a half have been in propulsion technology and, quite lately, in the hydrodynamic shape of the vessels. Ferries have always been accepted as being slow; suddenly, we expect major velocity improvements. Before the Industrial Revolution, the most advanced ferries used horses on a treadmill arrangement to pull the vessel across; afterwards the steam engine took over completely, and it reigned supreme for about a century. It was only in the second half of the twentieth century that deliberate efforts were made to redesign the vessel, making it more responsive to contemporary expectations of speed and comfort in urban transportation.

While technical developments have followed the same trends everywhere in the world where ferries have been employed, their local role, extent of operations, and level of performance have been quite different in various locations. Each place is certainly unique in what can be and what has been done with waterborne transportation. Since not too many of the individual cases can be described and analyzed in a single chapter, a deliberate selection had to be made; not surprisingly, the choice was New York.[1] Over an extended period, the largest waterborne system in operation provided basic services to this urban center, and there has been a rather impressive reemergence of ferry operations in recent

[1] The author may be accused of excessive ethnocentrism regarding the Big Apple, but we are what we are, and—given the choice—it is safer and wiser to write about things that are more familiar than to catch up in other sectors.

decades. Furthermore, the events in and around New York harbor have been recorded and examined in great detail—not only what has happened is known, but also why.[2]

Water was the determining element for the first settlement of New Amsterdam (1624), with a superb natural harbor, a huge hinterland that shaped the fortunes of the ever-growing city, and linkages to regional farming settlements in all directions. Wagon trails from the original city on Manhattan Island always led to the water's edge, where some means of crossing had to be provided. These places often became the sites of organized private ferries, operating with a government franchise and published tariffs, providing suitable accommodations and refreshments for the travelers. The Harlem River was not particularly wide, so it received the first bridge in 1693, eliminating crossing delays for the busy overland traffic to Boston, New Haven, and Albany.

The East River (actually a tidal strait) became the focus of early ferry operations, since many active villages of Long Island (later the city of Brooklyn) were on the other side. Crossings were made at several places by simple means, since the width is quite manageable and the tidal flows are predictable. The focus, however, was on the lower stretch, where settlements were concentrated and major roadways from both sides (later named for Robert Fulton) constituted important regional links. The first suburb (Brooklyn Heights) became established here since it was easily accessible by water; the armies of the Revolutionary War crossed at this location. Later, steam ferries providing high-volume service were placed here because of the established demand, only to be displaced by the first great bridge—Brooklyn Bridge, in 1883.

The Hudson River (actually an estuary) to the west was much more of a challenge—it is a mile wide, and the need for linkages to New Jersey was not very great in the early days. Nevertheless, a service was begun in 1661, from a location known as Communipaw (and Paulus Hook), which later became the large urban center

---

[2] Among the many books and articles, there are: B. J. Cudahy, *Over and Back: The History of Ferryboats in New York Harbor* (Fordham University Press, 1990); F. B. Roberts and J. Gillespie, *The Boats We Rode* (Quadrant Press, 1974, 101 pp.); A. G. Adams, *The Hudson Through the Years* (Lind Publications, 1983, 334 pp.); S. Grava, "Water Mode to the Rescue: Past and Future Ferry Service in New York City," *Transportation Quarterly,* July 1986, pp. 333–356; and K. Ascher, "Down to the Sea Again," *Portfolio,* Winter 1988, pp. 11–22.

of Exchange Place in today's Jersey City. Direct regular service to Staten Island—at least 5 mi (8 km) away across the Upper Bay— was instituted in 1755, when the communities there started to grow and wealthy New Yorkers established rural vacation retreats. Because there are no satisfactory parallel land paths to the center, the latter operations have continued vigorously without interruption to the present as the well-known Staten Island Ferry, the single highest-volume ferry service in North America.

Before the introduction of the steam engine, ferryboats were not only slow, but also unreliable and uncomfortable. Almost any type of floating device could be found providing service in New York Harbor, and these early operations were instrumental in defining a regional network with major routes converging on crossing points, establishing a permanent pattern. Even today there are city bus routes that run to some obscure waterfront locations, which would be difficult to explain except by remembering history and recalling that ferry terminals were once in operation there.

The vessels changed in the early nineteenth century with the introduction of the steam engine, and ferry services became popular and attractive means of transportation for the next century. They were instrumental in assisting the metropolitan development of the rapidly expanding industrial and service center. Experiments by Robert Fulton, who was a brilliant promoter of new technology, as well as the entrepreneurial efforts of a few other leading businessmen, culminated in the first steam ferry in regular operation in 1812 (a year after the gridiron plan for Manhattan was adopted), running between the city center and New Jersey. While a few boilers did explode in the early period, the boats were able to maintain schedules and provide reliable support to daily activities regardless of water obstacles.

A major milestone in the provision and management of public transport in the United States was the 1824 U.S. Supreme Court decision in *Gibbons vs. Ogden,* which broke the monopoly rights on ferry services throughout the harbor secured by the Fulton group. The Court declared that it is the prerogative of the national government to regulate interstate commerce (and that of state governments to control intrastate operations) for the benefit of the public at large, not that of selected private interests.

The "walking beam" sidewheelers—the drive system used a visible rocking beam atop the vessel that transferred power from the steam cylinder to the driving wheels—became a common sight in New York Harbor, as well as anywhere else where substantial ferry services were provided. Although hull length varied considerably, from 100 to 200 ft (30 to 61 m), the boat assumed a standard configuration that is still seen today—a symmetrical double-ended vessel with a flat deck that could accommodate wagons. Enclosed passenger compartments were on the sides or on the second deck; pilot houses on top allowed quick in-and-out operations; a tall smoke stack and a powerful steam whistle completed the ensemble.

The ferry terminal also soon evolved into a functional shape that responded well to service requirements. The larger facilities consisted of several *slips* on the waterside that allowed vessels to be positioned quickly and precisely for disembarking and boarding.[3] Movable bridges served as inclined ramps regardless of the elevation of the tides. The larger terminals were built on two levels, separating wheeled traffic from pedestrians. It is a curious fact that, while transit lines and major roadways marked these locations, the terminals themselves do not appear to have generated any significant land development around them. They did, however, attract clusters of services that catered directly to the travelers—hotels and certainly eating and drinking places. Patrons still had considerable difficulty in reaching their inland destinations, their workplaces and commercial districts, because of frequent local street traffic congestion.[4]

The nineteenth-century ferry operations across the East River to the growing communities of Brooklyn and Queens can be thought of as connections between the ends of major streets on both sides, extending all

Old Hudson River ferry (the *Binghamton*) used as a restaurant.

[3] *Slips* are curved walls of flexible timber piles that guide and hold the vessel in place. A good captain will bring in the boat without major bumps by sliding gently against the walls.

[4] Several of these terminals still exist, awaiting possible reuse. A few of the old boats can be seen in maritime museums or in use as theme restaurants.

along the waterway as far as southern Brooklyn. There were many such operations under the ownership and management of various private enterprises until the Union Ferry Company amalgamated most of them. Another major actor was the Long Island Rail Road (formerly a component of the Pennsylvania Railroad Company), which operated several ferry routes from its Hunter's Point terminal in Long Island City.

On the Hudson side, as railroads became the dominant national transportation mode, the early small private efforts were replaced as each major railroad company established and operated its own ferry service into Manhattan. The transcontinental and regional rail networks terminated on the west shore of the Hudson River, and passengers and goods could reach their final destination only across water. Freight cars were accommodated by an extensive system of roll-on/roll-off barges (*car floats* or *lighters*); passengers were channeled through the ever-more-elaborate and -prestigious ferry terminals.

By the end of the nineteenth century, these railroad terminals were as follows, reading downstream:

- Weehawken—New York Central Railroad,
- Hoboken—Delaware, Lackawanna and Western Railroad,
- Jersey City—Erie Railroad,
- Jersey City (Paulus Hook)—Pennsylvania Railroad
- Jersey City—Central Railroad of New Jersey

There were corresponding terminals on the Manhattan side for each railroad company. Particularly large and busy were those at the ends of Cortlandt Street, Chambers Street, and West 23rd Street. All traces of these major access nodes are obliterated today.

Thus, ferries became a basic component of daily life in Gotham. All writers and journalists who looked at the city noticed them (ranging from O. Henry to H. L. Mencken), including visitors from abroad (such as Charles Dickens), and they frequently recorded some reaction to them. This usually included exhilaration at the "sea voyage" and either admiration of the vitality of the city or dismay at the squalor of the immediate surroundings. Apparently, there were good reasons to be concerned about the quality of service much of the time.

Technical changes did come eventually; these included such items as the screw propeller (1888), the steel hull, and eventually

Crossing Brooklyn Ferry

1

Flood-tide below me! I see you face to face!

Clouds of the west-sun there half an hour high—I see you also face to face.

Crowds of men and women attired in the usual costumes, how curious you are to me!

On the ferry-boats the hundreds and hundreds that cross, returning home, are most curious to me than you suppose,

And you that shall cross from shore to shore years hence are more to me, and more in my meditations, than you might suppose.

9

Flow on, river! flow with the flood-tide, and ebb with the ebb-tide!

Frolic on, crested and scallop-edg'd waves!

Gorgeous clouds of the sunset! drench with your splendor me, or the men and women generations after me!

Cross from shore to shore, countless crowds of passengers!

Stand up, tall masts of Mannahatta! stand up, beautiful hills of Brooklyn!

Throb, baffled and curious brain! throw out questions and answers!

Suspend here and everywhere, eternal float of solution!

Walt Whitman, *Leaves of Grass,* 1856

the diesel engine (1926). But while the vessels became larger and more efficient and maneuverable, the form of operations and the shape of the boat were not affected by the new elements at all.

Several events took place toward the end of the nineteenth century that soon had a profound effect on ferry services in New York Harbor. The most dramatic of these was the progressive construction of the great bridges across the East River and the subsequent building of tunnels, first for rapid transit, but eventually also for automobiles. The Hudson River experienced the same events, only a few decades later. The general decline of intercity passenger rail service was also a major factor; many of the large westward services associated with ferry crossings faded away, and the Pennsylvania Railroad built its own tunnel.

Also significant was a growing public demand for government intervention in the ferry industry—first, because the quality of service was deteriorating; and second, because the growing loss of ridership made increasingly more routes unprofitable, and the patrons expected the municipal government to keep the services going. These demands became a political issue, and several candidates for high office adopted the cause and carried through with municipal acquisitions (or bailouts).

The first of these actions was the municipal takeover of the Staten Island ferry services in 1905. This was a successful and responsible transition—the operations are doing well even today. It is, incidentally, the only public transportation service that the City of New York runs directly. The rest of the story concerning the large and small private operations across the East River is much less inspirational, and the blows struck by each new bridge and tunnel were reflected immediately in the patronage volumes of the nearest ferries. The routes that were taken over by the city languished for a while with heavy subsidies, but eventually had to be abandoned as well.

Several interesting urban development consequences are associated with this revolutionary change toward continuous land-based wheeled and motorized transport. The slow marine vessels of that period, operating through cumbersome transfer points, could not even remotely compete with the speed and convenience offered by cars and subway trains. There was talk of support for the ferries, but no actual loyalty was shown by the patrons when other choices appeared. All these changes were possible only with the emerging dominance of motor vehicles, because horses pulling heavy wagons could not have really coped with the long and steep gradients of the new surface crossings.

In 1910, there were already four major bridges, one subway tunnel and one railroad tunnel across the East River; by 1925, five more subway tunnels had been added. The Long Island Rail Road's Hunter's Point ferry operation had closed by 1907, and the Brooklyn Union Ferry Company was dissolved in 1908. Service was maintained on several routes under municipal auspices, but by the start of World War II all of them were gone from the East River (Fig. 16.1).

Because the bridges had to maintain high navigational clearances and the tunnels had to be placed deep under the water's surface, they made "landfall" quite far inland, away from the water's edge. Since maritime and industrial activities were also migrating outside the city, the waterfront strip lost its functional attractiveness and utility. These districts were doomed to severe neglect and abandonment over many decades, which is changing only today as waterfront sites are being rediscovered.

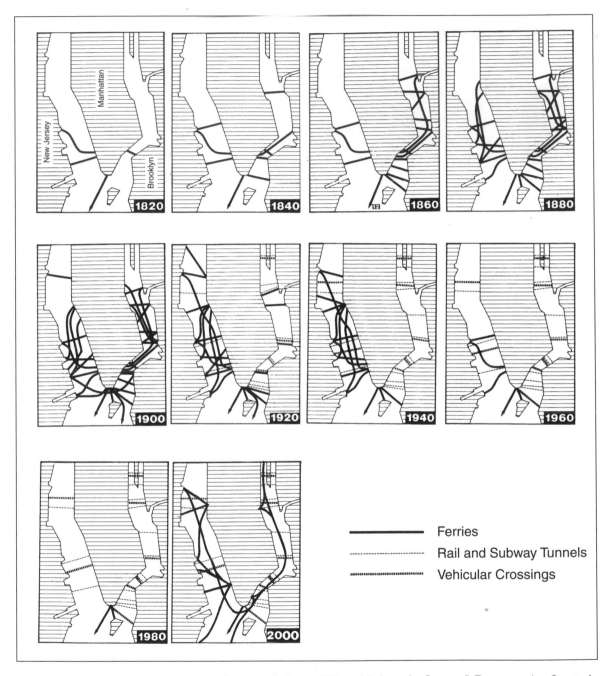

Figure 16.1   History of New York ferries. (*Source:* S. Grava, "Water Mode to the Rescue," *Transportation Quarterly,* July 1986.)

Recuerdo

We were very tired, we were very merry—
We had gone back and forth all night on the ferry.
It was bare and bright, and smelled like a stable—
But we looked into a fire, we leaned across a table,
We lay on a hill-top underneath the moon;
And the whistles kept blowing, and the dawn came soon.

Edna St. Vincent Millay, *A Few Figs from Thistles,* 1920

These patterns were repeated on the Hudson River, except that there were no efforts to provide government assistance for the railroad ferries. The Hudson & Manhattan rapid transit tunnel was opened in 1908, the Pennsylvania Railroad tunnel and Penn Station in 1910, the Holland vehicular tunnel in 1927, the George Washington Bridge in 1931, and the Lincoln Tunnel in 1941. The last Hudson River ferry of the historic period ran in 1967.

After a hiatus of about two decades, regular ferry service is back in New York harbor. This happened not because waterborne passenger transportation is intrinsically better, but because the conventional choices have become badly overloaded. Crossing the Hudson River by car during peak periods involves delays of at least 20 minutes and frequently much longer. There are many commuters who accept these conditions, but some do not. The latter constitute the core ridership for the new ferries. There are, however, also instances in which the geography makes a water crossing much more direct than a roundabout land route (for example, from Monmouth County, New Jersey, to Manhattan). As speeds attainable over water increase, these connections become very competitive. Another factor that is likely to carry more weight in the future is the return of development (offices and apartments) to the waterfront. This would tend to eliminate one land-access leg to a ferry service and thereby increase its utility.

The first of these new efforts was a rather limited service instituted in 1986 from Monmouth County, New Jersey, to Manhattan, covering a long distance at premium fares. This is seen as a distinct opportunity for service operators, generating significant competition among themselves. The size of the vessels and their speed have progressively increased, and business remains strong.

Edith Wharton described the Hudson River rail ferry crossing as a major romantic adventure in the 1870s. Was it possible for people at that time to foresee air travel?

His wife's dark blue brougham (with the wedding varnish still on it) met Archer at the ferry, and conveyed him luxuriously to the Pennsylvania terminus in Jersey City.

It was a somber snowy afternoon, and the gas-lamps were lit in the big reverberating station. As he paced the platform, waiting for the Washington express, he remembered that there were people who thought there would one day be a tunnel under the Hudson through which the trains of the Pennsylvania railway would run straight into New York. They were of the brotherhood of visionaries who likewise predicted the building of ships that would cross the Atlantic in five days, the invention of a flying machine, lighting by electricity, telephonic communication without wires, and other Arabian Night marvels.

"I don't care which of their visions comes true," Archer mused, "as long as the tunnel isn't built yet." In his senseless schoolboy happiness he pictured Madame Olenska's descent from the train, his discovery of her a long way off, among the throngs of meaningless faces, her clinging to his arm as he guided her to the carriage, their slow approach to the wharf among slipping horses, laden carts, vociferating teamsters, and then the startling quiet of the ferry-boat, where they would sit side by side under the snow, in the motionless carriage, while the earth seemed to glide away under them, rolling to the other side of the sun. It was incredible, the number of things he had to say to her, and in what eloquent order they were forming themselves on his lips. . . .

The clanging and groaning of the train came nearer, and it staggered slowly into the station like a prey-laden monster into its lair. Archer pushed forward, elbowing through the crowd, and staring blindly into window after window of the high-hung carriages. And then, suddenly, he saw Madame Olenska's pale and surprised face close at hand, and had again the mortified sensation of having forgotten what she looked like.

They reached each other, their hands met, and he drew her arm through his. "This way—I have the carriage," he said.

After that it all happened as he had dreamed. He helped her into the brougham with her bags, and had afterward the vague recollection of having properly reassured her about her grandmother and given her a summary of the Beaufort situation (he was struck by the softness of her: "Poor Regina!"). Meanwhile the carriage had worked its way out of the coil about the station, and they were crawling down the slippery incline to the wharf, menaced by swaying coal-carts, bewildered horses, dishevelled express-wagons, and an empty hearse—ah, that hearse! She shut her eyes as it passed, and clutched at Archer's hand.

"If only it doesn't mean—poor Granny!"

"Oh, no, no—she's much better—she's all right, really. There—we've passed it!" he exclaimed, as if that made all the difference. Her hand remained in his, and as the carriage lurched across the gangplank onto the ferry he bent over, unbuttoned her tight brown glove, and kissed her palm as if he had kissed a relic. She disengaged herself with a faint smile, and he said: "You didn't expect me today?"

(From *The Age of Innocence*, 1920)

The most prominent operation among several that were started in the 1980s is the expanding system of Arcorp Industries from Weehawken (site of one of the old rail ferry terminals, now named Port Imperial) to several locations in Manhattan. It was intended to be a special connection to a pending large real estate development on the western shore, but it works well even without that captive ridership. The service has a visible corporate identity as NY Waterway, and the key to its success is undoubtedly very convenient access systems at both ends, particularly on Manhattan, with dedicated feeder buses accommodating ferry customers throughout Midtown and Lower Manhattan. The vessels are reasonably fast front-loading boats that, through successive replacements, can now accommodate 400 passengers (Fig. 16.2). There are now a number of lines crossing the river from various locations, but the most direct linkage takes 3 to 4 minutes, with two vessels able to operate at 10-minute intervals (2100 passenger capacity per hour). The one-way fare varies among the several crossings, starting at $2 for a one-way trip, but it is always higher than the comparable regular transit choice and lower than the expenses of commuting by car. The total NY Waterway daily ridership of 30,000 trips (as of 2001) is not particularly large in comparison to that of other transit operations (the daily ridership on the subways is 3 million), but it does make a difference—it is estimated that the vehicular load on Hudson River bridges and tunnels may be eased by some 5 percent. (After September 11, 2001, when the PATH connection from New Jersey was knocked out, NY Waterway's service assumed a critical role.)

All these ferry operations have been implemented at no cost to the public for the acquisition of equipment and operations and maintenance. Local government agencies take some pride in this feature, while they maintain a most favorable attitude toward the ferry options. There is a form of assistance, however, in the provision of landing places and the public maintenance of those facilities. Even though a negotiable landing fee is charged for their use, major public capital investments have been made, and such programs continue. For example, Pier 11 on the East River is a city-owned terminal with all the necessary features that is being used by several ferry companies. Building other municipal facilities has been discussed. The next New York City project will be a multiuse facility at the end of West 39th Street. The NY Waterway terminal on the New Jersey side is a refurbished retired fer-

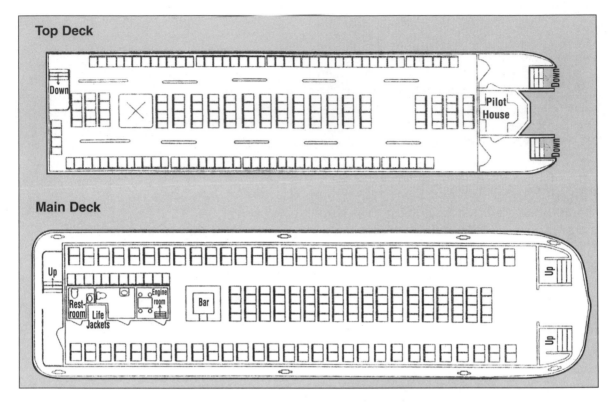

Figure 16.2   Ferryboat layout: NY Waterway vessel. (*Source:* Manufacturer's data.)

ryboat, but New Jersey Transit is expected to build a public ferry terminal in conjunction with the new shoreside light rail line.

Given the apparent opportunities to provide further ferry service in New York Harbor, extending along Long Island Sound into Connecticut and the North Shore, up the Hudson to any number of riverside communities, and deep into New Jersey via several bays and rivers (not to mention going out into the Atlantic Ocean), several attempts to institute waterborne operations have been made in the last three decades, besides the aforementioned NY Waterway and Monmouth County efforts. Many experiments have attempted long-distance waterborne service to compete with commuter rail, as well as shorter operations in town. Various types of vessels have been tried, including hydrofoils (for example, the Albatross, in 1963). The success rate has been low, most of these valiant efforts foundering on landside access constraints, unreliable schedules, maintenance issues, overall speed, and, of course, costs associated with pioneering small ventures. Even NY Waterway was not able to continue a service to a large development site across the East River, because the project has not yet really occurred. There have been no instances in the New York area in which financial assistance from public budgets has been provided, either for vessel acquisition or as an operating subsidy.

Service and private watercraft on the *klongs* (canals) of Bangkok.

Nevertheless, some services have worked (besides the indisputable competence of NY Waterway). These have included shuttle service from LaGuardia Airport, which on a good day was the most spectacular ride available anywhere in the city (and it was quick), and seasonal services to Yankee and Shea Stadiums. The LaGuardia service encountered administrative complexities, and is currently suspended, but it is being developed into a major airport access operation. The several "boutique" services at high fares from selected points in New Jersey are doing well, as has been mentioned previously. Undoubtedly, other waterborne services will be tried from time to time.

Types of Waterborne Operations

There is a great variety of movement on most waterways within urbanized regions, ranging from heavy cargo handling to recreational boating. Only a portion of these activities is relevant as an urban transportation mode—i.e., providing service for passengers and/or vehicles within cities and metropolitan areas. This strict urban

Ferry activity on the Golden Horn of Istanbul.

definition leaves out many long-distance ferry operations, such as, for example, those with large vessels providing overnight accommodations (on the Baltic Sea and the Mediterranean), those connecting remote island locations to central cities (some of which are listed in Table 16.1), those linking separate islands (as in Hawaii and the Caribbean), those crossing major bodies of water (such as the Long Island Sound or the Bay of Fundy), and those making long runs on coastal or inland waterways (such as on the lakes of Finland, along rivers in Russia, and along the coast of Alaska). The largest coordinated waterborne system in the United States—the many water routes of Puget Sound from Seattle and Tacoma—encompasses several of these types; the scale and the vessels are different as compared to regular urban services, but the system does provide for regular commuting as well as other activities.[5]

The operative definition of a *ferry service* for this discussion is, therefore: regular and frequent operations along and across waterways accommodating short trips by persons and vehicles. It is to be seen as an extension of transit systems or as filling in the gaps in a metropolitan highway network.

Thus, the first way of classifying ferry services is by whether only *passengers* are accommodated, or *vehicles* are also carried. (There are vessels designed for vehicle carriage alone, but they are always able to accommodate a few walk-on passengers.) The second significant distinction is whether the service *crosses* a waterway along the most direct alignment between points on opposite sides of a water obstacle (acts as a shuttle), or the service runs *along* a waterway connecting a string of embarkation points (acts as a transit line).

Within this context, large variations in the type of operations and in the vessels used are encountered. This is understandable, because there are not that many examples, and every existing service has been specifically designed as, or has evolved into a unique system that fits the local requirements. For example, there are the very large (6000-passenger) vessels on the Staten Island service in

[5] While the Puget Sound ferry operations represent the largest ferry system in North America in terms of annual passenger miles (there are more trips in New York harbor, but they are short), they are not discussed in this chapter because of their primarily nonurban role. Much information is available elsewhere; for example, two articles in *Transportation Research Record No. 1677*, 1999, pp. 93–116.

New York, the nimble but simple boats that zig-zag across the Bosporus in Turkey, the high-technology air-cushion vehicles in Hong Kong, the low-clearance waterbuses in Amsterdam and Hamburg, the *vaporetti* in Venice, the converted landing craft of the U.S. military services, and the motorized barges in Bangkok.

Vessels

Ferryboats can be grouped in the following generic classes according to the basic types of maritime vessels employed:

High-speed ferry vessel at Cape May, New Jersey.

- *Conventional displacement boats.* Single-hull vessels (monohulls) with hydrostatic (vessel-at-rest) buoyancy provided by the displaced water (the weight of which equals the weight of the fully loaded boat). This is the classical design, requiring little power input to move, but it has serious speed limitations because of the friction of its large surface area against the water and the accompanying creation of trailing water turbulence in the form of waves and eddies. The *wake* or *wash* (following wave) increases with speed, consumes power, and can be most disturbing to other vessels and shoreside facilities. For any given size of vessel, the most effective mitigation measure is to decrease the weight of the boat (reduce displacement) by the use of such materials as aluminum and glass-reinforced plastic composites.

- *Planing craft.* As a watercraft starts to move with any speed, there is a natural tendency to skim over the top of the water surface due to hydrodynamic lift forces. This reduces the wetted surface area and displacement, thus improving performance and allowing higher speeds. Speedboats and racing craft rely on this feature almost entirely, and they actually fly over the surface. Speeds in excess of 200 mph (320 kph) can be reached by special powerboats, obtained at the cost of a tremendous power output and an extremely rough ride that can shatter human spines. Obviously, these concerns are not of great relevance in a discussion of ferry-boats, except that the faster craft do take at least partial advantage of the planing effect. Air resistance comes into play at speeds exceeding 30 knots (34 mph; 54 kph).[6]

- *Catamarans.*   These vessels have two or more sharp and thin hulls that reduce friction and can take advantage of a planing effect, joined by a flat deck. The wake is considerably reduced, and the vessel has good stability against rolling. They are more costly than conventional craft, but appear to dominate the worldwide ferry market today because of speed and smoothness of ride. There are also "wave-piercing" catamarans, but that feature is significant only for craft that are likely to operate in open seas.

- *Hydrofoils.*   These boats are equipped with *foils* (underwater "wings") that lift the hull out of the water at high speeds and greatly reduce friction, as well as the wake. They do need distance to get up to the required speed; therefore, they are not suitable for service with frequent stops. There are two basic types: *surface-piercing foils*—V-shaped wings that protrude above the surface and have inherent stability; and *fully submerged foils* which are more efficient but require elaborate controls. All hydrofoils are by definition high-speed vessels, but they are expensive to purchase and operate and they are vulnerable to debris in the water.

Hydrofoil boat in operation between islands in the Caribbean.

Nevertheless, they operate all around the world where the seas are reasonably calm and extensive distances are involved.

- *Hovercraft.*   These are vessels that float atop a bubble of

[6] A *knot* is a speed of 1 nautical mile (6076 feet) per hour. Speed in knots = speed in mph ÷ 1.15.

compressed air (contained by flexible side skirts); thus, they experience no friction at all and are amphibious. They can move on any reasonably level surface; they are, however, difficult to control since they do not grip the surface and are affected by any wind. They require much power to replenish the escaping air, and they are noisy and kick up much spray and dust. The engineering is reasonably

Hovercraft in Hong Kong harbor.

advanced, and hovercraft have been used in regular service at the Montreal Expo, across the English Channel, and in a number of other places. They are a part of the regular waterborne service systems in Hong Kong.

- *Surface-effect ships.* These are partial hovercraft with rigid sidewalls that contain the air bubble at high speeds. The front and aft skirts are flexible; in effect, a catamaran platform is provided.

- *Submerged vessels.* Submarines have the advantage of generating no surface waves; they can be fully streamlined (like fish), and they move with great efficiency and speed (said to be more than 45 knots [52 mph; 83 kph] for naval submarines). Obviously, they make no sense whatsoever for ferry operations with frequent stops, but vessels with submerged buoyant cylinders—known as *stable semisubmerged platforms* (SSPs)—carrying a deck atop struts have been envisioned, developed, and tested (primarily by the U.S. Navy).

This is not the end of the classification possibilities. A major difference among ferryboats is whether they are single-direction vessels (i.e., having a prow and a stern and moving forward whenever possible) or bidirectional (i.e., having identical ends, able to move back and forth with equal ease, not requiring turn-arounds).[7]

At the present time, particularly looking at future possibilities, the *speed* of ferryboats is a major factor. This is the only transportation mode, leaving aside aircraft, which has substantially increased its basic speed capabilities in recent decades.[8] The conventional and traditional craft operate at 5 to 8 knots, which corresponds to a land speed of 6 to 9 mph (10 to 14 kph)—slower than a bus on a congested street. Modern vessels, however, can reach and exceed 20 knots (23 mph; 37 kph); the speed of the large Staten Island ferries is 15 to 17 knots (17 to 20 mph; 27 to 32 kph). It is an inescapable fact of hydrodynamics that any vessel of a given size and geometric configuration has a limiting speed beyond which power input requirements become unworkable. The figures given here represent the capabilities or normal displacement-type ferryboats, even though advanced designs that take advantage of partial planing can reach 30 knots (34 mph; 54 kph). The absolute ceiling is somewhere below 35 knots (40 mph; 64 kph) for semiplaning displacement vessels, involving considerable fuel consumption and the use of very powerful engines at that.

Much attention has been paid to fast ferries recently—any vessel that can exceed 25 knots (29 mph; 46 kph) at regular cruising speed. This is possible with catamarans, which can operate at 25 to 35 knots (29 to 40 mph; 46 to 64 kph) and routinely approach 40 knots (46 mph; 74 kph) or better as a maximum speed. Some claim to be able to reach 60 knots (69 mph; 110 kph) with special propulsion arrangements—the multihull ferry between Buenos Aires and Montevideo has a loaded speed of 57 knots (65 mph; 104 kph).

The consideration of maximum speed capabilities is only relevant for service with long line-haul runs, where head-to-head competition with land-based modes is an issue or measurable time

[7] Some specialists insist that a true *ferryboat* can only be a bidirectional vessel; everything else is a conventional marine vessel. Such a strict attitude will not be taken here: anything that ferries people or vehicles is a ferryboat. It is, however, a *boat, vessel,* or *craft;* it is not a *ship.*

[8] Urban rail moves just about as fast today as it did 100 years ago; street traffic creeps at about the same rate whether it is congested by horse wagons or motor vehicles.

savings can be accumulated on long runs. The problem is different on short runs: a marine vessel cannot get up to top cruising speed as quickly as a bus or metro train; it requires some time to leave the dock and overcome the initial inertia of a large mass; and, if the spacing between stops is short, it may have to slow down again before the theoretical maximum speed is even reached. Under these conditions, travel time is governed much more by the dispatch with which turnarounds at stops are accomplished, rather than the mechanical capability of the vessel.

Hydrofoils operate in the 40-knot (46-mph; 74-kph)-plus range, but higher speeds can be reached as well—up to 55 knots (60 mph; 100 kph). The ride quality is generally quite good, but significant waves will cause vibration in the craft. Hovercraft can maintain speeds up to about 40 or 45 knots (46 to 52 mph; 74 to 83 kph), but beyond that there are problems in controlling the air-support bubble. Surface-effect ships can operate at higher velocities.

Closely associated with the speed considerations is the question of the power plant in use and the method of propelling the craft. While some steam engines can still be found in a few remote locations, the engine of choice has mainly been a "marinized" diesel engine (or two). Current models are very reliable, easy to maintain, and offer good performance and fuel efficiency. They are not always very clean in terms of emissions, but those problems can be mitigated when necessary. Gas turbines are in increasing use because of their higher performance capabilities, especially for high-speed vessels and those with water-jet propulsion. Even though the technology is not fully perfected, fuel efficiency is not particularly good, and severe noise levels are generated, they are encountered with some frequency on the longer routes in Australia, Italy, and the United Kingdom.

The push against water to achieve motion is achieved by conventional screw propellers, or increasingly by water jets. The latter are cylindrical devices that suck in and expel water with force (similarly to the Jacuzzi pumps in bathtubs), and they have the advantages of higher efficiency and lower vulnerability to underwater entanglement. They also take up less space and therefore reduce the total draft of the vessel. Other choices are cycloidal propellers ("egg beaters" or vertical vanes on a rotating plate), surface-piercing propellers, and stern drive units. The propulsion devices are frequently mounted at both ends of the boat to achieve maneuverability, which is a highly desirable characteristic for ferryboats that have to operate in tight spaces. Some are able to turn around within their own length and move sideways.

The last, and perhaps most important, measure of a ferryboat is its passenger-carrying capacity. The usual vessels on the market today are able to accommodate 50 to 400 riders. Larger craft are certainly possible with monohull or catamaran design, and there are vessels rated to carry 600 to 650 passengers (plus cars), usually with water-jet propulsion. Because of the hydrodynamics in play, hydrofoils are limited to the smaller sizes. The 300-ft (91-m) Barberi-class ferries (1981 design) on the Staten Island run are rated for 6000 passengers (with seats for 4850), and the supporting Austen-class craft (1986 design) can carry 1200 patrons (see Table 16.2). Boats in long-distance service, particularly those operating on open seas, are larger yet because they frequently carry many vehicles, not necessarily more passengers.

Vertical Clearances

The *draft* of vessels (*draught* in British English, the classic maritime language)—the vertical dimension below the waterline—is an obvious critical concern in any navigational activity. Many harbors have to be dredged continuously to maintain adequate water depths for normal operations. It is not likely that such costs could

Table 16.2 Ferryboat Specifications

Model of Boat, Manufacturer	Type of Vessel	Length, ft (m)	Beam, ft (m)	Draft, ft (m)	Speed, knots	Passenger Capacity
SeaBus	Catamaran	112.5 (30.3)	41.5 (12.65)	6.7 (2.03)	13.3	400 seats 600 total
Barberi class, Equitable	Monohull	300 (91)	70 (21)	20 (6.1)	15	6000
Austen class, Derecktor	Monohull	207 (63)	41 (12.5)	15 (4.6)	15	1200
Sun Eagle, Crowther	Catamaran	96 (29.2)	31 (9.5)	4.9 (1.5)	32	190
INCAT, Gladding-Hearn	Catamaran	121 (37)	33 (10)	6 (1.9)	35	300 inside 350 total

be justified just for ferry operations, but it is possible to find vessels that can operate under almost any water-depth conditions (see Table 16.2).

The smaller boats may require not much more than 2 ft (0.6 m); rarely will a ferryboat draw more than 7 or 8 ft (2.1 or 2.4 m). The same is true for catamarans; the use of water jets instead of propellers will reduce the draft by about 2 ft for comparable vessels.

Hydrofoil boats, if the foils are not retractable, will have considerable draft when at rest, which may occur, of course, anywhere along a route. Hovercraft, on the other hand, require no depth because they skim over the surface, except that several feet of draft should be available in case they become incapacitated and must float on the surface.

The other side of the vertical clearance coin is the available height on top of the water surface, i.e., to the undersides of bridges. Again, it is not likely that any bridges will be rebuilt to accommodate ferry operations; if there are movable bridges that have to be opened on request to allow passage of watercraft, such a situation would not be workable with frequent ferry operations that would repeatedly interrupt surface traffic. Therefore, the choices are to use suitably low-profile vessels (such as the water buses operating under the historic bridges of Amsterdam), use craft with collapsible superstructures or masts that can be lowered temporarily (as is done on some sightseeing routes), or seek other paths between terminals (which usually represents a fatal flaw in the viability of ferry service).

Landings

The interface between water and land is a critical threshold in ferry operations and is frequently its weakest link. First, issues of personal danger to careless passengers will be present; second, the efficiency with which the transfers by patrons can be made will largely determine the quality and schedule responsiveness of the service.

In situations where the water level remains at a permanent elevation, it is a relatively simple matter to provide mooring accommodations and a gangplank to walk across. The principal operational choices are presented by the vessel:

- *Bow- or end-loaded* vessels require that the boat be positioned perpendicularly to the quaywall. If there is any current in the waterway, greater navigational skills will be required, or guide pilings will have to be provided (such as the funnel-like slips of some ferry facilities). The vessel will be kept in place during the minutes required for disembarking and loading by mooring ropes (or other devices) or by the propulsion systems pushing against the shore.

- *Side-loaded* vessels, i.e., boats having entries/exits on the broadsides, require mooring ropes, even if the dwell time is short. Lateral operations are the only choice if the currents are strong or unpredictable, allowing safe arrival and departure of the craft. Boats that are expected to operate sometimes in rough seas have to be built with watertight enclosures along the sides, which may complicate boarding operations somewhat.

Given the fact that shoreside facilities are frequently of different sizes and configurations, there is an advantage in having boats that can accommodate themselves to the various characteristics of landing places.

If the waterways experience tidal changes in water level, the landing facilities have to be able to provide reasonably level or mild gradients for the access gangways or ramps. This can be accomplished in two basic ways:

- Provide *movable bridges* that are fixed at the landside, but swivel up or down depending on the elevation of the water at any given time. This is done with higher-volume operations or when vehicles have to be driven on and off. Large ferryboats and converted military landing craft may have the movable ramp built into the vessel.[9]

- A simpler approach, suitable for relatively low volumes of passengers, is to moor a *barge* that floats up and down with the tides along the shoreline.[10] Its deck should be at the same level above water as that of the ferryboat (usually about 5 ft; 1.5 m), and long ramps or gangways supported by rollers at the bottom end adjust their gradient automati-

---

[9] The large ferryboat *Estonia* sank in a storm in the Baltic Sea in 1994 because the bow opening was not closed properly.

[10] The landing barge may be tied to the bulkhead wall, allowing it to move up and down, or it may be "nailed down" by vertical spud piles that are sunk in the river bottom but allow the barge to slide up and down along them.

cally to the vertical position of the barge. Loading facilities for high passenger flows may also be floating (such as in Vancouver), but they involve rather elaborate arrangements beyond a simple barge.

Again, since boats come in various sizes and shapes, the landing facilities should have enough flexibility to reasonably accommodate all of them.

Hovercraft facilities have a different configuration. The vessels ride up on a gently sloping apron and come to rest on land, thus providing direct passenger access from all sides.

As is the case with all terminals and transit stations, proper land access and passenger amenities have to be provided. These encompass not only normal street linkage and sidewalks, but preferably also bus routes with nearby loading bays, pickup and drop-off lanes for automobile passengers, and possibly park-and-ride lots. Passengers require at least a shelter for waiting since weather conditions on the water can become rather raw; with high-volume operations the passenger facility may have to be a full-scale station with a waiting room, ticket sales, concessions, and food services.

Since waterborne operations are already handicapped by relatively slow trip times, the guiding principle in all terminal design and operation has to be quickness and dispatch in moving the vessels in and out. There still appears to be some uncertainty about the extent to which full-access provisions under the Americans with Disabilities Act apply to ferry operations, particularly if no federal funds are involved.

If vehicles are accommodated, the terminal area has to provide marshaling lanes to arrange waiting cars, which then can be loaded on the vessel quickly by rows (see Figs. 16.3 and 16.4).

Repair, Maintenance, and Fueling

Ferryboats are found in a great variety of sizes and configurations, but their components and elements are almost always of standard design, only rarely incorporating unusual power or navigation devices. As such, they can be repaired, maintained, and cared for at normal maritime shipyards or service establishments, which are likely to exist in places with extensive waterway systems. Only very large operations may find some advantage in providing their own facilities.

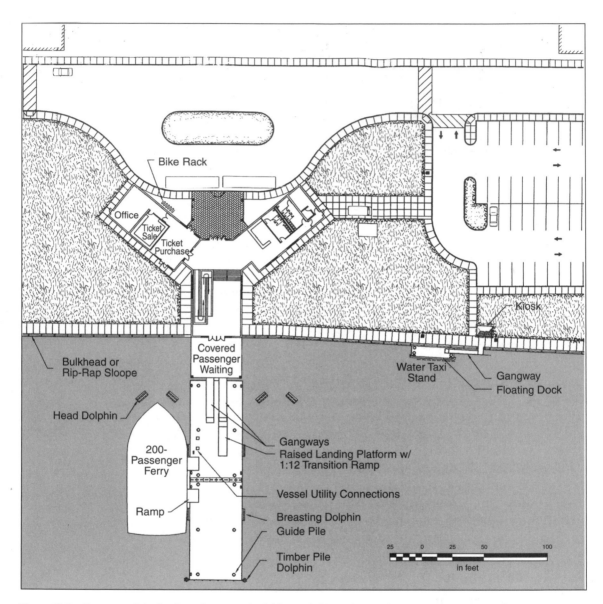

Figure 16.3 Conceptual design for a ferry terminal. (*Source:* Rhode Island Transportation Authority.)

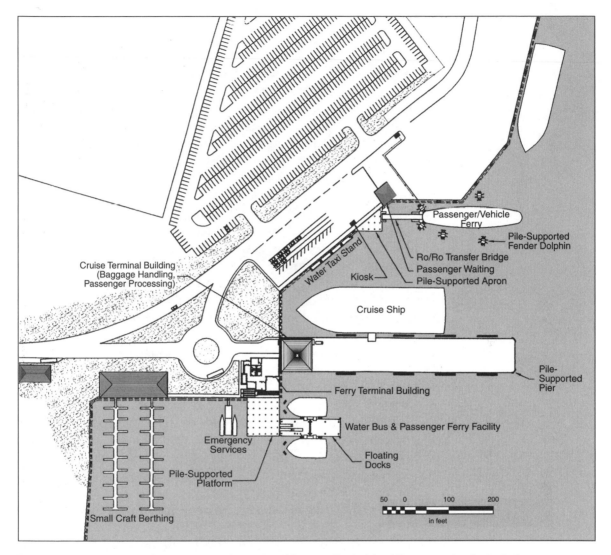

Figure 16.4 Conceptual design for a multiple terminal. (*Source:* Rhode Island Transportation Authority.)

Regulators

Any waterborne operation within the United States is controlled by a remarkable array of agencies, frequently with overlapping authority. This is not surprising, since serious issues of public safety and the efficiency of vital commercial activities are involved, and there is a long accumulated history regarding responsibilities and jurisdiction. Waterways serve not only maritime transportation; they are also relevant for recreation, flood control, power generation, potable water supply, marine ecology, and visual environmental quality.

The U.S. Army Corps of Engineers (ACE) has jurisdiction over all navigable waterways,[11] as established long ago by Congress. This agency controls channel locations and depths; it is responsible for any physical improvements and changes to the waterways. The channels are public rights-of-way, accessible to all, and any construction of facilities that involves altering the shoreline or bottom of the waterway requires ACE permission and probably an environmental impact statement. It is an elaborate process, not undertaken lightly.

The U.S. Coast Guard is responsible for the safety and proper performance of waterborne activity. The Coast Guard certifies all vessels as to their permitted use, crew requirements, and safety features, and it makes inspections and enforces applicable regulations. The register of record is *Merchant Vessels of the United States,* which includes ferryboats. There are *captains of ports* who oversee operations in busy areas.

To protect the American shipbuilding industry, which has been severely impacted by competition abroad, national legislation—the Jones Act, or Merchant Marine Act of 1920—requires that all vessels used in domestic cargo and passenger public service be built in the United States. While most of the advanced high-speed designs have been developed elsewhere (particularly in Australia), this is not a major issue because many American shipyards have been licensed to build them. But it does cause much unhappiness in other shipbuilding nations.

Next, there is the matter of ownership of waterside landing sites. If the operating enterprise has full rights on such properties with access from public streets, the problems may be minimal. Frequently, however, the shoreline may be under the jurisdiction of the municipal government or some other public body. In those instances, arrangements have to be made for use under a lease or joint operation permits. Fees will then usually be involved.

If a private enterprise provides service to the public by selling tickets or gaining compensation in any other way, it will have to obtain (and pay for) a franchise or license, as is the case for any other transportation operation. The issuing authority rests with

---

[11] If a boat can move along a waterway, the waterway is *navigable.*

the municipal or state government. The City of New York, for example, may grant *operating authority* through the city council and collect an annual fee for each specific service, or the Department of Transportation may issue a *temporary permit* with a monthly charge of $50. If public landing facilities or terminals are used by private companies, *landing rights* have to be secured, and most probably a *landing fee* will have to be paid—quite similar to the arrangements at airports and anywhere else where services for sale have to use somebody else's property.

The final addition to this list of control and financial elements is the inclusion of urban ferry operations under federal transportation assistance programs. This means that the acquisition of vessels, their operation, and the development of shoreside facilities are eligible for assistance under various programs, primarily as contained currently in the Transportation Equity Act for the 21st Century (TEA 21). It is interesting to note that ferry projects may be placed under the highway sector (Federal Highway Administration, Ferry Boat Discretionary Program) or the transit sector (Federal Transit Administration, Transit Ferry Boat Program). Up to 80 percent of the construction costs of ferries and terminals are eligible for federal assistance, including facilities that may be privately owned but provide a public service. In the states of Alaska, Washington, and New Jersey, as well as at some other locations, waterborne links are supported as components of the national highway system.

Reasons to Support Waterborne Modes

Use of Open Channels

Almost all the navigable waterways in American metropolitan areas carry much lower volumes of maritime traffic than they once did. Cargo facilities have moved to new consolidated port locations, taking advantage of large tracts of land outside urbanized districts for goods handling and unconstrained landside access. Total volumes have dropped, and the ships are much larger, thus freeing space on access channels even in busy ports. There is coastal and inland shipping by barge to many dispersed locations, and there certainly is recreational boating from marinas and other tie-up places, but much space on the water is available—frequently, it is the only unused or underutilized surface within the boundaries of contemporary cities.

Cost Efficiency of Connections

Most ferry operations have been instituted as the most economical means of providing necessary linkage between points (as compared to landbased crossings, such as bridges and tunnels). When demand volumes increase, it may be reasonable to build a fixed facility. On the other hand, when that facility becomes overloaded, the cycle may have to be repeated again, as has been shown by experience in several American communities.

Economy of Operations

If high speeds are not required, the physical propulsion of any marine vessel of any size consumes the least amount of power per unit of weight carried. A ferryboat is perhaps the ultimate high-occupancy vehicle available.

Safety

The safety record of properly managed ferry operations is remarkably good. Collisions do not happen when modern navigational aids are used and sensible practices are followed. There are also strict rules, inspections, and licensing requirements. This does not mean that the operations are foolproof, as attested by the badly overloaded and unseaworthy boats that capsize and lose hundreds of passengers now and then in some developing countries, but the quality of performance is subject to control.

Passenger Amenities

There is considerable pleasure in taking a boat ride, and this aspect should not be ignored in planning waterborne services. Even if the trip is made every day, the visual and psychological attraction does not disappear, and it can be an invigorating experience time and again. This is particularly true if the patrons keep in mind the quality of the alternative choices. To maintain the level of quality, however, the comfort and sanitary aspects of ferries should be under continuous monitoring. There are a number of waterborne operations of the "boutique" type, where exclusivity and high service levels are major demand-inducing features, but at a cost. Yet, these comfort features by themselves will not be enough to maintain patronage if the service cannot also show a travel time advantage.

Comfort features encompass climate control, airline-type seats, low noise levels, fully equipped toilets, perhaps the availability of telephones and TV monitors, some food and newspaper services, and perhaps a full bar. At the base level, passengers should be able to hold a coffee cup, read a newspaper, and maintain a conversation while the boat is moving.

Above all, the ride has to be smooth, because it will not do to have seasick passengers.

Flexible Configuration

Ferry operations usually do not involve major nonrecoverable investment. There is no construction expense associated with the right-of-way (assuming that the channel is maintained under normal harbor and waterway programs), the terminals are mostly simple facilities (parts of which can be moved around), and the vessels can be placed on the secondhand market if they are no longer needed. Route configurations can be changed easily, and this feature of ready experimentation is indeed a significant asset in developing a responsive service. One can break it in gradually, and progressive adjustments are possible and frequently advisable. Turnkey services can be chartered by engaging private operators.

Resiliency

Waterborne operations are much more immune to disaster events than are fixed land-based systems. For example, ferries carried the emergency loads immediately after the 1906 earthquake in San Francisco, and ferries were able to respond quickly in evacuating people after the September 11, 2001, World Trade Center attack. It is well to keep in mind that waterborne modes have vulnerabilities and recovery potentials that are different from (not necessarily greater than) those of land-based modes. Rail service, for example, will be little affected by a severe rainstorm or blizzard, while ferries will be. On the other hand, once the storm is over, boats can resume normal service immediately, while cleaning up and drying out may be necessary on land.

Goods Movement

Historically, ferries have carried much cargo, either as special operations (the lighters of New York Harbor) or as a part of routine service. They could do that again in urban situations where the highway systems are so congested that the economic performance of conventional distribution activities is severely impaired.

Reasons to Exercise Caution

Slowness

Because of friction against the hull and the need to overcome the tension of the air–water interface, it is difficult to reach high speeds in a boat, except with extraordinary power input and specially designed vessels. It is also not reasonable to expect quick acceleration and deceleration with any watercraft, compared to land vehicles on good pavement or rails. Nevertheless, the traditional image of the plodding slow boat no longer holds. Thus, ferries can be quite competitive in this realm if the corresponding land path is most circuitous or severely congested. This, of course, is not an infrequent situation on metropolitan networks today.

Door-to-Door Accessibility

It is still possible to moor one's gondola (actually, a motor boat these days) at most front doors in Venice, but even Amsterdam and Bangkok lost that ability some time ago with industrialization and automobilization. It is a fact that in American cities very few origin and destination points are directly on the waterfront. Historically, people-oriented activities were placed away from the neglected and unattractive shoreline, and this attitude has started to change only relatively recently. This means that almost all trips by ferry require access and distribution systems at both ends. Two additional operations are thus usually involved, with accompanying time consumption as well as the inconvenience of transfers and waiting. Since every traveler chooses modes and paths with total door-to-door time in mind, these considerations become critical (but not always fatal) factors in the successful operation of waterborne modes, as long as the total trip effort remains comparable to that of the other options.

Weather Conditions

Ferryboats are sturdy and reliable, but they are not completely immune to bad weather, particularly if open stretches of water have to be crossed and if the boats are small. Fog, extreme winds, and blizzards will interrupt service. Even if this does not happen frequently, standby alternate choices have to be considered— probably straining the capacity of parallel conventional transit services during such events.

Flotsam and Jetsam

A conventional boat moving slowly can push any floating debris out of the way without damage, but this may be a problem for fast vessels if the waterways are not clear. A hydrofoil, for example, that hits a log will not topple over but may lose the struts that support its foils. Ice used to be a significant problem in some cities, but with the general warming of the climate this tends to be a smaller issue today. However, drifting floes call for care in navigation.

Wake or Wash Disturbance

As has been mentioned before, vessels moving at any respectable speed produce bow and stern waves that travel with the boat and move out laterally. Such turbulence can erode or damage the shoreline and marine structures and throw smaller craft violently against bollards and quay walls. It interferes with navigation and can capsize boats. Moving at a suitable speed is not only a matter of marine courtesy; violators of proper practice are liable for damages. Thus, implicit speed limits exist on waterways, even if specific velocities are not stated, and they are enforced. This is a serious practical matter, because the theoretical maximum speeds that some vessels could attain are meaningless under real-life conditions on busy channels.

Environmental Impact

Due to the relatively limited presence of ferries on local waterways, their environmental impact has not yet received much attention. Marine engines can, however, produce considerable noise levels if they are not properly muffled. This is particularly the case with the more exotic power plants, such as gas turbines and aircraft-type propellers. Excessive noise conditions may affect passengers inside the boat, and care should be taken that normal conversation is not affected. Noise may also present problems outside the vessel, particularly regarding shoreside land uses. Specific standards have not yet been adopted, but a limit of 82 decibels at 50 ft (15 m) appears to be reasonable (New Jersey norm).

Exhaust emissions from diesel engines should also be of some concern, and it is likely that specific regulations will be developed or existing controls will be extended to cover this sector.

St. George ferry terminal in Staten Island, New York City.

Security

Given the times that we live in, the issue of security may loom larger for watercraft than for surface vehicles, particularly with vessels that carry trucks.

Matching the Technology to the Service

The use of fast ferries involves a major leap in all costs. This can usually be justified only with long routes and continuous utilization of the vessels with adequate payloads.

Application Scenarios

Designing a ferry system appears to be simple process. After a waterway is identified that seems to offer an opportunity to connect promising zones with sufficient trip origin and destination volumes, suitable landing places, can be found, a type of boat can be selected for the fleet, and a schedule that would attract ridership can be structured. Such a trial-and-error approach has been attempted repeatedly, assuming that waterborne operations have an intrinsic magnetism that will generate patrons. This improvised process usually does not work, because the novelty of the water trip wears off quickly, and commuters will continuously review the available choices.

In most instances, the necessity for convenient landside access at both ends of the trip is a critical element. For example, it has been suggested repeatedly that a ferry service along the New York side of the Hudson River should work well, because residential and commercial districts with intensive activity are quite close to the waterfront. The problem is sharply illustrated by a promising stop at the new Riverside South development (at the end of the streets of the West 60s), which is only some thousand feet inland horizontally. The obstacle is the fact that the regular street level is up to 60 ft (18 m) above water level, and climbing up the equivalent of a five-story building on the journey home will discourage most patrons. Minibus shuttle services could be provided, but, unless they operate at very frequent intervals, the total trip becomes cumbersome and time consuming (and also costly). The nearby parallel subway, regardless of its shortcomings, may remain the preferred choice.

The full planning process should, therefore, rely largely on a comparison of options, since it can be assumed that the ferry system is to be added to an existing transportation infrastructure in American communities.

First, it is a matter of delineating *tributary zones* for each stop. These consist of walk-to access (say 10 minutes or 2500 ft; 760 m), inland transit service (say within a 20-minute radius), and automobile access (say a 15-minute drive). It can then be assumed that the potential travelers during any given period (the peak A.M. period, for example) will compare the *total* travel time from their residences to their final destinations using the waterborne and the land-based service. Their choices will be heavily influenced by this time difference, but there are other considerations in play as well.

The other major component is the cost, which is difficult to compare to personal time—unless a value of time can be defined that allows direct mathematical calculation. (Value of time when traveling = 80 percent of salary rate when working?) Points in favor of the waterborne service can be added for the attractiveness of the experience; points may be subtracted if inclement weather is a common occurrence. Finally, even if everything comes out equal, people tend to stay with services that they are used to, not seek new challenges.

The conclusion from all of this has to be that planning for such rather unusual transit modes as ferries is not an exact science. Thus, experimentation is not an approach to be dismissed. Some further insights might be gained from a review of a few systems that are doing quite well at this time.

Vancouver, British Columbia

The SeaBus in Vancouver, British Columbia (Fig. 16.5), has received as much attention as any new transit system in North America, and for good reasons.[12] Two special vessels (the Otter and the Beaver) operate across Burrard Inlet between the city center and North Vancouver, a residential district. The run is about 2 mi (3.2 km), and it can be accomplished in 12 minutes, with a

[12] Numerous articles on SeaBus can be found in periodicals; current information is available on the Web at coastmountainbus.

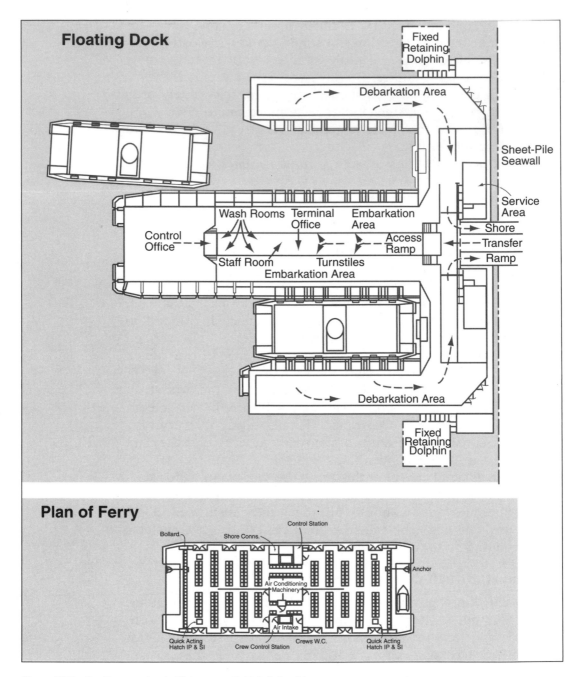

Figure 16.5 SeaBus service in Vancouver, British Columbia.

reasonable round-trip time of 30 minutes for each vessel on regular schedule. The carrying capacity of the boat is up to 600 passengers (400 seats). The normal volume on workdays is about 14,000 to 20,000 passengers, but the record for a single day is 42,000. This was achieved by maintaining 6 crossings per hour all day long during a festival; no fares were charged, and effective crowd control was implemented at both ends.

The remarkable efficiency of the SeaBus is undoubtedly due to the design of the vessel and its interface with the landside facilities. The boats are double-ended catamarans that can shuttle directly between the two terminals; they are fully enclosed, with six sliding doors on both sides. The propulsion system consists of four through-the-hull drive engines that can attain 13.5 knots (16 mph; 26 kph), but also stop the vessel in 2 boat-lengths. A single control cabin with 360-degree visibility allows one captain to operate the boat continuously (the crew is four people). Wheelchairs and bicycles are provided for.

When the ferry enters the slip (of which there are two in each terminal; see Fig. 16.5), one side opens, and passengers disembark in that direction along exiting walkways; the second set of doors opens after a short delay to admit the waiting patrons from the separated entrance areas. It is claimed that this process can load and unload 400 passengers in 37 seconds. A more realistic time in routine operations is 90 seconds—as good a performance as can be expected, even in rail rapid transit. The E-shaped terminals float and basically consist of pathways to keep the flows of entering and exiting passengers separated. Protection against the elements is provided, and the spaces are equipped with passenger services and amenities. There are direct connections to a network of bus services and the automated SkyTrain line at both ends of the water route.

When the SeaBus system was first implemented in 1977, the determining factor was economics: another bridge at that time would have cost at least $300 million, but the total bill for the ferry system was only

Vancouver SeaBus vessel entering its slip.

$35 million. A private ferry had operated on the same link many years previously, but it had been abandoned with the construction of bridges across the inlet. As North Vancouver continued to expand and demand increased, the restoration of ferry service in a modern format was an effective response. The bright orange color of the vessels and the blue strobe lights emphasize the proud visibility of this service.

San Francisco

The growth, demise, and rebirth of ferry services in the San Francisco Bay Area very much mirror the events in New York harbor. It started in 1850 with regular service (twice a week) from Oakland to San Francisco, although less formal operations existed even before that time. Services were instituted in many directions, connecting the growing communities on the bay. A tradition was established by equipping the boats as clubhouses, offering entertainment and relaxation for those who became used to the spectacular marine scenery. In the 1930s, some 50 ferries could be counted on the bay's waterways at any given time. As the Golden Gate Bridge (1937) and San Francisco–Oakland Bay Bridge (1936) were constructed and automobiles became dominant, the passenger ferries gradually were abandoned, and East Bay service was gone by 1939, although a special linkage to the Southern Pacific rail terminal in Oakland lasted until 1958. Indeed, since the toll income from the Bay Bridge was critical for the repayment of its construction bonds, the state legislature prohibited any competition by ferries. However, the ferries showed their utility during World War II by carrying cars and providing service during several transit strikes.

In the 1960s, traffic congestion on the bridges increased, and other commuting options were explored. After some experimentation, this eventually brought the opening of regular ferry operations under the Golden Gate Bridge Highway and Transportation District (in 1970), again connecting Sausalito and Larkspur in Marin County with downtown San Francisco. This service was not only successful functionally; it also represents one of the most attractive public transit rides in any American city. (The bar on the boats may be a contributing feature.) The crossing could be accomplished in 35 minutes (25 knots; 29 mph; 46 kph), but disturbance of shore-side facilities by the wake required slower operation. The original use of gas turbine engines had to be stopped as well. All this had some impact on ridership, but the volumes keep growing.

The private Red and White Fleet entered the field as well, with service from Sausalito and Tiburon, and in 1986 from Vallejo at the north end of the bay. The latter was started as a recreational service, but the 60-minute trip was found to be most competitive with regular highway travel. As ridership grew, the issue of public support for the vital service became the subject of extensive local debate. Along the way, experiments with hovercraft and hydrofoils took place, but they encountered obstacles due to the Jones Act and cumbersome local regulations. Likewise, haulage of trucks was considered as another means of balancing demand with the limited capabilities of land-based crossings.

It is of considerable significance that public agencies in the Bay Area continue to explore the expansion of waterborne systems, and the overall attitude is favorable. For example, in 1999 there was an effort under a blue-ribbon task force to structure water transit options across the bay. Out of this came a plan for the phased development of a system that would encompass some 35 to 40 terminals for more than 30 routes serviced by 120 vessels, to carry 25 to 30 million passengers each year. In addition, there were integrated recommendations for two separate cargo terminals and five remote check-in airline terminals to be connected by special ferries. Implementation has not happened yet, but the options have not been foreclosed, either.

A San Francisco Bay Area Water Transit Authority—the first of its type in the country—was established in 1999 to implement a long-range plan for a comprehensive system.[13] Much attention is being paid to the identification of responsive markets and environmental concerns. Cost-effectiveness is to be examined in a regional context in comparison to other modes, but operational subsidies for waterborne operations are expected to continue.

Capacity and Cost Considerations

When opportunities for new or expanded ferry services are examined, the capacity of the channels is almost never the issue—waterways, if they exist, are available for use at no charge, and they are mostly open. The determining factors are the size of the vessel and the throughput capability of the terminals.

Boats can be made in any size; the largest existing capacity is 6000 passengers. Since the boat market is not very large, it is not

[13] See www.watertransit.org.

likely that suitable vessels will be immediately available. There may be some secondhand craft idle, but most likely construction orders will have to be placed with shipyards that have experience with the appropriate types of vessels. Completely new designs are possible in such instances, tailor-made to the requirements of any given place, but manufacturers usually have prototypes that have already been developed and tested. Replication of these models will expedite the acquisition process and save some costs. Most of these boats are in the 100- to 400-passenger range.

There are a few notes regarding the passenger capacity of vessels. First, capacity is determined by the U.S. Coast Guard after rating any boat, and this volume has much to do with the entry/exit arrangements, interior layout, and other space considerations, not just the number of seats provided. Corresponding safety elements and facilities are also defined. The second consideration is that passenger capacity should be measured under two conditions:

- The *unconstrained capacity* that would be available in good weather, with some patrons standing and using unenclosed deck space
- The *all weather capacity* that would be available when passengers seek sheltered space and avoid external sections of the boat

The travel time between any two points, and consequently the total time-distance of any journey, will depend on the cruising speed that the vessel can sustain, accounting for slowing down when approaching a stop and leaving it. It has to be kept in mind that no boat should operate at its maximum full-throttle speed because of excessive fuel consumption and the strain that it generates on the mechanical systems, and that the speed may have to be considerably reduced when traveling on busy and tight waterways (due to the wake), probably below 25 knots (29 mph; 46 kph). The same caution applies in rough weather.

The overall service performance depends, of course, very much on the rapidity with which vessels can be processed at terminals and stops. This should not be a matter of physical capacity, because, if shoreside space is available, the necessary number of mooring berths and slips can be provided to accommodate vessels in parallel—just as a large bus terminal would have multiple loading lanes.

It is difficult and it may be misleading to generalize about cost estimates associated with waterborne services. Each situation is different; various service patterns and schedules will be in play, and the vessels probably will have to be built with local specifications in mind. The capital cost of typical vessels will range from $0.5 million to $10 million, and may be much more for advanced craft. The cost of terminals spans even a greater range. It may be nominal, if existing facilities can be used (there will be landing charges, however); it may also reach $10 million, as some studies have estimated in preliminary evaluations, or exceed that level if major stations have to be constructed.

The principal elements of operation and maintenance costs are fuel consumption and crew. The former is a direct function of the type of vessel used and its power plant, usually expressed as gallons of diesel fuel per hour. It will range over a wide span, depending on the size and type of boat. Lubricating oil expenses would be about 10 percent of the fuel costs. U.S. Coast Guard regulations require that public-service vessels be regularly dry-docked for full inspection and maintenance, thus taking them out of service. Total annual maintenance costs may represent about 5 percent of the acquisition cost of the vessel. There will be considerable insurance costs.

Crew size is also determined by the U.S. Coast Guard certification process, with limits on the allowed lengths and sequencing of shifts. The crew will include a licensed captain and several seamen, usually with an additional deckhand for each 100 passengers. Vessels moving at more than 30 knots (34 mph; 54 mph) will have to carry a licensed mate as well. Land and managerial staff represent further staffing requirements.

Conclusion

As a means of urban transportation, waterborne modes have the considerable benefit of taking advantage of the last sizable underutilized spaces left in urban areas. They also offer a pleasant way to travel most of the time, even on routine trips. The opportunities for providing effective service,

Front (boarding) end of a NY Waterway vessel on the Hudson River.

NY Waterway landing at Battery Park City in Lower Manhattan.

however, are usually quite limited. Undoubtedly, ferries remain effective as a means to overcome water obstacles, short of building a bridge or tunnel; they will quite probably have an increasing role in this sector as the conventional channels and crossings become overloaded.

Whether waterborne transportation can assume the role of regular transit along corridors at the metropolitan level remains an open question. The issue of whether ferries should be subsidized like any other form of public transit has not yet been explored. Evidence indicates that ferries are likely to be successful only under special circumstances, when there is a favorable convergence of many factors. It is possible, however, to identify such instances.

There is a need for a cautionary note: the fascination that waterborne modes generate among urbanists and planners today probably exceeds the amount of useful work that can be obtained from such services.

Bibliography

Roess, Roger P.: *Some Critical Aspects of Ferry Planning,* U.S. Department of Transportation, Documents Service Center, 1988.

Transportation Research and Training Center, Polytechnic Institute of New York, *Functional Design of Ferry Systems,* prepared for the Maritime Administration, U.S. Department of Transportation, July 1980. Comprehensive review of the field, reissued by the Technology Sharing Office in 1988.

Transportation Research Board: "Ferries of the 21st Century," *TR News,* no. 209 (special Issue), July–August 2000. Articles reviewing the current state of industry and prospects for the future; descriptions of cases.

Special Modes

The inventory of regular urban transportation modes, as explored in the previous chapters, is a rich one, but there is more yet. Over a time span of almost two centuries, there have been great necessities to find better means of mobility, and there has been urgency to do so quickly so that progress in development is not impeded. In many places growing districts have encountered physical obstacles that have had to be surmounted to capture new territories for continued expansion. Engineers and inventors of our technical age have been able to respond with any number of ideas, but, alas, not too many of the new devices have been able to survive the harsh tests of reality, prove themselves useful in actual applications, and persevere under changing conditions. Those that have been able to do all that are the principal urban transportation choices today. They are the workhorses.

But there is also a second group: modes that were dominant once but have been supplanted by better options, modes that respond to very specific needs without claiming much general applicability, and modes that appear to have great potential for service but have inherent limitations that are not always immediately obvious. These are the donkeys and the racehorses—interesting and sometimes useful beasts, but not all-purpose workers.

The purpose of this chapter, with a brief survey of each special mode, is to be comprehensive, i.e., to provide some information on every urban transportation possibility in the expectation that new ideas about feasible choices will emerge from time to time and reasonable judgments will have to be made again, perhaps referring to previous experience. The aim is certainly not to dismiss the unusual modes. After all, they do work, and they can be quite successful in special instances.

Inclines, Funiculars, and Cog Railways

Many cities, particularly those located on rivers with high bluffs along the floodplains, have encountered the need to move masses of people with some dispatch between different elevations within a short distance. These situations were not considered by the original settlers, but connections become vital as development expands. Faced with a hillside and having a need to transport people and goods up and down frequently, the logical solution is to stretch a rope and pull the payload along the slope. A wheel on top eases the movement, two containers at either end of the rope more or less balance each other, and, if they are placed on wheels and track, friction is reduced much more. A power source to pull the cable back and forth can be applied at any location. This is a *funicular,* as these devices are known around the world, or an *incline* or *inclined plane,* which are the names used more often in the United States.[1]

The classical funicular design consists of two passenger vehicles that have no engine or operator and are permanently attached to the same wire cable. The specific arrangement of the carrying ropes can take different configurations. The simplest format is to tie the two cars at the ends of a single cable; frequently, the loop is closed by tail ropes that maintain proper tension continuously. The operations are controlled from a single location, with the ascending and descending cars moving at the same time in opposite directions. Power input is only needed to overcome friction, accelerate or slow down the system, and balance any differences in weight between the two ends. In some instances, the

---

[1] Since funiculars at this time have entered the realm of exotic transportation modes, much information on them is not available in regular references. Specialized books do exist, however, covering every conceivable variation on rail or tram technology; a good source is the Web page *Funimag,* created and maintained by Michel Azema.

weight of water has been used to compensate for the weight of passengers. Each car may have a separate track, but more frequently there will be only a single guideway with a short section of double tracks in the middle to allow the vehicles to bypass each other. The steepness of the gradient is the distinguishing feature of funiculars, and it does not have to be uniform along the entire length. Slopes of 45 degrees are not unknown, and the record is said to be 52 degrees (in Australia). Beyond that, of course, these devices would become elevators.

The cars may have horizontal floors supported by a triangular understructure that reflects the gradient of the track, allowing level platforms for loading and unloading. More often the vehicles are built stepping along the slope, with each cabin at a different staggered elevation than the neighboring ones. The platforms also retain the general gradient, and passengers have to climb up and down at least the length of the platform. In some cases arrangements have been made to carry freight as well as to roll on loaded road vehicles.

Funiculars are effective only over relatively short distances, i.e., not much more than a mile, and not above vertical differences of a few thousand feet. They perform best as double-ended shuttles, but stations along the way are possible. Their capacity is limited by the size of the car (anywhere from a dozen to more than 100 passengers) and the time it takes to complete a round trip. The usual moving speed is up to about 20 mph (30 kph). Additional volume can only be provided by building a parallel line.

While several ancient transportation devices pulled by ropes can be identified in early periods, the historic examples were built mostly in the 1870s and 1880s and were powered by steam engines. Many other systems were added, utilizing electric motors, up until the time of World War I. They became quite common in cities with difficult topography and steep bluffs, but their utility faded as street networks expanded, as motorized traffic that had little trouble with steep grades became dominant, and as regular transit systems extended their reach. Nevertheless, a number of funiculars and inclines are still to be found inside cities, even though in most cases they have become primarily tourist and sightseeing services. Indeed, in a number of cases they have been retained as historic artifacts and working examples of earlier urban infrastructure.

The best known funicular probably is the line reaching Montmartre in Paris; the *tünel* in Istanbul is integrated as a link in the rail transit system. Haifa has a line well over 1 mi (1.75 km) long that has six symmetrically placed stations and utilizes rubber-tired vehicles. There are others in cities in Switzerland, Germany, and Austria, as well as scattered examples elsewhere in Europe, Asia, Africa, and Latin America—for example, Valparaiso, Chile, with 15 separate routes (*ascensores*).

In North America, a number of funiculars have operated or continue to operate in recreation areas; the most popular urban examples were found in Quebec City and Los Angeles until recently. Of the several inclines once operating in the Los Angeles area (including some private ones), Angel's Flight, connecting downtown to the top of Bunker Hill, had been restored as a major local feature, but is now (temporarily) closed after a serious accident. The largest inventory of some 15 inclines was developed in Pittsburgh, linking the riverfront work areas to residential districts on the bluff. Of these, the Monongahela and Duquesne inclines are maintained in service. Systems in Chattanooga, Dubuque, and Johnstown are the remaining examples.

While all these instances represent examples from the past, the fundamentally simple rope-based technology is by no means obsolete. It is being used with some recent automated guideway systems (see Chap. 15)—the access line to the Getty Museum complex in Los Angeles and the connection to Mud Island in Memphis. Other funiculars are likely to continue providing service in recreation areas, as will rope tows in ski resorts.

Cogwheel railroads have the same purpose as funiculars, but they utilize completely different technology. To move trains and cars up steep slopes (and maintain control on downgrades) where steel wheels would slip on the smooth rails,[2] a linear rack with large teeth is installed in the middle of the track, engaged by a geared (cogged) wheel on the powered vehicle that pulls up the train by rotation. There is no limit on the length of the route; the principal concern is safety devices to preclude runaway situations. The best examples are found outside cities at major recreation areas. However, there are also a few urban examples, some even of relatively

---

[2] Regular rail service can only operate with gradients less than 3 percent (an incline below 2 degrees); light rail transit can cope with a maximum of 6 percent.

recent origin (Lausanne, Zurich, Lyon, and Stuttgart).

There are a few instances in very hilly places where public vertical elevators have been placed in operation to overcome severe differences in elevation between city districts: Stockholm; Bahia, Brazil; and Lisbon, among others. Elevators have a single cab counterbalanced by inert weights.

Cable Cars

There was a brief period in the history of urban transportation

Cog railroad car climbing to peak (Hong Kong).

when powerful steam engines became workable, but they were not yet refined enough to be placed on individual vehicles. Given stationary power plants, the obvious way to transfer this energy to moving cars was to pull them with a rope. That is how cable cars operate: a rope loop runs under the pavement along a system of pulleys and rollers, there is a slot above through which a device from the vehicle grips the cable and is pulled along; the car remains in the proper position because it runs on rails that straddle the rope channel.[3]

Pulling things by rope is an ancient concept, and the ancestry of cable cars can be traced back to transport systems used in British mines in the eighteenth century. There were some early but not particularly successful transit applications— suburban rail lines in London in

The Katarina lift with a broad panoramic view (Stockholm).

[3] Cable cars obtain their propulsion power from a moving rope, but the vehicles can be individually attached and disengaged by the operator; funiculars are permanently tied to the rope.

the 1840s, steep access service to a ferry terminal in Hoboken, New Jersey, and an elevated line in Manhattan (1868 to 1870).[4] The key elements needed for a reasonably reliable system were the rope itself (so called, even though it is a steel cable), manufactured using the same technology that made the great suspension bridges possible, and a gripping device that can attach the vehicle to the rope and disengage again. Andrew Hallidie achieved this in 1869; he took out several patents, formed a construction trust, and built the first lines in San Francisco (1873). The intricacy of the mechanical arrangements spawned many inventions and patents, and most implementation efforts were associated with extensive litigation. The original transit service across the Brooklyn Bridge (1883) was a cable car.

Because one of the positive features of cable cars is their ability to function regardless of the steepness of grades, the leadership of San Francisco is easily understandable, but, in their eagerness to find a substitute for the horse, many other transit agencies quickly embraced the new technology during the next decade as well. Some 27 cities in North America built cable car lines, and they appeared also in the United Kingdom, France, Portugal, New Zealand, and Australia. Chicago developed the largest system in terms of passenger volume and size of fleet; Melbourne had the most route-miles. Cable cars enjoyed great initial success. Finally, mechanical power could be applied to street transit, it was a clean service, and it moved at twice the speed of horsecars (6 to 8 mph, or 10 to 13 kph). The commuting range was extended, making new territories developable—a critical concern at that time—and real estate values along routes boomed.

But there were problems that were characteristic of the mode itself and therefore not really amenable to mitigation and technological correction. The operating expenses were much lower than for comparable horse-pulled services, but the capital investment had to be extraordinarily high, so that only very high-demand corridors could be so equipped. A subsurface channel has to be built along the entire length of every route with elaborate cable support systems, and it has to be accessible for maintenance. The ropes

[4] Cable cars have their own enthusiasts and historians as well; the principal sources are G. W. Hilton, *The Cable Cars of America* (Howell-North, 1971) and B. J. Cudahy, "Cable Railways," *A Century of Service* (American Public Transit Association, 1982).

Special Modes 771

wear out quickly and have to be replaced several times each year on busy lines. Even a single loose strand of steel wire can snag the works entirely and require considerable effort to repair. The slots allow debris and water to enter the channel. The gripman has to have much physical strength and agility to start and stop the vehicle. There have been instances, not always publicized, of cars running away on steep grades. There are serious mechanical challenges when lines cross each other, and it is not possible to slow down when corners have to be negotiated. Cars are whipped around, which adds to the excitement of the ride but does not result in smooth performance. The worst problem is that most of the power is consumed in moving the rope, regardless of any payload. The cable weighs some 2.5 lb (1 kg) per foot, which means that on a 1-mi (1.6 km) route 13.2 tons have to be kept in motion, overcoming considerable friction before any cars are attached.

A superior replacement mode appeared very soon: the electric streetcar with a thin power supply wire became workable in 1888 (see Chap. 11). Cable car systems reached their peak in 1893 with a total of 305 mi (488 km) in the United States, but a decline soon started. Chicago closed its system by 1906 and Kansas City in 1913, leaving only operations in cities with major differences in elevation. Tacoma's service ended in 1938 and Seattle's in 1940.

That leaves, as everybody knows, San Francisco. The lines still in operation have assumed more of a role as a tourist attraction than regular transit service, but they offer arguably the most exhilarating ride on public transit anywhere in the world. Cable cars contribute dramatically to the special image of this city. The service is regarded with affection not only by local residents, but by just about everybody who experiences it, although it could not be considered a competitive urban transportation choice under normal circumstances. The San Francisco cable cars are one of the few instances where voluntary

San Francisco cable car being turned around at end of line.

private contributions toward the survival of public transit have been made; a referendum was passed in 1956 to preserve them, and soon thereafter they were designated a National Historic Landmark.

Aerial Tramways

Another way to overcome distances across rough and inaccessible terrain is to stretch cables from pylons and move suspended passenger containers or gondolas overhead. Even though the construction expenses of a bridge, roadway, or a track are saved, the capital improvements are substantial because a series of large columns, a power house, the wire cables, hanging vehicles, and intricate control and safety systems have to be built. The principal use of aerial tramways (*Seilbahn* in German, *téléphérique* in French, *teleferico* in Portugese) is, of course, in mountainous resort areas to carry skiers or sightseers. The technology is well developed and much tested, and standard systems are available from manufacturers. The same concept has been used repeatedly for a long time in freight handling on a small scale (in warehouses) or a large scale (a ropeway was the principal freight access to the Kathmandu Valley in Nepal from India).

Aerial tramways can also be found in urban environments, but primarily to carry tourists to nearby high elevations such as the Sugar Loaf in Rio de Janeiro or the Avila Mountain in Caracas. The only instance of aerial tramway use for regular public transit exists in New York City—the service to Roosevelt Island in the East River.

When major residential development was started on this heretofore institutional island, transit access was a primary issue.[5] A bridge was possible across the secondary channel from Queens, but could not be justified nor properly laid out from Manhattan. A subway extension was programmed, but opening could only be expected years in the future. A ferry connection could be made, but only with some difficulty because of currents and shore development. The ingenious, and presumably temporary, solution was to erect an Alpine cableway and call it a tram.

---

[5] The Queensboro Bridge crosses the island at a high elevation, and it had an elevator in the middle that was able to accommodate even vehicles. This was a cumbersome affair, and the elevator was removed in 1970.

The tram was opened in 1976, running from a terminal near major commercial activities and a subway station in Manhattan for 3100 ft (945 m) to a terminal on the island where local distribution buses are available (intended to be zero-emission vehicles). The two gondolas, each able to accommodate 125 passengers, operate simultaneously from both ends and make the crossing in about 4.5 minutes. Navigational clearance of 250 ft (76 m) is maintained atop the river. Service is usually at 15-minute intervals, and the regular Metro-Card can be used to pay the fare, even though a separate agency is responsible for the tramway. There have been no serious accidents as a total of more than 30 million passengers have been transported so far. Safety devices include multiple cables, standby power sources to bring home stalled cars, rescue cabins that can reach any point along the ropeway, and devices to lower passengers to the surface.

Aerial tramway to Roosevelt Island off Manhattan.

At this time, the subway connection has been completed, but there is no intention to remove the aerial tramway. It continues to provide useful and convenient service for many local residents, and it is an icon for the Roosevelt Island development. One of the popular activities for tourists who come to New York is to ride from the Big Apple to the Little Apple.

Airborne Modes

Since the 1920s, science fiction illustrations of future cities have tended to show the skies filled with aircraft maneuvering blithely among skyscrapers. Many movies imagining the future, whether bright or bleak, do the same, and personal flying scooters and large transit craft crowd the air space that, admittedly, is largely unused today. What keeps all these flying devices on paths straight as an arrow with no likelihood of crashes, chain collisions, and major unintended mayhem? Only the ignorance of the illustrators and scene designers who have never heard of wind or air currents, updrafts and downdrafts, air turbulence and eddies, not even of rain and fog.

A train maintains a fixed path because it runs along rails; motor vehicles stay (mostly) in lanes because the tires grip the pavement; even waterborne vessels cannot change positions or be dislocated abruptly because water provides some resistance. Air does not do much of that at all, and aircraft move together with the medium in which they are embedded. Local disturbances and turbulence have significant effects, and there is no such thing as a minor fender-bender involving airborne vehicles.

All these are reasons why aircraft are certified for specified types of operation, why pilots are fully licensed, why mandatory inspection and maintenance programs exist, and why strict regulations are observed in the use of air space. Significant vertical and horizontal separation has to be maintained between aircraft; no crowding can be tolerated; and variable weather conditions come strongly into play. There are rules as to when planes can fly and restrictions on how they can fly over built-up districts. At least, these are the conditions in countries that maintain responsible safety and management programs. Thus the use of airborne vessels carrying significant numbers of passengers as regular transportation modes inside cities remains a distant dream (or a nightmare).

Beyond the general safety issue, there is also the critical concern with noise. There is no airplane or helicopter available today that could operate at tolerable noise levels in proximity to people and normal urban activities on a regular schedule.

There is no need to discuss in detail the obstacles in urban use of conventional aircraft, or even short takeoff and landing (STOL) planes or lighter-than-air devices (blimps, dirigibles, zeppelins). They have their important roles, but not inside cities. That leaves helicopters or vertical takeoff and landing (VTOL) craft, which require some attention.

Helicopters are intriguing devices with interesting possibilities. Unfortunately, they are extremely energy inefficient (much air has to be moved to keep a heavy body airborne) and they generate extreme noise; they are also not inherently stable aerodynamically, and much skill and continuous attention is required to keep them flying properly. They are indispensable in police and emergency work, and better and more extensive systems will have to be developed for fire rescue from tall buildings. Civilian use, however, is a different story in American cities.

It is doubtful that even VIP service is appropriate in central areas, except on rare occasions; and private flying by individuals and members of corporations also raises quality of life issues for everybody who may be impacted by such operations. If nothing else, the noise will not be tolerated by the population, even in low-density areas. However, some of this activity under strict controls will have to be accepted. There is no way to envision public transit–type operations because of safety, noise, and premium cost considerations. Where would the landing pads be located?

That leaves a few remaining possibilities.[6] One of these is sightseeing—a very popular activity that can highlight the attractive features of any city from a new perspective. This is a matter of locating terminals where impacts on other uses are minimal, with flight paths routed over largely open space (preferably water) and the unavoidably high charges being acceptable to a reasonable volume of patrons.

A justifiable airborne service might also be fast linkage among major transportation centers. The most obvious connections would be between airports, where terminals have the advantage of already accommodating aircraft and providing noise-tolerant situations. Landing pads and glide paths are likely to present few difficulties; the selection of flightways will almost certainly require much care. If multimodal transportation centers are also located elsewhere in the city ("inland" from airports and waterways), the issues become more complex. A good case can be made for placing a heliport on top of such a structure, thereby achieving a higher degree on connectivity among various modes and nodes, but the safety and noise issues loom large. Reasonably secure and impact-tolerant alignments of flight paths would have to be found and designated because risk is involved, and there is a statistical probability that at some time a service craft will not complete its journey as scheduled.

Two cases can be briefly outlined at this point to illustrate the problems associated with helicopter operations inside cities.

In 1965, in the face of much controversy, a heliport was opened atop the 808-ft (246-m)-high tower then known as the Pan Am Building in the very core of Midtown Manhattan next to Grand Central Terminal. Regular service was made available to

---

[6] The complex specifications and standards associated with the placement, construction, and operation of heliports will not be outlined here. Material is available from the Federal Aviation Administration, trade associations, and, in many places, municipal ordinances.

Helicopter service operating from the Pan Am building in Midtown Manhattan.

the regional airports, and the spectacular ride as the helicopter left or approached the landing site amid the skyscrapers was worth the ticket charge. However, in 1977, the landing gear of one of the machines collapsed, and parts crashed to the street below. Four passengers and a pedestrian on the sidewalk below were killed. This was a watershed event in the history of urban helicopter use, and any regular activity was immediately banished from tall buildings within dense districts in New York. When another helicopter crashed at Newark airport in 1979, all further discussion of the helicopter option ceased.

Recent information from Latin America, on the other hand, indicates that cities there, particularly São Paulo, are experiencing a major boom in private helicopter use. This trend is driven primarily by security considerations, with kidnappings and attacks on automobiles reaching epidemic proportions on the massively congested streets. Wealthy people, who can afford to own these limousine replacements and to hire pilots and maintenance personnel, use them not only to commute to work but also to conduct other family business. The total helicopter fleet currently is said to exceed 500 machines (New York's inventory is 2000, Tokyo's 700), and there are some 200 helipads, many on top of tall buildings. Much of the flying occurs at night.

The private operations depend on numerous helipads that are now found throughout the cities and act as public landing sites. The air space is becoming seriously overcrowded, with controls not being particularly strict. Regrettably, it has to be assumed that corrective measures will be taken only when some serious mishaps show the need for careful control and judicious curtailment of activity. Working-class residents in these areas are not in a position to say much about upper-class noise.

Bibliography

Vuchic, Vukan: *Urban Public Transportation: Systems and Technology,* Prentice-Hall, 1981, 773 pp. Contains references to most of the special modes reviewed in this chapter. Beyond that, Web pages devoted to the various types of technology should be examined.

Intermodal Terminals

Background

All the chapters of this book are devoted to single and separate urban transportation modes, describing their capabilities and defining their actual roles in different types of communities. Yet, many reminders are included that several, if not many, modes are needed in every place to cover the mobility needs that exist locally and to provide access to the various urban activities. These modes have to act together, frequently supporting each other, to enable residents, workers, and visitors to operate within their own urban environments.

This "coming together" is expressed through transfer possibilities between modes, which may be as simple as motorists becoming train riders through regular park-and-ride lots to elaborate terminal structures where a multitude of patrons arrive and depart by diverse modes and can freely interchange among them.

The concept of the intermodal terminal thus becomes a convenient instrument to emphasize the need for system integration and to explore how this desirable state of coordination can best be achieved. The structure and all its associated access facilities will become a very visible service complex because of its centrality and role as the prime node of the entire metropolitan transportation system. Such a project has probably consumed quite a notice-

Intermodal terminal and trade center next to rail and metro stations in Stockholm.

able amount of public resources and has received well-publicized design and review attention within the community. This is where operational, safety, or efficiency problems within the larger network are likely to become most apparent.

The significance of intermodal[1] terminals has been specifically recognized by U.S. government transportation programs. The omnibus transportation act passed in 1991—the Intermodal Surface Transportation Efficiency Act (ISTEA)—contained language (as well as a separate Title V)[2] that purposefully encouraged and assisted the construction of such facilities. Perhaps this emphasis was in honor of the name of the act itself, but there were a number of projects highlighted under this designation.[3] Many involved simple service coordination, but there were also projects implementing true terminals of the form described in this chapter. The next, and current, act—the Transportation Equity Act for the 21st Century (TEA 21)—does not include such a specific designation, but similar work continues under other programs.

It can be assumed that the significance of the intermodal terminal as a key component of transportation systems is well established and does not need special promotion any more.

Airline terminals in airports are invariably intermodal facilities, their purpose being the transfer of passengers between the air mode and various land access possibilities. There is a trend toward providing airline terminals with rail service by building new light rail or even metro linkages. The Frankfurt Airport stands out by having placed a regular station of the federal rail system (*Bundesbahn*) below the concourse, but there are also others where linkages to long-distance networks have been pro-

---

[1] *Caution:* When "intermodal transportation" is mentioned without any qualifiers, the reference is usually to freight operations. This involves a complex situation of long standing regarding joint efforts by the rail and trucking industries.
[2] The National Commission on Intermodal Transportation prepared a summary report in 1994 suggesting policies for future action.
[3] See *Five Years of Progress: 110 Communities Where ISTEA is Making a Difference,* Surface Transportation Policy Project, 1996, pp. 9–17.

vided (Gatwick, Charles de Gaulle, Newark). Airport terminals, however, will not be included in this discussion because their principal planning and design concerns are not found in the intermodal area. They are to be recognized as primary gateways to cities and regions that have to have effective and direct linkages to the central intermodal terminal downtown.

Purpose and Design Criteria

The core purpose of an intermodal terminal is to expedite the interchange of travelers among modes and to tie together the local and regional systems (often including intercity operations). More precisely, these terminals function to:

- Allow entry/exit by travelers utilizing selected modes
- Provide for interchange between different routes of the same mode
- Provide for interchange among modes (local, regional, and intercity)
- Serve passengers and visitors (nontravelers) and provide space for commercial and service activities (revenue generation)
- Assist management and control of operations (ticketing, documentation, information service, supervision, staff accommodations)
- Handle various types of vehicles

The principal point about "terminals"[4] is that nothing really terminates here, except the movements of service vehicles; the essential users of the facility—passengers and some goods—continue their journey through the building and spaces, expecting as little delay and friction as possible as they move on. Of course, the terminal may also include secondary activities and establishments that are destination points in their own right and accommodations for patrons whose further travel is not immediate. By far the most important planning aspect of a terminal is the

---

[4] From the Latin *terminus*—a boundary, limit. Suggestions have been made by some transportation specialists that a better term would be "transportation interface areas," but that designation is a bit cumbersome and not likely to catch on.

arrangement of the multiple flow patterns to achieve internal connections with the best aggregate effectiveness, efficiency, and safety. Each mode has specific requirements and suitable settings that allow it to operate to its own best advantage, but its patrons have to be able to reach and transfer to the next service. Every individual trip usually consists of separate segments, each of which is accommodated by a different mode or route.

The terminal can be seen as a device (machine) that organizes and processes flows internally. But it is much more than that in any community. To emphasize its character as the principal node of transportation networks within an extended service area and the key feature of the overall system, as many modes as possible should converge on the site, public as well as private. There are also good reasons for the terminal, because of its active and intensive use, to be a proud civic building giving shape and orientation to the central district.

As has been discussed before, for transit to be used by city residents, communal operations have to offer, among a number of service qualities, convenience at the highest reasonable level. This is frequently interpreted as a "one-seat ride," i.e., an ability to move between origin and destination without changing modes or vehicles. (This is what the private automobile promises, assuming that convenient parking will be found.) Since that is usually not possible with public transit, the transfers should be as *seamless* as possible, i.e., involving the least amount of delay and inconvenience along the way.

That consideration should govern any transfer point, but it is particularly relevant to the design of the larger facilities, where interchanges can become quite complex with possible internal conflicts. It is not a question of the physical layout of the terminal alone, but should also involve such considerations as:

- *Schedule Coordination* among arriving and departing vehicles, even if they belong to different agencies, to minimize waiting for transfer

- *Unified Ticketing* so that patrons do not lose time for a routine transaction in the middle of the journey

- *Real-Time Reliable Travel Information* that allows travelers to manage their time or at least move forward with confidence that order prevails

It is a psychological fact, of some relevance to transportation planners, that travelers when in motion have a sense that progress is being made and some discomforts can be tolerated. When they are stopped, however, they not only become impatient because useful time is being lost, but they also have an opportunity to contemplate their surroundings and review their situation, usually finding deficiencies in both areas.

Multilevel local bus station with pedestrians above and U-Bahn below, Wandsek Markt in Hamburg, Germany.

Location

A simple transfer point can be placed anywhere modes cross if the only aim is to accommodate intrasystem interchanges between routes by travelers.[5] Indeed, even large facilities, such as pulse-scheduled bus transfer nodes, can be independent of the surroundings—as long as the operations are entirely "internal" and many linkages to surrounding land uses and neighboring activities need not be made. Such isolation and separation is likely to be rare, however, and the terminal becomes a facility that is closely related to the city components around it. This condition is best described as forming a *gateway*.

The gateway concept is an interesting feature associated with transportation in cities and deserves some exploration. It represents an entry/exit to a city or a district, functionally and symbolically. It is the node that receives the first leg of a longer outgoing trip (walk to the place where long-distance transport modes can be boarded), or, conversely, the place from which the last connection is made to the traveler's final destination in the city. If this is what many patrons do, to and from a central dis-

---

[5] A related example is provided by the air industry. Each airline operating under a hub system has to select a separate airport where it can dominate the local departure/arrival schedule and expedite its own internal connections, with only a secondary concern for linkage to the host city and region.

trict, the facility grows in its importance and role as an urban component.

There are two major consequences of this observation:

- High-intensity land uses and activities should surround the site as an existing inventory, thus allowing direct connection to intensive trip generators. This also means that suitable locations for a terminal in a high-demand real estate situation are not likely to be easily obtainable.

- The superior accessibility of the terminal should influence the land market in its vicinity, leading to new and intensive development. While this does not always happen, the probability suggests the need for constructive planning and land use management within the surrounding district.

The literal gateway of fortified cities during historical periods was a point where many paths converged, where controls over movements and persons were exercised, where many auxiliary activities clustered, and where the pulse of the city was most apparent. The gateway led to and from the important locations inside, and ceremonial movements (as well as hostile ones) focused on this node.

The central railroad station of the nineteenth century had the same role, in a form much closer to today's operational patterns. Multitudes of people used it every day, and it was also just about the only connector to the outside world. Traders, peasants, and people visiting family may have represented the bulk of the travel volume, but kings and opera stars also moved through the same facility and were received with proper respect and pomp. The central station up to the time of World War I was envisioned and built as a landmark structure (with a tower and a clock, almost without exception). Its ranking as prestige architecture was probably exceeded only by the cathedral and the ruler's residence in any given place. It was a building on which movements, avenues, and views focused, and the station square counted among the principal urban activity spaces of a city. The best restaurants and hotels were frequently on it or in the vicinity.

In the very large cities, served by several railroad companies, several or many prominent stations appeared because the independent operators had no inclination to cooperate and coexist in the same building. Also, since most of the travel was in and out of the city, with very little transfer of passengers among stations, there were few functional reasons to interconnect the radial rail lines. The new stations entered into the city as deeply as possible, i.e., reaching the edge of the contiguous high-density development as it existed at the time the stations were built. No expense was spared to express a corporate identity (a term coined much later) and to show the technological advancement and civic good taste of the service operator/building owner. The results are the chains of stations surrounding the cores of London, Paris, and Moscow, with fewer examples in many other places.

Old Union Station of St. Louis converted into a commercial center.

Given the central access needs of intermodal terminals, particularly if they are of the gateway type serving intense activity districts, the logical conclusion is that they have to be as central as possible in terms of their location.[6] It is not likely that the "100 percent corner" of any core area can be considered for this purpose, but the argument remains strong that patrons should be able to reach their final destinations conveniently on foot. The major constraints on an absolutely central location are the likelihood of the prevailing congestion there, as well as the fact that the terminal has to be serviced by a multitude of access tracks and vehicular lanes that consume much space but should be able to operate freely. The search for a site thus has to concentrate on a location that penetrates inside as much as possible, but has some free space in the back that can accommodate efficient vehicular access.

---

[6] It is possible to engage in a theory-based exercise searching for an optimum location. It is more likely, however, that in any given place the opportunities for sites will be relatively few and the constraints rather limiting. Under those circumstances, a pragmatic approach will be more appropriate. See, however, J. B. Schneider, "Selecting and Evaluating Intermodal Stations for Intercity High Speed Ground Transportation," *Transportation Quarterly*, April 1993, pp. 221–245.

Another consideration reflecting contemporary and foreseeable urban development trends is the probability that the single-center (mononuclear) metropolitan structure no longer governs. This suggests that the principal terminal may have to be regarded as the center of a solar system of nodes, all linked together with some efficiency. The airport will certainly be one of these "planets," but so will be several large suburban activity centers (edge cities) on the periphery.

Grand Central Terminal in New York City

The argument can be made that the direct ancestor of today's intermodal terminal is the railroad station of the earlier period. Grand Central Terminal (GCT) in New York City is a case in point.* It was opened in 1913 (replacing the earlier Grand Central Station), "nailing down" its location at the northern edge of the Midtown commercial district at that time. It turned its face and all entrances single-mindedly to the south, which became a serious problem decades later since commercial buildings continued to be erected, covering many blocks to the north. (The needs of these commuters were accommodated only a few years ago by providing the functional but rather tight pedestrian tunnels of the North End Access project.)

The operations of the terminal have undergone a dramatic change from its original purpose to today's operational needs. It was originally built for travelers who embarked on long-distance trips, some across the continent, with much ceremony and luggage. There were some hundred trains each day with sleepers and restaurant cars, although suburban service was also in operation to the growing communities within the service area in the northern part of the region. Today, only a few long-distance trains are left, but there are about 500 trains each day that carry commuters to and from their residences—with repeat trips every day, no luggage, frequently tired and stressed even in the morning, and worried about each minute on the schedule. A completely different clientele with a different purpose has to be accommodated in the same building and spaces. Surprisingly enough, it works quite well, particularly after a recent restoration and modernization project.

The rail station was built with two levels of track to provide for the anticipated train volume (a total of 67 tracks), and the entire complex was placed atop the new subway system. The original rapid transit line (1904) made a sharp turn from Park Avenue (north-south) to 42nd Street (east-west), which was later converted into the Lexington express and local lines (north-south) and the multitrack Shuttle below 42nd. Two interconnected subway stations are imme-

* It is probably the most thoroughly documented transportation facility found anywhere. Among the many books and publications, a useful comprehensive summary is D. Nevins (ed.), *Grand Central Terminal: City within the City,* Municipal Art Society of New York, 1982, 145 pp.

diately below GCT. A major crosstown streetcar line ran in front of the terminal, now replaced by buses, and there has been a thought to restore surface rail on 42nd Street for many years. Another trolley line in a tunnel below Park Avenue reached the terminal with a station at 42nd Street (closed for a long time). A number of city bus lines are in operation today on the avenues along both sides as well, but none of these services have been integrated physically with the terminal, except that stops are placed along the curbs at several locations, all of which are within a short walking distance.

Express buses, likewise, stop at many places within this district nearby, but they too are not components of the terminal itself. A number of taxi stands and service lanes have been provided, however, as a part of the building or along the curb. There is no parking within the terminal, but commercial garages are found on surrounding blocks. For a brief period, a heliport was in operation from the roof above the 59th floor of the (then) Pan Am building, until a severe crash forced the termination of this fast service to the regional airports. There was also the East Side Airport Terminal across the street, which provided convenient check-in facilities and bus and taxi service to and from the airports. It too was abandoned as of the late 1960s.

A significant feature of GCT is the viaduct for cars that wraps around the building at the second level and connects Park Avenue South and North, thereby precluding much worse street traffic congestion than is the case anyway. The many levels of the building, especially their arrangement inside, is a major positive feature that has allowed so many intense activities to operate on a relatively small site, bordered by a concentration of tall buildings as can be found in any city.

The key characteristic of the entire complex, however, is its ability to accommodate people—not only travelers but also local office workers and visitors who enter the building for any number of purposes, even if it is only to seek a shortcut between points. Some 500,000 people enter and leave GCT every day, many to take advantage of the many shops and eating places (at all price levels) that now have turned the terminal into a major commercial and recreational enterprise. The superb architecture, the many convenient portals, and above all the great hall that is one of the most splendid rooms in the city (374 × 118 × 125 ft or 114 × 36 × 38 m in size). It brings visual enjoyment to most who enter it and can lift the spirit of just about any stressed urbanite.

As old as the terminal is, it is by no means finished yet. A major and most expensive project (East Side Access) is currently under way to bring Long Island rail commuters, who now can only reach Penn Station on the West Side, into GCT and the heart of the office district as well. Beyond that, another project (Access to the Region's Core) contemplates a direct rail connection between the two rail stations allowing much operational flexibility and choices in accessibility. It probably will have to wait a while, given the current financial constraints.

A view into Grand Central Terminal in Manhattan, with its many levels of operation.

Components and Overall Configuration

If the principal point of an intermodal terminal is to bring together many modes, it would be appropriate to start organizing the facility with the heaviest mode that has the most exacting and capital-intensive access needs. In most places this will be the railroad. For many cities in North America this has been an appropriate point of beginning because the historical railroad stations are underutilized, frequently neglected, and seeking a good contemporary role, and they tend to be well located next to the activity core (but not always).

An interesting example of using the railroad station as a base is the Union Station Intermodal Transportation Center of Worcester, Massachusetts, a recently renovated facility. The old building, much admired for its architectural quality ("a poem in stone"), had been abandoned for years, reaching a deplorable state of disrepair, to a large extent because the business core had moved several blocks away from it over the decades. New development projects, however, have been placed in the vicinity recently that generate urban activity and have allowed restoration and rejuvenation of the building as a transportation center. It includes Amtrak and Massachusetts Bay Transportation Authority's rail commuter service; local, express, and inner-city bus access; and airport and taxi service. Upgraded pedestrian and bicycle paths make the splendid facade very visible again—a true civic icon.

The highest priority in an intermodal terminal belongs to *public transit*—the principal reason why the terminal is considered in the first place. The connectivity among these modes and the various routes is the key requirement. Likewise, access by nonmechanical modes (*pedestrians* and *bicycles*) is of high importance to ensure user convenience and to minimize personal energy expenditure. These paths are somewhat easier to accommodate because they take less space and have much flexibility in their placement, but they require careful attention nevertheless to make the entire complex work well from the perspective of the users.

The old rail station of Worcester, Massachusetts, before recovery as a multimodal terminal.

Then there are the various specialized motor vehicle services that include *taxis, paratransit* operations, *airport limousines, sightseeing buses,* and any other similar *special-purpose connectors*. Above all, there are the considerations of *emergency* and *service vehicle* access, which is crucial in all instances, even though actual entry may be limited and infrequent. Some terminals may include the handling of goods, at least the shipment of parcels, which will require access by *motor trucks*.

The last item on the list is *private cars*. Such operations cannot be avoided because many patrons will be dropped off or picked up by friends and relatives, and reasonably well located access and loading lanes are expected. A *parking garage* either below or above grade may also be a desirable convenience, with short-time if not long-time parking provisions; it will almost certainly represent a good source of income for the entire complex. This is usually a controversial issue, related to local policy toward car use and presence in central districts. Similarly, it might be appropriate and convenient to accommodate *car rental* agencies too, since many long-distance travelers are likely to use these facilities in our automobile-oriented communities.

A *heliport* on top of the building may be considered for premium and emergency service, provided the safety issues are surmountable. Not optional is the requirement that all paths and spaces be navigable by *people with handicaps* and *wheelchair users* so that they can move through the complex without any obstacles and impediments (Americans with Disabilities Act).

In some instances, if the terminal is truly a gateway, it may be useful to consider the possibility that on some occasions ceremo-

A network of passages connecting various modes in Hong Kong.

nial VIP travel may be a factor. Can a motorcade for the president or other head of state be accommodated? Can an entertainment star or official hero be properly greeted upon arrival?

Besides dealing with all the functional transportation elements, an important part of the design program is the definition of the extent to which the intermodal terminal will also encompass auxiliary services and activities. The range here is from a few kiosks that sell newspapers and refreshments to travelers to a full-fledged array of shopping and entertainment establishments, possibly with office floors added. Again, this is a question of gaining revenues, but also of creating an activity center at the city scale that has attractiveness beyond the immediate transportation functions. The current trends are very much in this direction. Many airport terminals, for example, are deliberately becoming shopping centers as destination nodes for the general public. Another type of activity that has been incorporated in many terminals is a visitors' center that can provide useful information to the many patrons who need guidance about the city and its facilities.

Any terminal, beyond the many options just outlined has to include a series of standard elements (as discussed further in the section on space allocation). This list encompasses:

- Corridors and concourses for unimpeded movement.
- Waiting spaces for long-distance passengers.
- Ticket and information booths.
- Rest rooms.
- Luggage storage.
- Staff accommodations.
- Control and management room.

- Customer convenience establishments (at least to buy reading material, have breakfast or a snack, and possibly places to develop film, drop off laundry, buy staple goods, etc. It has even been suggested that the terminal might include day care centers where commuters can "park" their children for the day or a shorter period.)

A final design consideration relates to the quality of architecture for the terminal inside and out and its accessways reaching out into the surrounding zones. The argument can be made that the tradition of the classical railroad stations should be continued by creating significant buildings as landmarks of civic importance and points of districtwide orientation (in a contemporary or contextual image, of course). There is also the general concern of functional attractiveness to maintain patronage. Since everybody has seen modern airport terminals, with their advanced and frequently high-tech design, it is reasonable for the patrons to expect the same visual and functional design attainment. The seedy downtown bus terminal that can still be found in many places is exactly the type of facility that is not needed.

Among the examples in the United States of major intermodal terminals is the reorganized South Station of Boston, which is served by rapid transit, commuter and intercity rail (including the substantially upgraded Northeast Corridor service), suburban and local buses with a large marshaling area, taxis, and connectors to Logan Airport. Similarly, Philadelphia's 30th Street Station accommodates Amtrak and regional rail, the subway, NJ Transit buses and rail, SEPTA local feeder buses, and taxis and shuttles.

The Journal Square Transportation Center in Jersey City, New Jersey, was built as a new facility (1975) above transit air rights by the Port Authority of New York and New Jersey. It encompasses a 10-story office tower and a 600-car garage, as well as a PATH rail transit station and more than 20 bus route connections.

The Fullerton, California, and Oxnard, California, transportation centers have received much attention in publications. The latter encompasses as its principal elements commuter rail and Amtrak, long-distance and local buses, and a substantial park-and-ride facility. One Gateway Center in Los Angeles (note the inclusion of the term *gateway*) has most of the listed facilities, plus heavy and light rail transit. Interesting intermodal terminals are also found in Indianapolis; Cincinnati; Cleveland; Akron; Harrisburg, Pennsylvania; and a number of other cities at this time. Cities that have taken significant steps in organizing good intermodal connections (usually between rail and bus modes) include Sacramento, St. Louis, Chicago, Baltimore, Atlanta, Miami, and others.[7]

Internal Flow Patterns

An intermodal terminal is an arrangement of flow paths for people and vehicles, with some attached spaces where they may remain stationary for a brief period and some auxiliary service establishments, most of it enclosed by a building skin and a roof. The design process is exactly that: to arrange the flow patterns for maximum effectiveness and then put a building shell around it all. There will be, of course, further additions and embellishments, but all those efforts should be secondary to the principal purpose.[8]

The types of terminal users will be the following:

- *Daily Commuters,* who are frequently in a hurry, will not linger, have a ticket or pass already, and know their paths very well.

- *Commuters and Local Travelers,* who are not under any time pressure and might wish to take advantage of local services.

- *Long-Distance Travelers,* who have to buy tickets, check the schedule, deal with luggage, buy supplies, get information and become oriented, and may need to wait for a connection.

- *Nontravelers,* who may be companions and well-wishers of travelers or receivers of guests, and thereby become a part of the patronage volume. The practice of accompanying travelers is no longer very strong in the United States, but in developing countries it happens occasionally that a whole village will say goodbye to somebody going a long way,

[7] G. M. Smerk, "Intermodal Makes Sense," *Bus Ride,* April 1995, p. 36.
[8] The planning and design process is described in considerable detail and with an emphasis on technical precision that will not always be attainable in actual practice by L. Goodman in Chap. 7, "Transportation Interface Areas," *Transportation Planning Handbook* (Institute of Transportation Engineers, 1992). There are separate sections on bus terminals (pp. 216–228), rail terminals (pp. 236–245), and waterborne terminals (pp. 249–251), which can be regarded as components of a multimodal terminal.

with all appropriate ceremony and festivities. (This requires space.)

- *Service Users,* who are attracted to the facility because of the nontransportation commercial and entertainment opportunities provided.

- *Seekers of Shelter,* who include the homeless, nonsettled newcomers, the idle, and the socially disoriented. They appear in every transportation terminal, with very few exceptions, because the facility may be open 24 hours, inexpensive services would be available, there are communal spaces, and anybody can blend in among many strangers. The issue is a complex and delicate one because basic human rights are involved, and exclusion from a space in open public use just because of somebody's personal appearance is not quite acceptable. Yet, these persons should not be in the terminal, and their accommodation is a challenge both for transportation and social service agencies in any community.

Returning to design issues, there are certain criteria that need to be satisfied in a general sense before precise plans can be prepared, including:

- Efficient access by all service vehicles from the outside, ensuring as much directness and safety as possible for in and out movements, and the provision of adequate and convenient loading spaces for each mode

- Maintenance of smooth pedestrian flows inside the facility, minimizing walking distances for most patrons, providing adequate size corridors and channels, removing any constraints to flow, and precluding unsafe conditions

- Provision of information that is understandable and helpful to all users, equipping the facility with signs and orientation devices that allow each patron to find his or her way without confusion

- Provision of human amenities that range from toilets to works of art, certainly including basic commercial services

The key design tool to organize these considerations is undoubtedly a flowchart that should trace the movement paths of all users with their logical starting, ending, and intervening

Access ramps from the rear of the Port Authority Bus Terminal in New York City.

points. The schematic structure of paths can be amplified with information on the expected volumes of people, as well as a time scale for movements and processing (ticket buying, waiting, queuing, etc.). A significant fact is that some arrivals/departures will be individual (random entry/exit by separate persons) and others will be bunched (batch entry/exit as, for example, the arrival of a fully loaded train).

Once the flow patterns are defined, the next step should be a careful estimate of space and corridor width requirements, as outlined in the next section. It has to be observed first, though, that the transportation functions within the terminal should be planned so that all patrons and vehicles are processed as expeditiously as possible, i.e., be kept in motion, minimizing their time inside the facility, while the desirable condition from the point of view of the merchants and shopkeepers in the terminal is to slow down the flow, making the customers linger and thereby be tempted to spend some money in the stores and eating places.

Space Allocation

The design process, assuming somewhat ideal conditions with much information available, would follow this sequence:

1. Identify all modes that will operate through or touch the terminal.

2. Establish an operational schedule for the future (the design period), i.e., how many vehicles of each mode will arrive and depart every hour of the day (perhaps every 15 min during the peak periods).

3. Estimate how many passengers will exit from and board those vehicles (per hour, per 15 min, per vehicle). The total numbers for these transportation service users (public and

private) will have to come from comprehensive regionwide or citywide studies that estimate overall movement patterns and volumes by mode. From this information, the portion of movements associated with the planned interchange node will have to be extracted. If an elaborate planning process for a major facility is called for, a simulation model of all vehicle and user operations is possible.[9] A shortcut in arriving at useful results under this task may be offered, if some terminal or transfer activity already exists, by assembling these present data and extrapolating volumes in the future.

4. Define the origin and destination points of all the passengers entering and leaving the terminal, on foot or in a vehicle, i.e., their linkages to zones external to the terminal and their connections among modes within the terminal. (Note: the layout of the terminal is not yet determined, and therefore the internal geography is abstract and schematic at this stage.) There are likely to be significant differences in these patterns hour by hour. This information should give the volumes of specific connecting movements among all the portals within the terminal (doors to the outside and all internal locations where entry into/exit from vehicles takes place). This defines the flow linkages (with quantified volumes), which requires careful functional analysis (as outlined in the previous section) to shape the most efficient overall pattern.

5. Each of the flows has to be characterized as to the speed with which the patrons will move along these paths, including the time that will be spent at zero speed (waiting or completing various transactions). This allows sizing of the corridors and spaces to be utilized by each flow, i.e., determination of needed widths and floor space to accommodate the moving or stationary volumes. (The time-space method of pedestrian space design would be utilized, as discussed in Chap. 2.) Of course, in many instances the various flows

[9] See D. E. Whitney and J. C. Brill, "Development of an Intermodal Transit Simulation and Its Application to the Frankford Transportation Center," *Transportation Research Record,* no. 1623, 1998, pp. 71–79, which examines the implications of location and design alternatives, their effect on surface transit operations, and traffic flow in and around the facility.

will overlap, and the cumulative space requirements will have to be obtained, allowing for some friction among movements.

6. The analysis so far has only dealt with the prime customers of the terminal—actual travelers. To the extent that other activities are to be provided in the building, which need to be located at least in their approximate positions at this time, the anticipated volumes of shoppers, visitors, and well-wishers have to be added. This may increase the internal space requirements considerably, but it should not interfere with the efficiency of the functional movements of travelers.

7. All this represents a precise design program for the facility. It can be turned over to architects, who have undoubtedly participated all along, to design a building around the functional elements. In addition, there will be operational and support spaces, room for mechanical and electrical equipment, staff accommodations, and other behind-the-scenes components.

This description of the planning/design process is what the process *should* be. It is elaborate, and it can be precise; it is not intended to frighten anybody, but rather to explain what the key considerations have to be. In real life, particularly with smaller facilities, many shortcuts and assumptions will be made. This is to be expected and it is acceptable, as long as the designers have sufficient experience and judgment regarding the functional needs of the terminal.

Given the urban situation we live in, and recognizing the likely attraction of persons with questionable intent to the facility, as noted previously, significant design attention has to be devoted to security and making the terminal safe for all users. This encompasses good visibility throughout, the absence of secluded and unsupervised spaces, and the presence of surveillance systems.

It must also be pointed out that a major functional planning task exists outside the terminal building—the efficient and safe arrangement of all the access lanes and track for the various modes that will service the facility, as well as the placement of the loading berths. Priorities have to be established, and the threat of congestion, confusion, and mutual interference is present unless much care is taken in allocating vehicular space properly. The entire complex has to be fitted into the surrounding urban environment.

The latter consideration will most likely be of major local significance generating considerable attention. This is a matter of:

Interior of a bus terminal in Buffalo, New York.

- *Community Concerns* related to the observance of local development controls and conformance with land use and development policies as formulated by the community. There will be aspects of job generation and overall economic activity, including the advancement of the city's standing in the larger business world. Visual integration of the facility will be examined.

- *Environmental Concerns,* which may be positive in an overall sense for the project, but generate localized traffic overloads and accompanying air quality impacts, as well as excessive noise levels.

Administration and Management

An intermodal terminal, by definition, is a joint enterprise. It has to accommodate not only a series of regular transit operations, which may be under the jurisdiction of separate agencies, but also any number of private and auxiliary operations under diverse ownerships. Key issues are who will be in charge, who will make the project happen, and who will run it.

It is possible to consider a commercial effort, particularly today when privatization of many infrastructure elements is deemed a desirable policy (it minimizes the need for allocating scarce public resources). Under such a scenario, an investor builds the facility, gains revenue from the rental of office, retail, and garage space, and collects fees from transport operators for the use of internal space and berths. The private owner manages the facility and is responsible for maintenance, having signed leases or contracts with the various tenants. Nevertheless, it has to be recognized in making all the arrangements that a public purpose is embedded in the effort. Undoubtedly, government participation will be necessary in structuring the access facilities, which will extend into the public street network. It is also conceivable that eminent domain powers may have to be employed in securing an appropriate and adequate site. The possible need for some element of public subsidy is not excluded.

It is more likely for a public agency to take the lead and assume implementation and operational responsibilities. It can be done by the municipality itself, but the more common approach would be for the principal metropolitan transit agency or authority to do the work. It has the necessary powers and instrumentalities, and it is likely to be the largest user of the facility anyway. Or, a special-purpose authority can be identified or established for this task.

A public investment would be necessary, but with reasonable prospects to make the construction and operation self-sufficient. This does not exclude the possibility of augmenting the balance sheet by some public assistance, depending on the degree of general public benefit assumed to be associated with the facility. The public agency would collect rental and concession fees from all the tenants, as would be the case with a commercial effort. Thus, the financial and administrative arrangements would not be much different from those widely practiced with airport terminals.

An example of long standing is the Port Authority Bus Terminal in Manhattan (Eighth Avenue and 41st Street, opened in 1950). While it is not exactly a true intermodal terminal (local buses and paratransit remain outside), it was a public effort to organize and rationalize rather chaotic long-distance bus operations scattered over many blocks. The city took the initiative in securing a large site as close to the Midtown core as possible, but all responsibility was turned over to the Port Authority. There was some reluctance by the various private bus companies to move in and thereby submerge their corporate identities, but they were persuaded to do so. The multilevel building extending over two blocks, with a parking garage on top, processes some 200,000 passengers each day with considerable efficiency. The most successful operating feature is a complex arrangement of ramps in the rear, connecting directly to the Lincoln Tunnel, which are able to handle a multitude of buses without delays near the facility itself. The control of taxis in front of the building has always been a headache.

The last point to be made regarding intermodal terminals is to repeat that they are not only devices to integrate transportation services for the benefit of the riding public, allowing them to reach destinations with dispatch and comfort; they also are, or can be, major urban elements that help structure the city pattern and give points of orientation. They should be the highest-visibility transportation component in any community.

Front of the Port Authority Bus Terminal on Eighth Avenue in Manhattan.

Bibliography

The basic references on rail (American Railway Engineering Association) and buses (Giannopoulos) contain sections on their respective stops, stations, and terminals. Information on terminals is also found in the books by Gray and Hoel and by Vuchic.

American Railway Engineering Association: *Manual for Railroad Engineering,* 1994.

Edwards, John D. (ed.): *Transportation Planning Handbook,* Institute of Transportation Engineers/Prentice-Hall, 1992, 525 pp. Chapter 7 covers terminals of all kinds, for passengers and freight, with emphasis on the facilities for each specific mode.

Giannopoulos, G. A.: *Bus Planning and Operation in Urban Areas: A Practical Guide,* Aldershot, United Kingdom: Avebury/Gower, 1989, 370 pp.

Gray, George E., and Lester A. Hoel (eds.): *Public Transportation* (2nd ed.), Prentice-Hall, 1992, 750 pp.

Vuchic, Vukan R.: *Urban Public Transportation: Systems and Technology,* Prentice-Hall, 1981, 673 pp.

CHAPTER 19

Conclusion—The Road Ahead

The twenty-first century should be, and it certainly can be, better in urban transportation than the twentieth century was. The last 100 years were basically devoted to the exploitation and further development of modes, technologies, and service approaches that were created in the nineteenth century. All the rail types, bicycles, and even the automobile are now more than 100 years old. We used them all in North America, and to a large extent built cities that are not particularly attractive, are certainly very inefficient, and are plagued by various quite familiar ills. Sprawl, pollution, and loss of many traditional urban amenities can be ascribed to the inappropriate and unwise use of the automobile, which remains immensely popular. Communal transit that should bring a higher degree of livability to cities has become the neglected and unloved orphan among mobility choices.

At the same time, travel distances between cities have shrunk to a few hours in most cases, and any civilized place on the globe can be reached within a day. The cheap transport of goods back and forth among production stations has cut costs of finished products and generated an interlinked global economy, for better or worse. The automobile, as a personal device and a family appliance, has revolutionized lifestyles, given most of us on this continent unprecedented mobility, and usually pleases each owner and user. Yet, we may not particularly enjoy the long rides that are now associated with any trip purpose; there may not be too many interesting places to go to. The public realm and the organized communities bear the brunt of excessive spatial and environmental loads.

But enough moaning. There are hundreds of books, TV shows, newspaper articles, and panel discussions that have painted the contemporary picture fully and quite well (and repeatedly).

The really important question is, What happens now? Or more specifically, what are the current and the next generations going to do to shape the urban environment? How will it be made not only workable, but also more pleasant? We do not know, of course, what will happen, but the crucial conclusion is that the means and the knowledge are on hand to do well. It is possible to make urban transportation much better than what we have seen and experienced so far—at any period in history, near or distant.

Much of this happy and promising state is due to transportation technology. We are no longer limited by steam engines or even electric motors, by 4-ft 8.5-in tracks or even rubber tires, by painted stripes on the pavement or even mindless red/green traffic signals. This is the first instance in urban history when service systems can be purposefully assembled from available components to do a specific task with a precise focus (once the needs are defined). Just as a car buyer can go to a dealer and select—following his or her own taste and demands—the elements of an automobile for quick delivery, public transit vehicles can now be ordered in any size with any type of propulsion system, guided by any number of controls, and responding to a wide array of service tasks. The traditional concept of a bus, streetcar, or metro becomes very blurred because we are no longer looking at a lineup of discrete and separable units, but rather at a continuous spectrum of transportation opportunities shading into each other: with or without tracks, long or short, with or without external power pickup, with space for a dozen passengers or a hundred dozen, manually operated or fully automatic, in a tunnel or up in the air, fast with few stops or accommodating patrons through many points, and so on. Communal urban transportation can range from a self-driven two-seater that one picks up along the curb (perhaps even a bicycle) to a double-deck express train racing along in a tunnel carrying thousands in considerable comfort.

The vehicular pollution problem that has haunted us for decades should be a thing of the past. It is possible today, without delay, to switch to clean fuels or rely on clean engines. Granted, it will not happen overnight because higher costs are involved and the elements need to be upgraded further, but the obstacles are a lack of will and vested interests, not technology. Yes, the congestion problem caused by persons sitting within 3 ft^2 in an automobile but preempting at least 300 ft^2 of roadway space still remains. That will not be solved by technology—not even by smaller cars. But there are other means available to us.

Even more significant than the advances in mechanical technology are the breakthroughs in electronic and communications technology. Instead of having individual transportation units scurrying around the metropolitan space on their own without anybody knowing exactly how the total "chaotic aggregate" behaves at any given time, we should institute system management—and we can. If a vehicle operator encounters obstacles on the street today, the driver seeks a better path by trial and error or simply accepts delay stoically as a part of the contemporary urban condition. However, if that driver had had full information on what prevails in that hour and what is to be expected, the modal choice might have been more effective and the routing certainly more efficient. This is possible with a comprehensive network of sensors, centralized intelligence that can do short-range forecasting anticipating conditions and can manage controls, and various communication systems making this information instantaneously and directly available to operators and travelers.

We are on the verge of doing exactly that on a citywide scale. The engineering has been substantially done, and the systems of intelligent transportation are operational. It is now a question of implementation and deployment. Here, too, new expenditures are involved because basically a new infrastructure system has to be created and built. The benefits to be gained, however, in greater overall efficiency and peace of mind for urban residents should outweigh the costs. This might be the most important program that we can institute in the first half of the twenty-first century to cross a threshold in urbanization. The ability to manage all transport operations at a metropolitan scale with purpose and positive intent is an incredible opportunity within reach.

There is also a side benefit that politicians do not yet wish to talk about because it sounds like additional taxes—the levying of user charges (or "congestion management charges," if that sounds more acceptable). This would be not only a defensible policy in providing equitable allocation of scarce circulation space, but also a means to generate revenues that could be spent to upgrade the total system for the largest aggregate public good.

The most important new situation, however, is probably a widely held public recognition that constructive actions are necessary and that positive if sometimes painful steps have to be taken. The inefficiencies and problems associated with urban transportation should not be tolerated, and the overwhelming majority of the population knows that. If overall mobility and accessibility can be enhanced without sacrificing too much individual comfort and convenience, corrective and upgrading actions should be politically acceptable at all levels—provided, of course, that credible and understandable programs are advanced.

To the extent that the urban environment is a living organism, a number of problem-correcting events have occurred without any planning or centralized guidance. The urban field concept, as a structure of multiple centers and specialized activity locations dispersed over large metropolitan space, is a natural market reaction, not so much to walk away in development from historical deficiencies as to take advantage of contemporary operational advantages. Factories and distribution centers will never return to the central districts in multistory structures. The much maligned shopping centers serve the retail and entertainment needs of families and individuals very well. Separate institutional campuses, research centers, and edge city concentrations of office buildings have great internal advantages, even though they are associated with longer access trips by workers and visitors. Families overwhelmingly prefer individual homes (if they can afford them), but there is a market for clustered high-density living as well. The challenge is not only to find the most effective means of transportation to service this pattern, but, more importantly, to develop mobility systems that foster more efficient use of space and resources. Good transport systems have the ability to organize and shape land use patterns. This cannot be legislated; it has to be achieved by providing communal service of superb quality, because anything less will not do in a prosperous society that allows individual choice.

The private automobile will not go away. The attractiveness of a personal device that extends one's ability to reach a large number of destinations in comfort and privacy is incontrovertible. Yet, it is necessary and logical to develop a vehicle that is responsive to the larger societal requirements. It should also be politically acceptable to use it wisely and to curtail its presence in situations where the impacts on the community exceed any individual benefits. This can be done effectively, and we know how to do it.

Let us seize the opportunities to do well with transportation services. The well-being of our communities depends on it.

Rail Transit Definitions

This glossary does not include nonguideway, rubber tire modes (paratransit, buses, trolleybuses) or automated guideway transit and waterborne and special modes.

The definitions are the author's own interpretations (with the advice of Winfred Salter), attempting to generate a coherent structure. Great reliance has been placed on established definitions published by the Federal Transit Administration, the American Public Transportation Association, the Transportation Research Board, and Professor Vukan Vuchic.

cable car A vehicle on rails pulled by a continuously moving steel cable loop under the pavement within the trackway, with a central power plant. Examples include the San Francisco cable car service.

commuter rail Rail passenger service operating at the metropolitan level, usually between the center and adjacent suburbs, using either locomotive-hauled or self-propelled railroad cars. Stations are relatively far apart, platforms may be high or low, and the right-of-way will be largely reserved and segregated, but possibly having some protected at-grade crossings. May provide peak period service only. Examples include the operations by Metra in the Chicago metropolitan area.

elevated (el) (American)   Early rapid transit systems placed on elevated viaducts atop streets. Examples include the els of New York and Chicago, many of which are still in operation.

fixed guideway transit   Any transit mode with operations confined to a permanent channel or trackway. A general term encompassing heavy rail transit, light rail transit, and other rail modes. Examples include BART and the Los Angeles Blue Line.

heavy rail transit (HRT)   Transit service using multicar trains, with self-propelled high-acceleration/deceleration vehicles powered by electricity, operated on exclusive rights-of-way, with high platforms in stations. Examples include the HRT or Metro line of Baltimore, in contrast to its Central LRT and MARC commuter rail service.

heritage/vintage trolley or tram   Services intended almost entirely for tourists and visitors, attempting to capture a historical image through the use of restored or replicated equipment. Examples include the McKinney Avenue tramway in Dallas. The vehicles may run on rails, but frequently are motor vehicles (on a truck chassis) with a passenger compartment somewhat resembling a historical tram.

light rail transit (LRT)   Transit service using rail cars singly or in short trains, powered by electricity usually supplied by overhead wires, operated on exclusive rights-of-way, on non-exclusive rights-of-way (with grade crossings), or in mixed street traffic, with stations close together. Other characteristics are rapid acceleration of vehicles, automatic or manual control systems, and high or low platforms at stations, under a variety of right-of-way conditions (fully grade separated, predominantly reserved, designated by pavement markings, or mixed traffic). Examples include the new light rail transit lines of Cleveland, with the old Shaker Heights streetcar service upgraded to LRT level.

mass transportation   See **transit.**

metro   Short for *metropolitan railway,* frequently used popular designation of heavy rail transit, particularly in Europe. Examples include the Métro system of Paris.

monorail   Guided transit vehicles operating on or suspended from a single rail, beam, or tube. Examples include the Seattle Monorail.

premetro/light metro   Light rail transit systems designed for later or progressive conversion into rapid or metro/heavy rail service, i.e., eventually with exclusive right-of-way and high platforms. Examples include the premetro of Brussels.

public transit   See **transit**.

rail bus   Self-propelled rail vehicles utilizing a diesel engine, usually in suburban service. Examples include the Merseyrail diesel service component in Liverpool.

rail transit   All modes that employ rail technology and run on steel tracks, including all railroad operations at the city or metropolitan level, but not intercity and other long-distance service. Examples include the Orange, Red, and Blue Lines of Boston, as well as the several branches of the light rail transit Green Line.

rapid transit   All modes characterized by fast running speed and few delays, usually on an exclusive right-of-way. Buses under very special arrangements can be included in this group, but not streetcars. Examples include the Red Line service in Los Angeles, but not the Blue Line; the Green Line is almost rapid transit. The El Monte busway qualifies as well.

regional rail   See **commuter rail**.

regional transit   Rail or bus operations at the metropolitan level, usually from the suburbs to the center, with few stops in the middle portions of routes, in contrast to city or short-haul transit. Examples include the commuter rail and express bus operations in the New York region leading to Midtown Manhattan.

S-Bahn (German)   Short for *Schnell-Bahn* ("fast rail") regional rail service. Examples include the S-Bahn of Berlin, in contrast to the U-Bahn and other city services.

sneltram (Dutch)   "Fast tramway," an upgraded streetcar mode, almost light rail transit. Examples include the fast city and interurban trams of Holland.

Stadtbahn (German)   "City rail," an upgraded streetcar mode, the equivalent of light rail transit. Examples include the Stadtbahn network of Hannover.

streetcar   Transit service using rail cars singly or in short trains, powered by electricity supplied by overhead wire, operated usually on city streets in mixed traffic, with stations close together. Examples include the five streetcar or tramway lines of Philadelphia, not yet converted to light rail transit.

suburban rail   See **commuter rail**.

subway (American)   A rail rapid transit service or an entire system that is predominantly but not entirely underground. Examples include the New York City subway network.

supertram   An upgraded streetcar mode, light rail transit for all practical purposes. A British promotional designation. Examples include the Sheffield Supertram.

tramway (British)   Another name for streetcar. Examples include the Blackpool tramway.

transit   Urban transport services open to everybody, usually upon payment of a fixed charge, following a defined route, having specified stations or stops, and running according to a set schedule. Examples include all the services provided by the Metropolitan Transportation Authority of New York, including buses.

trolley   Another name for streetcars, referring primarily to simple technology. Examples include the legendary Toonerville Trolley. Trolleybuses are also frequently referred to by this abbreviated designation, but they should not be.

tubes/underground (British)   Heavy rail transit. Examples include the tubes of London or the underground of Glasgow.

U-Bahn (German)   Short for *Untergrund Bahn* ("underground rail") heavy rail transit. Examples include the U-Bahn of Munich.

Summary of Performance and Service Characteristics

The criteria for evaluating and judging various modes as to their suitability in various situations are outlined in Chap. 1. Modes are classified from the perspective of individual users, as appropriate to communities at large, and responding to several broad objectives that are effective at the national scale. The bulk of the book is devoted to a reasonably detailed review of every mode that could be considered for implementation in a given community. It is now possible to provide a general summary, which is presented in the following table. This has been done as an overall approach and characterization, recognizing that in each specific instance a service can be structured to operate at a higher or lower level than shown in the general summary.

Responses by Modes to Transportation Needs/Demands/Expectations

	Private and Individual Means			Single Party for-Hire Ride		Shared Ride, Paratransit	
	Chauf-feur Driven Limo	Self-Drive Auto-mobile	Rented Auto-mobile	Taxi (by Street Hail)	Taxi (by Arrangement)	Carpool, Vanpool	Special Clients Dial-a Ride
For Users							
Time/effort prior to start of trip?	Alert driver	Reach parking space	Pick up vehicle	Reach street	Call and response time	Assemble and distribute riders	Assemble and distribute riders
Time needed to complete trip?	Usually short, except for congestion			Usually short		Rather long	Long
Door-to-door access? One seat ride?	Abso-lutely	Yes	Yes	Yes	Yes	Some-times	Usually
Parking require-ments?	None for user	High	High	None	None	None	None
Privacy, comfort level?	Extremely high	High	High	High, except for driver presence		Good, with known riders	Adequate with similar riders
Reliability, safety, and security?	Very high	High	High	Reason-able	Good	Good	Reason-able
Costs to user?	Extremely high	High	Very high	Very high	Very high	Reason-able	Usually subsidized
Interconnectivity, acceptable trans-fer conditions?	Extremely high	High, if parking available		Very high	Very high	Limited	Usually none
Ability to carry luggage and goods?	Very high	High	High	High	High	Limited	Some
Social status?	Extremely high	High	High	High	High	Limited	N/A

| Shared Ride, Paratransit (Publicly Available) | | Rubber-Tired Transit | | Rail Transit | | | | | |
Dial-a-Ride	Jitney	Buses	Trolley buses	Light Rail	Mono-rail	Heavy Rail/Metro	Commuter Rail	AGT	Water-borne
For Users									
Call and response time	Reach street	Reach stop	Reach stop	Reach stop	Walk or use feeder to station		Feeder service	Reach station	Feeder service
Acceptable	Reasonable	Somewhat slow		Reasonably fast	Fast	Fast	Fast	Reasonable	Slow
Usually	No	Possibly	Rarely	Seldom	Probably not		No	Seldom	No
None	None	None	None	None	Possibly	None	Most likely	None	Possibly
Reasonable	Limited	Limited	Limited	Acceptable	Acceptable	Low	Good	Good	Good
Adequate	Limited	Reasonable	Reasonable	Good	Good	Acceptable	Superior	Superior	Superior
Acceptable	Low	Low, subsidized		Low, subsidized		Low, subsidized	Acceptable, subsidized	Acceptable	Acceptable
Acceptable	Acceptable	Usually constrained		Can be superior		Superior	Limited	Good	Constrained
Limited	Very limited	Limited	Limited	Limited	Possible	Limited	Good	Limited	Possible
Acceptable	Low	Inferior	Low	High	High	Limited	High	Very high	High

Responses by Modes to Transportation Needs/Demands/Expectations *(continued)*

	Private and Individual Means			Single Party for-Hire Ride		Shared Ride, Paratransit	
	Chauffeur Driven Limo	Self-Drive Automobile	Rented Automobile	Taxi (by Street Hail)	Taxi (by Arrangement)	Carpool, Vanpool	Special Clients Dial-a Ride
For Communities							
Need for capital funds?	N/A, except for the construction of roadways and traffic controls					N/A	N/A
Reasonable O&M costs?	Maintenance of streets and traffic controls					Some	Possible subsidies
Environmental impacts?	Very high	Very high	Very high	High	High	Reasonable	Acceptable
Efficiency of networks and land use?	Inferior	Inferior	Inferior	Acceptable		Good	N/A
Boost to local economy?	Significant	Neutral	Some	High	High	Some	Vital
Civic image, political support?	High	Neutral	High	High	High	Limited	Vital
For the Nation							
Conservation of fuel resources?	Poor	Very poor	Poor	Acceptable	Acceptable	Positive	Good
Equity of service availability?	Very poor	Poor	Very poor	Limited	Acceptable	Good	Very good
Advancement of technology?	Limited	Some	Limited	Limited	Limited	Limited	Some
Good use of national wealth?	No	No	Reasonable	Reasonable	Reasonable	Yes	Absolutely

| Shared Ride, Paratransit | | Rubber-Tired Transit | | Rail Transit | | | | | |
| Publicly Available | | | | | | | | | |
Dial-a-Ride	Jitney	Buses	Trolley buses	Light Rail	Mono-rail	Heavy Rail/Metro	Commuter Rail	AGT	Water-borne
For Communities									
N/A		Limited	Considerable	Considerable	Very high	Very high	Considerable	Very high	Acceptable
Some	Some	Yes	Yes	Good	Yes	Yes	Yes	Acceptable	High
Reasonable	High	Considerable	Low	Very low	Very low	Very low	Low	Very low	Low
Good	Good	Reasonable	Reasonable	Good	Inferior	Very good	Very good	Good	Limited
Some	Neutral	Limited	Limited	Significant	Limited	Major	Major	Significant	Limited
Some	Limited	Inferior	Limited	Major	Major	Considerable	Considerable	Major	Significant
For the Nation									
Acceptable	Good	Inferior	Good	Superior	Superior	Superior	Good	Good	Limited
Good	Very good	Good	Good	Acceptable	Acceptable	Good	Limited	Limited	Limited
Some	Limited	Limited	Good	Good	Limited	High	Limited	Very high	High
Yes	Yes	Yes	Doubtful	Yes	Doubtful	Yes	Yes	Limited	Yes

Suggested Places for Field Inspection

The preceding chapters make reference to any number of transportation and transit systems around the world, particularly those in North America. These are described in greater or lesser detail and illustrations of networks and vehicles are provided. Yet, reading about a system is not the same as actually seeing it and riding on it. Therefore, as a further guide for those who are truly interested in a specific mode and seek real familiarity with operational patterns and service quality, selected candidates are listed here that would be worthwhile to see—preferably over an extended period, but a short visit would be instructive, too. The selection, limited arbitrarily to only three examples per mode, attempts to identify successful instances that provide a general overview of a specific mode or illustrate significant characteristics. These choices may not necessarily be endorsed by every transportation specialist, but some care has gone into making the list. Any investigator who wishes to go further or is concerned with specific elements or features should be able to receive guidance for visits from the various other examples mentioned in the chapters on the different modes.[1]

---

[1] See also recent special issues of *Passenger Transport* (The Weekly Newspaper of the Public Transportation Industry, APTA) associated with annual and speciality conferences, containing articles prepared by various agencies that outline their programs and accomplishments.

Walking

- A number of American cities are blessed with natural or historic environments that make them most attractive as places to walk in. Planners and designers can take little credit for New Orleans, Honolulu, Litchfield, Aspen, and several other similar localities. Places with large, deliberately created pedestrian systems are the *waterfront areas* of *Boston* and *Manhattan*.

- Most of the downtown pedestrian zones and transit malls in North America are pleasant enough, but sometimes not easily distinguishable from each other. Exceptions are the large underground system of *Montreal* and the transit mall and skyway arrangements of *Minneapolis*.

- Precise pedestrian time-space design methods have been employed in a number of recent transportation projects, but the specific results are hardly apparent to an outside observer. It is well to note, however, that the recent additions to both *Grand Central Terminal* and *Pennsylvania Station* in *Manhattan* (as well as other large projects across the country) have employed these study procedures.

Bicycles

- The community in the United States that has done as much about bicycle systems as any other and has been well documented is *Davis, California*.

- To see bicycles in active use and in great volumes, the distinct and separate university towns with large student enrollments are the places to go: *Ann Arbor, Michigan, Madison, Wisconsin, Berkeley, California,* and a few others. The same observation applies to European cities.

- The highest concentration of bikers, bordering on congestion (particularly on a nice weekend), is to be found in *Central Park, New York*. The other choice is still any city in *China*.

Automobiles

- The *Twin Cities,* and the *State of Minnesota* in general, appear to be in the forefront of developing advanced (intelligent transportation) systems to manage automobile operations at a metropolitan level and to guide their behavior in a responsible and effective way.

- To experience fully the pleasures and tribulations of a lifestyle dominated by car culture, the place is still *Los Angeles*. The staggering investment placed in a highway network that reaches almost all places and has generated an urban structure and an inventory of facilities never seen before in any urban area can only be experienced behind the wheel.

- The communities that have taken serious steps to minimize automobile dependency are the several neotraditional developments at several locations, such as *WaterColor* (in the Florida panhandle), *Kentlands* (in Gaithersburg, Maryland), and *Mashpee Commons* (in Massachusetts). These are basically test cases to gauge whether Americans will use alternate modes in daily operations.

Paratransit

- While just about every community in North America has some paratransit service, the most active concentration appears to be found in southern California. *Orange County* might be the place to go with its extensive dial-a-ride service and other auxiliary operations.

- The islandwide van service at a high level of organization and support by major terminals in *Puerto Rico* is of considerable interest, with demand-responsive operations assuming a dominant transit role.

- The spontaneous, self-generated local services (frequently on the borderline of legality) are best seen in the minority districts of the larger cities—*New York, Chicago, Miami,* and *Los Angeles*. Other than that, there are the hundreds of examples in all cities of the developing world.

Taxis

- *Los Angeles* has been able to structure a regulatory system that works well.

- *Fairfax County, Virginia* is a good example of a midsize community with a responsive taxi service.

- *Chicago* has a large system with many components that operates as well as can be expected.

Buses

It is not easy to select just three bus systems for special recognition. Many places are doing interesting things with service-upgrading programs. These tend to be localities that appear in other modal areas on this larger list. There are, however, several that have gained deserved visibility for their efforts.

- *Las Vegas* is looking forward to advanced vehicles and forms of operation.
- The *Pace Suburban Bus Division of RTA* of the *Chicago* region has a solid and responsive record.
- *Long Beach, California,* has found innovative ways to expand the role of bus transit.

Bus Rapid Transit

- *Houston* has developed an extensive radial express bus system that provides rapid, frequent, and comfortable service over rather long distances at a premium fare. The many other HOV lane projects around the country are usually separate efforts. (For sheer volume, the record holder is still the contraflow express lane at the *Lincoln Tunnel* leading into *Manhattan*.)
- The *El Monte* busway in *Los Angeles,* a pioneering effort, is the core of a service system with a wide reach, and has combined fast express movements with local service utilizing off-line stops.
- *Ottawa* has built a busway system that serves as the principal transportation mode for the metropolitan area. Various routes and services are accommodated, and the network expands into the surrounding districts.

Trolleybuses

In the United States, the options are not many. There are the five cities (*Boston, Dayton, Philadelphia, San Francisco,* and *Seattle*) that have maintained trolleybus service in spite of national trends. *Seattle* has done more than the others by ordering new equipment and building a downtown tunnel. Other than that, there are many cities in *Eastern Europe* and *Germany* that have extensive systems—not necessarily using advanced technological or operational features, but providing steady and reliable service.

Light Rail Transit

- The *San Diego* system was an example and inspiration for most of the subsequent LRT efforts in the country. It still provides excellent service, and there have been continuous expansion programs.

- The *Hudson-Bergen* line along the western shore of the Hudson River in *New Jersey* is expected to play a significant role in a region that is transit oriented, but congested nevertheless. It should become a key element in the total integrated movement network, and it is as advanced in its technology as can be seen anywhere.

- The *Green Line* in *Los Angeles* is of considerable interest as an LRT system that is as capital intensive and has as large a service capability as any other in the country, and verges on being a metro (heavy rail). The other end of the scale would be the *McKinney Avenue* trolley in *Dallas*.

Monorail

In the United States, the options are not too many—either the *Seattle* monorail shuttle or the *Disney* visitor services (and the *Newark* airport system, if it is taken out of the AGT class). Other than that, a trip to *Japan* is indicated to see several operating systems.

Heavy Rail Transit (Metro)

- The *Washington, D.C. (WMATA)* system is not only highly visible in the nation's capitol, but its geographic extent and popularity among the public are significant features. The quality of design, construction, and operation are exemplary; coordination with land development programs in station vicinity is a strong element.

- *Atlanta (MARTA)* offers a robust rail service that is largely built in a traditional format, but contains advanced features and good design touches. It has much potential for further expansion.

- The *San Francisco Bay Area (BART)* system is the first "modern" effort of the current era. It set the stage for most subsequent activities in other places, and it remains a rare example of a central rail network reaching out within a large service region (but not able to reach all sectors yet).

- There are also noteworthy examples of single lines of advanced design being added to existing systems. This certainly includes the *Jubilee Line* in *London* and the *Météor Line* in *Paris,* with a more modest example being the *B and Q Lines* from *Manhattan* into *Queens.*

Commuter Rail

- The *Long Island Rail Road,* extending from Penn Station in Manhattan to Montauk Point 115 mi away, is not only the largest system in the United States, but also one with diverse equipment and several innovative features instituted recently. A ride on any of the longer branches is suggested, as well as observation of the operations at Jamaica Station during a peak hour when a maximum number of trains is processed.

- *Metra* out of *Chicago* shows many instances of adapting facilities built a long time ago to contemporary needs and achieving a superior service quality with advanced rolling stock. Of particular interest may be the *Burlington Northern Santa Fe* line from Union Station to Aurora.

- The *Metrolink* system centered on downtown *Los Angeles* is significant as an example of relatively recent reactivated operations in a region known for its automobile dominance. The well-designed new stations, passenger facilities, and parking lots deserve attention.

Automated Guideway Transit

- The *University of West Virginia* system in *Morgantown* still deserves a visit as a pioneer endeavor, and also because it operates well and comes as close to personal rapid transit as we are likely to get.

- Most of the airport AGT systems are of interest. They all serve the same purpose, but respond with different configurations and hardware. The *Atlanta* and *Newark* systems are important because of the dominant role that they play within their own territories.

- As a service in full public operation providing basic transportation, the *Miami* system is notable for its size and good record of performance. It is effective and attractive. (Alternate choices might be *Scarborough* outside Toronto and *Vancouver, British Columbia*.)

Waterborne Modes

- The urban ferry system with the most impressive record and level of technological advancement is undoubtedly the *SeaBus* of *Vancouver, British Columbia,* although on a modest scale.

- Operations on a large scale both old and new, with significant expected growth, are found in *New York harbor,* with services by the *Staten Island ferry* and *NY Waterway* leading the way.

- The waterborne system centered on *Seattle* and extending into *Puget Sound* is an example with considerable variety, intensive use, but an emphasis on service beyond the city scale.

Integrated and Coordinated Systems

- *Toronto* can claim the title of overall champion in North America as a place that has succeeded in organizing a large and diverse multimodal system that operates most effectively. The various components tend to be rather conventional in nature, but they are brought to a high level of efficiency, and they are mutually supported by other services and elements.

- *Portland, Oregon,* continues to lead the way in practically every development and transportation sector by selecting and implementing advanced systems and concepts. Its inventory of pioneering equipment and planning ideas responding to service needs is unsurpassed on this continent. (Its international counterpart would be *Curitiba, Brazil.*)

- All the transportation systems of *New York City* and its region do not work perfectly,[2] but there are many of them. While the inventory does not match everything that is found in *Hong Kong,* there are some unique examples and special situations. Within the very large transportation universe, many instances can be found where advanced and innovative approaches are taken. The choices range from the first parkways to the latest ferry vessels; from illegal jitney operations in low-income areas to premium limousine services in some districts; from the old els to advanced bimodal commuter locomotives.

---

[2] Much to the regret and chagrin of this author. If a mode or a vehicle works in New York, it will work anywhere.

Index